New Directions in Civil Engineering

Series Editor
W. F. CHEN
Purdue University

Zdeněk P. Bažant and Jaime Planas
Fracture and Size Effect in Concrete and Other Quasibrittle Materials

W.F. Chen and Seung-Eock Kim
LRFD Steel Design Using Advanced Analysis

W.F. Chen and E.M. Lui
Stability Design of Steel Frames

W.F. Chen and K.H. Mossallam
Concrete Buildings: Analysis for Safe Construction

W.F. Chen and S. Toma
Advanced Analysis of Steel Frames: Theory, Software, and Applications

W.F. Chen and Shouji Toma
Analysis and Software of Cylindrical Members

Y.K. Cheung and L.G. Tham
Finite Strip Method

Hsai-Yang Fang
Introduction to Environmental Geotechnology

Yuhshi Fukumoto and George C. Lee
Stability and Ductility of Steel Structures under Cyclic Loading

Ajaya Kumar Gupta
Response Spectrum Method in Seismic Analysis and Design of Structures

C.S. Krishnamoorthy and S. Rajeev
Artificial Intelligence and Expert Systems for Engineers

Boris A. Krylov
Cold Weather Concreting

Pavel Marek, Milan Guštar and Thalia Anagnos
Simulation-Based Reliability Assessment for Structural Engineers

N.S. Trahair
Flexural-Torsional Buckling of Structures

Jan G.M. van Mier
Fracture Processes of Concrete

S. Vigneswaran and C. Visvanathan
Water Treatment Processes: Simple Options

Finite Strip Method

Y.K.Cheung
L.G. Tham

CRC Press
Boca Raton Boston London New York Washington, D.C.

Library of Congress Cataloging-in-Publication Data

Cheung, V. K.
Finite strip method / by V.K. Cheung, L.G. Tham.
p. cm. -- (New directions in civil engineering)
Includes bibliographical references and index.
ISBN 0-8493-7430-8 (alk. paper)
1. Structural analysis (Engineering). 2. Finite strip method.
I. Tham, L.G. II. Title. III. Series.
TA646.C477 1998
624.1'.71'015194—dc21
for Library of Congress 97-24697
CIP

International Standard Book Number 0-8493-7430-8
Library of Congress Card Number 97-24697
Printed in the United States of America 1 2 3 4 5 6 7 8 9 0
Printed on acid-free paper

PREFACE

Since the publication of the first book on the finite strip method in 1976, tremendous developments of the method have been made. Interpolation functions have been generalized such that it permits the usage of functions other than the trigonometrical series for the series component of the shape functions. The availability of these shape functions has not only greatly widened the applications of this method in structural analysis but also extended the applications to fields beyond structural engineering. Recently, researchers have further carried out non-linear analyses using this method, and current publications have demonstrated the activities and interests in this area. Furthermore, finite strips formulated by carrying out appropriate Fourier or Hankel transformations have been widely adopted with particular reference to the geotechnical field. In view of the recent advancement, it was decided that an updated text on the finite strip method should be written such that "state of the art" reference on this subject could be made available to both researchers as well as practising engineers.

In this book, we attempt to provide a concise introduction to the theory of the finite strip method including the classical strip as well as the newly developed spline strip and computed shape function strip. The applications of various topics in structural engineering with special reference to practical structures such as bridges, box girder bridges, tall buildings, etc. are extensively discussed. References to the applications in the geotechnical field are also made. Recent development of the applications of the method in nonlinear analyses is also briefly described.

A detailed introduction on the theory of the finite strip method is made in Chapters 1 and 2. Chapters 3 to 5 cover the applications of the method in fundamental problems including plate bending, plane stress/strain as well as vibration and stability problems. The extension of the method to three-dimensional analysis is explored in Chapter 6. The domain transformation which has proved to be a valuable method for mapping complex geometry into regular domain is discussed in Chapter 7. The recent applications of finite strip method in various areas of engineering significance are fully covered in Chapters 8, 9 and 10. Development in nonlinear applications is reserved to Chapter 11, whereas the next chapter concentrates on the transformation approach of the method for semi-infinite domains.

The source code of a program for the static analysis of folded plate structures is listed in Appendix I. The program is written for the IBM SP2 but it can also easily be adapted for the micro-computer environment.

ACKNOWLEDGMENTS

The authors would like to express their sincere gratitude to Associate Professor J. C. Small of University of Sydney and Associate Professor S. C. Fan of Nanyang Technological University for reading parts of the manuscripts and making helpful suggestions, and to Ms. K. Wong for preparing the figures.

Permission from our former graduate students for us to use material from their theses are acknowledged :

Dr. F. T. K. Au,
Dr. C. R. C. Delcourt,
Associate Professor S. C. Fan,
Dr. D. J. Guo,
Dr. J. Kong,
Mr. H. Y. Szeto,
Associate Professor S. Swaddiwudhipong, and
Professor D. S. Zhu.

The contribution of the late Dr. W. Y. Li to the development of the spline finite strip method is also acknowledged. We wish to pay tribute to him by making extensive references to materials from his thesis.

Our sincere thanks are also extended to American Society of Civil Engineers, Elsevier Science, John Wiley & Sons Limited, Structural Engineering International and Thomas Telford Publishing for their permission to use material from their journals.

CONTENTS

CHAPTER FOUR

PLANE STRESS ANALYSIS

CHAPTER FIVE

VIBRATION AND STABILITY

CHAPTER SIX

MODELLING OF THREE-DIMENSIONAL SOLIDS : FINITE PRISM AND FINITE LAYER METHODS

CHAPTER SEVEN

DOMAIN TRANSFORMATION : TREATMENT FOR ARBITRARILY SHAPED STRUCTURES

CHAPTER EIGHT

APPLICATIONS TO SHELL STRUCTURES AND BRIDGES

CHAPTER NINE

APPLICATIONS TO TALL BUILDINGS

CHAPTER TEN

APPLICATIONS TO LAYERED SYSTEMS IN STRUCTURAL AND GEOTECHNICAL ENGINEERING

CHAPTER ELEVEN

NON-LINEAR ANALYSIS

CHAPTER TWELVE

TRANSFORMATION APPROACH: FOURIER AND HANKEL TRANSFORMS

APPENDIX

CHAPTER ONE

INTRODUCTION

Rapid developments of various numerical techniques in engineering analyses have occurred in the last three decades as faster and bigger computers are made available in rapid succession. It is widely acknowledged that the finite element method[1] has dominated the field, but various other methods, such as the finite strip and the boundary element[2] methods, continue to have their own roles to play and have not been outclassed in their more specialized areas.

The finite element method, being the most versatile tool, requires discretisation in every dimension of the problems, and, therefore, will generally require more unknowns for approximation than the other methods. The advent of super-computers has provided engineers and scientists with the means to handle millions of unknowns, and problems which had once been considered to be intractable because of their complex nature and size have been solved successfully by the finite element method. Though the cost of the solutions has decreased considerably, it is still far from being considered cheap. The cost also multiplies by orders of magnitude when a more refined, or a higher dimensional analysis is required. Furthermore, the availability of such powerful machines is still limited to the privileged few, and this is particularly true in developing countries. In such countries, most computations have to be carried out either on less powerful machines or personal micro-computers. Very often the problem size of an accurate analysis might be so overwhelming as to overtax machines of these grades, so that the problem either will have to be solved roughly, or some additional lengthy, time-consuming subroutines have to be written to lower the core requirements. It is also well known that for many problems having regular geometric shapes and simple boundary conditions a full finite element analysis is very often both extravagant and unnecessary. Therefore, alternative methods that can reduce the computational effort and core requirements, while at the same time retaining to a great extent the versatility of the finite element analysis, are evidently desirable.

The finite strip method for two-dimensional problems, which has been developed since the late sixties, is one of the alternative methods. In the three-dimensional regime, the method is referred to as either the finite prism method or the finite layer method depending on the form of the functions chosen for describing the variables for the analyses.

The philosophy of the finite strip methods is similar to the Kantorovich[3] method which is used extensively for reducing partial differential equations to ordinary or partial differential equations of a lower order. In the finite strip method, the reduction is achieved either by assuming that the separation of variable approach can be applied in expressing the interpolation functions of the unknowns (*separation of variable approach*) or by carrying out suitable transformations (*transformation approach*).

Early formulations of the finite strips have been based on the former approach. The methods initially call for the use of simple polynomials in some directions and continuously differentiable smooth trigonometrical as well as hyperbolical series in the other directions. The general form of the displacement function is given as a product of polynomials and series. Broadly speaking, all series, which can satisfy *a priori* the boundary conditions at the ends of the strips (prisms or layers), can also be used. For example, exponential and decaying power series have been employed in the analyses of field problems and stress problems with one or two boundaries of infinite extent. Such an approach can be referred to as the *classical* finite strip method, and a brief comparison between this method and the finite element method is listed in Table 1.1.

The recently developed spline finite strip and computed shape function strip have further increased the flexibility in the choice of the interpolation functions by allowing polynomials to be used in all directions.

The spline finite strip, which has been devised to overcome the difficulties experienced by the classical finite strip method in dealing with multi-span or column-supported structures, has chosen the B-3 spline functions to replace the trigonometrical series and hyperbolic series in the interpolation functions for plates and shell analyses. Since the B-3 spline functions are, in fact, the solution of a beam under a point load, the spline strips also show considerable improvements in convergence in simulating the action of a point load. In conjunction with the domain transformation concept, the method can handle arbitrarily shaped structures easily. In addition, the localized property of the spline functions has allowed all characteristic matrices or vectors of the analyses to be formed section by section. Integration for the formation of these matrices or vectors can be carried out by explicit formulae (for regular domain) or numerical integration schemes (for arbitrarily shaped structures). Apart from being nearly as versatile as the standard finite element method, the method can also achieve a higher order of continuity (C^2 continuity) with a smaller number of degrees of freedom.

Two forms of computed shape function strips (transverse and longitudinal strips) were made available recently for structural analysis. The transverse (longitudinal) component of the shape function of a transverse (longitudinal) strip is obtained by analyzing a typical transverse (longitudinal) section. In the analysis, a unit load is placed sequentially on selected points (joints) of the section, and deflections of the joints due to such loads are computed by a simple frame stress analysis program. These deflections constitute the basis of the interpolation parameters along the transverse (longitudinal) direction. As

TABLE 1.1
Comparison between finite element and classical finite strip methods.

Finite element	Finite strip
Applicable to any geometry, boundary conditions and material variation. Extremely versatile and powerful.	In static analysis, more often used for structures with two opposite simply supported ends and with or without intermediate elastic supports. In dynamic analysis it is used for structures with all boundary conditions and with discrete supports.
Usually large number of equations and matrix with comparatively large bandwidth. Can be very expensive and at times impossible to work out solution because of limitation in computing facilities.	Usually much smaller number of equations and matrix with narrow bandwidth, especially true for problems with an opposite pair of simply supported ends. Consequently much shorter computing time for solution of comparable accuracy.
Large quantities of input data and easier to make mistakes. Requires automatic mesh and load generation schemes.	Very small amount of input data because of the small number of mesh lines involved due to the reduction in dimensional analysis
Large quantities of output. As a rule all nodal displacements and element stresses are printed. Also many lower order elements will not yield correct stresses at the nodes and stress averaging or interpolation techniques must be used in the interpolation of results.	Easy to specify only those locations at which displacements and stresses are required and then output accordingly.
Requires a large amount of core and is more difficult to program. Very often, advanced techniques such as mass condensation, subspace iteration or Lanczos method have to be resorted to for eigenvalue problems in order to reduce core requirements.	Requires smaller amount of core and is easier to program. Because only the lowest few eigenvalues are required (for most cases anyway), the first two or three terms of the series will normally yield sufficiently accurate results for the classical finite strip method. Matrix can usually be solved by standard eigenvalue subroutines.

in the finite element method, simple polynomials in terms of the joint deflections are used to interpolate for the deflection between the joints. Standard polynomials are chosen for the other components of the shape functions.

Applications of Fourier and Hankel transforms in the analysis of geotechnical problems are examples of the transformation approach. The transformations reduce the governing differential equations for each strip (layer) into lower order equations for which solutions can be sought. The undetermined coefficients of the solutions can then be determined by conditions based on physical relations (e.g. flexibility relations).

Though the underlying principle is the same for the two approaches, the formulations and procedures are slightly different. For clarity, the transformation approach will be discussed in detail in Chapter 12 after allowing for a more systematic coverage on the separation of variable approach in the previous eleven chapters. Nevertheless, the following common procedures are adopted for both approaches:

(i) The continuum is divided into two- (strips) or three-dimensional (prisms, layers) subdomains in which one opposite pair of sides (2-dimensional) or one or more opposite pairs of faces (3-dimensional) of such a subdomain are in coincidence with the boundaries of the domain (Figure. 1.1). The geometry of the domain is preferably constant along one or two coordinate axes so that the width of a strip or the cross-section of a prism or layer will not change from one end to the other. Recent advances show that two-dimensional (three-dimensional) problems defined in irregular geometrical domain can be dealt with by transforming the domain into a unit square (cube) before the discretisation. For convenience, only the strip will be discussed in the general formulation, which is in any case also applicable to prisms and layers.

(ii) The strips are assumed to be connected to one another along a discrete number of nodal lines which coincide with the longitudinal boundaries of the strips. In some cases it is also possible to use internal nodal lines to arrive at a higher order strip.
The degrees of freedom (DOF) at each nodal line, called nodal parameters, are normally connected with the displacements and the partial derivatives of the displacements. They can also include non-displacement terms such as forces and strains (including direct strains, shear strain, bending and twisting curvatures) in stress analysis; pore pressure in consolidation and seepage analyses; as well as temperature in thermal analysis, etc.
Due to the use of continuous functions in the longitudinal direction, the DOF of the strip nodal line is usually less than that at the element node. For example, in plate bending, w and θ_x exist at each strip nodal line while w, θ_x, θ_y exist at each element node.

(iii) An interpolation function (or functions), in terms of the nodal parameters, is (are) chosen to represent the variables. Consequently, the stress/strain field (in the case of stress problems) and the velocity field (in the case of field problems) can be derived.

(iv) Based on the chosen functions, it is possible to obtain the respective characteristic matrices through variational or energy principles. In the case of static analysis of structural problems, the stiffness and load matrices which equilibrate the various external loadings acting on the strip are formed through minimum total potential energy principles. The principle states that

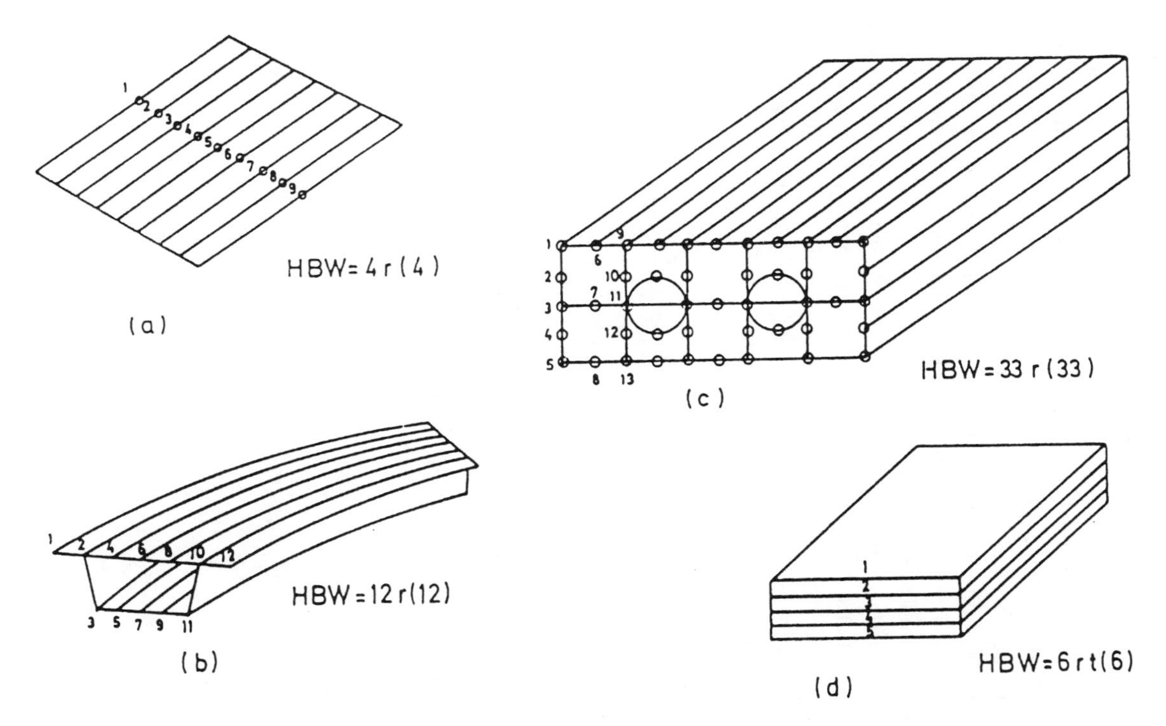

Figure 1.1 Some structures and typical mesh divisions: (a) Encastre slab (plate strips). (b) Curved box girder bridge (shell strips). (c) Voided slab bridge with circular holes (quadrilateral finite prisms). (d) Thick, multi-layer plate (finite layer). HBW = half bandwidth; r,t = number of terms of series; ( ) =simply supported case.

of all compatible displacements satisfying the boundary conditions, those which satisfy the equilibrium conditions make the total potential energy assume a stationary value.
In mathematical form we have

$$\left[\frac{\partial \phi}{\partial \{\delta\}}\right] = 0 \qquad (1.1)$$

in which the total potential energy ϕ is defined as the sum of the potential energy of external forces W and the strain energy U.
Other approaches may be used to formulate the solutions as well. For example, flexibility approach has been used to formulate the structural problems, and weighted residual method is widely used in the formulation for the field problems.

(v) The characteristic matrices of all strips are assembled to form a set of the overall equations for the whole domain under consideration. Since the bandwidth and the size of the matrix are usually small, the equations can be solved easily by any standard band matrix solution technique to yield the nodal parameters.

REFERENCES

1. Zienkiewicz, O. C., *The finite element method*, Fourth Edition, McGraw-Hill, London, 1989.
2. Brebbia, C. A. and Dominguez, J., *Boundary element: an introductory course*, Computational Mechanics Pub, Southampton, 1989.
3. Kantorovich, L. V. and Krylov, V. I., *Approximate method of higher analysis*, Interscience Pub, New York, 1958.

CHAPTER TWO

INTERPOLATION SHAPE FUNCTIONS

2.1 CHOICE OF INTERPOLATION SHAPE FUNCTIONS

It has been pointed out in the previous chapter that the choice of suitable interpolation functions for a strip based on the separation of variable approach is the most crucial part of the analysis, and great care must be exercised at such a stage. A wrongly chosen interpolation function might not just produce obviously ridiculous answers but might even lead to results which converge to the wrong answer for successively refined meshes.[1]

A classical strip reduces a two-dimensional problem to a one-dimensional one with the shape functions for the interpolation variables in the analysis given as

$$\delta = \sum_{m=1}^{r} f_m(x)\, Y_m(y) \tag{2.1a}$$

Similarly, three-dimensional problems can be reduced to two-dimensional or one-dimensional ones in cases of finite prism and finite layer respectively. The shape functions for the finite prisms are

$$\delta = \sum_{m=1}^{r} f_m(x,z)\, Y_m(y) \tag{2.1b}$$

while those for the finite layers are

$$\delta = \sum_{n=1}^{t} \sum_{m=1}^{r} f_{mn}(x)\, Y_m(y)\, Z_n(z) \tag{2.1c}$$

In the above expressions, $f_m(x)$, $f_m(x,z)$ and $f_{mn}(x)$ are polynomial expressions with undetermined constants for the m-th and mn-th terms. $Y_m(y)$ and $Z_n(z)$ are series which satisfy the end conditions in the y and z directions respectively.

Recently developed spline and computed shape function strips have allowed the replacement of the series component by polynomials. The interpolation variables are then written as

$$\delta = f(x)\ \phi_m(y) \tag{2.1d}$$

where m represents the node number and mode number for the spline and computed shape function strips respectively.

To ensure convergence to correct results the following simple requirements have to be satisfied:

(i) The series part (Y_m and Z_n) or ϕ_m (after modifications) of the interpolation function should satisfy *a priori* the end conditions (for vibration problems only the displacement conditions have to be satisfied). For example, for a simply supported plate strip in bending, the displacement function should be able to satisfy the conditions of both deflection w and normal curvature $\partial^2 w/\partial y^2$ being equal to zero at the two ends.

(ii) The polynomial part of the interpolation function ($f_m(x)$, $f_m(x,z)$ and $f_{mn}(x)$) must be able to represent a state of constant 'strain' in the transverse (x) direction or x-z plane. If this is not obeyed, then there is no guarantee of convergence as the mesh is further and further refined. The methods for the checking of the constant 'strain' conditions are discussed in detail in Appendix 2.A.

(iii) The interpolation function must satisfy the compatibility requirement along the boundaries with the neighbouring strips, and this might include the continuity of the partial derivatives of the interpolated variables as well as the variables themselves.
The above statement in the stress analysis context can also be rephrased as:
"*the displacement function should be chosen in such a way that the strains which are required in the energy formulations should remain finite at the interface between the strips*".
Thus for in-plane and three-dimensional elasticity the strains involved are the first partial derivatives, and only the displacements need to be continuous. On the other hand, for bending problems the strains involved are the second partial derivatives of the deflection (curvatures), and therefore both the deflection and its first derivatives have to be continuous along the interface.
If such conditions are complied with, then there will be no infinite strains at the interface and therefore no contribution to the energy formulation from the interface, which can be considered as a narrow strip of area converging to zero. Only in this way can we be assured that a simple summation of the total potential energy of all strips will in fact be equal to the potential energy of the elastic body in question. The total potential energy of such a finite strip representation will always provide an approximate energy greater than the true one and, therefore, give a bound to the absolute total potential energy of the elastic system. A detailed mathematical discussion of this condition was presented in a paper by Tong and Pian.[(2)]
Similar argument for field problems, which are described by second order differential equations, leads to the fact that only the variables, such as

excess pore water pressure and temperature, have to be continuous along the boundaries.

2.2 AVAILABLE INTERPOLATION FUNCTIONS FOR CLASSICAL FINITE STRIPS

Since an interpolation function is always made up of two parts [see Eqs. 2.1a to 2.1c], it would be more convenient for us to discuss each part separately. The polynomial component (shape function) is governed by the shape of the cross-section (e.g. line, triangle, etc.) together with the nodal arrangement within the cross-section, and the series part is determined by the end conditions.

2.2.1 SERIES PART OF INTERPOLATION FUNCTIONS

Depending on the boundary conditions as well as the nature of the problems, different series have been sought. Some of the choices are listed as follows:

(1) Vibration eigenfunctions
(2) Buckling eigenfunctions
(3) Exponential functions
(4) Decaying power series

2.2.1.1 Vibration eigenfunctions

The most commonly used series for stress analysis are the basic functions (or eigenfunctions) which are derived from the solution of the beam vibration differential equation

$$\frac{\partial^4 Y}{\partial y^4} = \frac{\mu^4}{l^4} Y \tag{2.2}$$

where l is the length of the strip and μ is a parameter.

The general solution of Eq. 2.2 is

$$Y(y) = C_1 \sin(\frac{\mu y}{l}) + C_2 \cos(\frac{\mu y}{l}) + C_3 \sinh(\frac{\mu y}{l}) + C_4 \cosh(\frac{\mu y}{l}) \tag{2.3}$$

with the coefficients C_1, etc., to be determined by the end conditions. These have been worked out explicitly in the literature[3] for the various end conditions and are listed below:

(a) Both ends simply supported ($Y(0) = Y''(0) = 0$, and $Y(l) = Y''(l) = 0$)

$$Y_m(y) = \sin(\frac{\mu_m y}{l}) \qquad (\mu_m = \pi, 2\pi, 3\pi, \ldots m\pi) \tag{2.4}$$

(b) Both ends clamped ($Y(0) = Y'(0) = 0$, and $Y(l) = Y'(l) = 0$)

$$Y_m(y) = \sin(\frac{\mu_m y}{l}) - \sinh(\frac{\mu_m y}{l}) - \alpha_m[\cos(\frac{\mu_m y}{l}) - \cosh(\frac{\mu_m y}{l})] \tag{2.5}$$

$$\alpha_m = \frac{\sin\mu_m - \sinh\mu_m}{\cos\mu_m - \cosh\mu_m} \qquad (\mu_m = 4.7300,\ 7.8532,\ 10.9960,..\ \frac{2m+1}{2}\pi)$$

(c) One end simply supported and the other end clamped ($Y(0) = Y''(0) = 0$, and $Y(l) = Y'(l) = 0$)

$$Y_m(y) = \sin(\frac{\mu_m y}{l}) - \alpha_m \sinh(\frac{\mu_m y}{l}) \tag{2.6}$$

$$\alpha_m = \frac{\sin\mu_m}{\sinh\mu_m} \qquad (\mu_m = 3.9266,\ 7.0685,\ 10.2102,...\ \frac{4m+1}{4}\pi)$$

(d) Both ends free ($Y''(0) = Y'''(0) = 0$, and $Y''(l) = Y'''(l) = 0$)

$$Y_1(y) = 1, \qquad \mu_1 = 0 \tag{2.7a}$$

$$Y_2(y) = 1 - \frac{2y}{l} \qquad \mu_2 = 1 \tag{2.7b}$$

$$Y_m(y) = \sin(\frac{\mu_m y}{l}) + \sinh(\frac{\mu_m y}{l}) - \alpha_m[\cos(\frac{\mu_m y}{l}) + \cosh(\frac{\mu_m y}{l})] \tag{2.7c}$$

$$\alpha_m = \frac{\sin\mu_m - \sinh\mu_m}{\cos\mu_m - \cosh\mu_m} \qquad (\mu_m = 4.7300,\ 7.8532,\ ..\ \frac{2m-3}{2}\pi,\ m=3,4,\ ....)$$

(e) One end clamped and the other end free ($Y(0) = Y'(0) = 0$, and $Y''(l) = Y'''(l) = 0$)

$$Y_m(y) = \sin(\frac{\mu_m y}{l}) - \sinh(\frac{\mu_m y}{l}) - \alpha_m[\cos(\frac{\mu_m y}{l}) - \cosh(\frac{\mu_m y}{l})] \tag{2.8}$$

$$\alpha_m = \frac{\sin\mu_m + \sinh\mu_m}{\cos\mu_m + \cosh\mu_m} \qquad (\mu_m = 1.875,\ 4.694,\ ....\ \frac{2m-1}{2}\pi)$$

(f) One end simply supported and the other end free ($Y(0) = Y''(0) = 0$, and $Y''(l) = Y'''(l) = 0$)

$$Y_1(y) = \frac{y}{l}, \qquad \mu_1 = 1 \tag{2.9a}$$

$$Y_m(y) = \sin(\frac{\mu_m y}{l}) + \alpha_m \sinh(\frac{\mu_m y}{l})] \tag{2.9b}$$

$$\alpha_m = \frac{\sin\mu_m}{\sinh\mu_m} \qquad (\mu_m = 3.9266,\ 7.0685,\ 10.2102,...\ \frac{4m-3}{4}\pi)$$

(g) Continuous beam[4]

To derive the series solution for a continuous beam consisting of S spans (Figure 2.1), the eigenfunctions can be computed by using a general stiffness approach. The vibration of the s-th span of such a beam, with reference to the local coordinate system, is defined by:

$$\frac{d^4 Y_s}{dy^4} - \frac{m_s \omega^2}{E_s I_s} Y_s = 0 \tag{2.10}$$

Y_s is the amplitude of the mode and ω is the natural frequency of the beam. I_s, m_s, and E_s are respectively the second moment area of the section, the mass per unit length and Young's modulus.

It can be shown readily that the general solution (basis function) for Eq. 2.10 is

$$Y_{sm} = A_{sm}\sin G_s\mu_m y_s + B_{sm}\cos G_s\mu_m y_s + C_{sm}\sinh G_s\mu_m y_s + D_{sm}\cosh G_s\mu_m y_s \quad (2.11)$$

where $\mu_m = \left[\frac{m\omega_m^2}{EI}\right]^{1/4}$ and $G_s = \left[\frac{m_s}{E_s I_s}\right]^{1/4}\left[\frac{m}{EI}\right]^{-1/4}$. E, I and m are the reference values of the corresponding variables and are usually taken as the parameters of the first span.

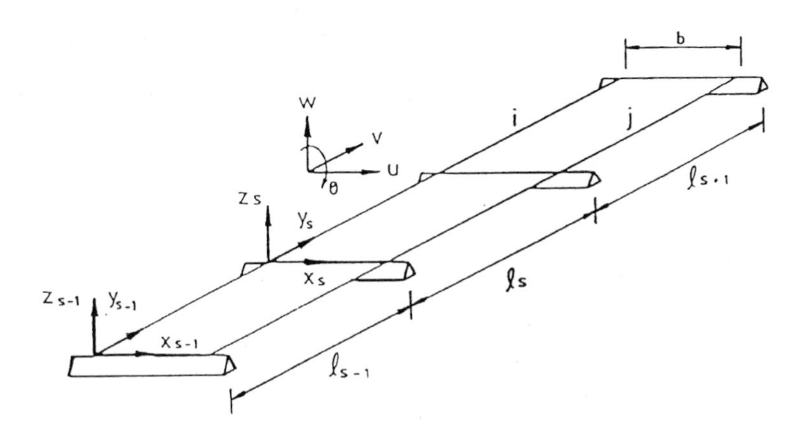

Figure 2.1 A typical continuous finite strip.

The undetermined coefficients, μ_m, A_{sm}, B_{sm}, C_{sm}, D_{sm} are determined by considering the boundary conditions at both end supports and at all intermediate supports. They can be determined by the procedures described as follows.

Inserting the coordinates of the two end supports of the s-th span, the end deflections and rotations can be written as

$$\begin{Bmatrix} Y_{sm}(0) \\ \theta_{sm}(0) \\ Y_{sm}(l_s) \\ \theta_{sm}(l_s) \end{Bmatrix} = \begin{bmatrix} 0 & 1 & 0 & 1 \\ -\beta & 0 & -\beta & 0 \\ s_1 & c_1 & s_2 & c_2 \\ -\beta c_1 & \beta s_1 & -\beta c_2 & \beta s_s \end{bmatrix} \begin{Bmatrix} A_{sm} \\ B_{sm} \\ C_{sm} \\ D_{sm} \end{Bmatrix} \quad (2.12)$$

The two end moments for this particular span can also be written in terms of the undetermined coefficients as:

$$\begin{bmatrix} M_{sm}(0) \\ M_{sm}(l_s) \end{bmatrix} = E_s I_s \beta^2 \begin{bmatrix} 0 & -1 & 0 & 1 \\ s_1 & c_1 & s_2 & c_s \end{bmatrix} \begin{Bmatrix} A_{sm} \\ B_{sm} \\ C_{sm} \\ D_{sm} \end{Bmatrix} \tag{2.13}$$

In the above equations (Eqs. 2.12 and 2.13) the following definitions are adopted:

$$\theta_{sm} = -\frac{dY_{sm}}{dy_s}, \; M_{sm} = -E_s I_s \frac{d^2 Y_{sm}}{d^2 y_s}, \; \beta = G_s \mu_m$$

$s_1 = \sin\beta l_s$, $s_2 = \sinh\beta l_s$, $c_1 = \cos\beta l_s$, $c_2 = \cosh\beta l_s$

Taking into account the fact that the intermediate supports are rigid, end moments and slopes for each span (except the first and last spans) are related as:

$$\begin{Bmatrix} M_{sm}(0) \\ M_{sm}(l_s) \end{Bmatrix} = \begin{bmatrix} \alpha(l_s) & \gamma(l_s) \\ \gamma(l_s) & \alpha(l_s) \end{bmatrix} \begin{Bmatrix} \theta_{sm}(0) \\ \theta_{sm}(l_s) \end{Bmatrix} \tag{2.14}$$

in which $\alpha(l_s) = E_s\, I_s\, \beta(s_1\, c_2 - s_2\, c_1)/(1 - c_1\, c_2)$, and

$\gamma(l_s) = E_s\, I_s\, \beta(s_2 - s_1)/(1 - c_1\, c_2)$.

Having established the moment equilibrium and compatibility conditions at all intermediate supports, the following matrix equation can be obtained:

$$\begin{bmatrix} \gamma(l_1) & \alpha(l_1)+\alpha(l_2) & \gamma(l_2) & 0 & . & 0 & 0 & 0 & 0 \\ & \gamma(l_2) & \alpha(l_2)+\alpha(l_3) & \gamma(l_3) & . & 0 & 0 & 0 & 0 \\ & & & & . & . & . & . & . \\ & & & & & . & . & . & . \\ & & & & & & \gamma(l_{s-1}) & 0 & 0 \\ & & & & & \gamma(l_{s-1}) & \alpha(l_{s-1})+\alpha(l_s) & 0 & 0 \end{bmatrix} \begin{Bmatrix} \theta_{1m}(0) \\ \theta_{2m}(0) \\ \theta_{3m}(0) \\ . \\ . \\ \theta_{sm}(0) \\ \theta_{sm}(l_s) \\ Y_{1m}(0) \\ Y_{sm}(l_s) \end{Bmatrix} = \begin{Bmatrix} 0 \\ 0 \\ 0 \\ . \\ . \\ 0 \\ 0 \\ 0 \\ 0 \end{Bmatrix} \tag{2.15}$$

Incorporating the boundary conditions at the end supports (two at each end), this equation is reduced to a square matrix. As examples, the reduced matrix equations for a three span beam with simply supported, free and clamped ends respectively are given as follows:[(4)]

(a) simply supported ends

$$\begin{bmatrix} \alpha(l_1) & \gamma(l_1) & 0 & 0 \\ \gamma(l_1) & \alpha(l_1)+\alpha(l_2) & \gamma(l_2) & 0 \\ 0 & \gamma(l_2) & \alpha(l_2)+\alpha(l_3) & \gamma(l_3) \\ 0 & 0 & \gamma(l_3) & \alpha(l_3) \end{bmatrix} \begin{Bmatrix} \theta_{1m}(0) \\ \theta_{2m}(0) \\ \theta_{3m}(0) \\ \theta_{3m}(l_3) \end{Bmatrix} = \begin{Bmatrix} 0 \\ 0 \\ 0 \\ 0 \end{Bmatrix} \quad (2.16)$$

(b) free ends

$$\begin{bmatrix} \alpha(l_1) & -\beta\eta_1 & \gamma(l_1) & 0 & 0 & 0 \\ -\beta\eta_1 & \beta^2\xi_1 & -\beta\zeta_1 & 0 & 0 & 0 \\ \gamma(l_1) & -\beta\zeta_1 & \alpha(l_1)+\alpha(l_2) & \gamma(l_2) & 0 & 0 \\ 0 & 0 & \gamma(l_2) & \alpha(l_2)+\alpha(l_3) & \gamma(l_3) & \beta\zeta_3 \\ 0 & 0 & 0 & \gamma(l_3) & \alpha(l_3) & \beta\eta_3 \\ 0 & 0 & 0 & \beta\zeta_3 & \beta\eta_3 & \beta^2\xi_3 \end{bmatrix} \begin{Bmatrix} \theta_{1m}(0) \\ Y_{1m}(0) \\ \theta_{2m}(0) \\ \theta_{3m}(0) \\ \theta_{3m}(l_3) \\ Y_{3m}(l_3) \end{Bmatrix} = \begin{Bmatrix} 0 \\ 0 \\ 0 \\ 0 \\ 0 \\ 0 \end{Bmatrix} \quad (2.17)$$

where $\zeta_s = E_s I_s \beta (c_2 - c_1)/(1 - c_1 c_2)$; $\eta_s = E_s I_s \beta s_2 s_1/(1 - c_1 c_2)$

$\xi_s = E_s I_s \beta (s_1 c_2 + c_1 s_2)/(1 - c_1 c_2)$

(c) clamped ends

$$\begin{bmatrix} \alpha(l_1)+\alpha(l_2) & \gamma(l_2) \\ \gamma(l_2) & \alpha(l_2)+\alpha(l_3) \end{bmatrix} \begin{Bmatrix} \theta_{2m}(0) \\ \theta_{3m}(0) \end{Bmatrix} = \begin{Bmatrix} 0 \\ 0 \end{Bmatrix} \quad (2.18)$$

In the next step, the non-trivial solution of μ_m is obtained by equating the determinant of this matrix to zero. The solution can be obtained either by the Newton-Raphson method[4] or 'modified regula falsi' method.[4] Attention should be drawn here to the fact that for the case of a beam clamped at both ends, Eq. 2.18 will miss out on the roots corresponding to the situation in which the whole displacement vector is equal to zero, that is, the continuous structure degenerates into a series of single span substructures with both end clamped. The eigenvalues of such beams are, of course, common knowledge and they can be inserted into the correct positions in the ascending array of μ_m. The position of missing roots can also be determined by using a method proposed by Wittrick and Williams[6] in which a 'sign count' of the matrix is made.

Having μ_m determined, the eigenfunction coefficients (A_{sm} to D_{sm}) can be determined by utilizing the moment equilibrium, the compatibility condition and the zero deflection conditions at intermediate and end supports,

a) for all spans

$$Y_s(0) = 0$$

$$Y_{s-1}(l_{s-1}) = 0$$

b) for all intermediate supports

$$Y'_{s-1}(l_{s-1}) = Y'_{s}(l_{s})$$

$$EI_{s-1}Y''_{s-1}(l_{s-1}) = EI_{s}Y''_{s}(l_{s})$$

c) for simple end support

$$Y''_1(0) = 0$$

d) for clamped end support

$$Y'_1(0) = 0$$

If A_{1m} is taken as a unit reference constant, the other coefficients can be determined accordingly. Referring to the simply-supported three span example once again, the independent unknowns can be chosen to be[4]

$$[A_{1m}, B_{1m}, C_{1m}, A_{2m}, B_{2m}, C_{2m}, A_{3m}, B_{3m}, C_{3m}]$$

and they can be determined by solving the following matrix equation:

$$\begin{Bmatrix} 1 \\ 0 \\ 0 \\ 0 \\ 0 \\ 0 \\ 0 \\ 0 \\ 0 \end{Bmatrix} = [ABC] \begin{Bmatrix} A_{1m} \\ B_{1m} \\ C_{1m} \\ A_{2m} \\ B_{2m} \\ C_{2m} \\ A_{3m} \\ B_{3m} \\ C_{3m} \end{Bmatrix} \quad (2.19)$$

The matrix [ABC] is given explicitly in Table 2.1.

The above functions [Cases (a) to (g)] are primarily used for bending strips. For two- and three-dimensional elasticity problems, which will be discussed in more detail in Chapter 4, both Y_m and Y_m' will be used for the u, v (and w) displacements. Only one other function has been employed successfully in plane analysis and will be discussed later.

Some of these basic functions also possess the valuable properties of orthogonality, i.e. for $m \neq n$

$$\int_0^l Y_m Y_n \, dy = 0 \quad (2.20a)$$

$$\int_0^l Y''_m Y''_n \, dy = 0 \quad (2.20b)$$

It will be observed that these integrals appear in all subsequent formulations, and the utilization of these properties results in a significant saving in computational effort.

TABLE 2.1
Matrix [ABC] (simply supported three span beam)[4].

$$
\begin{bmatrix}
1 & 0 & 0 & 0 & 0 & 0 & 0 & 0 & 0 \\
\sin\mu_m l_1 & \cos\mu_m l_1 - \cosh\mu_m l_1 & \sinh\mu_m l_1 & 0 & 0 & 0 & 0 & 0 & 0 \\
0 & 0 & 0 & \sin\mu_m l_2 & \cos\mu_m l_2 - \cosh\mu_m l_2 & \sinh\mu_m l_2 & 0 & 0 & 0 \\
0 & 0 & 0 & 0 & 0 & 0 & \sin\mu_m l_3 & \cos\mu_m l_3 - \cosh\mu_m l_3 & \sinh\mu_m l_3 \\
\cos\mu_m l_1 & -\sin\mu_m l_1 - \sinh\mu_m l_1 & \cosh\mu_m l_1 & -1 & 0 & -1 & 0 & 0 & 0 \\
0 & 0 & 0 & \cos\mu_m l_2 & -\sin\mu_m l_2 - \sinh\mu_m l_2 & \cosh\mu_m l_2 & -1 & 0 & -1 \\
-\sin\mu_m l_1 & -\cos\mu_m l_1 - \cosh\mu_m l_1 & \sinh\mu_m l_1 & 0 & -2 & 0 & 0 & 0 & 0 \\
0 & 0 & 0 & -\sin\mu_m l_2 & -\cos\mu_m l_2 - \cosh\mu_m l_2 & \sinh\mu_m l_2 & 0 & -2 & 0 \\
0 & -2 & 0 & 0 & 0 & 0 & 0 & 0 & 0
\end{bmatrix}
$$

The existence of so many functions should in no way be interpreted as a need for a matching number of computer subroutines. In actual practice all the relevant functions are stored in the same subroutine, and integrals involving different combinations of the functions are integrated numerically. Consequently problems with many different types of boundary conditions can be solved by the same program.

2.2.1.2 Buckling eigenfunctions

In stability analysis, more accurate results can be obtained if the buckling mode shape is used for the interpolation series. For example, the functions for a structure clamped at the base and free at the tip can be chosen to be:[7]

$$Y_m = 1 - \cos\frac{(2m-1)\pi y}{2} \tag{2.21}$$

2.2.1.3 Exponential functions

Attempts have been made to model the behaviour of layered medium of infinite extent in the horizontal plane by exponential functions,[8]

a) rectangular coordinate system

$$Y_m(x) = \exp\left(-\frac{mx}{L}\right) \tag{2.22}$$

b) circular coordinate system

$$Y_m(r) = \exp\left(-\frac{mr}{R}\right) \tag{2.23}$$

where L and R are length constants and are sufficiently large in magnitude so that the infinite extent of the medium can be modelled approximately. In most cases, only one term is necessary for the analysis.

Exponential series in the time variable have also been used in the solution of the parabolic differential equation.[9] For example, the temperature field is defined as:

a) one-dimensional problem

$$T(x,\tau) = \sum_{m=1}^{r} f_m(x)\left[1 - \exp\left(-\frac{m^2\pi^2\tau}{L_x^2}\right)\right] \tag{2.24a}$$

b) two-dimensional problem

$$T(x,y,\tau) = \sum_{m=1}^{r}\sum_{n=1}^{r} f_{mn}(x,y)\left[1 - \exp\left(-\frac{m^2\pi^2\tau}{L_x^2} - \frac{n^2\pi^2\tau}{L_y^2}\right)\right] \tag{2.24b}$$

c) three-dimensional problem

$$T(x,y,z,\tau) = \sum_{m=1}^{r}\sum_{n=1}^{r}\sum_{k=1}^{r} f_{mnk}(x,y,z)\left[1 - \exp\left(-\frac{m^2\pi^2\tau}{L_x^2} - \frac{n^2\pi^2\tau}{L_y^2} - \frac{k^2\pi^2\tau}{L_z^2}\right)\right] \tag{2.24c}$$

where τ is the temporal variable; L_x , L_y and L_z are characteristic lengths in the corresponding directions.

2.2.1.4 Decaying power series

Instead of exponential series, decaying power series may also be chosen to model a medium of infinite extent in two directions,[10] such as layered soils. In order to reflect the difference in behaviour between the near and far fields, the series should consist of two parts:

a) near field functions ($r < R$): series of r^m

$$Y_1^{\,1}(r) = 1 \tag{2.25a}$$

$$Y_2^{\,1}(r) = \frac{r}{R} \tag{2.25b}$$

$$Y_3^{\,1}(r) = 3\left(\frac{r}{R}\right) - 4\left(\frac{r}{R}\right)^2 \tag{2.25c}$$

$$Y_m^{\,1}(r) = \sum_{k=1}^{m-1} (-1)^{k-1} \frac{(m+k-1)!}{(m-k-1)!(k+1)!(k-1)!}\left(\frac{r}{R}\right)^k \tag{2.25d}$$

b) far field function ($r > R$): series of $1/r^m$

$$Y_m^{\,2}(r) = \left(\frac{R}{r}\right)^m \tag{2.25e}$$

in which R is distance of the interface of the far and near field from the centre.

Enforcing the continuity along the interface, it is possible to relate the two sets of interpolation parameters. For example, if L terms are chosen for the far field functions, we will have

$$\Sigma Y_m^{\,1}(R)\,\delta_m^{\,1} = \Sigma Y_m^{\,2}(R)\,\delta_m^{\,2} \tag{2.26a}$$

$$\Sigma \frac{d}{dr} Y_m^{\,1}(R)\,\delta_m^{\,1} = \Sigma \frac{d}{dr} Y_m^{\,2}(R)\,\delta_m^{\,2} \tag{2.26b}$$

$$\Sigma \frac{d^{L-1}}{dr^{L-1}} Y_m^{\,1}(R)\,\delta_m^{\,1} = \Sigma \frac{d^{L-1}}{dr^{L-1}} Y_m^{\,2}(R)\,\delta_m^{\,2} \tag{2.26c}$$

where $\delta_m^{\,1}$ and $\delta_m^{\,2}$ are the interpolation parameters for the near field and far fields respectively.

In matrix form,

$$\mathbf{A}\,\Delta^1 = \mathbf{B}\,\Delta^2 \tag{2.27}$$

where Δ^1 and Δ^2 are the vectors containing $\delta_m^{\,1}$ and $\delta_m^{\,2}$.

The elements ($\mathbf{a}_{ij}$ and $\mathbf{b}_{ij}$) of $\mathbf{A}$ and $\mathbf{B}$ are defined as follows:

$$\mathbf{a}_{ij} = \mathbf{r}_{ik}\,\mathbf{a}_{kl}^{*}\,\mathbf{c}_{jl} \text{ and } \mathbf{b}_{ij} = \mathbf{r}_{ik}\,\mathbf{b}_{kl}^{*}$$

$$\mathbf{r}_{ij} = \begin{cases} \left(\frac{1}{R}\right)^{i-1} & i = j \\ 0 & i \neq j \end{cases} \quad ; \quad \mathbf{a}_{ij}^{*} = \begin{cases} \frac{(i-1)!}{(j-1)!} & i \geq j \\ 0 & i < j \end{cases}$$

$$\mathbf{c}_{ij} = \begin{cases} 1 & i = j = 1 \\ 0 & i = 1 \neq j \text{ or } i \neq j = 1 \text{ or } i < j \\ (-1)^{j-2} \dfrac{(i+j-2)!}{(i-j)!(j-2)!j!} & i, j > 1 \text{ and } i \geq j \end{cases}$$

$$\mathbf{b}_{ij}^{*} = (-1)^{i-1} \frac{(i+j-2)!}{(i-1)!}$$

On simplification, the n-th term of the series can be written as

$$Y_m = \Gamma_{r_0 R}\, Y_m^1 + \Gamma_{R\infty} \sum_{l=1}^{L} e_{lm}\, Y_m^2 \tag{2.28}$$

where $e_{lm} = \sum_{k=1}^{L} b_{lk} a_{km}$

$$\Gamma_{pq} = \begin{cases} 1 & p \leq r \leq q \\ 0 & \text{elsewhere} \end{cases}$$

2.2.2 SHAPE FUNCTION PART OF THE INTERPOLATION FUNCTION

A polynomial shape function is a polynomial associated with a nodal interpolation parameter, and it describes the corresponding interpolation field within the cross-section of a strip when the nodal interpolation parameter in question is given a unit value. Such shape functions are, in fact, derived by specifying a normalized unit value of the relevant interpolation component at its own node, and a value of zero for the same interpolation component at all other nodes.

For example, the shape functions for a strip with three nodes (Eq. 2.1a) can be written as

$$w = \sum_{m=1}^{r} [C_1 \quad C_2 \quad C_3] \begin{Bmatrix} \delta_1 \\ \delta_2 \\ \delta_3 \end{Bmatrix}_m Y_m = \sum_{m=1}^{r} Y_m \sum_{k=1}^{3} [C_k]\{\delta_k\}_m \tag{2.29}$$

in which $\{\delta_k\}_m$ is a vector representing the m-th term nodal displacement parameters (deflection) at nodes 1, 2, and 3, and C_1 , C_2 , C_3 are the shape functions associated with δ_1, δ_2, δ_3 respectively.

In order to satisfy the stated criterion, it can be shown readily that

at $x=0$ $\quad [C_1\ C_2\ C_3] = [1\ 0\ 0]$

at $x=b/2$ $\quad [C_1\ C_2\ C_3] = [0\ 1\ 0]$

and at $x=b$ $\quad [C_1\ C_2\ C_3] = [0\ 0\ 1]$

These three relations suggest that the shape functions should be second degree polynomials. The methods for the derivation of such shape functions can be found in most standard finite element texts.

The main purpose of using the shape function directly instead of a simple polynomial with the undetermined constants is twofold: to avoid the lengthy process of relating the undetermined constants to the nodal displacement parameters, and to make sure that there is compatibility of displacement along the adjoining strips (prisms). The first point is rather obvious and requires no discussion. The second point can be best explained by noting that the displacements along any interface of adjoining strips (prisms) are uniquely determined by the displacement parameters at the node (or nodes) common to the adjoining strips (prisms), since by definition the shape function for the displacement parameters of any other node will take up zero value at the said interface. Many shape functions are available, and some of the more common ones are listed below.

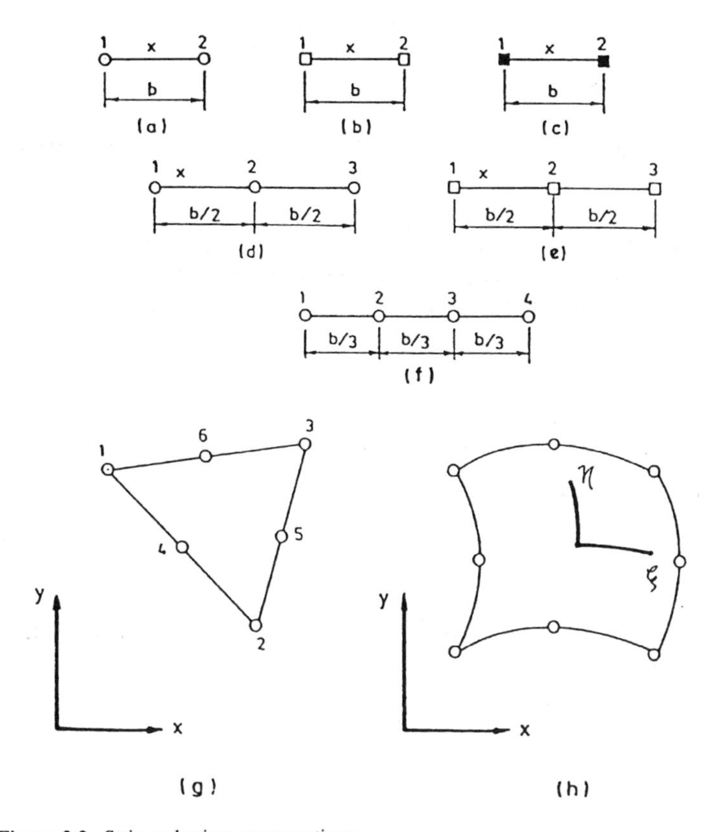

Figure 2.2 Strip and prism cross-sections.

Nodal DOF : ○ = f; □ = f, $\frac{\partial f}{\partial x}$; ■ = f, $\frac{\partial f}{\partial x}$; $\frac{\partial^2 f}{\partial x^2}$

(a) Straight line with two nodes (Figure 2.2a) and with variables (zero derivatives) as nodal parameters:

$$C_1 = (1-\bar{x}) \quad , \quad C_2 = \bar{x} \tag{2.30}$$

where $\bar{x} = x/b$

(b) Straight line with two nodes (Figure 2.2b); variables and first derivatives:

$$[C_1] = [(1-3\bar{x}^2+2\bar{x}^3)\, ,\, x(1-2\bar{x}+\bar{x}^2)\,] \tag{2.31a}$$
$$[C_2] = [(3\bar{x}^2-2\bar{x}^3)\, ,\, x(\bar{x}^2-\bar{x})\,] \tag{2.31b}$$

(c) Straight line with two nodes (Figure 2.2c); variables, first and second derivatives:

$$[C_1] = [(1-10\bar{x}^3+15\bar{x}^4-6\bar{x}^5)\, ,\, x(1-6\bar{x}^2+8\bar{x}^3-3\bar{x}^4),\; x^2(0.5-1.5\bar{x}+1.5\bar{x}^2-0.5\bar{x}^3)] \tag{2.32a}$$
$$[C_2] = [(10\bar{x}^3-15\bar{x}^4+6\bar{x}^5)\, ,\, x(-4\bar{x}^2+7\bar{x}^3-3\bar{x}^4),\; x^2(0.5\bar{x}-\bar{x}^2+0.5\bar{x}^3)] \tag{2.32b}$$

(d) Straight line with three nodes (Figure 2.2d); variables only:

$$C_1 = (1-3\bar{x}+2\bar{x}^2) \quad , \quad C_2 = (4\bar{x}-4\bar{x}^2) \quad , \quad C_3 = (-\bar{x}+2\bar{x}^2) \tag{2.33}$$

(e) Straight line with three nodes (Figure 2.2e); variables and first derivatives:

$$[C_1] = [(1-23\bar{x}^2+66\bar{x}^3-68\bar{x}^4+24\bar{x}^5)\, ,\, x(1-6\bar{x}+13\bar{x}^2-12\bar{x}^3+4\bar{x}^4)] \tag{2.34a}$$
$$[C_2] = [(16\bar{x}^2-32\bar{x}^3+16\bar{x}^4)\, ,\, x(-8\bar{x}+32\bar{x}^2-40\bar{x}^3+16\bar{x}^4)] \tag{2.34b}$$
$$[C_3] = [(7\bar{x}-34\bar{x}^3+52\bar{x}^4-24\bar{x}^5)\, ,\, x(-\bar{x}+5\bar{x}^2-8\bar{x}^3+4\bar{x}^4)] \tag{2.34c}$$

(f) Straight line with four nodes (Figure 2.2f); variables only

$$C_1 = -\frac{9}{2}\,(\bar{x}-\frac{1}{3})\,(\bar{x}-\frac{2}{3})\,(\bar{x}-1) \tag{2.35a}$$

$$C_2 = \frac{27}{2}\,\bar{x}\,(\bar{x}-\frac{2}{3})\,(\bar{x}-1) \tag{2.35b}$$

$$C_3 = -\frac{27}{2}\,\bar{x}\,(\bar{x}-\frac{1}{3})\,(\bar{x}-1) \tag{2.35c}$$

$$C_4 = \frac{9}{2}\,\bar{x}\,(\bar{x}-\frac{1}{3})\,(\bar{x}-\frac{2}{3}) \tag{2.35d}$$

(g) Triangle with six nodes (Figure 2.2g); variables only:

(1) Corner nodes: $C_i = (2L_i - 1)L_i \qquad (i=1,2,3)$ (2.36a)

(2) Midside nodes: $C_4 = 4L_1 L_2,\ C_5 = 4L_2 L_3,\ C_6 = 4L_3 L_1,$ (2.36b)

in which L_i, etc., are area coordinates.

A useful formula for integrating area coordinate quantities over the area of a triangle is

$$\iint L_1^a L_2^b L_3^c \, dxdy = \frac{a!b!c!}{(a+b+c+2)!} 2\Delta \tag{2.37}$$

in which Δ refers to the area of the triangle.

(h) Serendipity and Lagrange elements: variable only. For example 8-noded Serendipity elements (in natural coordinates) (Figure 2.2h):

(1) Corner nodes:

$$C_i = \frac{1}{4}(1+\xi_o)(1+\eta_o)(\xi_o+\eta_o-1) \tag{2.38a}$$

(2) Midside nodes ($\xi_i = 0$)

$$C_i = \frac{1}{2}(1-\xi^2)(1+\eta_o) \tag{2.38b}$$

TABLE 2.2

Family of Serendipity and Lagrange elements ($\xi_o = \xi\xi_i$, $\eta_o = \eta\eta_i$).

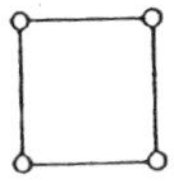	$\frac{1}{4}(1+\xi_o)(1+\eta_o)$
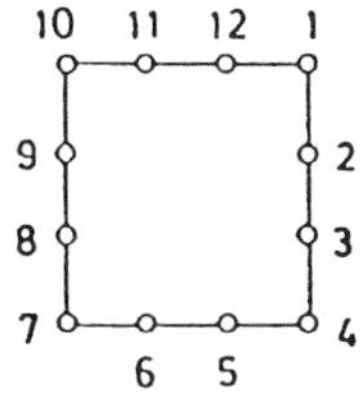 10 11 12 1 / 9 2 / 8 3 / 7 4 / 6 5	$\frac{1}{32}(1+\xi_o)(1+\eta_o)[9(\xi^2+\eta^2)-10]$ (Corner nodes) $\frac{9}{32}(1+\xi_o)(1-\eta^2)(1+9\eta_o)$ (Nodes : 2,3,8,9) $\frac{9}{32}(1+9\xi_o)(1-\xi^2)(1+\eta_o)$ (Nodes : 5,6,11,12)
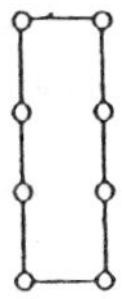	$\frac{1}{32}(1+\xi_o)(1+\eta_o)(9\eta^2-1)$ (Corner nodes) $\frac{9}{32}(1+\xi_o)(1-\eta^2)(1+9\eta_o)$ (Side Nodes)

3) Midside nodes ($\eta_i = 0$)

$$C_i = \frac{1}{2}(1+\xi_o)(1-\eta^2) \tag{2.38c}$$

with $\xi_o = \xi\xi_i$, $\eta_o = \eta\eta_i$, and ξ_i and η_i being the ξ and η coordinates of the i-th node.

Table 2.2 lists a number of other members of the element families that have been called up for use in the finite strip analysis. In addition, such elements can also be used for domain (geometric) transformation[11].

(i) Higher order quadrilateral elements;[12] interpolation parameters include the variables as well as the derivatives of the variables. Some examples of such elements are given in Table 2.3.

TABLE 2.3
Higher order quadrilateral elements ($\xi_o = \xi\xi_i$, $\eta_o = \eta\eta_i$).

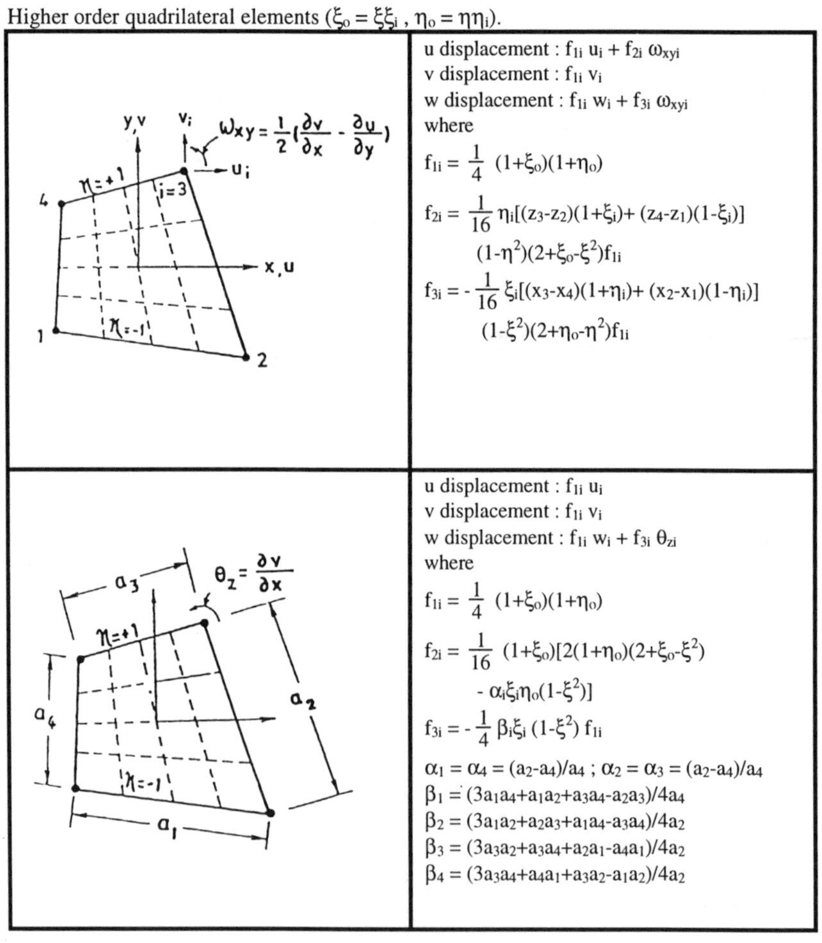

Element	Shape functions
	u displacement : $f_{1i}\, u_i + f_{2i}\, \omega_{xyi}$ v displacement : $f_{1i}\, v_i$ w displacement : $f_{1i}\, w_i + f_{3i}\, \omega_{xyi}$ where $f_{1i} = \frac{1}{4}(1+\xi_o)(1+\eta_o)$ $f_{2i} = \frac{1}{16}\eta_i[(z_3-z_2)(1+\xi_i)+(z_4-z_1)(1-\xi_i)]$ $(1-\eta^2)(2+\xi_o-\xi^2)f_{1i}$ $f_{3i} = -\frac{1}{16}\xi_i[(x_3-x_4)(1+\eta_i)+(x_2-x_1)(1-\eta_i)]$ $(1-\xi^2)(2+\eta_o-\eta^2)f_{1i}$
	u displacement : $f_{1i}\, u_i$ v displacement : $f_{1i}\, v_i$ w displacement : $f_{1i}\, w_i + f_{3i}\, \theta_{zi}$ where $f_{1i} = \frac{1}{4}(1+\xi_o)(1+\eta_o)$ $f_{2i} = \frac{1}{16}(1+\xi_o)[2(1+\eta_o)(2+\xi_o-\xi^2)$ $-\alpha_i\xi_i\eta_o(1-\xi^2)]$ $f_{3i} = -\frac{1}{4}\beta_i\xi_i(1-\xi^2)f_{1i}$ $\alpha_1 = \alpha_4 = (a_2-a_4)/a_4$; $\alpha_2 = \alpha_3 = (a_2-a_4)/a_4$ $\beta_1 = (3a_1a_4+a_1a_2+a_3a_4-a_2a_3)/4a_4$ $\beta_2 = (3a_1a_2+a_2a_3+a_1a_4-a_3a_4)/4a_2$ $\beta_3 = (3a_3a_2+a_3a_4+a_2a_1-a_4a_1)/4a_2$ $\beta_4 = (3a_3a_4+a_4a_1+a_3a_2-a_1a_2)/4a_2$

2.3 LONGITUDINAL FUNCTIONS FOR SPLINE FINITE STRIPS[13]

The spline finite strip was initially developed for the analysis of plates and shells. In such a model, a strip is divided into a number of sections by knots (nodes) (Figure 2.3), and variables are expressed in terms of the knot values.

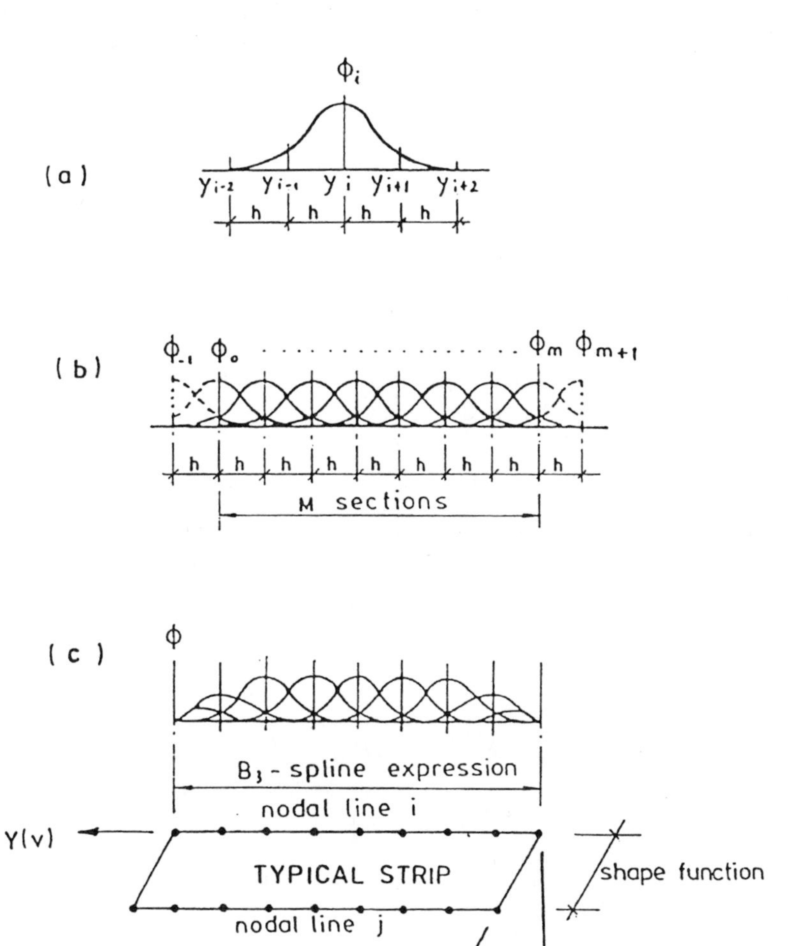

Figure 2.3 (a) Typical B-3 spline. (b) Basis of B-3 spline expression. (c) Displacement functions for typical strip.[13]

The shape functions, which consist of two parts, are written as the product of spline functions and suitable piecewise continuous polynomials (see Section 2.2.2). A variety of spline functions[14] are available, but it has been demonstrated that the B-3 spline function is the most appropriate choice for most plate and shell problems.

Each local spline function has non-zero values over four consecutive sections with the section knot $y=y_i$ as the centre. The functions with equal spacings are

$$\phi_i = \frac{1}{6h^3}\begin{cases} 0 & y < y_{i-2} \\ (y-y_{i-2})^3 & y_{i-2} \le y \le y_{i-1} \\ h^3 + 3h^2(y-y_{i-1}) + 3h(y-y_{i-1})^2 - 3(y-y_{i-1})^3 & y_{i-1} \le y \le y_i \\ h^3 + 3h^2(y_{i+1}-y) + 3h(y_{i+1}-y)^2 - 3(y_{i+1}-y)^3 & y_i \le y \le y_{i+1} \\ (y_{i+2}-y)^3 & y_{i+1} \le y \le y_{i+2} \\ 0 & y < y_{i+2} \end{cases} \tag{2.39a}$$

Spline functions of unequal spacings were also adopted in analyses, and the explicit forms of the local B-3 splines in the normalized variable η $(-1 \le \eta \le 1)$ are:

$$\phi_i = \begin{cases} 0 & \eta < \eta_{i-2} \\ A_i(\eta-\eta_{i-2})^3 & \eta_{i-2} \le \eta \le \eta_{i-1} \\ A_i(\eta-\eta_{i-2})^3 + C_i(\eta-\eta_{i-1})^3 & \eta_{i-1} \le \eta \le \eta_i \\ B_i(\eta_{i+2}-\eta)^3 + D_i(\eta_{i+1}-\eta)^3 & \eta_i \le \eta \le \eta_{i+1} \\ B_i(\eta_{i+2}-\eta\ )^3 & \eta_{i+1} \le \eta \le \eta_{i+2} \\ 0 & \eta < \eta_{i+2} \end{cases} \tag{2.39b}$$

where $A_i = [h_{i-1}\,(h_{i-1} + h_i\,)(h_{i-1}+h_i + h_{i+1}\,)]^{-1}$

$B_i = [h_{i+2}\,(h_{i+2} + h_{i+1}\,)(h_{i+2} + h_{i+1} + h_i\,)]^{-1}$

$C_i = -(h_{i-1} + h_i + h_{i+1} + h_{i+2}\,)[h_{i-1}\, h_i\, (h_{i+1} + h_i)(h_{i+2} + h_{i+1} + h_i\,)]^{-1}$

$D_i = -(h_{i-1} + h_i + h_{i+1} + h_{i+2})[h_{i+2}\, h_{i+1}\, (h_{i+1} + h_i\,)(h_{i+1} + h_i + h_{i-1}\,)]^{-1}$

$h_i = \eta_i - \eta_{i-1}$.

Therefore, the displacement function for the deflection of a lower order plate strip (Figure 2.3) is:

$$w = [[C_1]\ [C_2]]\begin{bmatrix} [\phi] & & & \\ & [\phi] & & \\ & & [\phi] & \\ & & & [\phi] \end{bmatrix}\{\delta\} \tag{2.40}$$

where C_1 and C_2 are given by Eq. 2.31,

$\{\delta\} = [(\alpha_w)_j^T\ (\alpha_{wr})_j^T\ (\alpha_w)_{j+1}^T\ (\alpha_{wr})_{j+1}^T]^T$

$[\,\bar{\phi}\,] = [\phi_{-1}\ \phi_o\ .....\phi_M\ \phi_{M+1}\,]$

$(\alpha_w)_j$ and $(\alpha_{wr})_j$ are the vectors containing the interpolation parameters for displacement and corresponding rotation, that is
$(\alpha_w)_j = [\alpha_{w,-1,j}\ \alpha_{w,o,j}\ .....\ \alpha_{w,M+1,j}\ ]^T$, and
$(\alpha_{wr})_j = [\alpha_{wr,-1,j}\ \alpha_{wr,o,j}\ .....\ \alpha_{wr,M+1,j}\ ]^T$

Fan[15] suggested that the spline function can be modified in such a way that instead of the α's, the variable (w) and its first as well as second derivatives (w_y and w_{yy}) can be chosen to be the interpolation parameters at the ends of the strip. Mathematically, $\{\alpha\}$ is modified as

$$\{\alpha\} = [T]\ \{\beta\} \tag{2.41}$$

where $\{\beta\} = \{w_1\ w_{y1}\ w_{yy1}\ \alpha_2\ ....\alpha_{M-2}\ w_{yyM}\ w_{yM}\ w_M\ \}$

$$[T] = \begin{bmatrix} 1 & -h & \frac{h^2}{3} & & & & & \\ 1 & 0 & \frac{-h^2}{6} & & & & & \\ 1 & h & \frac{h^2}{3} & & & & & \\ 0 & 0 & 0 & 1 & & & & \\ & & & & \cdot & & & \\ & & & & & \cdot & & \\ & & & & 1 & 0 & 0 & 0 \\ & & & & & 1 & -h & \frac{h^2}{3} \\ & & & & & 1 & 0 & \frac{-h^2}{6} \\ & & & & & 1 & h & \frac{h^2}{3} \end{bmatrix}$$

Eq. 2.40 then becomes

$$w = [[C_1]\ [C_2]] \begin{bmatrix} [\bar{\phi}] & & & \\ & [\bar{\phi}] & & \\ & & [\bar{\phi}] & \\ & & & [\bar{\phi}] \end{bmatrix} [T]\{\delta\} \tag{2.42}$$

where C_1 and C_2 are given by Eq. 2.31,
$\{\delta\} = [(\alpha_w)_j^T\ (\alpha_{wr})_j^T\ (\alpha_w)_{j+1}^T\ (\alpha_{wr})_{j+1}^T]^T$
$(\alpha_w)_j = [w_{1,j}\ w_{y1,j}\ w_{yy1,j}\ \alpha_{w,1,j}\ \alpha_{w,2,j}\ .....\ \alpha_{w,M-1,j}\ w_{yyM,j}\ w_{yM,j}\ w_{M,j}\ ]^T$
$(\alpha_{wr})_j = [w_{r1,j}\ w_{ry1,j}\ w_{ryy1,j}\ \alpha_{wr,1,j} .....\ \alpha_{wr,M-1,j}\ w_{ryyM,j}\ w_{ryM,j}\ w_{rM,j}]^T$
$[\phi] = [\phi_{-1}\ \phi_o\ .....\phi_M\ \phi_{M+1}\ ]$

2.4 COMPUTED SHAPE FUNCTIONS

2.4.1 LONGITUDINAL COMPUTED SHAPE FUNCTION STRIPS[16]

Figure 2.4a shows a typical strip with varying rigidity along the longitudinal direction. In order to generate the longitudinal component of the shape functions, a unit width of the strip is taken out as a beam with the same

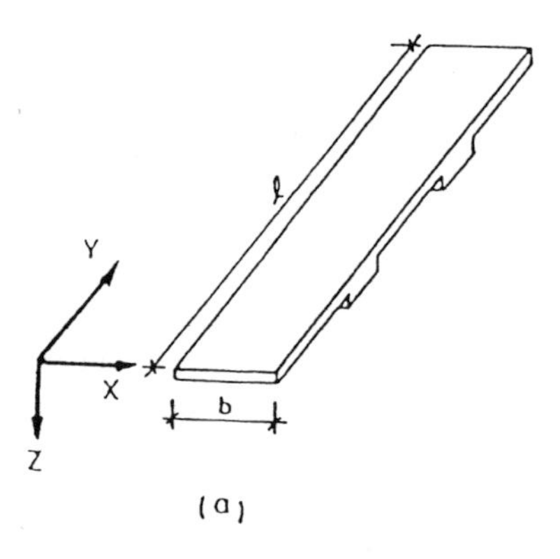

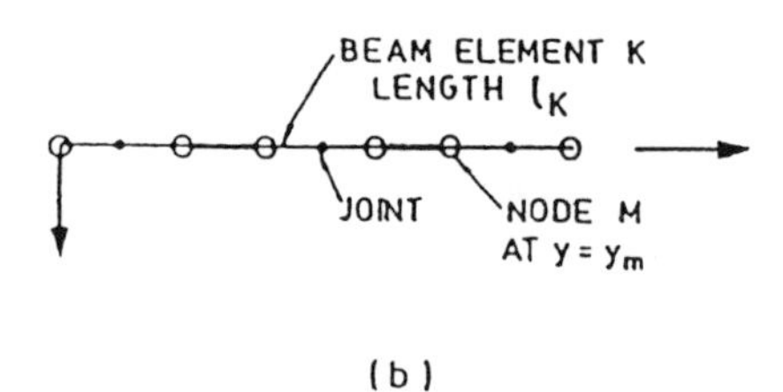

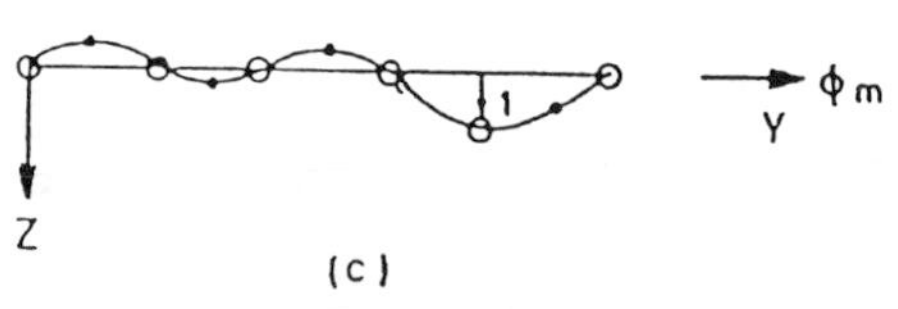

Figure 2.4 (a) A strip with varying longitudinal rigidity. (b) An equivalent beam of the strip in (a). (c) Deflection profile for a shape function Y_m.

variation of the longitudinal rigidity as the original strip (Figure 2.4b). The beam is divided into p beam elements which are not necessarily of equal length. A number of joints, say r, are assigned as nodes for the purpose of computing the longitudinal shape functions. Corresponding to such joints, a total of r computed shape functions can be obtained. These shape functions can be easily found by means of a simple continuous beam computer program. For example, the m-th mode of the shape function can be derived by determining the exact deflection pattern of a continuous beam subjected to a concentrated load at the node m (Figure 2.4c). Letting the deflection and rotation of the k-th beam with nodes k and k+1 be denoted by w_k , θ_k , w_{k+1} and θ_{k+1} respectively, one can show readily that the deflection of the beam can be written as

$$w = (1 - 3\bar{y}^2 + 2\bar{y}^3)w_k + y(1 - 2\bar{y} + \bar{y}^2)\theta_k + (3\bar{y}^2 - 2\bar{y}^3)w_{k+1} + y(\bar{y}^2 - \bar{y})\theta_{k+1}$$

$$= [N(\bar{y})]\{\alpha_m\}_k \tag{2.43}$$

where $[N(\bar{y})] = [(1-3\bar{y}^2+2\bar{y}^3)\ \ y(1-2\bar{y}+\bar{y}^2)\ \ (3\bar{y}^2-2\bar{y}^3)\ \ y(\bar{y}^2-\bar{y})]$

$\{\alpha_m\}_k = \{w_k\ \theta_k\ w_{k+1}\ \theta_{k+1}\}$

Therefore,

$$\phi_m(y) = \sum_{k=1}^{p} [N(\bar{y})]\{\alpha_m\}_k \tag{2.44}$$

The transverse components of the shape functions can again be chosen from Section 2.2.2.

2.4.2 TRANSVERSE COMPUTED SHAPE FUNCTION STRIPS[17]

The shape function can be obtained by interchanging x and y for Eq. 2.1d, that is

$$\delta = f(y)\ \phi_m(x) \tag{2.45}$$

f(y) is polynomial component (Section 2.2.2). ϕ_m is the computed part, and its derivation is based on the principle developed in the previous section. For example, to derive a three-cell box girder bridge with the cross-section as shown in Figure 2.5a, a plane frame (Figure 2.5b) can be used to represent the

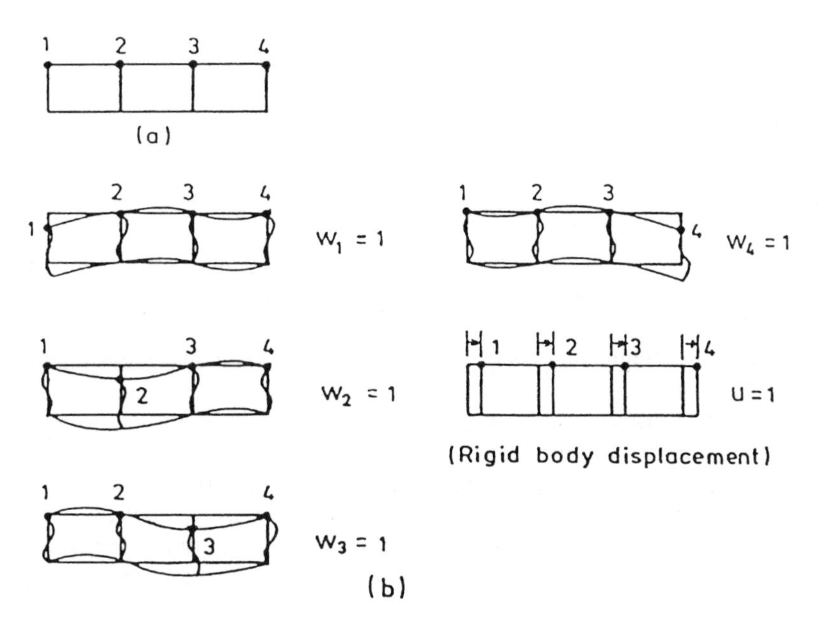

Figure 2.5 (a) Cross-section of a three-celled box girder bridge. (b) Five shape functions corresponding to prescribed deflections w_1, w_2, w_3, w_4 and rigid body displacement u.

behaviour of the bridge in the transverse direction. Following the same procedure outlined in Section 2.4.1, we can select the four top flange web junctions as nodes where vertical deflections can be prescribed one after the other. For the first shape function, node '1' is given a unit vertical deflection while nodes '2'-'4' are restrained from deflection. Similarly a unit vertical deflection is in turn prescribed at nodes '2'-'4', and a total of four shape functions is obtained. In addition, a fifth shape function consisting of a unit horizontal rigid body displacement of the frame may be included. This shape function is obtained by prescribing a unit horizontal displacement at any one node. These five shape functions are shown in Figure 2.5b. The shape functions can thus be written as:

$$\phi_m(x) = \sum_{k=1}^{p} [N(\bar{x})] \{\alpha_m\}_k \qquad (2.46)$$

In fact any degree of freedom at a node, e.g. vertical deflection, horizontal deflection and rotation, can be included in the small set of degrees of freedom, each of which is in turn given a unit displacement.

APPENDIX 2.A
CONSTANT STRAIN CONDITIONS

The constant strain conditions can be tested in one of the following two ways:

(a) If a simple polynomial of the form $a_o + a_1x + a_2x^2 + \ldots$ is used as f(x), then constant 'strain' will exist if the polynomial is complete up to or above the order in which a constant term will actually be obtained when the necessary differentiation for computing strains are carried out. For example, in case of a bending strip, the polynomial must be complete up to the quadratic term at least. If the polynomial is chosen as $a_o + a_1x + a_3x^3 + a_4x^4$, then the transverse bending curvature $\kappa_x = -\partial^2w/\partial x^2$ will be equal to $-6a_3x - 12a_4x^2$. This would naturally imply that κ_x is always zero at x=0, and that κ_x will never converge towards the correct solution.

(b) If the shape functions are used, it is not possible to decide very easily whether the polynomial is complete or not, and a different approach has to be resorted to.

The nodal displacement parameters of a deformed strip, in general, will take up arbitrary values which bear no fixed relationship to each other. However, for a constant strain situation such as cylindrical bending, pure shear, etc., the nodal displacement parameters will be related to each other in some specific manner on taking up some prescribed values. If such a set of parameters compatible with a constraint condition is substituted into the displacement function, then constant strain should in fact be obtained if the shape functions are correctly formulated and well behaved.

For example, the displacement function of a bending plate strip is given by

$$w = \sum_{m=1}^{r} [(1-3\bar{x}^2+2\bar{x}^3)w_{1m}+(x-2x\bar{x}+x\bar{x}^2)\theta_{1m}+ (3\bar{x}^2-2\bar{x}^3)w_{2m}+(-x\bar{x}+x\bar{x}^2)\theta_{2m}]Y_m(y) \quad (2.A.1)$$

in which $\bar{x} = x/b$ and b is the width of the strip.

For $w_{1m} = w_{2m} = 0$ and $\theta_{1m} = -\theta_{2m}$, a state of constant curvature in the x-direction can be expected. If these nodal displacement parameters are substituted into (1.1), then

$$w = \sum_{m=1}^{r} (x-x\bar{x})\theta_{1m}Y_m(y) \quad (2.A.2)$$

and

$$\frac{\partial^2 w}{\partial x^2} = \sum_{m=1}^{r} \frac{-2}{b}\theta_{1m}Y_m(y) \quad (2.A.3)$$

Thus a state of constant bending strain in the x direction in fact exists.

REFERENCES

1. Clough, R., The finite element method in structural mechanics, Chapter 7 in *Stress Analysis* (ed. O. C. Zienkiewicz and G. S. Holister), John Wiley, 1965.
2. Tong, P. and Pian, T. H., The convergence of finite element method in solving linear elastic problems, *Int J Solids Struct*, 3, 865-879, 1967.
3. Vlazov, V., *General theory of shells and its application in engineering*, NASA TT F-99, April 1964.
4. Delcourt, C. R. C., *Linear and geometrical non-linear analysis of flat-walled structures by the finite strip method*, PhD thesis, Department of Civil Engineering, University of Adelaide, 1978.
5. Hildebrand, F. B., *Introduction to numerical analysis*, McGraw-Hill, 1956.
6. Wittrick, W. H. and Williams, F. W., A general algorithm for computing natural frequencies of elastic structures. *The Quarterly J of Mechanics and Applied Mathematics*, Part 3, 263-284, 1971.
7. Swaddiwudhipong, S., *A unified approximate analysis of tall building using generalised finite strip method*, PhD thesis, Department of Civil Engineering, University of Hong Kong, 1979.
8. Li, Y. S. and Qi, T. X., *Spline Function Method*. Science Press, 1979 (in Chinese).

9. Tham, L. G., *Numerical solutions for time-dependent problems*, PhD thesis, Department of Civil Engineering, University of Hong Kong, 1981.
10. Guo, D. J., *Infinite layer method and its application to the analysis of pile systems*, PhD thesis, Department of Civil & Structural Engineering, University of Hong Kong, 1988.
11. Zienkiewicz, O. C., *Finite element method*, 4th Edition, Mc-Graw Hill, London, 1989.
12. Cheung, M. S. and Chan, M. Y. T., Three-dimensional finite strip analysis of elastic solids, *Comp and Struct*, 9, 629-638, 1978.
13. Fan, S. C., *Spline finite strip method in structural analysis*, PhD thesis, Department of Civil Engineering, University of Hong Kong, 1982.
14. Cao, Zhiyuan and Cheung, Y. K., A semi-analytical method for structure-external fluid dynamic interaction problems. *Acta Mechanica Sinica*, 17, 389-399, 1985 (in Chinese).
15. Fan, S. C. and Luah, M. H., New spline finite element for analysis of shells of revolution, *J of Eng Mech, ASCE*, 116(3), 709-726, 1990.
16. Cheung, Y. K., Au, F. T. K. and Kong, J., Structural analysis by the finite strip method using computed shape functions, *Proc Int Conf on Computational Methods in Engineering, Singapore*, 2, 1140-1145, 1992.
17. Cheung, Y. K. and Au, F. T. K., Finite strip analysis of right box girder bridges using computed shape functions, *Thin-Walled Struct*, 13(4), 275-298, 1992.
18. Cheung, Y. K. and Yeo, M., *A practical introduction to finite element analysis*, Pitman International Text, 1979.

CHAPTER THREE

PLATE BENDING PROBLEMS

3.1 INTRODUCTION

The finite strip approach has considerable advantages over the conventional finite element method for the plate bending problems considered here in which the geometry is fairly simple and does not change in one direction, but nevertheless such structures are so frequently used in practice that a special and more economical treatment is warranted. Since less nodal degrees of freedom are required by the present method, the size as well as the bandwidth of the matrices are greatly reduced, and consequently such matrices can be handled by small computers or solved in a much shorter time.

The first paper on the finite strip method was presented by Cheung[1] on thin plate bending problems[2] using a simply supported rectangular strip. A paper on the formulation of the simply supported rectangular strip was published independently at a later date by Powell and Ogden.[3] The method was subsequently generalized by Cheung to include other support conditions.[4] In the above-mentioned publications, a lower order strip using cubic polynomial function [shape function (b), Eq. 2.31] has been used throughout. Whilst the deflections obtained have been almost always very good, discontinuity of moments at strip boundaries and existence of some residual transverse moments at a free edge can be expected for coarse meshes. Of course, by increasing the number of strips used in the analysis such discontinuities and residual moments will diminish and even disappear altogether.

An alternative way to achieve higher accuracy without increasing the number of strips is to use higher order strips, and these can be formulated either by establishing higher order nodal line compatibilities [shape function (c), Eq. 2.32] or by introducing internal nodal lines [shape function (e), Eq. 2.34], in the same way as was done in finite element analysis.[5]

The use of shape function (c), which involves a quintic polynomial for a higher order bending strip, was originally suggested by Cheung,[6] although the actual formulation were carried out by Loo and Cusens,[7] who also worked out a second higher order strip using one additional internal nodal line.[8] The inclusion of end conditions other than either of the two ends simply supported or clamped was presented in a paper by Cheung and Cheung for vibration analysis.[9] Re-formulation of the problem by the least square method was attempted by Kwok et al.[10] to demonstrate that the weighted residual approach can also be used in the formulation of this problem.

Attempts to carry out analysis of skew plates by finite strips were reported by Brown[11,12] and Mukhopadhyay.[13] On the other hand, the method was also applied to the analyses of moderately thick plates. Mawenya and Davies[14] used the quadratic strips [shape function (d), Eq. 2.33] to solve for the bending of moderately thick plates. The shear locking phenomenon associated with this formulation was also addressed. The work of Hinton[15] and Onate[16] demonstrates that the accuracy of the linear and quadratic strips [functions (a) and (d); Eqs. 2.31 and 2.33] can be improved if the stiffness terms are integrated by the reduced/selective integration schemes.[17] Onate also studied the cubic strips [shape function (f), Eq. 2.35]. Applications of the strain distribution technique to reduce the locking of linear strips were investigated by Chulya.[18] Extensions to vibration and stability analyses were reported by Benson[19] and Dawe.[20]

To improve the convergence in modelling the action of point loads and to be able to model multi-span problems accurately and conveniently, the spline finite strip was devised by Cheung, Fan and Wu[21] for stress analysis. They expressed the shape functions as products of B-3 spline functions and piecewise continuous polynomials. The applications of lower order spline strip to static as well as dynamic analyses of rectangular plates were fully explored by them.[22,23] Higher order spline strips were proposed by Loo,[24] while bending analysis of skew plates was studied first by Chen et al.,[25] and later by Li et al.[26,27] Incorporating the domain transformation concept,[27] the method was successfully extended to the analyses of arbitrarily shaped structures.[28,29,30] Analysis for plates undergoing large deformation was reported by Cheung and Zhu.[31] Based on the Reddy plate theory,[56] Kong and Cheung [32] analyzed moderately thick plates and laminated plates using the spline strips. Choosing the deflection and its first and second derivatives as the interpolation parameters of the end knots, Fan[33] proposed a new technique for modifying the spline function to satisfy the boundary conditions. Using the X-spline function, Yang and Chong[34,35] and then Chong and Chen[36,37,38] proposed a modified finite strip method to solve some irregular shaped plate problems, including static, free vibration and buckling analysis.

Based on computed shape functions, Cheung and his collaborators[39,40,41,42] developed a new form of finite strip and carried out analyses for plates and flat slabs.

3.2 FORMULATION FOR THIN PLATE BENDING

Thin plate bending strips are developed through the Principle of Minimum Total Potential Energy. First, the deflection of the plates are defined in terms of the shape functions [N] and nodal displacement parameters $\{\delta\}$ as:

$$w = [N]\{\delta\} \tag{3.1}$$

Following the standard procedure, the strain $\{\varepsilon\}$ and stress $\{\sigma\}$ vectors have to be defined and they are respectively:

$$\{\varepsilon\} = \begin{Bmatrix} \kappa_x \\ \kappa_y \\ 2\kappa_{xy} \end{Bmatrix} = \begin{Bmatrix} -\dfrac{\partial^2 w}{\partial x^2} \\ -\dfrac{\partial^2 w}{\partial y^2} \\ 2\dfrac{\partial^2 w}{\partial x \partial y} \end{Bmatrix} = \begin{Bmatrix} -\dfrac{\partial^2 [N]}{\partial x^2} \\ -\dfrac{\partial^2 [N]}{\partial y^2} \\ 2\dfrac{\partial^2 [N]}{\partial x \partial y} \end{Bmatrix} \{\delta\} = [B]\{\delta\} \tag{3.2}$$

and

$$\{\sigma\} = [D]\{\varepsilon\} = [D][B]\{\delta\} \tag{3.3}$$

In the above equations, [B] is the strain matrix and [D] is the property matrix.

Furthermore, it can be shown readily that the total potential energy is:

$$\begin{aligned} \phi &= U + W \\ &= 1/2 \int \{\delta\}^T [B]^T [D] [B] \{\delta\} \, d(\text{area}) - \int \{\delta\}^T [N]^T \{q\} \, d(\text{area}) \end{aligned} \tag{3.4}$$

Minimizing this energy with respect to the deflection parameters, we have the equilibrium equation:

$$[K^e]\{\delta\} - \{F^e\} = \{0\} \tag{3.5}$$

where $[K^e]$ is the stiffness matrix,
$\{F^e\}$ is the load vector.

The explicit forms of the characteristic matrices ($[B]$,$[K^e]$ and $\{F^e\}$) depend on the shape functions of the strips, and they will be described in detail in the forthcoming sections.

3.3 CLASSICAL FINITE STRIPS

3.3.1 LOWER ORDER RECTANGULAR STRIP (LO2)

(a) Displacement function

Consider the strip shown in Figure 3.1b. Each nodal line is free to move up and down in the z-direction and to rotate about the y-axis, with the result that there are two DOF per nodal line and a total of four DOF for the whole strip. The deflection can be written as

$$w = [N]\{\delta\} = \sum_{m=1}^{r} [N]_m \{\delta\}_m = \sum_{m=1}^{r} Y_m [[C_1]\ [C_2]] \{\delta\}_m \tag{3.6}$$

in which $[C_1]$ and $[C_2]$ are given by Eq. 2.31, and $\{\delta\}_m^T = \{w_{1m}\ \theta_{1m}\ w_{2m}\ \theta_{2m}\}^T$ are the deflection and rotation parameters at the two longitudinal edges for the m-th term of the series.

The above function assures compatibility of displacement values as well as its first derivatives at the interfaces of the strips, and therefore a convergent solution is expected. For static analysis, Eqs. 2.4 to 2.9 and 2.11 have been used as Y_m (the series part of the displacement function) successfully.

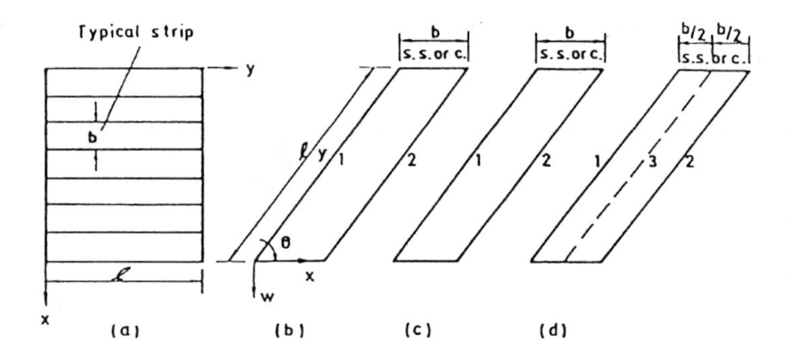

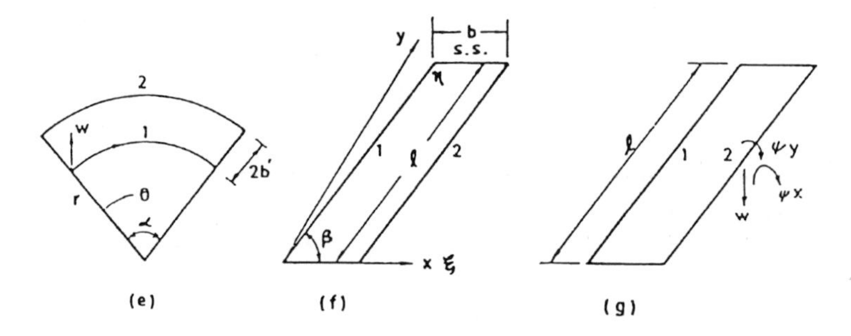

Figure 3.1 (a) Top view of a structure divided into finite strips. (b) LO2 rectangular bending strip. (c) HO3 rectangular bending strip. (d) HO2 rectangular bending strip. (e) LO2 curved bending strip. (f) HO2 skew bending strip. (g) Mindlin plate strip.

(b) Strains (curvatures)

Having the displacement functions defined, it is possible to obtain the bending strains

$$\{\varepsilon\} = \sum_{m=1}^{r} \begin{Bmatrix} -\dfrac{\partial^2[N]}{\partial x^2} \\ -\dfrac{\partial^2[N]}{\partial y^2} \\ 2\dfrac{\partial^2[N]}{\partial x \partial y} \end{Bmatrix} \{\delta\}_m = \sum_{m=1}^{r} [B]_m \{\delta\}_m \tag{3.7}$$

The strain matrix $[B]_m$ is obtained through the appropriate differentiations.

$[B]_m =$

$$\begin{bmatrix} \frac{6}{b^2}(1-2\bar{x})Y_m & \frac{2}{b}(2-3\bar{x})Y_m & \frac{6}{b^2}(-1+2\bar{x})Y_m & \frac{2}{b}(-3\bar{x}+1)Y_m \\ -(1-3\bar{x}^2+2\bar{x}^3)Y_m'' & -x(1-2\bar{x}+\bar{x}^2)Y_m'' & -(3\bar{x}^2-2\bar{x}^3)Y_m'' & -x(\bar{x}^2-\bar{x})Y_m'' \\ \frac{2}{b}(-6\bar{x}+6\bar{x}^2)Y_m' & 2(1-4\bar{x}+3\bar{x}^2)Y_m' & \frac{2}{b}(6\bar{x}-6\bar{x}^2)Y_m' & 2(3\bar{x}^2-2\bar{x})Y_m' \end{bmatrix}$$

(c) Stresses (moments)

The stresses (moments) are related to the strains by

$$\{\sigma\} = \begin{Bmatrix} M_x \\ M_y \\ M_{xy} \end{Bmatrix} = [D]\{\varepsilon\} = [D] \sum_{m=1}^{r} [B]_m \{\delta\}_m \tag{3.8}$$

The property matrix [D] is:

(i) for isotropic plates

$$[D] = \frac{Et^3}{12(1-\upsilon^2)} \begin{bmatrix} 1 & \upsilon & 0 \\ \upsilon & 1 & 0 \\ 0 & 0 & \frac{(1-\upsilon)}{2} \end{bmatrix} \tag{3.9}$$

(ii) for orthotropic plates

$$[D] = \begin{bmatrix} D_x & D_1 & 0 \\ D_1 & D_y & 0 \\ 0 & 0 & D_{xy} \end{bmatrix} \tag{3.10}$$

where $D_x = \dfrac{E_x t^3}{12(1-\upsilon_x\upsilon_y)}$; $D_y = \dfrac{E_y t^3}{12(1-\upsilon_x\upsilon_y)}$;

$$D_1 = \frac{\upsilon_y E_x t^3}{12(1-\upsilon_x\upsilon_y)} = \frac{\upsilon_x E_y t^3}{12(1-\upsilon_x\upsilon_y)} \quad ; \; D_{xy} = \frac{Gt^3}{12}$$

(d) Total potential energy

The total potential energy, as stated previously, is the sum of the elastic strain energy stored in the body and the potential energy of the loads. Thus

$$\phi = U + W = \frac{1}{2} \sum_{m=1}^{r} \sum_{n=1}^{r} \{\delta\}_m^T [B]_m^T [D] [B]_n \{\delta\}_n d(\text{area}) - \sum_{m=1}^{r} \{\delta\}_m^T [N]_m^T \{q\} d(\text{area}) \tag{3.11}$$

(e) Minimization procedure

The principle of minimum total potential energy requires that

$$\frac{\partial \phi}{\partial \{\delta\}} = \{0\} \tag{3.12}$$

Performing the partial differentiation, we obtain the equilibrium equations which are expressed in terms of the stiffness matrix $[K^e]$ and the load vector $\{F^e\}$,

$$[K^e]\{\delta\} - \{F^e\} = \{0\} \tag{3.13}$$

in which
$$[K^e] = \begin{bmatrix} [K]_{11} & [K]_{12} & \cdot & \cdot & \cdot & [K]_{1r} \\ & [K]_{22} & \cdot & \cdot & \cdot & [K]_{2r} \\ & & & \cdot & \cdot & \cdot \\ & & & & \cdot & \cdot \\ & & & & & \cdot \\ & & & & & [K]_{rr} \end{bmatrix}, \quad \{F^e\} = \begin{Bmatrix} \{F\}_1 \\ \{F\}_2 \\ \cdot \\ \cdot \\ \cdot \\ \{F\}_r \end{Bmatrix}$$

where

$$[K]_{mn} = \int [B]_m^T [D] [B]_n \, d(\text{area}) \tag{3.14}$$
$$\{F\}_m = \int [N]_m^T \{q\} \, d(\text{area}) \tag{3.15}$$

(f) Stiffness matrix

The explicit form of the stiffness matrix for a lower order strip is listed in Table 3.1. For a simply supported strip, we have,

$$Y = \sin\left(\frac{m\pi y}{l}\right) = \sin k_m y$$

Therefore, $Y' = k_m \cos k_m y$ and $Y'' = -k_m^2 \sin k_m y$

Consequently all the integrals I_1 to I_5 and also $[K]_{mn}$ of Table 3.1 are equal to zero for $m \neq n$. This can be proved with the help of Eq. 2.20 as follows:

$$I_1 = \int_0^l Y_m Y_n \, dy = 0 \;;\; I_2 = \int_0^l Y_m'' Y_n \, dy = -k_m^2 \int_0^l \sin k_m y \sin k_n y \, dy = 0$$

$$I_3 = \int_0^l Y_m Y_n'' dy = -k_n^2 \int_0^l \sin k_m y \sin k_n y \, dy = 0 \;;\; I_4 = \int_0^l Y_m'' Y_n'' dy = 0$$

$$I_5 = \int_0^l Y_m' Y_n' dy = k_m k_n \int_0^l \cos k_m y \cos k_n y \, dy = 0$$

where $k_m = \dfrac{m\pi y}{l}$ and $k_n = \dfrac{n\pi y}{l}$ respectively.

Because of this orthogonal property, the stiffness matrix takes up the form

$$[K^e] = \begin{bmatrix} [K]_{11} & 0 & \cdot & \cdot & \cdot & 0 \\ & [K]_{22} & \cdot & \cdot & \cdot & 0 \\ & & & \cdot & \cdot & \cdot \\ & & & & \cdot & \cdot \\ & & & & & \cdot \\ & & & & & [K]_{rr} \end{bmatrix} \tag{3.16}$$

and all the terms of the series are thus uncoupled. The final form of the stiffness matrix $[K]_{mm}$ for a simply supported strip is given in Table 3.2.

Eq. 3.11 to Eq. 3.16 are quite general and shall be applied to the

TABLE 3.1
Bending stiffness matrix for a rectangular strip with any end condition.

$$
[K]_{mn} = \frac{1}{420b^3}\left[\begin{array}{c:c:c:c}
5040D_xI_1 & 2520bD_xI_1 & -5040D_xI_1 & 2520bD_xI_1 \\
-504b^2D_1I_2 & -462b^3D_1I_2 & +504b^2D_1I_2 & -42b^3D_1I_2 \\
-504b^2D_1I_3 & -42b^3D_1I_3 & +504b^2D_1I_3 & -42b^3D_1I_3 \\
+156b^4D_yI_4 & +22b^5D_yI_4 & +54b^4D_yI_4 & -13b^5D_yI_4 \\
+2016b^2D_{xy}I_5 & +168b^3D_{xy}I_5 & -2016b^2D_{xy}I_5 & +168b^3D_{xy}I_5 \\
\hdashline
2520bD_xI_1 & 1680b^2D_xI_1 & -2520bD_xI_1 & 840b^2D_xI_1 \\
-462b^3D_1I_2 & -56b^4D_1I_2 & +42b^3D_1I_2 & +14b^4D_1I_2 \\
-42b^3D_1I_3 & -56b^4D_1I_3 & +42b^3D_1I_3 & +14b^4D_1I_3 \\
+22b^5D_yI_4 & +4b^6D_yI_4 & +13b^5D_yI_4 & -3b^6D_yI_4 \\
+168b^3D_{xy}I_5 & +224b^4D_{xy}I_5 & -168b^3D_{xy}I_5 & -56b^4D_{xy}I_5 \\
\hdashline
-5040D_xI_1 & -2520bD_xI_1 & 5040D_xI_1 & -2520bD_xI_1 \\
+504b^2D_1I_3 & +42b^3D_1I_3 & -504b^2D_1I_2 & +462b^3D_1I_2 \\
+504b^2D_1I_2 & +42b^3D_1I_2 & -504b^2D_1I_3 & +42b^3D_1I_3 \\
+54b^4D_yI_4 & +13b^5D_yI_4 & +156b^4D_yI_4 & -22b^5D_yI_4 \\
-2016b^2D_{xy}I_5 & -168b^3D_{xy}I_5 & +2016b^2D_{xy}I_5 & -168b^3D_{xy}I_5 \\
\hdashline
2520bD_xI_1 & 840b^2D_xI_1 & -2520bD_xI_1 & 1680b^2D_xI_1 \\
-42b^3D_1I_3 & +14b^4D_1I_3 & +462b^3D_1I_3 & -56b^4D_1I_2 \\
-42b^3D_1I_2 & +14b^4D_1I_2 & +42b^3D_1I_2 & -56b^4D_1I_3 \\
-13b^5D_yI_4 & -3b^6D_yI_4 & -22b^5D_yI_4 & +4b^6D_yI_4 \\
+168b^3D_{xy}I_5 & -56b^4D_{xy}I_5 & -168b^3D_{xy}I_5 & +224b^4D_{xy}I_5
\end{array}\right]
$$

$$I_1 = \int_0^l Y_mY_n dy \;;\; I_2 = \int_0^l Y_m'' Y_n dy \;;\; I_3 = \int_0^l Y_mY_n'' dy$$

$$I_4 = \int_0^l Y_m'' Y_n'' dy \;;\; I_5 = \int_0^l Y_m' Y_n' dy \qquad (\text{for } m \neq n,\ I_1 = I_4 = 0)$$

formulation of all other classical finite strips for two-dimensional and three-dimensional problems.

(g) Consistent load vector

The load vector is given by Eq. 3.15 as

$$\{F\}_m = \int_0^l\int_0^b \begin{Bmatrix} [N_1]_m^T \\ [N_2]_m^T \end{Bmatrix} \{q\}\, dxdy \tag{3.17}$$

For uniformly distributed load q

$$\{F\}_m = q\int_0^b \begin{Bmatrix} [C_1]_m^T \\ [C_2]_m^T \end{Bmatrix} dx \int_0^l Y_m dy = q\begin{Bmatrix} \frac{b}{2} \\ \frac{b^2}{12} \\ \frac{b}{2} \\ -\frac{b^2}{12} \end{Bmatrix} \int_0^l Y_m dy \tag{3.18}$$

in which $[C_1]$ and $[C_2]$ are, as before, given by Eq. 2.31.

TABLE 3.2
Stiffness matrix of simply supported rectangular strip.

$$[K]_{mm} = \left[\begin{array}{c:c:c:c}
\begin{array}{c}\frac{13lb}{70}k_m^4D_y+\frac{12l}{5b}k_m^2D_{xy}\\+\frac{6l}{5b}k_m^2D_1+\frac{6l}{b^3}D_x\end{array} & & & \\ \hdashline
\begin{array}{c}\frac{3l}{5}k_m^2D_1+\frac{l}{5}k_m^2D_{xy}\\+\frac{3l}{b^2}D_x+\frac{11lb^2}{420}k_m^4D_y\end{array} & \begin{array}{c}\frac{lb^3}{210}k_m^4D_y+\frac{4lb}{15}k_m^2D_{xy}\\+\frac{2lb}{15}k_m^2D_1+\frac{2l}{b}D_x\end{array} & & \\ \hdashline
\begin{array}{c}\frac{9lb}{140}k_m^4D_y-\frac{12l}{5b}k_m^2D_{xy}\\-\frac{6l}{5b}k_m^2D_1-\frac{6l}{b^3}D_x\end{array} & \begin{array}{c}\frac{13lb^2}{840}k_m^4D_y-\frac{l}{5}k_m^2D_{xy}\\-\frac{l}{10}k_m^2D_1-\frac{3l}{b^2}D_x\end{array} & \begin{array}{c}\frac{13lb}{70}k_m^4D_y+\frac{12l}{5b}k_m^2D_{xy}\\+\frac{6l}{5b}k_m^2D_1+\frac{6l}{b^3}D_x\end{array} & \\ \hdashline
\begin{array}{c}-\frac{13lb^2}{840}k_m^4D_y+\frac{l}{5}k_m^2D_{xy}\\+\frac{l}{10}k_m^2D_1+\frac{3l}{b^2}D_x\end{array} & \begin{array}{c}-\frac{3lb^3}{840}k_m^4D_y-\frac{lb}{15}k_m^2D_{xy}\\-\frac{lb}{30}k_m^2D_1+\frac{l}{b}D_x\end{array} & \begin{array}{c}-\frac{11lb^2}{420}k_m^4D_y-\frac{l}{5}k_m^2D_{xy}\\-\frac{3l}{5}k_m^2D_1-\frac{3l}{b^2}D_x\end{array} & \begin{array}{c}\frac{lb^3}{210}k_m^4D_y+\frac{4lb}{15}k_m^2D_{xy}\\+\frac{2lb}{15}k_m^2D_1+\frac{2l}{b}D_x\end{array}
\end{array}\right]$$

$$k_m = \frac{m\pi}{l}$$

For a concentrated load P at a point $x=x_c$, $y=y_c$, it is only necessary to substitute the coordinates of the load into the first part of Eq. 3.17 and also to dispense with the integration. Letting $x_c /b = \overline{x}_c$, we have

$$\{F\}_m = P\,Y_m(y_c) \begin{Bmatrix} (1-3\overline{x}_c^2+2\overline{x}_c^3) \\ x_c(1-2\overline{x}_c+\overline{x}_c^2) \\ (3\overline{x}_c^2-2\overline{x}_c^3) \\ x_c(\overline{x}_c^2-\overline{x}_c) \end{Bmatrix} \tag{3.19}$$

Since the distribution of the transverse moment is linear from one edge to the other of the strip, it is not possible to obtain the maximum moment under the point load. Therefore, if circumstances permit, it is more desirable to actually insert a nodal line under the point load.

TABLE 3.3
Stiffness matrix for a HO3 bending strip.[43,44]

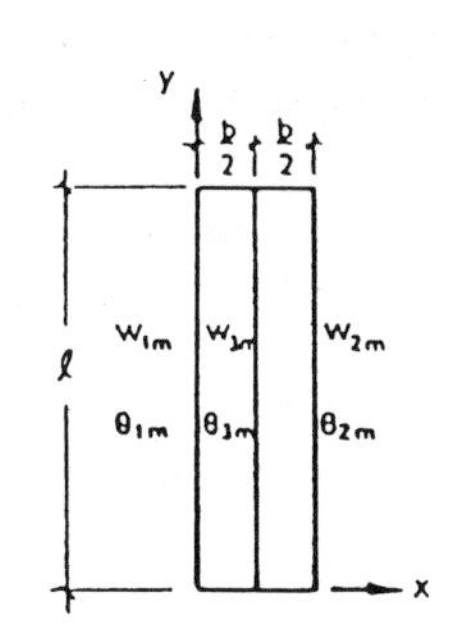

$$\begin{bmatrix} S1 & & & & & \\ S2 & S7 & & & \text{SYM} & \\ S3 & S8 & S11 & & & \\ S4 & S9 & 0 & S12 & & \\ S5 & -S6 & S3 & -S4 & S1 & \\ S6 & S10 & -S8 & S9 & -S2 & S7 \end{bmatrix}$$

where

$$S1 = \frac{5092B}{35b^3} + \frac{278C}{105b} + \frac{523bD}{3465} + \frac{278E}{105b} \quad ; \quad S7 = \frac{332B}{35b} + \frac{2bC}{45} + \frac{2b^3D}{3465} + \frac{2bE}{45}$$

$$S2 = \frac{1138B}{35b^2} + \frac{59C}{105} + \frac{19b^2D}{2310} + \frac{13E}{210} \quad ; \quad S8 = -\frac{128B}{5b^2} - \frac{8C}{105} + \frac{2b^2D}{315} - \frac{8E}{105}$$

$$S3 = -\frac{512B}{5b^3} - \frac{256C}{105b} + \frac{4bD}{63} - \frac{256E}{105b} \quad ; \quad S9 = \frac{64B}{7b} - \frac{4bC}{315} - \frac{b^3D}{1155} - \frac{4bE}{315}$$

$$S4 = \frac{384B}{7b^2} + \frac{8C}{21} - \frac{8b^2D}{693} + \frac{8E}{21} \quad ; \quad S10 = \frac{38B}{35b} - \frac{bC}{126} - \frac{b^3D}{4620} - \frac{bE}{126}$$

$$S5 = -\frac{1508B}{35b^3} - \frac{22C}{105b} + \frac{131bD}{6930} - \frac{22E}{105b} \quad ; \quad S11 = \frac{1024B}{5b^3} + \frac{512C}{105b} + \frac{128bD}{315} + \frac{512E}{105b}$$

$$S6 = \frac{242B}{35b^2} - \frac{C}{70} - \frac{29b^2D}{13860} - \frac{E}{70} \quad ; \quad S12 = \frac{256B}{7b} + \frac{128bC}{315} + \frac{32b^3D}{3465} + \frac{128bE}{315}$$

in which $B = \frac{lD_x}{2}$; $C = lk_m^2 D_1$; $D = \frac{lk_m^2 D_y}{2}$; $E = 2lk_m^2 D_{xy}$; $k_m = \frac{m\pi}{l}$

3.3.2 HIGHER ORDER STRIP WITH ONE INTERNAL NODAL LINE (HO3) (Figure 3.1c)

This strip uses an additional internal nodal line which, for the sake of convenience, is usually placed midway between the two longitudinal edges. Since each nodal line includes two displacement parameters, a deflection and a rotation, it is apparent that the half bandwidth of the final stiffness matrix is 50% higher than that for a lower order strip. However, due to the fact that the internal nodal displacement parameters are not connected with anything else apart from those of the same strip, it is possible to eliminate these two variables before assembly through the process of static condensation.

The displacement function used in this case is simply the product of shape function (e) of Section 2.2.2 and the basic function series

$$w = \sum_{m=1}^{r} [N]_m \{\delta\}_m = \sum_{m=1}^{r} Y_m [[C_1][C_2][C_3]] \{\delta\}_m \tag{3.20}$$

where $[C_1]$, $[C_2]$ and $[C_3]$ are now given by Eq. 2.34, and $\{\delta\}_m = \{w_{1m}\ \theta_{1m}\ w_{2m}\ \theta_{2m}\ w_{3m}\ \theta_{3m}\}^T$.

Thus, the stiffness matrix and load vectors can be formed by the procedure as described in the previous section. Explicit forms of these matrices for a simply supported strip are given in Tables 3.3 and 3.4.[43,44]

TABLE 3.4
Load vectors due to applied loads for a HO3 strip.[43,44]

(a) Concentrated load

$$\{F\}_m = \begin{Bmatrix} 1-\dfrac{23x_c^2}{b^2}+\dfrac{66x_c^3}{b^3}-\dfrac{68x_c^4}{b^4}-\dfrac{24x_c^5}{b^5} \\ x_c-\dfrac{6x_c^2}{b}+\dfrac{13x_c^3}{b^2}-\dfrac{12x_c^4}{b^3}+\dfrac{4x_c^5}{b^4} \\ \dfrac{16x_c^2}{b^2}-\dfrac{32x_c^3}{b^3}+\dfrac{16x_c^4}{b^4} \\ -\dfrac{8x_c^2}{b}+\dfrac{32x_c^3}{b^2}-\dfrac{40x_c^4}{b^3}+\dfrac{16x_c^5}{b^4} \\ \dfrac{7x_c^2}{b^2}-\dfrac{34x_c^3}{b^3}+\dfrac{52x_c^4}{b^4}-\dfrac{24x_c^5}{b^5} \\ -\dfrac{x_c^2}{b}+\dfrac{5x_c^3}{b^2}-\dfrac{8x_c^4}{b^3}+\dfrac{4x_c^5}{b^4} \end{Bmatrix} P \sin k_m y_c$$

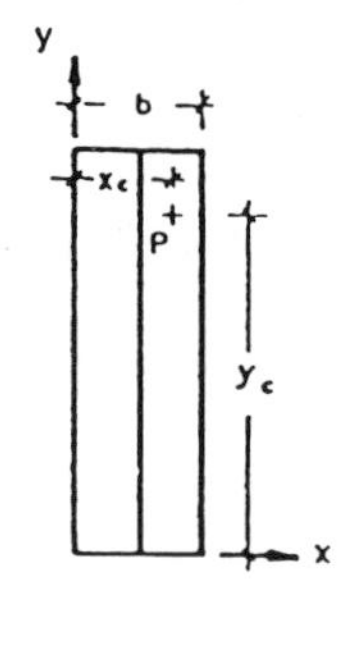

(b) Patch load

$$\{F\}_m = \begin{Bmatrix} x_o-\dfrac{23x_o^3}{3b^2}+\dfrac{33x_o^4}{2b^3}+\dfrac{68x_o^5}{5b^4}+\dfrac{4x_o^6}{b^5} \\ \dfrac{x_o^2}{2}-\dfrac{2x_o^3}{b}+\dfrac{13x_o^4}{4b^2}-\dfrac{12x_o^5}{5b^3}+\dfrac{2x_o^6}{3b^4} \\ \dfrac{16x_o^3}{3b^2}-\dfrac{8x_o^4}{b^3}+\dfrac{16x_o^5}{b^4} \\ -\dfrac{8x_o^3}{3b}+\dfrac{8x_o^4}{b^2}-\dfrac{8x_o^5}{5b^3}+\dfrac{8x_o^6}{3b^4} \\ \dfrac{7x_o^3}{3b^2}-\dfrac{17x_o^4}{2b^3}+\dfrac{52x_o^5}{5b^4}-\dfrac{4x_o^6}{b^5} \\ -\dfrac{x_o^3}{3b}+\dfrac{5x_o^4}{4b^2}-\dfrac{8x_o^5}{5b^3}+\dfrac{2x_o^6}{3b^4} \end{Bmatrix} qC_m$$

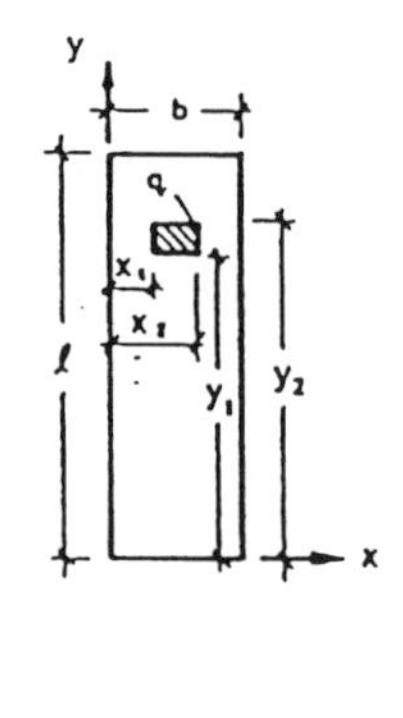

in which $C_m = (\cos k_m y_1 - \cos k_m y_2)/k_m$; $x_o^n = x_2^n - x_1^n$

3.3.3 HIGHER ORDER RECTANGULAR STRIP WITH TWO NODAL LINES (HO2) (Figure 3.1d)

In this strip, the transverse curvature is also used as a nodal displacement parameter in addition to the standard deflection and rotation variables. As a result, continuous curvatures and consequently moments will now exist at the interface between the neighbouring strips, and therefore more accurate results can be expected when compared with the lower order strip analysis for the same number of strips. Two points must be borne in mind, however. The first point is that due to the increase in nodal displacement parameters the half-bandwidth of the final stiffness matrix is also increased by 50%. This is not really serious, since it has been observed from experience that the increase in accuracy more than compensates the increase in computational effort. The second point is that such a strip cannot be used, at least not in a straightforward manner, for plates with abrupt changes in thickness or material properties in the transverse direction because at the line where such abrupt change occurs, the curvatures should always be discontinuous. The nature of this disadvantage is more serious since the scope of application of such higher order strips will be more or less limited to constant thickness plates. The nodal displacement parameters for the strip shown in Figure 3.1d include the displacement and its first and second partial derivatives with respect to x. A

TABLE 3.5
Stiffness matrix for a bending simply supported strip with curvature compatibility.[43,44]

$$\begin{bmatrix} S1 & & & & & \\ S4 & S2 & & & \text{SYM} & \\ S5 & S6 & S3 & & & \\ S10 & S7 & S8 & S1 & & \\ -S7 & S11 & S9 & -S4 & S2 & \\ S8 & -S9 & S12 & S5 & -S6 & S3 \end{bmatrix}$$

where

$$S1 = \frac{120B}{7b^3} + \frac{20C}{7b} + \frac{18lbD}{462} + \frac{10E}{7b} \quad ; \quad S7 = -\frac{60B}{7b^2} - \frac{3C}{7} + \frac{15lb^2D}{4620} - \frac{3E}{14}$$

$$S2 = \frac{192B}{35b} + \frac{16bC}{35} + \frac{52b^3D}{3465} + \frac{8bE}{45} \quad ; \quad S8 = -\frac{3B}{7b} - \frac{bC}{42} + \frac{18lb^3D}{55440} - \frac{bE}{84}$$

$$S3 = \frac{3bB}{35} + \frac{b^3C}{315} + \frac{b^5D}{9240} + \frac{b^3E}{630} \quad ; \quad S9 = \frac{4B}{35} - \frac{b^2C}{105} - \frac{13b^4D}{13860} - \frac{b^2E}{210}$$

$$S4 = \frac{60B}{7b^2} + \frac{10C}{7} + \frac{31lb^2D}{4620} + \frac{3E}{14} \quad ; \quad S10 = -\frac{120B}{7b^3} - \frac{20C}{7b} + \frac{25bD}{231} - \frac{10E}{7b}$$

$$S5 = \frac{3B}{7b} + \frac{bC}{42} + \frac{281b^3D}{55440} + \frac{bE}{84} \quad ; \quad S11 = \frac{108B}{35b} - \frac{bC}{35} - \frac{19b^3D}{1980} - \frac{bE}{70}$$

$$S6 = \frac{11B}{35} + \frac{b^2C}{30} + \frac{23b^4D}{18480} + \frac{b^2E}{60} \quad ; \quad S12 = \frac{bB}{70} + \frac{b^3C}{630} + \frac{b^5D}{11088} + \frac{b^3E}{1260}$$

in which $B = \frac{lD_x}{2}$; $C = \frac{lk_m^2 D_1}{2}$; $D = \frac{lk_m^2 D_y}{2}$; $E = 2lk_m^2 D_{xy}$; $k_m = \frac{m\pi}{l}$

TABLE 3.6
Load vectors due to applied loads for a bending simply supported strip with curvature compatibility.[43,44]

(a) Concentrated load

$$\{F\}_m = \begin{Bmatrix} 1-\dfrac{10x_c^3}{b^3}+\dfrac{15x_c^4}{b^4}-\dfrac{6x_c^5}{b^5} \\ x_c-\dfrac{6x_c^3}{b^2}+\dfrac{8x_c^4}{b^3}-\dfrac{3x_c^5}{b^4} \\ \dfrac{x_c^2}{2}-\dfrac{1.5x_c^3}{b}+\dfrac{1.5x_c^4}{b^2}-\dfrac{x_c^6}{2b^3} \\ \dfrac{10x_c^3}{b^3}-\dfrac{15x_c^4}{b^4}+\dfrac{6x_c^5}{b^5} \\ -\dfrac{4x_c^3}{b^2}+\dfrac{7x_c^4}{b^3}-\dfrac{3x_c^5}{b^4} \\ \dfrac{x_c^3}{2b}-\dfrac{x_c^4}{b^2}+\dfrac{x_c^6}{2b^3} \end{Bmatrix} P \sin k_m y_c$$

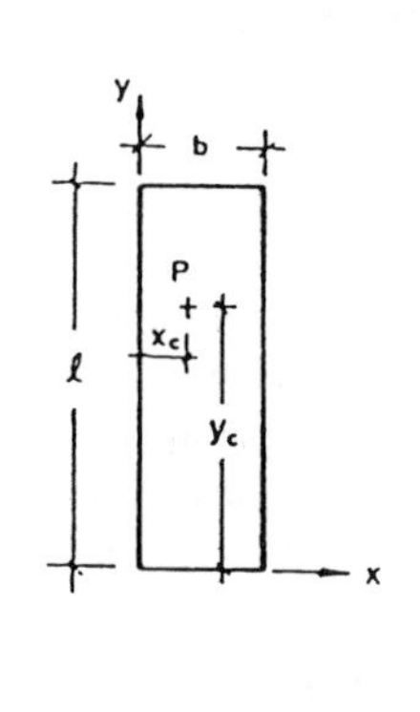

(b) Patch load

$$\{F\}_m = \begin{Bmatrix} x_o-\dfrac{2.5x_o^4}{b^3}+\dfrac{3x_o^5}{b^4}-\dfrac{x_o^6}{b^5} \\ \dfrac{x_o^2}{2}-\dfrac{1.5x_o^4}{4b^2}+\dfrac{1.6x_o^5}{b^3}-\dfrac{x_o^6}{2b^4} \\ \dfrac{x_o^3}{6}-\dfrac{3x_o^4}{8b}+\dfrac{0.3x_o^5}{b^2}-\dfrac{x_o^6}{12b^3} \\ \dfrac{2.5x_o^4}{b^3}-\dfrac{3x_o^5}{b^4}+\dfrac{x_o^6}{b^5} \\ -\dfrac{x_o^4}{b^2}+\dfrac{1.4x_o^5}{b^3}-\dfrac{x_o^6}{2b^4} \\ \dfrac{x_o^4}{8b}-\dfrac{x_o^5}{5b^2}+\dfrac{x_o^6}{12b^3} \end{Bmatrix} qC_m$$

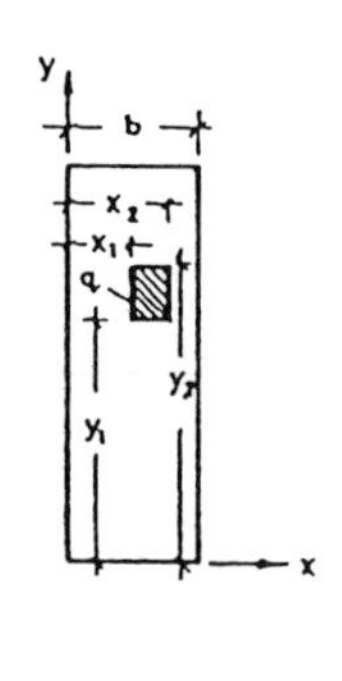

in which $C_m = (\cos k_m y_1 - \cos k_m y_2)/k_m$; $x_o^n = x_2^n - x_1^n$

suitable displacement function can be obtained as a product of the shape function (c) given in Section 2.2.2 and the basic function series as follows:

$$w = \sum_{m=1}^{r} [N]_m \{\delta\}_m = \sum_{m=1}^{r} Y_m [[C_1][C_2]] \{\delta\}_m \tag{3.21}$$

where $[C_1]$ and $[C_2]$ are now given by Eq. 2.32, and $\{\delta\}_m = \{w_{1m}\ \theta_{1m}\ \chi_{1m}\ w_{2m}\ \theta_{2m}\ \chi_{2m}\}^T$ (in which $\chi_{im} = (\partial^2 w/\partial x^2)_{im}$ is the curvature parameter at nodal line i for the m-th term of the series.)

With the displacement function established, the procedure outlined in Section 3.3.1 can be repeated to obtain the stiffness and load matrices. The general form of $[K]_{mn}$ and $\{F\}_m$ has been worked out by Cheung[45] but will not be given here because of its complexity. However, for a simply supported strip, $[K]_{mm}$ and $\{F\}_m$ are quite simple and are listed in Tables 3.5 and 3.6.[43,44]

3.3.4 EXAMPLES

(i) Simply supported isotropic slab under uniform load [4,43] (Figure 3.2)

The accuracy of the lower order strip LO2 is studied in this example. A series of calculations for half a slab was carried out for different mesh divisions and using different number of terms. The rate of convergence is compared with that of the finite element method in Table 3.7. It can be seen that while convergence of deflections is extremely rapid, more strips and terms are required for the moments to converge to the exact answer. Note that the comparison of bandwidths and total number of equations will be much more disadvantageous for the finite element method if the whole slab has to be analyzed, e.g. in the case of asymmetric loading.

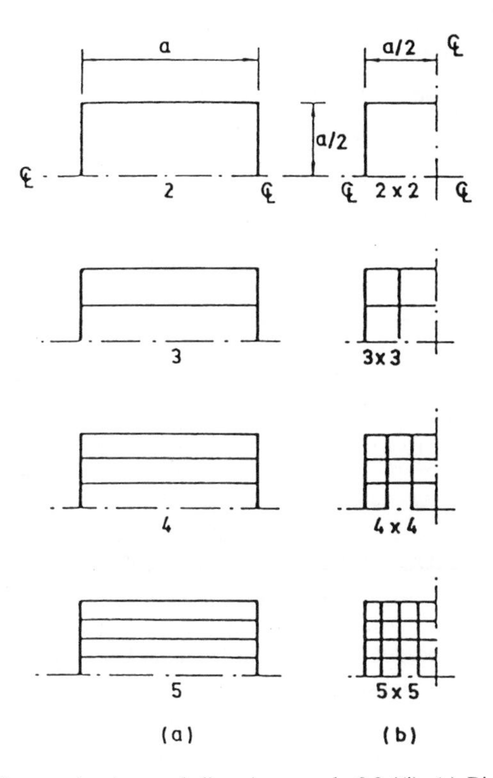

Figure 3.2 Diagram showing mesh lines in example 3.3.4(i): (a) Division into strips. (b) Division into elements[4,43].

TABLE 3.7
Convergence test using uniformly loaded simply supported slab.[4,43]

$\upsilon = .3$	Finite strip method				Finite element method		
No of lines for slab (2 equations per line)	Total number of terms	w_{max}	$M_{x\ max}$	$M_{y\ max}$	Nodal mesh for quarter slab (3 equations per nodes)	w_{max}	M_{max}
2	1 3 4	0.00414 0.00409 0.00409	0.0561 0.0549 0.0549	0.0540 0.0504 0.0504	2x2	0.00350	0.0669
3	1 3 4	0.00411 0.00406 0.00406	0.0502 0.0490 0.0489	0.0520 0.0485 0.0481	3x3	0.00395	0.0502
4	1 3 4	0.00411 0.00406 0.00406	0.0496 0.0484 0.0481	0.0518 0.0483 0.0479	4x4	0.00401	0.0483
5	1 3 4	0.00411 0.00406 0.00406	0.0494 0.0482 0.0481	0.0517 0.0482 0.0478	5x5	0.00403	
Exact[2]		0.00406	0.0479	0.0479		0.00406	0.0479
Multiplier		qa^4/D	qa^2			qa^4/D	qa^2

(ii) Square isotropic slab with all edges clamped and with a central concentrated load P [43,46]

In order to demonstrate the improved accuracy achievable by higher order strips, the slab is analyzed by dividing half of it, first of all, into five LO2 strips and then into only two HO2 strips. From Table 3.8 it is observed that the very coarse HO2 mesh actually produces results that are slightly more accurate. In this example, all the terms of the series are coupled together.

TABLE 3.8
Static analysis of slab with four clamped sides.[43,46]

$\upsilon = 0.3$	Central concentrated load, P					
	w_{max}		$-M_{x\ max}$		$-M_{y\ max}$	
	HO2	LO2	HO2	LO2	HO2	LO2
m=1	0.00511	0.00510	-0.11838	-0.11562	-0.14385	-0.14381
m=3	0.00030	0.00031	-0.00949	-0.00634	+0.05122	+0.05242
m=5	0.00011	0.00011	+0.00331	+0.00134	-0.04690	-0.04421
m=7	0.00003	0.00003	-0.00198	-0.00061	+0.02677	+0.02638
Σ	0.00555	0.00560	-0.12654	-0.12129	-0.11276	-0.10922
Exact	0.00560		-0.12570			
Multiplier	Pa^2/D		P			

It should be pointed out that from the computation point of view, no gain in efficiency has been achieved in this example through the use of HO2 because of the 50% increase in bandwidth, which more than offsets the small reduction in the total number of equations.

3.4 SPLINE BENDING STRIPS[47]

The spline strip which will be presented here is a lower order strip. Higher order strips can also be formulated without any difficulty, if desired.

3.4.1 FORMATION OF STRIP STIFFNESS MATRIX

As in the case of the classical finite strip, each node of the lower order bending strip requires two degrees of freedom (deflection and rotation) to satisfy the continuity requirements. The displacement function for the lower order strip as shown in Figure 3.3 is given as:

$$\{f\} = w = [[C_1]\,[C_2]] \begin{bmatrix} [\phi]_1 & & & \\ & [\phi]_2 & & \\ & & [\phi]_3 & \\ & & & [\phi]_4 \end{bmatrix} \{\delta\} \qquad (3.22)$$

where $[\phi]_i$ (i=1,2,3 and 4) are the corresponding B-3 spline functions.

The symbols have already been defined in Section 2.3. In short form

$$w = [N]\,\{\delta\} = [C]\,[\Phi]\,\{\delta\} \qquad (3.23)$$

Following the procedure outlined in Section 3.2, the stiffness matrix can be worked out by hand and expressed in an explicit form as follows:

$$[K^e] = \int [B]^T [D][B]\, dv$$

$$= \int \left\{ [\Phi]^T\ [\Phi'']^T\ [\Phi']^T \right\} \begin{bmatrix} -[C'']^T & 0 & 0 \\ 0 & -[C]^T & 0 \\ 0 & 0 & 2[C']^T \end{bmatrix} \begin{bmatrix} D_x & D_1 & 0 \\ D_1 & D_y & 0 \\ 0 & 0 & D_{xy} \end{bmatrix}$$

$$\begin{bmatrix} -[C''] & 0 & 0 \\ 0 & -[C] & 0 \\ 0 & 0 & 2[C'] \end{bmatrix} \begin{Bmatrix} [\Phi] \\ [\Phi''] \\ [\Phi'] \end{Bmatrix} dv$$

$$= \int \{ [\Phi]^T [C_1][\Phi] + [\Phi]^T [C_3][\Phi''] + [\Phi'']^T [C_2][\Phi] + [\Phi'']^T [C_4][\Phi''] + 4[\Phi']^T [C_5][\Phi'] \}\, dy \qquad (3.24)$$

in which $[C_1] = \int_0^b [C'']^T D_x [C'']\, dx$; $[C_2] = \int_0^b [C'']^T D_1 [C]\, dx$

$[C_3] = \int_0^b [C]^T D_1 [C'']\, dx$; $[C_4] = \int_0^b [C]^T D_y [C]\, dx$

$[C_5] = \int_0^b [C']^T D_{xy}\ [C']\, dx$

The integration in the above equation can be worked out easily while the integration for coupling matrices $\int \Phi^k\, \Phi^l\, dy$ in Eq. 3.24 may be reduced to summation operations for equally spaced spline knots. The stiffness matrix is tabulated in Table 3.9.[47]

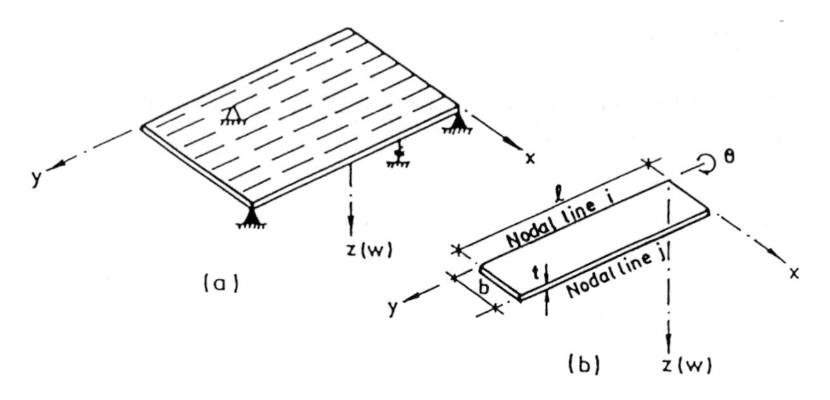

Figure 3.3 Spline finite strip model: (a) Plates as an assembly of strips. (b) Typical strip.

TABLE 3.9
Bending stiffness matrix of a rectangular strip.[47]

$\frac{1}{420b^3}\times$

$5040D_xI_{1_{11}}$ $-504b^2D_1I_{2_{11}}$ $-504b^2D_1I_{3_{11}}$ $+156b^4D_yI_{4_{11}}$ $+2016b^2D_{xy}I_{5_{11}}$	$2520bD_xI_{1_{12}}$ $-462b^3D_1I_{2_{12}}$ $-42b^3D_1I_{3_{12}}$ $+22b^5D_yI_{4_{12}}$ $+168b^3D_{xy}I_{5_{12}}$	$-5040D_xI_{1_{13}}$ $+504b^2D_1I_{2_{13}}$ $+504b^2D_1I_{3_{13}}$ $+54b^4D_yI_{4_{13}}$ $-2016b^2D_{xy}I_{5_{13}}$	$2520bD_xI_{1_{14}}$ $-42b^3D_1I_{2_{14}}$ $-42b^3D_1I_{3_{14}}$ $-13b^5D_yI_{4_{14}}$ $+168b^3D_{xy}I_{5_{14}}$
	$5040b^2D_xI_{1_{22}}$ $-504b^3D_1I_{2_{22}}$ $-504b^4D_1I_{3_{22}}$ $+156b^6D_yI_{4_{22}}$ $+2016b^4D_{xy}I_{5_{22}}$	$-2520bD_xI_{1_{23}}$ $+42b^3D_1I_{2_{23}}$ $+42b^3D_1I_{3_{23}}$ $+13b^5D_yI_{4_{23}}$ $-168b^3D_{xy}I_{5_{23}}$	$840b^2D_xI_{1_{24}}$ $+14b^3D_1I_{2_{24}}$ $+14b^4D_1I_{3_{24}}$ $-3b^6D_yI_{4_{24}}$ $-56b^4D_{xy}I_{5_{24}}$
	SYM	$5040D_xI_{1_{33}}$ $-504b^2D_1I_{2_{33}}$ $-504b^2D_1I_{3_{33}}$ $+156b^4D_yI_{4_{33}}$ $+2016b^2D_{xy}I_{5_{33}}$	$-2520bD_xI_{1_{34}}$ $+462b^3D_1I_{2_{34}}$ $+42b^3D_1I_{3_{34}}$ $-22b^5D_yI_{4_{34}}$ $-168b^3D_{xy}I_{5_{34}}$
			$5040b^2D_xI_{1_{44}}$ $-504b^3D_1I_{2_{44}}$ $-504b^4D_1I_{3_{44}}$ $+156b^6D_yI_{4_{44}}$ $+2016b^4D_{xy}I_{5_{44}}$

where $I_{1_{ij}} = \int_0^l t[\phi]_i^T[\phi]_j dy$; $I_{2_{ij}} = \int_0^l t[\phi]_i^{''T}[\phi]_j dy$; $I_{3_{ij}} = \int_0^l t[\phi]_i^T[\phi]_j^{''} dy$

$I_{4_{ij}} = \int_0^l t[\phi]_i^{''T}[\phi]_j^{''} dy$; $I_{5_{ij}} = \int_0^l t[\phi]_i^{'T}[\phi]_j^{'} dy$

3.4.2 LOAD VECTORS

i) For a patch load linearly distributed along the y-direction (Figure 3.4) from y_1 to y_2. The load vector {F} can be expressed as

$$\{F^e\} = \int_o^b \int_{y_1}^{y_2} [\Phi]^T [N]^T \{q_1 + (q_2 - q_1)(y - y_1)\}\, dydx$$

$$= q_1\left(\int_{y_1}^{y_2} [\Phi]^T dy\right) \begin{Bmatrix} \frac{b}{2} \\ \frac{b^2}{12} \\ \frac{b}{2} \\ -\frac{b^2}{12} \end{Bmatrix} + (q_2 - q_1)\left(\int_{y_1}^{y_2} [\Phi]^T (y - y_1) dy\right) \begin{Bmatrix} \frac{b}{2} \\ \frac{b^2}{12} \\ \frac{b}{2} \\ -\frac{b^2}{12} \end{Bmatrix} \quad (3.25)$$

ii) For a concentrated load on a nodal line

It is recommended to locate a nodal line passing through the concentrated load. Otherwise, its effect can only be approximately transferred to its two boundary nodal lines through the cubic interpolation polynomial of the assumed displacement function. For a concentrated force P, moments M_x and M_y at locations y_1, y_2 and y_3 of the nodal line i respectively (Figure 3.5) the load vector will be

$$\{F^e\} = [\Phi(y_1)]^T \begin{Bmatrix} P \\ 0 \\ 0 \\ 0 \end{Bmatrix} + [\Phi(y_2)]^T \begin{Bmatrix} 0 \\ M_x \\ 0 \\ 0 \end{Bmatrix} + [\Phi'(y_3)]^T \begin{Bmatrix} M_y \\ 0 \\ 0 \\ 0 \end{Bmatrix} \quad (3.26)$$

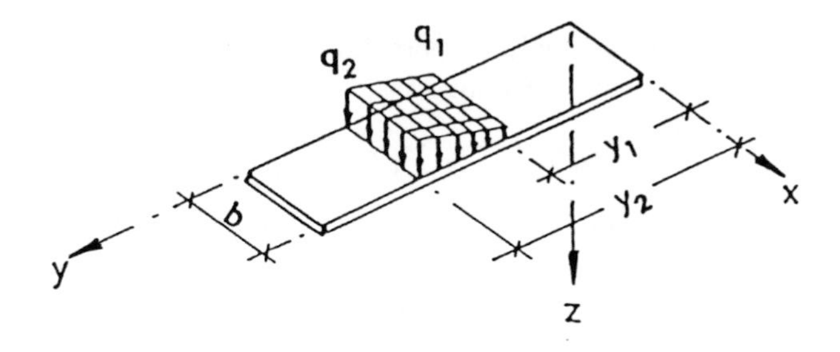

Figure 3.4 Linearly distributed patch load on rectangular plate strip.

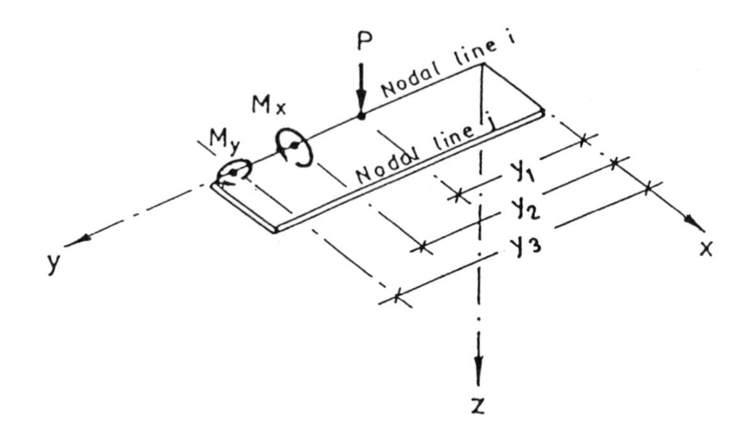

Figure 3.5 Concentrated loads on nodal line of rectangular plate strip.

Each of the three components in Eq. 3.26 will form a column vector with four non-zero entries only because there are only four local splines contributing non-zero values to it at the specified location y_i as shown in Figure 3.5.

It should be noted that the load vector for a concentrated load on a nodal line common to two strips should never be added to the overall load vector more than once.

3.4.3 TREATMENT FOR BOUNDARY CONDITIONS

In the strip direction, the B-3 spline function must satisfy the geometrical boundary conditions at the two ends, and this can be achieved by amending the spline functions.

For example, a strip simply supported at one end (say $y = y_0$) can be modified by imposing the conditions that the deflection is zero while the rotation is not, that is

$$\alpha_{-1}\,\phi_{-1}(y_0) + \alpha_0\,\phi_0(y) + \alpha_1\,\phi_1(y) = 0 \tag{3.27}$$

It is obvious that α_{-1} can be written in terms of the other α's. If the lengths of the three consecutive sections are the same, we have:

$$\alpha_{-1} = -4\,\alpha_0 + \alpha_1 \tag{3.28}$$

The spline functions are modified to:

$$\delta = \alpha_0\,[\phi_0(y) - 4\,\phi_{-1}(y)] + \alpha_1\,[\phi_0(y) - \phi_{-1}(y)] + \sum_{i=2}^{n+1} \alpha_i\,\phi_i(y)$$

or

$$\delta = \alpha_0\, \tilde{\phi}_0(y) + \alpha_1\, \tilde{\phi}_1(y) + \sum_{i=2}^{n+1} \alpha_i\, \phi_i(y) \tag{3.29}$$

The function $\tilde{\phi}_0(y)$ and $\tilde{\phi}_1(y)$ are referred to as the amended boundary spline. Note that α_{-1} has been eliminated. Other boundary conditions can be treated by this approach, and the results are listed in Table 3.10.

TABLE 3.10
Modified spline functions at end support.[27]

Boundary Condition	Modified local spline functions		
	$\tilde{\phi}_{-1}$	$\tilde{\phi}_0$	$\tilde{\phi}_1$
Free	$\tilde{\phi}_{-1}$	$\tilde{\phi}_0$	$\tilde{\phi}_1$
Simply supported	Eliminated	$\tilde{\phi}_0 - 4\tilde{\phi}_{-1}$	$\tilde{\phi}_1 - \tilde{\phi}_{-1}$
Clamped	Eliminated	Eliminated	$\tilde{\phi}_1 - \frac{1}{2}\tilde{\phi}_0 + \tilde{\phi}_{-1}$
Sliding clamped supported	Eliminated	$\tilde{\phi}_0$	$\tilde{\phi}_1 + \tilde{\phi}_{-1}$
Continuous	$\tilde{\phi}_{-1} - \frac{1}{4}\tilde{\phi}_0$	Eliminated	$\tilde{\phi}_1 - \frac{1}{4}\tilde{\phi}_0$

3.4.4 ELASTIC SPRING SUPPORTS

Elastic spring support with stiffness k_{si} (or $k_{\theta xi}$, $k_{\theta yi}$) may be treated as a fictitious load of magnitude $k_{si}\delta_i$ (or $k_{\theta xi}\theta_{xi}$, $k_{\theta yi}\theta_{yi}$), in which δ_i is the displacement (or rotation) at the location (x_i, y_i) where the elastic support is placed. As in the case of concentrated loading, it is advisable to locate a nodal line passing through the support at $(0, y_i)$ as shown in Figure 3.6.

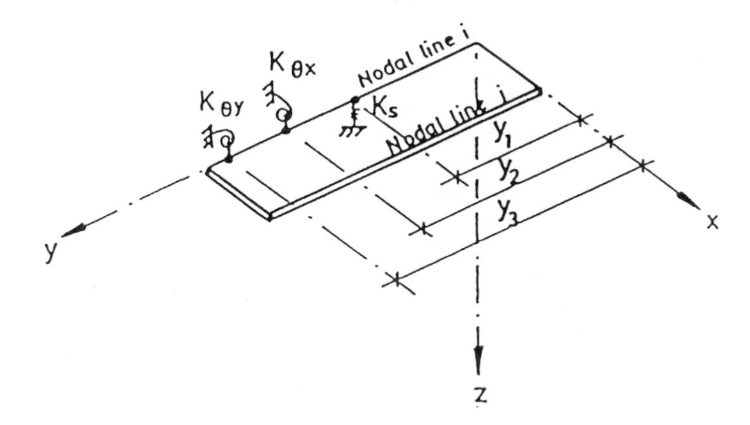

Figure 3.6 Elastic supports at nodal line of rectangular plate strip.

In the energy formulation, additional potential energy is thus introduced, that is

$$\Pi = -\{f(0,y_1)\}^T k_s \{f(0,y_1)\} - \{\frac{d}{dx} f(0,y_2)\}^T k_{\theta x} \{\frac{d}{dx} f(0,y_2)\} -$$

$$\{\frac{d}{dy} f(0,y_3)\}^T k_{\theta y} \{\frac{d}{dy} f(0,y_3)\} \quad (3.30)$$

Hence, the total potential energy (Eq. 3.4) becomes

$$\phi = \frac{1}{2}\int \{\delta\}^T [B]^T [D] [B] \{\delta\} d(area) - \int \{\delta\}^T [N]^T \{q\} d(area) + \Pi \quad (3.31)$$

After taking the first variation, it will become

$$\{[K] + [K_{es}]\} \{\delta\} = \{F\} \quad (3.32)$$

The additional matrix $[K_{es}]$ is derived from the elastic supports and is given by

$$[K_{es}] = [\Phi(y_1)]^T \begin{bmatrix} k_S & 0 & 0 & 0 \\ 0 & 0 & 0 & 0 \\ 0 & 0 & 0 & 0 \\ 0 & 0 & 0 & 0 \end{bmatrix} [\Phi(y_1)] + [\Phi(y_2)]^T \begin{bmatrix} 0 & 0 & 0 & 0 \\ 0 & k_{\theta x} & 0 & 0 \\ 0 & 0 & 0 & 0 \\ 0 & 0 & 0 & 0 \end{bmatrix} [\Phi(y_2)]$$

$$[\Phi'(y_3)]^T \begin{bmatrix} k_{\theta y} & 0 & 0 & 0 \\ 0 & 0 & 0 & 0 \\ 0 & 0 & 0 & 0 \\ 0 & 0 & 0 & 0 \end{bmatrix} [\Phi'(y_3)] \quad (3.33)$$

Each of the three components in Eq. 3.33 will form a matrix with a non-zero 4x4 submatrix only due to the localization of B-3 spline representation.

3.4.5 EXAMPLES

(i) Square plates with various support conditions

Fan[47] studied the versatility of the spline strips by considering plates with different support conditions. Some examples are quoted here for reference, and they include plates simply supported

1) at four corners (Table 3.11)
2) at mid-point of the edges (Table 3.12)
3) at one corner and two adjacent edges (Table 3.13)

The results converge fairly rapidly to the analytical or other published ones.

(ii) Square runway slab on Winkler's foundation[47]

By symmetry, analyses were performed only on one quarter of a slab with discretisation meshes as shown in Figure 3.7. Results of vertical deflections along two axes and the diagonal line are shown in Figure 3.8 and compared with finite element solutions obtained by Chen et al.[48] It can be seen that results obtained either by coarse (5 strips x 5 sections) or finer mesh (12 strips x 12 sections) analyses are in good agreement with the results published by Chen.[48]

TABLE 3.11
Corner supported plate subjected to central point load.[47]

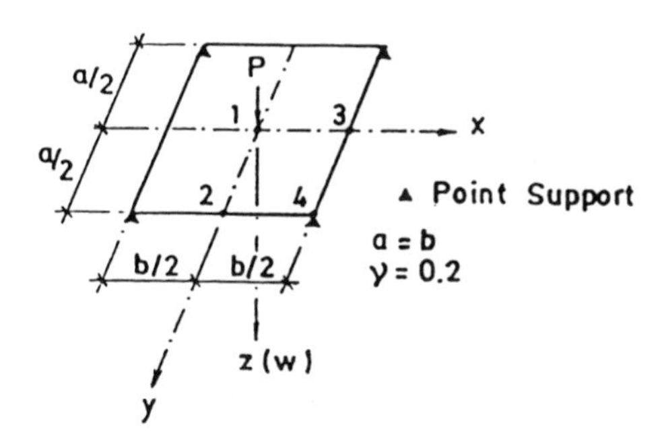

Discretisation (no of strips)x(no of sections)	Vertical deflection		Moment	
	W_1	W_2	M_{x2}	M_{xy4}
4 x 4	0.040180	0.022702	0.21360	0.11999
6 x 6	0.040202	0.022702	0.21206	0.12125
8 x 8	0.040208	0.022702	0.21148	0.12195
10 x 10	0.040213	0.022702	0.21126	0.12240
12 x 12	0.040215	0.022702	0.21109	0.12271
Multiplication factor	P/D		P	

TABLE 3.12
Edge mid-point supported square plate subjected to uniformly distributed load.[47]

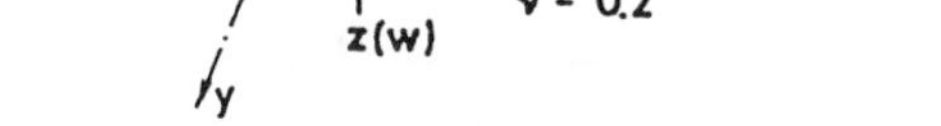

Discretisation (no of strips)x(no of sections)	Vertical deflection		Moment	
	W_2	W_3	M_{y3}	M_{x3}
4 x 4	0.13083	0.455556	0.040766	0.040617
6 x 6	0.13185	0.45662	0.040923	0.040890
8 x 8	0.13222	0.45701	0.040991	0.040973
10 x 10	0.13240	0.45718	0.041018	0.041009
12 x 12	0.13250	0.45729	0.041041	0.041029
Multiplication factor	$(qa^4/D) \times 10^{-2}$		qa^2	

TABLE 3.13
Square plate with two adjacent sides simply supported and one corner support subjected to uniform load.[47]

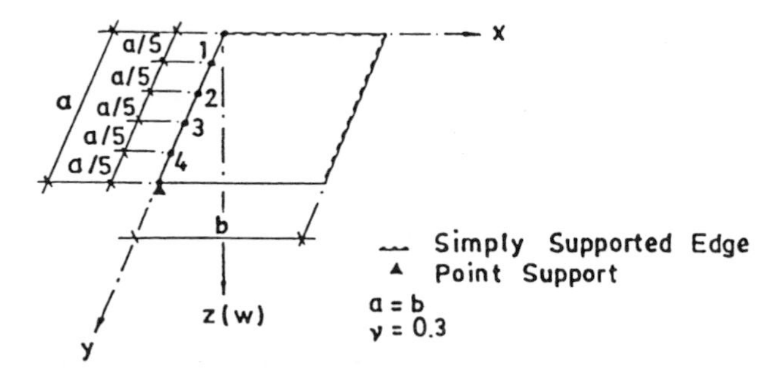

Discretisation (no of strips)x(no of sections)	Vertical deflection			
	W_1	W_2	W_3	W_4
4 x 4	0.08082	0.13031	0.13201	0.08395
6 x 6	0.08085	0.13039	0.13203	0.08406
8 x 8	0.08086	0.13040	0.13205	0.08405
10 x 10	0.08086	0.13040	0.13205	0.08405
12 x 12	0.08086	0.13040	0.13205	0.08405
Multiplication factor	$(qa^4/D) \times 10^{-1}$			

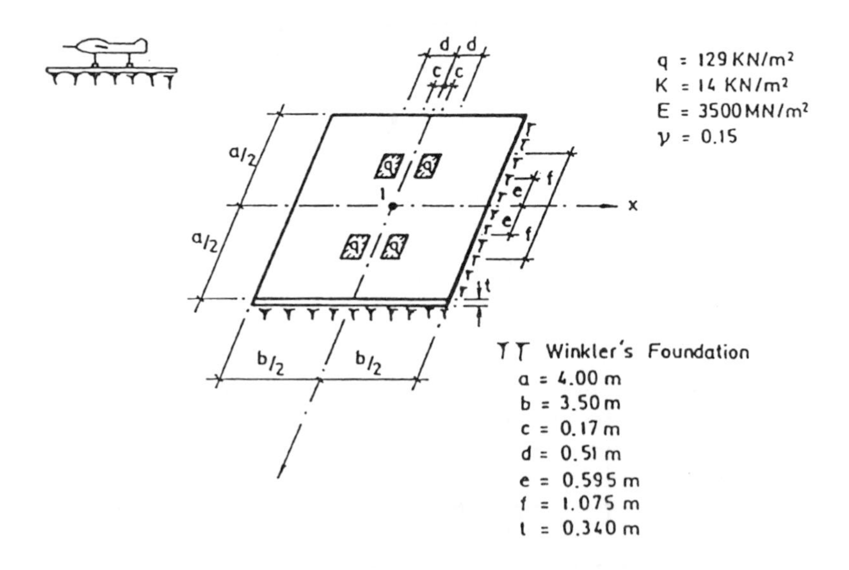

Figure 3.7 Runway slab on Winkler's foundation subjected to static patch load.[47]

		0.35	0.70	1.05	1.40	1.75
	0.650 * [0.652] (0.659)	0.627 * [0.628] (0.627)	0.563 * [0.565] (0.564)	0.474 * [0.476] (0.474)	0.375 * [0.377] (0.375)	0.274 * [0.276] (0.275)
0.40	0.649 * [0.651] (0.649)	0.625 * [0.627] (0.626)				
0.80	0.630 * [0.632] (0.630)		0.536 * [0.539] (0.536)			
1.20	0.567 * [0.570] (0.567)			0.395 * [0.397] (0.396)		
1.60	0.481 * [0.485] (0.482)				0.245 * [0.248] (0.245)	
2.00	0.399 * [0.402] (0.400)					0.097 * [0.098] (0.096)

* : Chen[48] [ ] : Spline strip (mesh 5x5) ( ) : Spline strip (mesh 12x12)

Figure 3.8 Runway slab. Comparison of vertical deflection.[47]

3.5 COMPUTED SHAPE FUNCTION (COMSFUN) STRIP

A strip shown in Figure 3.9 with varying rigidity along the y-axis is taken as an example. In Figure 3.9, a unit width of the strip is taken out as a fictitious beam with the same variation of the longitudinal rigidity as the original plate strip. In accordance with the boundary conditions at each end of the strip, the appropriate conditions have to be imposed on the fictitious beam (Table 3.14). A total of r computed shape functions are computed from the fictitious beam. The displacement functions for the strip can be written as:

$$w = \sum_{m=1}^{r} [N(x)]\phi_m(y)\{\delta\}_m \tag{3.34}$$

where $\{\delta\}_m = \{w_1 \ \ (\partial w/\partial x)_1 \ \ w_2 \ \ (\partial w/\partial x)_2\}^T$. $[N(x)]$ and $\phi_m(y)$ are the usual beam functions along the x-axis and the longitudinal COMSFUN.

The stiffness matrix and load vector can be formed accordingly and detailed discussion can be found in Reference 42.

3.5.1 MIXED STRIPS

Because there is a change in rigidity across the transverse direction (Figure 3.10), the longitudinal shape functions of the panel strip are different from those of the column strip. A mixed strip is, therefore, introduced to satisfy the

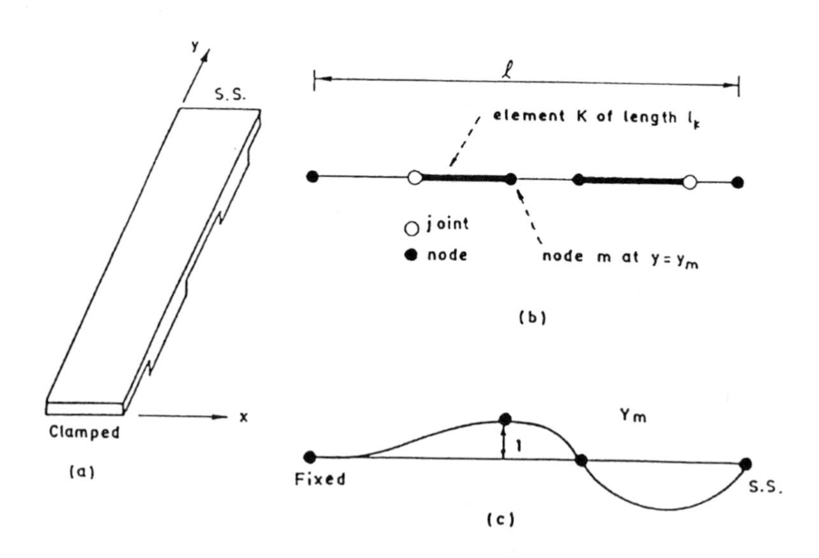

Figure 3.9 (a) A thin plate finite strip with varying longitudinal rigidity. (b) The fictitious beam and its discretisation. (c) COMSFUN ϕ_m corresponding to node m.

TABLE 3.14
Boundary conditions for plate bending problems.[42]

A	B	C
clamped	sliding-clamped	clamped
sliding-clamped	sliding-clamped	clamped
simply supported	free	simply-supported
free	free	simply-supported

A - actual boundary condition at y = 0 of a strip
B - boundary condition at y = 0 of the beam when computing the shape function for y = 0
C - boundary condition at y = 0 of the beam when computing the shape function for each of the internal nodes and the end nodes at y = b

compatibility requirement along the nodal lines between adjacent strips. The displacement function of a mixed strip is given by:

$$w = \sum_{m=1}^{r} [(N_1(x) \quad N_2(x))(\phi_1)_m(y) \quad (N_3(x) \quad N_4(x))(\phi_2)_m(y)]\{\delta\}_m \quad (3.35)$$

where $\phi_1(y)$ and $\phi_2(y)$ denote respectively the COMSFUN of nodal lines '1' and '2'. The characteristic matrices can follow the same track as for the standard strip.

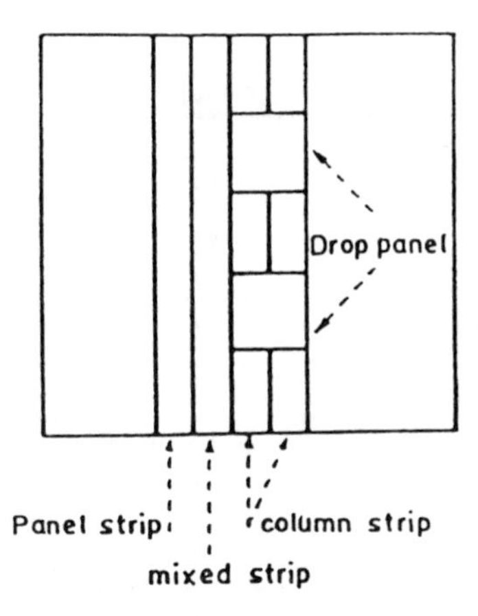

Figure 3.10 Classification of strips for analyzing flat slabs with drop panels.

3.5.2 ELASTIC SPRING SUPPORTS

For elastic springs corresponding to w and ($\partial w/\partial x$), its stiffness can be added directly to their corresponding location along the diagonal in the global stiffness matrix. However, for a point spring corresponding to a rotation about the x-axis ($\partial w/\partial y$), the spring stiffness can be incorporated by using the mixed strip because the spline functions have to be modified to satisfy the support condition.

3.5.3 EXAMPLES

Examples of square plates (a x a) were presented by Kong[(42)] to demonstrate the application of the computed shape function strips. For each plate, the finite strip mesh consists of eight strips in one direction and it adopts various number of nodes in the other orthogonal direction.

The plates are subjected to uniformly distributed load, and the boundary conditions are as follows:

a) Case A: all edges simply supported
b) Case B: four corners simply supported
c) Case C: all edges fixed.

Results of bending moments and deflections at the centre of the plates are tabulated in Table 3.15. They compare favourably with the exact solutions.[(2)] Good convergence characteristics can also be observed.

3.6 CURVED PLATE STRIPS

In modern construction, many structures are curved in plans because of aesthetical or functional reasons, and sector plates bounded by two radial lines and one or two (concentric) circular arcs are especially popular. These curved structures are usually either plain curved slabs or are made up of a system of

TABLE 3.15
Deflections and moments at the centre of square plates ($\upsilon = 0.3$).[42]

Case	No of nodes	Deflection	M_x	M_y
A	3	0.406	0.481	0.478
	5	0.406	0.481	0.481
	7	0.406	0.481	0.480
	Exact[2]	0.406	0.479	0.479
B	3	0.252	0.115	0.125
	5	0.254	0.113	0.114
	7	0.254	0.113	0.113
	Exact[2]	0.249	0.109	0.109
C	3	0.126	0.234	0.236
	5	0.127	0.233	0.233
	7	0.127	0.233	0.231
	Exact[2]	0.127	0.229	0.229
Multiplier		$qa^4 \times 10^{-2}/D$	$qa^2 \times 10^{-1}$	

slabs and multiple stiffening girders. These multiple girders are usually arranged orthogonally along the radii and arcs of concentric circles, and the entire system can be treated as an equivalent cylindrical orthotropic plate.

In the previous discussions, rectangular strips using a Cartesian coordinate system have been developed and applied successfully to the analysis of rectangular orthotropic plates. It is immediately apparent that the same approach can be applied to the analysis of cylindrical orthotropic curved plates by using curved strips and a polar coordinate system. In fact it will be demonstrated that any rectangular plate can be analyzed by the curved strip program by simply adopting a very small subtended angle together with a large radius.

As in the case of spline strips, only the formulation of the lower order strips will be presented here.

3.6.1 DISPLACEMENT FUNCTION

A comparison of the curved strip shown in Figure 3.1e and its counterpart in a Cartesian coordinate system (Figure 3.1b) will show clearly the various terms which correspond to one another in the two coordinate systems. Such a comparison is listed in Table 3.16, and from these data the displacement function for the curved strip is readily deduced.

The displacement function can be further simplified by introducing the dimensionless variables
$b' = (r_2 - r_1)/2$ and $R = (r - r_1)/b'$, so that at $r = r_1$, $R = 0$, and at $r = r_2$, $R = 2$.

Thus, finally,

a) classical strip

w =

$$\sum_{m=1}^{r} \Theta_m \left[(1-\frac{3R^2}{4}+\frac{R^3}{4}) \quad b'(R-R^2+\frac{R^3}{4}) \quad (\frac{3R^2}{4}-\frac{R^3}{4}) \quad b'(\frac{R^3}{4}-\frac{R^2}{2}) \right] \begin{Bmatrix} w_{1m} \\ \psi_{1m} \\ w_{2m} \\ \psi_{2m} \end{Bmatrix}$$

$$= \sum_{m=1}^{r} [N]_m \{\delta\}_m \Theta_m \tag{3.36}$$

TABLE 3.16
Conversion of displacement function for bending strip.

Rectangular strip	$w = \sum_{m=1}^{r} [(1-\frac{3x^2}{b^2}+\frac{2x^3}{b^3}),\ (x-\frac{2x^2}{b}+\frac{x^3}{b^2}),\ (\frac{3x^2}{b^2}-\frac{2x^3}{b^3}),\ (\frac{x^3}{b^2}-\frac{x^2}{b})] Y_m [w_{1m} \theta_{1m} w_{2m} \theta_{2m}]^T$				
Curved strip	$w = \sum_{m=1}^{r} [(1-3\bar{r}^2+2\bar{r}^3),\ (r-2r\bar{r}+r\bar{r}^2),\ (3\bar{r}^2-2\bar{r}^3),\ (r\bar{r}^2-r\bar{r})] \Theta_m [w_{1m} \psi_{1m} w_{2m} \psi_{2m}]^T$ where $\bar{r} = \frac{r-r_1}{r_2-r_1}$				
Conversion	x	y	l	b	Y_m	$\theta = \partial w/\partial x$
	$r-r_1$	θ	α	$r_2 - r_1$	Θ_m	$\psi = \partial w/\partial r$

b) spline strip

w =

$$\left[(1-\frac{3R^2}{4}+\frac{R^3}{4}) \quad b'(R-R^2+\frac{R^3}{4}) \quad (\frac{3R^2}{4}-\frac{R^3}{4}) \quad b'(\frac{R^3}{4}-\frac{R^2}{2}) \right]$$

$$\begin{bmatrix} [\phi]_1 & 0 & 0 & 0 \\ 0 & [\phi]_2 & 0 & 0 \\ 0 & 0 & [\phi]_3 & 0 \\ 0 & 0 & 0 & [\phi]_4 \end{bmatrix} \{\delta\} \tag{3.37}$$

or

$$w = [C]\,[\Phi]\,\{\delta\} = [N]\,\{\delta\} \tag{3.38}$$

To satisfy the boundary conditions, the B-3 spline functions again have to be modified as in the case of the rectangular strips (Table 3.10).

3.6.2 STIFFNESS AND LOAD MATRICES

The curvatures of a plate in polar coordinates are given by[2]

$$\{\varepsilon\} = \begin{Bmatrix} -\kappa_r \\ -\kappa_\theta \\ 2\kappa_{r\theta} \end{Bmatrix} = \begin{Bmatrix} -\dfrac{\partial^2 w}{\partial r^2} \\ -\dfrac{1}{r}(\dfrac{1}{r}\dfrac{\partial^2 w}{\partial \theta^2} + \dfrac{\partial w}{\partial r}) \\ -\dfrac{2}{r}(\dfrac{\partial^2 w}{\partial r \partial \theta} - \dfrac{1}{r}\dfrac{\partial w}{\partial \theta}) \end{Bmatrix} \tag{3.39}$$

In the case of a classical strip, the strains are

$$\{\varepsilon\} = \sum_{m=1}^{r} \begin{bmatrix} -\dfrac{\partial^2 [N]_m}{\partial r^2} \\ -\dfrac{1}{r}(\dfrac{1}{r}\dfrac{\partial^2 [N]_m}{\partial \theta^2} + \dfrac{\partial [N]_m}{\partial r}) \\ -\dfrac{2}{r}(\dfrac{\partial^2 [N]_m}{\partial r \partial \theta} - \dfrac{1}{r}\dfrac{\partial [N]_m}{\partial \theta}) \end{bmatrix} \{\delta\}_m = \sum_{m=1}^{r} [B]_m \{\delta\}_m \tag{3.40}$$

$[B]_m$ is obtained by carrying out the appropriate partial differentiations as indicated by Eq. 3.40 , and its explicit form is given in Table 3.17.

Similarly, the strain matrix for a spline strip is

TABLE 3.17
Strain matrix of a LO2 curved bending strip (classical strip).

$$\left[\begin{array}{c|c} (-\frac{3R}{2b'^2} + \frac{3}{2b'^2})\Theta_m & (\frac{2}{b'} - \frac{3R}{2b'})\Theta_m \\ \hline -\frac{1}{r^2}(1 - \frac{3}{4}R^2 + \frac{1}{4}R^3)\Theta_m'' & -\frac{b'}{r^2}(R - R^2 + \frac{R^3}{4})\Theta_m'' \\ +\frac{1}{r}(\frac{3R}{2b'} - \frac{3R^2}{4b'})\Theta_m & +\frac{1}{r}(2R - 1 - \frac{3R^2}{4})\Theta_m \\ \hline \frac{2}{r}(\frac{3R}{2b'} - \frac{3R^2}{4b'})\Theta_m' & +\frac{2}{r}(2R - 1 - \frac{3R^2}{4})\Theta_m' \\ +\frac{2}{r^2}(1 - \frac{3}{4}R^2 + \frac{1}{4}R^3)\Theta_m' & +\frac{2b'}{r^2}(R - R^2 + \frac{R^3}{4})\Theta_m' \end{array}\right.$$

$$\left.\begin{array}{c|c} (\frac{3R}{2b'^2} - \frac{3}{2b'^2})\Theta_m & (\frac{1}{b'} - \frac{3R}{2b'})\Theta_m \\ \hline -\frac{1}{r^2}(\frac{3}{4}R^2 - \frac{1}{4}R^3)\Theta_m'' & -\frac{b'}{r^2}(\frac{R^3}{4} - \frac{R^2}{2})\Theta_m'' \\ +\frac{1}{r}(\frac{3R^2}{4b'} - \frac{3R}{2b'})\Theta_m & +\frac{1}{r}(R - \frac{3R^2}{4})\Theta_m \\ \hline \frac{2}{r}(\frac{3R^2}{4b'} - \frac{3R}{2b'})\Theta_m' & \frac{2}{r}(R - \frac{3R^2}{4})\Theta_m' \\ +\frac{2}{r^2}(\frac{3}{4}R^2 - \frac{1}{4}R^3)\Theta_m' & +\frac{2b'}{r^2}(\frac{R^3}{4} - \frac{R^2}{2})\Theta_m' \end{array}\right]$$

$$[B]=\begin{bmatrix} -[C''] & 0 & 0 \\ \frac{-1}{r}[C'] & \frac{-1}{r^2}[C] & 0 \\ 0 & 0 & \frac{2}{r^2}[C]-\frac{2}{r}[C'] \end{bmatrix}\begin{Bmatrix} [\Phi] \\ [\Phi''] \\ [\Phi'] \end{Bmatrix} \tag{3.41}$$

One can show readily that the explicit form of [B] for the lower order spline strip can be obtained by changing Θ_m of Table 3.17 to $[\Phi]$, that is

$$\left[\begin{array}{c:c} (-\frac{3R}{2b'^2}+\frac{3}{2b'^2})[\Phi] & (\frac{2}{b'}-\frac{3R}{2b'})[\Phi] \\ \hdashline -\frac{1}{r^2}(1-\frac{3}{4}R^2+\frac{1}{4}R^3)[\Phi]'' & -\frac{b'}{r^2}(R-R^2+\frac{R^3}{4})[\Phi]'' \\ +\frac{1}{r}(\frac{3R}{2b'}-\frac{3R^2}{4b'})[\Phi] & +\frac{1}{r}(2R-1-\frac{3R^2}{4})[\Phi] \\ \hdashline \frac{2}{r}(\frac{3R}{2b'}-\frac{3R^2}{4b'})[\Phi]' & +\frac{2}{r}(2R-1-\frac{3R^2}{4})[\Phi]' \\ +\frac{2}{r^2}(1-\frac{3}{4}R^2+\frac{1}{4}R^3)[\Phi]' & +\frac{2b'}{r^2}(R-R^2+\frac{R^3}{4})[\Phi]' \end{array}\right.$$

$$\left.\begin{array}{c:c} (\frac{3R}{2b'^2}-\frac{3}{2b'^2})[\Phi] & (\frac{1}{b'}-\frac{3R}{2b'})[\Phi] \\ \hdashline -\frac{1}{r^2}(\frac{3}{4}R^2-\frac{1}{4}R^3)[\Phi]'' & -\frac{b'}{r^2}(\frac{R^3}{4}-\frac{R^2}{2})[\Phi]'' \\ +\frac{1}{r}(\frac{3R^2}{4b'}-\frac{3R}{2b'})[\Phi] & +\frac{1}{r}(R-\frac{3R^2}{4})[\Phi] \\ \hdashline \frac{2}{r}(\frac{3R^2}{4b'}-\frac{3R}{2b'})[\Phi]' & \frac{2}{r}(R-\frac{3R^2}{4})[\Phi]' \\ +\frac{2}{r^2}(\frac{3}{4}R^2-\frac{1}{4}R^3)[\Phi]' & +\frac{2b'}{r^2}(\frac{R^3}{4}-\frac{R^2}{2})[\Phi]' \end{array}\right] \tag{3.42}$$

The bending and twisting moments for a plate with cylindrical orthotropic material are:

$$\{\sigma\}=\begin{Bmatrix} M_r \\ M_\theta \\ M_{r\theta} \end{Bmatrix}=[D]\,[B]\,\{\varepsilon\} \tag{3.43}$$

and the property matrix

$$[D]=\begin{bmatrix} D_r & D_1 & 0 \\ D_1 & D_\theta & 0 \\ 0 & 0 & D_{r\theta} \end{bmatrix} \tag{3.44}$$

where D_r, D_θ are bending rigidities for directions of r and θ, and $D_{r\theta}$ is the twisting rigidity:

$$D_r = \frac{E_r t^3}{12(1-\upsilon_r\upsilon_\theta)}\ , \ D_\theta = \frac{E_\theta t^3}{12(1-\upsilon_r\upsilon_\theta)}\ , \ D_{r\theta} = \frac{G_{r\theta} t^3}{12}\ , \ D_1 = \upsilon_r D_\theta = \upsilon_\theta D_r\ .$$

For isotropic material, we have the familiar plate constants

$$D_r = D_\theta = D = \frac{Et^3}{12(1-\upsilon^2)}, \quad \upsilon_r = \upsilon_\theta = \upsilon\ , \ D_{r\theta} = \frac{(1-\upsilon)}{2} D.$$

For a classical strip, the above equation is modified to:

$$\{\sigma\} = [D] \sum_{m=1}^{r} [B]_m \{\delta\}_m \tag{3.45}$$

All the ingredients are now ready, and it is possible to compute the stiffness matrix and load vector through the following equations:

a) classical strip

$$[K]_{mn} = \int_0^\alpha \int_{r_1}^{r_2} [B]_m^T [D] [B]_n \, r dr d\theta \tag{3.46}$$

$$\{F\}_m = \int_0^\alpha \int_{r_1}^{r_2} 2\, [N]_m^T q \, r dr d\theta \tag{3.47}$$

TABLE 3.18

Stiffness matrix $[K]_{mn}$ of a LO2 curved strip.[47]

$$[S]_{mn} = \left[\begin{array}{l:l:l:l}
\begin{array}{l} D_r B_{11}B_{11} \\ +D_1 B_{21}B_{11} \\ +D_1 B_{11}B_{21} \\ +D_\theta B_{21}B_{21} \\ +D_{r\theta} B_{31}B_{31} \end{array} &
\begin{array}{l} D_r B_{11}B_{12} \\ +D_1 B_{21}B_{12} \\ +D_1 B_{11}B_{22} \\ +D_\theta B_{21}B_{22} \\ +D_{r\theta} B_{31}B_{32} \end{array} &
\begin{array}{l} D_r B_{11}B_{13} \\ +D_1 B_{21}B_{13} \\ +D_1 B_{11}B_{23} \\ +D_\theta B_{21}B_{23} \\ +D_{r\theta} B_{31}B_{33} \end{array} &
\begin{array}{l} D_r B_{11}B_{14} \\ +D_1 B_{21}B_{14} \\ +D_1 B_{11}B_{24} \\ +D_\theta B_{21}B_{24} \\ +D_{r\theta} B_{31}B_{34} \end{array} \\ \hdashline
\begin{array}{l} D_r B_{12}B_{11} \\ +D_1 B_{22}B_{11} \\ +D_1 B_{12}B_{21} \\ +D_\theta B_{22}B_{21} \\ +D_{r\theta} B_{32}B_{31} \end{array} &
\begin{array}{l} D_r B_{12}B_{12} \\ +D_1 B_{22}B_{12} \\ +D_1 B_{12}B_{22} \\ +D_\theta B_{22}B_{22} \\ +D_{r\theta} B_{32}B_{32} \end{array} &
\begin{array}{l} D_r B_{12}B_{13} \\ +D_1 B_{22}B_{13} \\ +D_1 B_{12}B_{23} \\ +D_\theta B_{22}B_{23} \\ +D_{r\theta} B_{32}B_{33} \end{array} &
\begin{array}{l} D_r B_{12}B_{14} \\ +D_1 B_{22}B_{14} \\ +D_1 B_{12}B_{24} \\ +D_\theta B_{22}B_{24} \\ +D_{r\theta} B_{32}B_{34} \end{array} \\ \hdashline
\begin{array}{l} D_r B_{13}B_{11} \\ +D_1 B_{23}B_{11} \\ +D_1 B_{13}B_{21} \\ +D_\theta B_{23}B_{21} \\ +D_{r\theta} B_{33}B_{31} \end{array} &
\begin{array}{l} D_r B_{13}B_{12} \\ +D_1 B_{23}B_{12} \\ +D_1 B_{13}B_{22} \\ +D_\theta B_{23}B_{22} \\ +D_{r\theta} B_{33}B_{32} \end{array} &
\begin{array}{l} D_r B_{13}B_{13} \\ +D_1 B_{23}B_{13} \\ +D_1 B_{13}B_{23} \\ +D_\theta B_{23}B_{23} \\ +D_{r\theta} B_{33}B_{33} \end{array} &
\begin{array}{l} D_r B_{13}B_{14} \\ +D_1 B_{23}B_{14} \\ +D_1 B_{13}B_{24} \\ +D_\theta B_{23}B_{24} \\ +D_{r\theta} B_{33}B_{34} \end{array} \\ \hdashline
\begin{array}{l} D_r B_{14}B_{11} \\ +D_1 B_{24}B_{11} \\ +D_1 B_{14}B_{21} \\ +D_\theta B_{24}B_{21} \\ +D_{r\theta} B_{34}B_{31} \end{array} &
\begin{array}{l} D_r B_{14}B_{12} \\ +D_1 B_{24}B_{12} \\ +D_1 B_{14}B_{22} \\ +D_\theta B_{24}B_{22} \\ +D_{r\theta} B_{34}B_{32} \end{array} &
\begin{array}{l} D_r B_{14}B_{13} \\ +D_1 B_{24}B_{13} \\ +D_1 B_{14}B_{23} \\ +D_\theta B_{24}B_{23} \\ +D_{r\theta} B_{34}B_{33} \end{array} &
\begin{array}{l} D_r B_{14}B_{14} \\ +D_1 B_{24}B_{14} \\ +D_1 B_{14}B_{24} \\ +D_\theta B_{24}B_{24} \\ +D_{r\theta} B_{34}B_{34} \end{array}
\end{array}\right]$$

($B_{ij}\,B_{kl}$ refers to product of the m-th term and n-th term coefficients in the strain matrix)

$$[K]_{mn} = \int_0^\alpha \int_{r_1}^{r_2} [S]_{mn} \, r dr d\theta$$

Due to the presence of $1/r^n$ terms it is more expedient to integrate the above expressions numerically both with respect to dr and dθ. Thus the stiffness matrix listed in Table 3.18 is still in an incomplete form, and numerical integration using Gaussian quadrature or Simpson's rule remains to be performed.

b) spline strip

$[K^e] = \int [B]^T [D][B]\, d(\text{area})$

$$= \int \left\{ [\Phi]^T \quad [\Phi'']^T \quad [\Phi']^T \right\} \begin{bmatrix} -[C'']^T & 0 & 0 \\ \frac{-1}{r}[C']^T & \frac{-1}{r^2}[C]^T & \\ 0 & 0 & \frac{2}{r^2}[C]^T - \frac{2}{r}[C']^T \end{bmatrix}$$

$$\begin{bmatrix} D_r & D_1 & 0 \\ D_1 & D_\theta & 0 \\ 0 & 0 & D_{r\theta} \end{bmatrix} \begin{bmatrix} -[C''] & 0 & 0 \\ \frac{-1}{r}[C'] & \frac{-1}{r^2}[C] & 0 \\ 0 & 0 & \frac{2}{r^2}[C] - \frac{2}{r}[C'] \end{bmatrix} \begin{Bmatrix} [\Phi] \\ [\Phi''] \\ [\Phi'] \end{Bmatrix} d(\text{area}) \tag{3.48}$$

The explicit form of the stiffness matrix was established by Fan.[47]

The load vector for uniformly distributed patch load q over the area of the strip from θ_1 to θ_2 as shown in Figure 3.11 is

$$\{F^e\} = \frac{q}{60}\int_{\theta_1}^{\theta_2} [\Phi]^T d\theta \begin{Bmatrix} 3b(10r_1 + 3b) \\ b^2(5r_1 + 2b) \\ 3b(10r_1 + 7b) \\ -b^2(5r_1 + 3b) \end{Bmatrix} \tag{3.49}$$

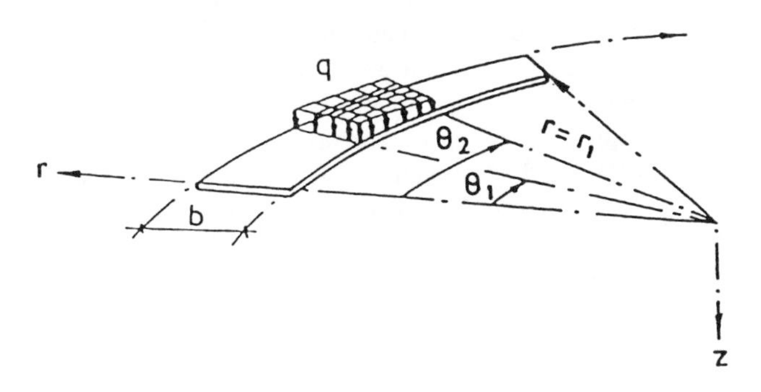

Figure 3.11 Uniformly distributed patch load on a circular plate strip.

While for concentrated loads P, M_r and M_θ at θ_1, θ_2 , θ_3 ($r=r_2$) as shown in Figure 3.12, the load vector is

$$\{F\} = [\Phi(\theta_1)]^T \begin{Bmatrix} 0 \\ 0 \\ P \\ 0 \end{Bmatrix} + [\Phi(\theta_2)]^T \begin{Bmatrix} 0 \\ 0 \\ 0 \\ M_r \end{Bmatrix} + [\Phi'(\theta_3)]^T \begin{Bmatrix} 0 \\ 0 \\ \dfrac{M_\theta}{r_2} \\ 0 \end{Bmatrix} \quad (3.50)$$

The elastic spring supports can be treated in a way similar to that for rectangular plate strips. The additional stiffness matrix [K_s] due to elastic supports k_s , $k_{\psi r}$ and $k_{\psi\theta}$ at θ_1 , θ_2, θ_2 ($r=r_2$) as shown in Figure 3.13 is

$$[K_{es}] = [\Phi(\theta_1)]^T \begin{bmatrix} 0 & 0 & 0 & 0 \\ 0 & 0 & 0 & 0 \\ 0 & 0 & k_s & 0 \\ 0 & 0 & 0 & 0 \end{bmatrix} [\Phi(\theta_1)] + [\Phi(\theta_2)]^T \begin{bmatrix} 0 & 0 & 0 & 0 \\ 0 & 0 & 0 & 0 \\ 0 & 0 & 0 & 0 \\ 0 & 0 & 0 & k_{\psi r} \end{bmatrix} [\Phi(\theta_2)] +$$

$$[\Phi'(\theta_3)]^T \begin{bmatrix} 0 & 0 & 0 & 0 \\ 0 & 0 & 0 & 0 \\ 0 & 0 & \dfrac{k_{\psi\theta}}{r^2} & 0 \\ 0 & 0 & 0 & 0 \end{bmatrix} [\Phi'(\theta_3)] \quad (3.51)$$

Each of the three components in Eq. 3.52 will form a matrix with a non-zero 4x4 submatrix only due to the localization of B-3 spline representation.

3.6.3 EXAMPLES

A clamped semi-circular plate of radius 'a' under a uniformly distributed load was divided into eight equal classical strips and four terms of the series were used in the analysis. The first strip had an inner radius equal to zero and also its two radial edges forming part of a straight line; nevertheless, no numerical difficulty was experienced because no singularity had occurred when Gaussian quadrature was used in the numerical integration of the strip stiffness. This is due to the fact that the end points have never been used as integration points in such a scheme.

The finite strip results give a maximum centre deflection of $0.002021qa^4/D$ at $r = 0.4859a$ and a maximum negative bending moment of $M = -0.0697qa^2$ at $r = a$, while the corresponding values from an analytical solution by Woinowsky-Krieger[49] are found to be $0.002022qa^4/D$ and $-0.0731qa^2$. Note that the use of equal width strips is merely for convenience, and better results are usually achievable for the same number of strips if strips of unequal widths are used so that regions of steep stress gradient are served by finer meshes.

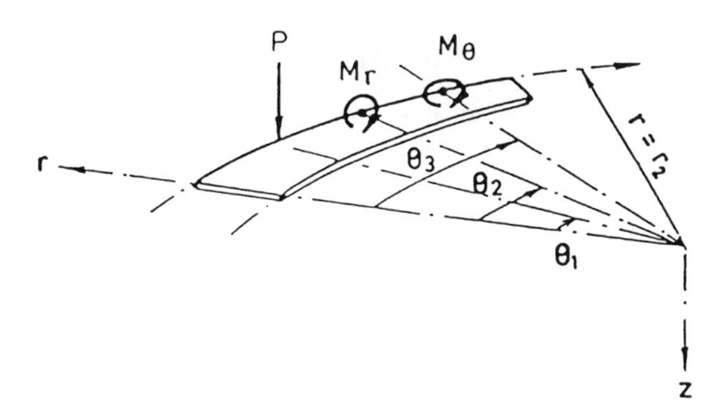

Figure 3.12 Concentrated loads at nodal line of circular strip.

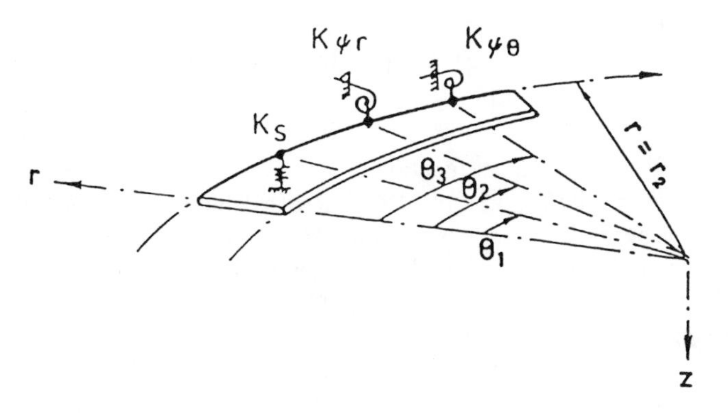

Figure 3.13 Elastic supports at nodal line of circular plate strip.

3.7 SKEW PLATES

The stiffness matrix for a skew plate strip can be developed in the same way as for a rectangular strip except that a skew coordinate (ξ,η) system should now be used. The relationship between the skew coordinate system and the Cartesian coordinate system is given by the following transformation (Figure 3.1f):

$$\begin{Bmatrix}\xi\\ \eta\end{Bmatrix}=\begin{bmatrix}1 & -\cot\beta\\ 0 & \operatorname{cosec}\beta\end{bmatrix}\begin{Bmatrix}x\\ y\end{Bmatrix} \tag{3.52}$$

from which another transformation matrix relating the curvatures in the two sets of coordinate systems can be derived:

$$\{\varepsilon\}=\begin{Bmatrix}-\dfrac{\partial^2 w}{\partial x^2}\\ -\dfrac{\partial^2 w}{\partial y^2}\\ 2\dfrac{\partial^2 w}{\partial x\partial y}\end{Bmatrix}=[T]\begin{Bmatrix}-\dfrac{\partial^2 w}{\partial \xi^2}\\ -\dfrac{\partial^2 w}{\partial \eta^2}\\ 2\dfrac{\partial^2 w}{\partial \xi\partial \eta}\end{Bmatrix}==[T]\{\bar{\varepsilon}\} \tag{3.53}$$

where [T] is the strain transformation matrix given by

$$[T]=\begin{bmatrix}1 & 0 & 0\\ \cot^2\beta & \operatorname{cosec}^2\beta & \cot\beta\operatorname{cosec}\beta\\ 2\cot\beta & 0 & \operatorname{cosec}\beta\end{bmatrix} \tag{3.54}$$

Therefore, the strain matrix is
a) classical strip

$$[B]_m=[T]\,[\,\overline{B}\,]_m \tag{3.55}$$

b) spline strip

$$[B]=[T]\,[\,\overline{B}\,] \tag{3.56}$$

where $[\,\overline{B}\,]_m$ and $[\,\overline{B}\,]$ are the strain matrices expressed in the skew coordinates.

Since the displacement function is now written in skew coordinates, it is not possible to use Eqs. 3.14 or 3.24 directly to obtain the stiffness matrix of a skew strip because all the strain components in $[B]_m$ and [B] are given in Cartesian coordinates. An intermediate step of coordinate transformation is therefore necessary. For example, the stiffness matrix of the classical strip can be shown to be

$$[K^e]=\sum_{m=1}^{r}\sin\beta\int_0^b\!\!\int_0^l [\,\overline{B}\,]_m^T\,[T]^T\,[D]\,[T]\,[\,\overline{B}\,]_m\,d\xi d\eta \tag{3.57}$$

and the load vector is:

$$\{F^e\}=\sum_{m=1}^{r}\sin\beta\int_0^b\!\!\int_0^l [C]_m\,Y_m\,q\,d\xi d\eta \tag{3.58}$$

The shape functions are to be given in terms of the skew coordinate system.

In the case of plates clamped along two opposite edges, the Y_m of such a strip can be chosen to be (Eq. 2.5)

$$Y_m = \sin\frac{\mu_m\eta}{l} - \sinh\frac{\mu_m\eta}{l} - \alpha_m\left(\cos\frac{\mu_m\eta}{l} - \cosh\frac{\mu_m\eta}{l}\right) \qquad (3.59)$$

and ($\mu_m = 4.730, 7.852, 10.9960, \ldots.. (2m+1)\pi/2$)

$$[C] = \left[(1-3\frac{\xi^2}{b^2}+2\frac{\xi^3}{b^3}) \quad (-\xi+2\frac{\xi^2}{b}+\frac{\xi^3}{b^2}) \quad (3\frac{\xi^2}{b^2}-2\frac{\xi^3}{b^3}) \quad (-\frac{\xi^2}{b}+\frac{\xi^3}{b^2})\right]$$

Clamped isotropic parallelogram plates under uniform load were studied by Mukhopadhyay[13] for aspect ratios of 1 and 2 and skew angles varying from 45° to 75° at an increment of 15° . The deflections of the centre point of the plates are presented in Table 3.19.

TABLE 3.19
Maximum deflection ($xqa^4/Dx10^{-3}$) of skew plates.

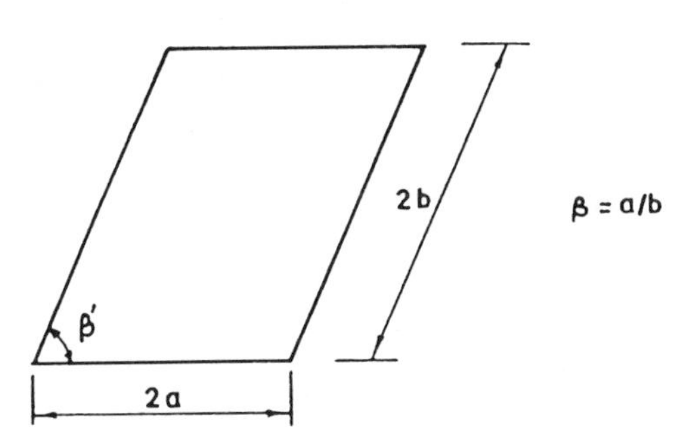

Case	Finite strip[13]	Morley[50]
β' = 75°		
β = 1	17.282	17.968
β = 2	2.221	2.221
β' = 60°		
β = 1	10.228	12.304
β = 2	1.367	1.450
β' = 45°		
β = 1	4.5049	6.032
β =2	0.568	0.653

Note that the chosen interpolation shape functions impose an unwanted restriction of quadrant symmetry on the solution. Thus the solution, though reasonably accurate, is restrictive in character. Furthermore, it is difficult to satisfy fully the boundary conditions of simply supported edges. If the ξ and η terms are simply replaced by x and y respectively, we are only able to have $w=0$, $\partial^2w/\partial\xi^2 = 0$ and $\partial^2w/\partial\eta^2 = 0$ but the normal curvature does not vanish. However, according to Brown and Ghali,[12] satisfactory results can be obtained by using the HO2 (although not for LO2, which has been used successfully for vibration problems) type of displacement function for uniformly loaded slabs with up to about 45° skew (see Table 2,3,4 of reference 13). For skew angles greater than 45°, the results start to deviate more strongly from the other results given by series solution, finite difference, and finite element.

Tham et al.[26] reported examples on the applications of spline finite strip in the analyses of skew plates. They carried out a series of analyses on parallelogram plates with skew angle varying from 30° to 90° (right plates). Two opposite edges of the plates were simply supported whereas the other two edges were free. The Poisson's ratio of the plates was 0.31. In the analysis, a 9x6 (node x strip) mesh was chosen. The deflection (Figure 3.14) as well as principal moments and their direction (Figure 3.15) compared fairly well with those of Ramstad.[51]

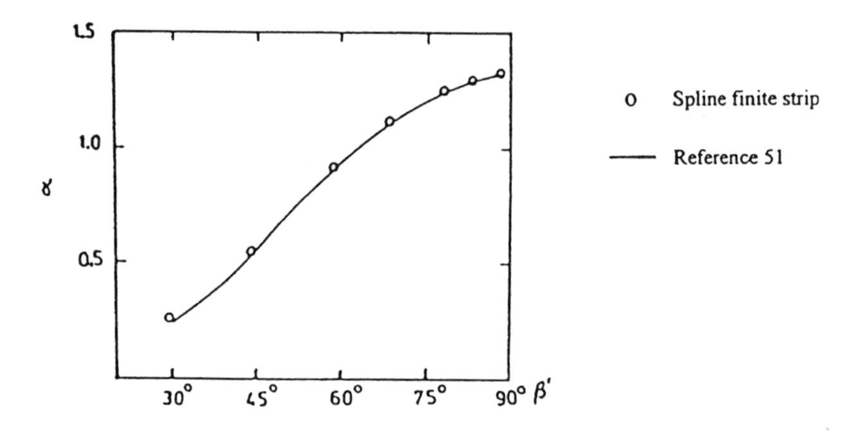

[Central deflection = $\frac{\alpha}{100D} qa^4 \sin\beta'$ (a = width; β' =skew angle)]

Figure 3.14 Central deflection ($\upsilon = 0.31$).[27]

3.8 ANALYSIS OF MODERATELY THICK PLATES

The formulation discussed so far has invariably been based on the thin plate theory which has ignored shear deformation and assumed that plane section will remain plane after bending (Kirchhoff normality assumption). However, it is well known that shear deformation becomes increasingly important as the thickness/span ratio increases. Relaxing the normality assumption, Mindlin[52,53] proposed a theory for the bending of moderately thick plates. The original normal line (Figure 3.1g) has now become a curved line which is closely approximated in an average sense by a straight non-normal line. It therefore requires not only the deflection but also the rotations to describe the deformations of the plate. The displacements can be written as

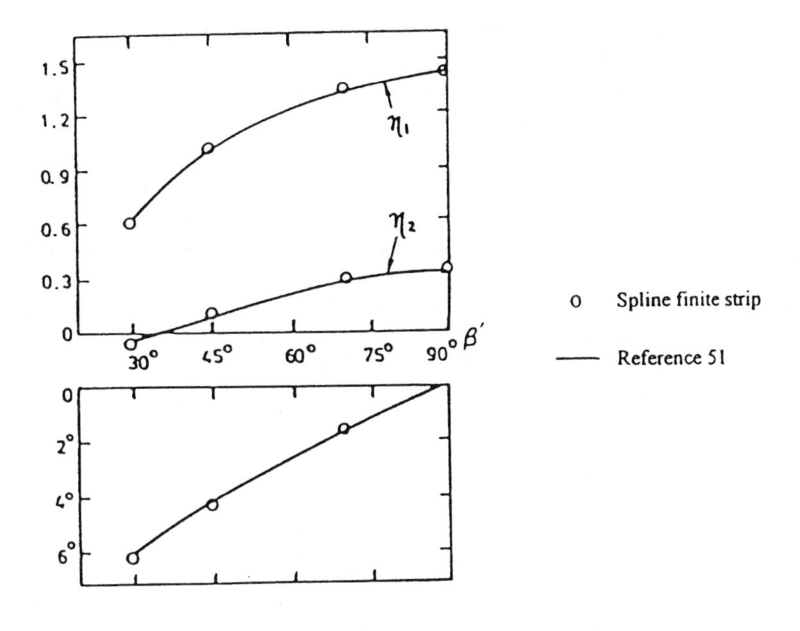

$[M_i = \eta_i q a^2 \sin\beta']$

Figure 3.15 Principal moments and direction at centre for parallelogram plate under uniformly distributed load.[27]

$$\begin{Bmatrix} u \\ v \\ w \end{Bmatrix} = \begin{Bmatrix} -z\psi_x(x,y) \\ -z\psi_y(x,y) \\ w(x,y) \end{Bmatrix} \tag{3.60}$$

where w, ψ_x and ψ_y are defined in Figure 3.1g.

Substituting Eq. 3.60 into the strain-displacement equations, the following relationships are obtained:

$$\begin{Bmatrix} \varepsilon_x \\ \varepsilon_y \\ \gamma_{xy} \\ \gamma_{xz} \\ \gamma_{yz} \end{Bmatrix} = \begin{Bmatrix} z\dfrac{\partial\psi_x}{\partial x} \\ z\dfrac{\partial\psi_y}{\partial y} \\ z(\dfrac{\partial\psi_y}{\partial x} + \dfrac{\partial\psi_x}{\partial y}) \\ \dfrac{\partial w}{\partial x} - \psi_x \\ \dfrac{\partial w}{\partial y} - \psi_y \end{Bmatrix} \tag{3.61}$$

The expression of the strain energy can be shown to be:

$$U = \frac{1}{2}\iint \{D_x[\frac{\partial \psi_x}{\partial x}]^2 + 2D_1 \frac{\partial \psi_x}{\partial x}\frac{\partial \psi_y}{\partial y} + D_y[\frac{\partial \psi_y}{\partial y}]^2 + D_{xy}[\frac{\partial \psi_x}{\partial y} + \frac{\partial \psi_y}{\partial x}]^2$$
$$+ S_x[\frac{\partial w}{\partial x} - \psi_x]^2 + S_y[\frac{\partial w}{\partial y} - \psi_y]^2\}dxdy \tag{3.62}$$

in which S_x and S_y are the shear moduli; D_x , D_y , D_1 and D_{xy} are the bending moduli.

The above relations suggest that the interpolation functions for the deflection and rotations only have to satisfy zero order continuity. Therefore, the displacement field can be chosen to be given in terms of the functions given in Section 2.2.2. For example, the deflection and rotations of a three-noded classical strip will be given as:

$$\psi_x = \sum_{m=1}^{r}\sum_{i=1}^{3} C_i\, \psi_{xi}^m\, Y_m = \sum_{m=1}^{r} [C]\, \psi_{xi}^m\, Y_m \tag{3.63a}$$

$$\psi_y = \sum_{m=1}^{r}\sum_{i=1}^{3} C_i\, \psi_{yi}^m\, \tilde{Y}_m = \sum_{m=1}^{r} [C]\, \psi_{yi}^m\, \tilde{Y}_m \tag{3.63b}$$

$$w = \sum_{m=1}^{r}\sum_{i=1}^{3} C_i\, w_i^m\, Y_m = \sum_{m=1}^{r} [C]\, w_i^m\, Y_m \tag{3.63c}$$

where $[C] = [C_1\ \ C_2\ \ C_3\ ]$ and C_i are given by Eq. 2.33.

Following standard procedures, the typical stiffness submatrix can be shown to be

$$\begin{aligned}[K]_{mn} &= \iint [B]_m^T\, [D]\, [B]_n\, dxdy \\ &= \iint [B_b\,]_m^T\, [D_b\,]\, [B_b\,]_n\, dxdy + \iint [B_s\,]_m^T\, [D_s\,]\, [B_s\,]_n\, dxdy \\ &= [K_b\,]_{mn} + [K_s\,]_{mn}\end{aligned} \tag{3.64}$$

($[K_b\,]_{mn}$ = bending stiffness matrix ; $[K_s\,]_{mn}$ = shear stiffness matrix) where

a) for a rectangular strip

$$[D] = \begin{bmatrix} D_x & D_1 & 0 & & \\ D_1 & D_y & 0 & & \\ 0 & 0 & D_{xy} & & \\ & & & S_x & 0 \\ & & & 0 & S_y \end{bmatrix} = \begin{bmatrix} D_b & \\ & D_s \end{bmatrix} \tag{3.65}$$

(D_x , D_1 , D_y and D_{xy} are given by Eqs. 3.9 and 3.10)

$[B]_m = [B_1^m\ \ B_2^m\ \ B_3^m\ ]$

$$B_i^m = \begin{bmatrix} -\frac{\partial C_i}{\partial x} Y_m & 0 & 0 \\ 0 & C_i \tilde{Y}_m' & 0 \\ -C_i Y_m' & -\frac{\partial C_i}{\partial x}\tilde{Y}_m & 0 \\ -C_i Y_m & 0 & -\frac{\partial C_i}{\partial x} Y_m \\ 0 & -C_i \tilde{Y}_m & C_i Y_m' \end{bmatrix} = \begin{bmatrix} B_{ib}^{\ m} \\ B_{is}^{\ m} \end{bmatrix} \tag{3.66}$$

b) for a circular strip

$$[D] = \begin{bmatrix} D_r & D_1 & 0 & & \\ D_1 & D_\theta & 0 & & \\ 0 & 0 & D_{r\theta} & & \\ & & & S_r & 0 \\ & & & 0 & S_\theta \end{bmatrix} = \begin{bmatrix} D_b & \\ & D_s \end{bmatrix} \quad (3.67)$$

(D_r , D_1 , D_θ and $D_{r\theta}$ are given by Eq. 3.37)

$[B]_m = [B_1^m \ B_2^m \ B_3^m]$

$$B_i^m = \begin{bmatrix} -\frac{\partial C_i}{\partial r}\Theta_m & 0 & 0 \\ -\frac{C_i}{r}\Theta_m & \frac{6C_i}{5r}\tilde{\Theta}_m' & 0 \\ -\frac{6C_i}{5r}\Theta_m' & (-\frac{\partial C_i}{\partial r}+\frac{C_i}{r})\tilde{Y}_m & 0 \\ -C_i\Theta_m & 0 & -\frac{\partial C_i}{\partial r}\Theta_m \\ 0 & -C_i\tilde{\Theta}_m & \frac{6C_i}{5r}\Theta_m' \end{bmatrix} = \begin{bmatrix} B_{ib}{}^m \\ B_{is}{}^m \end{bmatrix} \quad (3.68)$$

Again applying the Principle of Minimum Total Potential Energy, one can easily derive the stiffness matrix for such strips. The explicit form of the submatrix for a rectangular strip is tabulated in Table 3.20. Realizing that locking may stiffen the plates significantly, reduced/selective integration schemes[17] must be used for the integration of the terms of the stiffness matrix with C_i and its derivatives.

TABLE 3.20

Submatrix $[K]_{mn}$ for a moderately thick plate.

$$[K]_{mn} = \left[\begin{array}{c|c|c} D_x[C_3]I_1 + D_{xy}[C_1]I_5 + S_x[C_1]I_1 & D_1[C_2]^T I_2 - D_{xy}[C_2]I_3 & -S_x[C_2]I_1 \\ \hline D_1[C_2]I_6 - D_{xy}[C_2]^T I_7 & D_y[C_1]I_4 + D_{xy}[C_3]I_8 + S_y[C_1]I_8 & S_y[C_1]I_7 \\ \hline -S_x[C_2]^T I_1 & S_y[C_1]I_3 & S_x[C_3]I_1 + S_y[C_1]I_5 \end{array}\right]$$

in which $[C_1] = \int [C]^T [C] dx$; $[C_2] = \int [C]^T [\frac{\partial C}{\partial x}] dx$; $[C_3] = \int [\frac{\partial C}{\partial x}]^T [\frac{\partial C}{\partial x}] dx$

$I_1 = \int Y_m Y_n dy$; $I_2 = \int Y_m \tilde{Y}_n' dy$; $I_3 = \int Y_m' \tilde{Y}_n dy$; $I_4 = \int \tilde{Y}_m' \tilde{Y}_n' dy$

$I_5 = \int Y_m' Y_n' dy$; $I_6 = \int \tilde{Y}_m' Y_n dy$; $I_7 = \int \tilde{Y}_m Y_n' dy$; $I_8 = \int \tilde{Y}_m \tilde{Y}_n dy$

Using the Principle of Virtual Work, the load vector for a rectangular strip nodal line force of intensity P can be obtained. A typical node *i* acted upon by the m-th term of the load will yield the following nodal force

$$F_{im} = \int\int [\overline{N}_{im}]^T P dy \qquad (3.69)$$

where $$\overline{N}_{im} = \begin{bmatrix} 0 & & \\ & 0 & \\ & & C_i Y_m \end{bmatrix}$$

In the case of simply supported plates, we have

$$Y_m = \sin\left(\frac{m\pi y}{l}\right) \; ; \; \tilde{Y}_m = \cos\left(\frac{m\pi y}{l}\right) \qquad (3.70)$$

and it can be shown easily that the stiffness matrix for such plates will be decoupled and each sub-matrix can be solved individually as in the cases of thin plates.

Several examples of square isotropic plates of width 'a' were studied by various investigators. The plates were simply supported along four edges. The computed central deflections are given in Table 3.21.

Table 3.22 tabulates the bending moments for the mid-points of the plates with other support conditions. The plates were simply supported along two opposite edges and clamped or free on the other two edges. Note that the values are for the moments about the axis parallel to the simply supported edges.

The thick finite strip model was also employed to solve a curved plate problem (Figure 3.16) by Benson.[19] The results are compared to those obtained by Coull[55] in Table 3.23 and Figures 3.17 to 3.18.

TABLE 3.21
Central deflection of simply supported isotropic square plates (υ = 0.3)(E = Young's modulus; t = thickness; q = applied load).

	Thickness-span ratio		
	0.01	0.1	0.2
Mindlin Finite Strip[14]	0.0444	0.0467	
Modified Finite Strip	0.0444	0.0464	0.0521
Kong[42]		0.0467	0.0535
Thin Plate[2]	0.04437	0.04437	0.04437
Multiplier		qa^4/Et^3	

TABLE 3.22
Bending moments of square plates under various support conditions.[14]

Thickness-span ratio	SS-CC-SS-CC	SS-FF-SS-FF
2	0.042	0.1378
2.5	0.038	0.1365
5	0.029	0.1338
10	0.026	0.1321
Multiplier	qa^2	

TABLE 3.23
Comparison of deflection across the mid-span of a perspex model due to 1lb load at: (a) the centre. (b) the outer edge. (c) the inner edge.[19]

	Radius (in)	Finite Strip[19]	Coull[55]
Case (a)	7	0.147	0.155
	11	0.342	0.342
	13	0.462	0.457
Case (b)	7	0.193	0.194
	11	0.589	0.578
	13	0.909	0.876
Case (c)	7	0.147	0.169
	11	0.158	0.163
	13	0.193	0.195

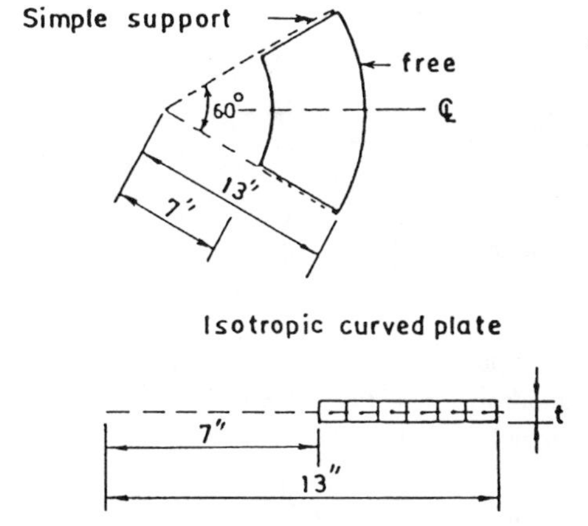

Figure 3.16 Cross-section of plate idealization to finite strips.[19]

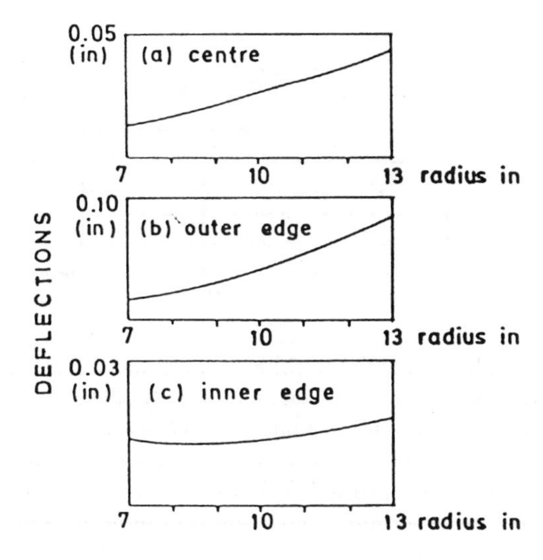

Figure 3.17 Radial distribution of mid-span deflections due to unit point loads.[19]

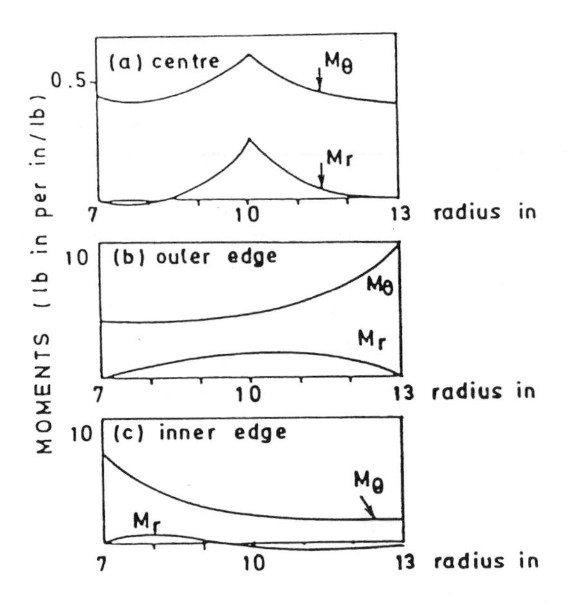

Figure 3.18 Distribution of tangential and radial moments at mid-span due to unit point load.[19]

A detailed study to compare the convergence and accuracy of the linear [(Section 2.2.2.1(c)), quadratic (Section 2.2.2.1(d)) and cubic (Section 2.2.2.1(f))] strips was carried out by Onate and Suarez.[16] By varying the number of integration points, the stiffness matrices were formed by using different numerical integration schemes (Table 3.24).

TABLE 3.24
Integration Schemes.[16]

$$\int_{-1}^{+1}\int_{-1}^{+1} f(\xi,\eta)d\xi d\eta = \sum_{i=1}^{N}\sum_{j=1}^{N} f(\xi_i,\eta_j)W_iW_j$$

Element	Exact		Selective		Reduced	
	K_b	K_s	K_b	K_s	K_b	K_s
Linear	2	2	2	1	1	1
Quadratic	3	3	3	2	2	2
Cubic	4	4	4	3	3	3

(ξ_I, η_j) : integration point coordinates ; W_i and W_j : weighting factor

Tables 3.25 and 3.26 show the results for simply supported square plates under uniformly distributed load. In the analysis, nine zero harmonic terms were used. It is noted that the locking for linear strips can be very serious if the analysis was carried out using exact integration. However, significant improvement can be attained by employing reduced/selective integration schemes. The results also show that the locking effect is less serious for the quadratic and the cubic strips.

TABLE 3.25
Convergence study of central deflection with number of strips and degrees of freedom (thickness/span = 0.1).[16]

(a) Full integration

Strip type	No of strips	Degrees of freedom	Central deflection ($x10^{-5}$)
Linear	1	3	128
	3	9	356
	5	15	399
Quadratic	1	6	416
	3	18	427
	5	30	427
Cubic	1	9	428
	3	27	427
	5	45	427
Analytical[2]			427
Multiplier			qL^4/D

(b) Reduced and selective integration

Strip type	Number of strips	Degrees of freedom	Central deflection ($x10^{-5}$)
Linear	1	3	400
	3	9	429
	5	15	428
Quadratic	1	6	435
	3	18	427
	5	30	427
Cubic	1	9	427
	3	27	427
	5	45	427
Analytical[2]			427
Multiplier			qL^4/D

(q = uniformly distributed load; D = rigidity; L = span)

TABLE 3.26
Convergence study of central deflection with number of strips and degrees of freedom (thickness/span = 0.001).[16]

a) Full integration

Strip type	No of strips	Degrees of freedom	Central deflection ($x10^{-5}$)
Linear	1	3	0.02
	3	9	0.16
	5	15	0.45
Quadratic	1	6	379
	3	18	405
	5	30	406
Cubic	1	9	409
	3	27	406
	5	45	406
Analytical[2]			406
Multiplier			qL^4/D

(b) Reduced and selective integration

Strip type	Number of strips	Degrees of freedom	Central deflection ($x10^{-5}$)
Linear	1	3	373
	3	9	407
	5	15	407
Quadratic	1	6	414
	3	18	406
	5	30	406
Cubic	1	9	406
	3	27	406
	5	45	406
Analytical[2]			406
Multiplier			qL^4/D

(q = uniformly distributed load; D = rigidity; L =span)

Chulya and Mullen[18] employed the assumed strain distribution technique to improve the convergence of the linear strips. For each strip (Figure 3.19), the average transverse shear strain can be defined in terms of the strains at four selected points A, B, C and D as:

$$\bar{\gamma}_{xz} = 0.5\,(\gamma_{xz}^{A} + \gamma_{xz}^{C})$$
$$\bar{\gamma}_{yz} = 0.5\,(\gamma_{yz}^{B} + \gamma_{yz}^{D})$$

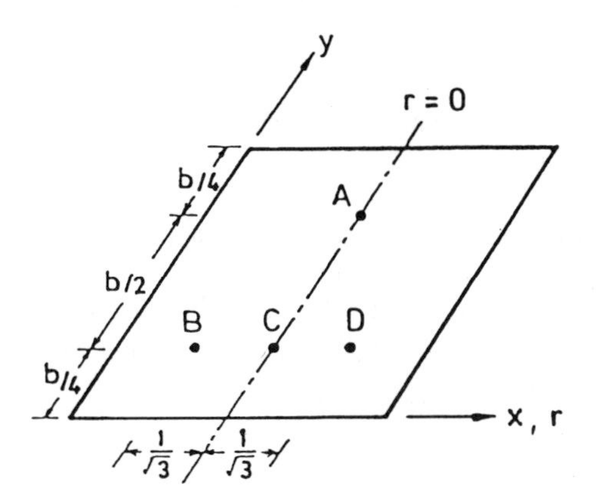

Figure 3.19 Selected points for assumed shear strains in each strip.[18]

One can easily show that such average strains can be written in terms of the nodal parameters. The strain energy expression is then modified as:

$$U = \frac{1}{2}\iint \{D_x[\frac{\partial\psi_x}{\partial x}]^2 + 2D_1\frac{\partial\psi_x}{\partial x}\frac{\partial\psi_y}{\partial y} + D_y[\frac{\partial\psi_y}{\partial y}]^2 + D_{xy}[\frac{\partial\psi_x}{\partial y} + \frac{\partial\psi_y}{\partial x}]^2 + S_x\,\bar{\gamma}_{xz}^{\,2} + S_y\,\bar{\gamma}_{yz}^{\,2}\}dxdy \qquad (3.71)$$

Again following the standard procedure, the stiffness matrix can be derived easily, and the solutions can then be obtained accordingly.

Results of parametric study of the bending of square plates with the span/thickness ratio varying from 5 to 10 indicate that such an approach greatly reduces the locking effects and improves the rate of convergence.[18]

To demonstrate the application of spline strip in moderately thick plate analysis, two thick isotropic plates were analyzed by Kong.[42] The Reddy plate theory[56] was employed in the study. The plates were simply supported along all edges and subjected to a uniformly distributed load. The deflections and moments at the centre as well as twisting moments at the corner are tabulated in Table 3.27. They are in fairly good agreement with those obtained by Kant[57].

TABLE 3.27

Results of simply supported thick plates ($\upsilon = 0.3$).[42]

	Deflection (xqa^4/D)	Moment (xqa^2)		Twisting Moment (xqa^2)
		M_x	M_y	M_{xy}
a) span/thickness=5				
Kong[42]	0.00490	0.0479	0.0480	0.0318
Kant[57]	0.00480	0.0484	0.0485	0.0299
b) span/thickness=10				
Kong[42]	0.00427	0.0479	0.0480	0.0318
Kant[57]	0.00424	0.0480	0.0480	0.0317

q = magnitude of uniformly distributed load ; D = bending rigidity ; a =span

3.9 LEAST SQUARE FORMULATION OF PLATE BENDING PROBLEMS

Kwok et al.[10] demonstrated that the Least Square approach can also be used to solve the plate bending problems by finite strips.

Assuming the approximate solution for the plate deflection to be $\overline{w}(x,y)$, one can show readily that the residual for the plate biharmonic equation is:

$$E = \frac{\partial^4 \overline{w}}{\partial x^4} + 2\frac{\partial^2 \overline{w}}{\partial x \partial y} + \frac{\partial^4 \overline{w}}{\partial y^4} - \frac{q(x,y)}{D} \tag{3.72}$$

where q(x,y) is the applied load.

If the plates are rectangular in shape and simply supported along two opposite edges, one can write

$$\overline{w}(x,y) = \sum_{m=1}^{r} [C]\{\delta\}_m \sin(\frac{m\pi y}{l}) \tag{3.73}$$

$$q(x,y) = \sum_{m=1}^{r} q_m(x) \sin(\frac{m\pi y}{l}) \tag{3.74}$$

Substituting Eqs. 3.73 and 3.74 into Eq. 3.72, the residual corresponding to the m-th term is obtained:

$$E_m = [\frac{\partial^4 [C]}{\partial x^4}\{\delta\}_m + 2\frac{\partial^2 [C]}{\partial x \partial y}\{\delta\}_m + \frac{\partial^4 [C]}{\partial y^4}\{\delta\}_m - \frac{q_m(x)}{D}] \sin(\frac{m\pi y}{l}) \tag{3.75}$$

or

$$E_m = L_m \delta_m - \frac{q_m(x)}{D}$$ (Note that $\sin(\frac{m\pi y}{l})$ has been deleted).

The Least Square approach can be used to minimize the residual with respect to the deflections $\{\delta\}$. The relation thus derived can be shown to be

$$\frac{\partial \int E_m^2 dx}{\partial \{\delta\}_m} = \frac{\partial \int (L_m \delta_m - \frac{q_m}{D})^2 dx}{\partial \{\delta\}_m} = 0 \tag{3.76}$$

This results in a matrix equation of the form

$$(\int L_m^T L_m \, dx) \{\delta\}_m = \int (L_m^T q_m(x)/D) \, dx \tag{3.77}$$

or in the form of standard stiffness relationship,

$$[K]_m \{\delta\}_m = \{F\}_m \tag{3.78}$$

where $[K]_m$ and $\{F\}_m$ are the stiffness and load matrix respectively.

In the collocation Least Square approach, Eq. 3.76 is only forced to vanish at a number of chosen (collocation) points and Eq. 3.77 is modified to:

$$(\Sigma L_m^T L_m) \{\delta\}_m = \Sigma(L_m^T \frac{q_m(x)}{D}) \tag{3.79}$$

Since L_m involves the fourth order differential of deflection, the interpolation functions must be continuous to the third order. Therefore, the interpolation parameters include:

$$\{w_{m1} \; w_{m1}' \; w_{m1}'' \; w_{m1}''' \; w_{m2} \; w_{m2}' \; w_{m2}'' \; w_{m2}''' \}$$

and the corresponding shape functions can be derived by the methods outlined in Chapter 2.

The results[10] (deflections and moments) of a simply supported plate obtained by this approach are tabulated in Table 3.28. The plate, which was subjected to a uniformly distributed load, was modelled by only two strips. It can be seen that the agreement between the analytical method and the present method is quite remarkable. Also tabulated in the table are the results of the conventional finite strip method, in which the plate was divided into six strips so that the number of unknowns is the same for both cases. While the present method yields somewhat more accurate results, it must be kept in mind that the bandwidth of the matrix used here is also bigger.

TABLE 3.28
Maximum deflection and bending moments in uniformly loaded plates with simply supported edges ($\upsilon = 0.3$; aspect ratio = 1.0).[10]

m	Deflection	M_x	M_y
1	0.00410936	0.0492024	0.0516683
3	-0.00005057	-0.0015517	-0.0045534
5	0.00000418	0.0003155	0.0010349
7	-0.00000079	-0.0001158	-0.0003830
9	0.00000023	0.0000540	0.0001828
3 terms	0.004063	0.04797	0.04815
5 terms	0.004062	0.047904	0.047949
Analytical[2]	0.00406	0.0479	0.0479
Classical finite strip	0.00406	0.0484	0.0484
Multiplier	qa^4/D	qa^2	

REFERENCES

1. Cheung, Y. K., Finite strip method in the analysis of elastic plates with two opposite ends simply supported, *Proc Inst Civ Eng*, 40, 1-7, 1968.
2. Timoshenko, S. P. and Woinowsky-Krieger, S., *Theory of plates and shells*, Second Edition, McGraw-Hill, 1959.
3. Powell, G. H. and Ogden, D. W., Analysis of orthotropic bridge decks, *J of Struct Div, ASCE*, 95(5), 909-923, 1969.
4. Cheung, Y. K., Finite strip method analysis of elastic slabs, *J of Eng Mech, ASCE*, 94(6), 1365-1378, 1968.
5. Ergatourdis, J. G., Irons, B. M., and Zienkiewicz, O. C., Curved, isoparametric, quadrilateral elements for finite element analysis, *Int J Solids Struct*, 4, 31-42, 1968.
6. Cheung, Y. K., Orthotropic right bridges by the finite strip method, Concrete Bridge Design, *ACI Publications SP-26*, 182-205, 1971.
7. Loo, Y. C. and Cusens, A. R., A refined finite strip method for the analysis of orthotropic plates, *Proc Inst Civ Eng*, 40, 85-91, 1971.
8. Loo, Y. C. and Cusens, A. R., Developments of finite strip method in the analysis of bridge decks, *Developments in Bridge Design and Construction* (ed. Rockey et al.), Crosby Lockwood, 1971.
9. Cheung, Y. K. and Cheung, M. S., Flexural vibrations of rectangular and other polygonal plates, *J of Eng Mech, ASCE*, 97(2), 391-411, 1971.
10. Kwok, W. L., Cheung, Y. K. and Delcourt, C., Application of least square collocation technique in finite element and finite strip formulation, *Int J for Num Meth in Eng*, 11, 1391-1404, 1977.
11. Brown, T. G. and Ghali, A., Finite strip analysis of skew plates, *Proc Mcgill-EIC Specialty Conf on the Finite Element Method in Civil Engineering*, Montreal, June, 1972.
12. Brown, T. G. and Ghali, A., Semi-analytical solution of skew plates in bending, *Proc Inst Civ Eng*, 57, Part 2, 165-175, 1972.
13. Mukhopadhyay, M., Finite strip method of analysis of clamped skew plates in bending, *Proc Inst Civ Eng*, 61, Part 2, 189-195, 1976.
14. Mawenya, A. S. and Davies, J. D., Finite strip analysis of plate bending including transverse shear effects, *Build Sci*, 9, 175-180, 1974.
15. Hinton, E. and Zienkiewicz, O. C., A note on a simple thick finite strip, *Int J for Num Meth in Eng*, 11, 905-909, 1977.
16. Onate, E. and Suarez, B., A comparison of the linear, quadratic and cubic Mindlin strip elements for the analysis of thick and thin plates, *Comp and Struct*, 17, 427-439, 1983.
17. Zienkiewicz, O. C., Taylor, R. L. and Too, J. M., Reduced integration technique in general analysis of plates and shells, *Int J for Num Meth in Eng*, 3, 575-586, 1971.
18. Chulya, A. and Mullen, R. L., Assumed strain distributions for a finite strip plate bending element using Mindlin-Reissner plate theory, *Comp and Struct*, 33(2), 513-521, 1989.

19. Benson, P. R. and Hinton, E., A thick finite strip solution for static, free vibration and stability problems, *Int J for Num Meth in Eng*, 10, 665-678, 1976.
20. Dawe, D. J., Finite strip models for vibration of Mindlin plates, *J of Sound and Vib*, 59, 441-452, 1978.
21. Cheung, Y. K., Fan, S. C. and Wu, C. Q., Spline finite strip in structural analysis, *Proc of the Int Conf on finite element method*, Shanghai, August, 1982, 704-709.
22. Wu, C. Q., Cheung, Y. K. and Fan, S. C., *Spline finite strip in structural analysis*, Guangdong Scientific and Technology Press, 1986 (in Chinese).
23. Fan, S. C. and Cheung, Y. K., Flexural free vibration of rectangular plates with complex support conditions, *J of Sound and Vib*, 93(1), 81-94, 1984.
24. Loo, Y. C., Tam, W. S. and Byun, Y. J., Higher order spline finite strip method, *Int J Structures*, 5(1), 45-69, 1985.
25. Chen, M. J., Tham, L. G. and Cheung, Y. K., Spline finite strip for parallelogram plate, *Proc Int Conf on Accuracy Estimates and Adaptive Refinements in Finite Element Computation*, Lisbon, Portugal, 1, 95-104, 1984.
26. Tham, L. G., Li, W. Y., Cheung, Y. K. and Chen, M. J., Bending of skew plates by spline-finite-strip method, *Comp and Struct*, 22(1), 31-38, 1986.
27. Li, W. Y., *Spline finite strip analysis of arbitrarily shaped plates and shells*, PhD thesis, Department of Civil & Structural Engineering, University of Hong Kong, 1988.
28. Li, W. Y., Cheung, Y. K. and Tham, L. G., Spline finite strip analysis of general plates, *J of Eng Mech, ASCE*, 112(1), 43-54, 1986.
29. Cheung, Y. K., Tham, L. G. and Li, W. Y., Application of spline-finite-strip method in the analysis of curved slab bridge, *Proc Inst Civ Eng*, 81, Part 2, 111-124, 1986.
30. Cheung, Y. K., Tham, L. G. and Li, W. Y., Free vibration and static analysis of general plates by spline finite strip, *Comp Mech*, 3, 187-197, 1988.
31. Cheung, Y. K. and Zhu, Dashan, Large deflection analysis of arbitrary shaped thin plates, *Comp and Struct*, 26(5), 811-814, 1987.
32. Kong, J. and Cheung, Y. K., Application of spline finite strip to the analysis of shear-deformable plates, *Comp and Struct*, 46, 985-988, 1993.
33. Fan, S. C. and Luah, M. H., New spline finite element for analysis of shells of revolution, *J of Eng Mech, ASCE*, 116(3), 709-726, 1990.
34. Yang, H. Y. and Chong, K. P., Modified finite strip method, *Proc Int Conf on finite element methods*, Shanghai, 824-829, 1982.
35. Yang, H. Y. and Chong, K. P., Finite strip method with X-spline functions, *Comp and Struct*, 18(1), 127-132, 1984.
36. Chen, J. L. and Chong, K. P., Vibration of irregular plates by finite strip method with spline functions, *Engineering Mechanics in Civil Engineering, Proc 5th Eng Mech Div of ASCE*, 1, 256-260, 1984.

37. Chen, J. L. and Chong, K. P., Buckling of irregular plates by finite strip method with spline functions, *Collection of Technical Papers - AIAA/ASME/ASCE/AHS 26th Structures, Structural Dynamics and Materials Conf*, Part 1, 147-151, 1985.
38. Chong, K. P. and Chen, J. L., Buckling of irregular plates by splined finite strips, *J of AIAA*, 24(3), 534-536, 1986.
39. Cheung, Y. K. and Au, F. T. K., Finite strip analysis of right box girder bridges using computed shape functions, *Thin-Walled Struct*, 13(4), 275-298, 1992.
40. Cheung, Y. K., Au, F. T. K. and Kong, J., Structural analysis by the finite strip method using computed shape functions, *Proc Int Conf on Computational Methods in Engineering*, Singapore, 2, 1140-1145, 1992.
41. Cheung, Y. K. and Kong, J., Analysis of flat slabs by a new finite strip method, *Proc of Aust Struct Eng Conf,* Sept 21-23, 1994, Sydney, Australia, 259-264.
42. Kong, J., *Analysis of plate type structures by finite strip, finite prism and finite layer method*, PhD thesis, Department of Civil and Structural Engineering, University of Hong Kong, 1994.
43. Cheung, Y. K., *Finite strip method in structural analysis*, Pergamon Press, 1976.
44. Loo, Y. C., *Developments and applications of the finite strip method in the analysis of right bridge decks*, PhD thesis, Department of Civil Engineering, University of Dundee, 1972.
45. Cheung, M. S., *Finite strip analysis of structures*, PhD thesis, Department of Civil Engineering, University of Calgary, 1971.
46. Cheung, M. S. and Cheung, Y. K., Static and dynamic behaviour of rectangular strips using higher order finite strips, *Building Sci*, 7(3), 151-158, 1972.
47. Fan, S. C., *Spline finite strip in structural analysis*, PhD thesis, Department of Civil Engineering, University of Hong Kong, 1982.
48. Chen, S. J. and Peng, Y. C., On spline finite element computation of plates on elastic foundations for various base-models, *J of Zhongshan University*, 2, 1980 (in Chinese).
49. Woinowsky-Krieger, S., Clamped semicircular plate under uniform bending load, *J of Appl Mech*, 22, 1955.
50. Morley, L. S. D., Bending of clamped rectilinear plates, *Q J Mech Appl Maths*, 17, 293-317, 1964.
51. Ramstad, H., Parallelogram elements in bending accuracy and convergence of results, *Tech Report, Division of Structural Mech*, The Tech University of Norway, Trondheim, 1967.
52. Mindlin, R. D., Influence of rotary inertia and on shear flexural motions of isotropic elastic plates, *J of Appl Mech*, 18, 31-38, 1951.

53. Mindlin, R. D., Schacknow, A. and Deresiewicz, H., Flexural vibrations of rectangular plates, *J of Appl Mech*, 23, 430-436, 1956.
54. Reissner, E., The effect of transverse shear deformation on the bending of elastic plates, *J of Appl Mech*, 12, 69, 1945.
55. Coull, A. and Das, P. C., Analysis of curved bridge decks, *Proc Inst Civ Eng*, 37, 75-85, 1967.
56. Reddy, J. N., A simple higher order theory for laminated composite plates, *J of Appl Mech*, 51,745-752, 1984.
57. Kant, T., Numerical analysis of thick plates, *Comput Meth Appl Mech Eng*, 31, 1-18, 1982.

CHAPTER FOUR

PLANE STRESS ANALYSIS

4.1 INTRODUCTION

In this chapter solutions for plane stress problems in two-dimensional elasticity will be presented, with plane strain problems similarly tackled by a suitable conversion of elastic constants. The scope of direct application for the plane stress strips in structural analysis is somewhat limited but may include deep beams, layered or sandwich beams. Nevertheless, they are of considerable importance when used in conjunction with bending strips in which combined flat shell strips can be formed for the analysis of straight or curved stiffened plates, folded plate structures, box girder bridges and tall building shear cores. Furthermore, the plane strain strips are applied widely in solutions of geotechnical problems (Chapter 10) in which the plane strain assumption may be considered to be acceptable.

Various types of classical strips with two simply supported or two clamped ends have been formulated and each of them will be discussed in detail in subsequent paragraphs. The simply supported strip is by far the most important and useful one, and, similar to the simply supported bending strip, the series used in the displacement functions also decouples, and therefore each term of the series can be analyzed separately.

In Section 2.2.1, it has been mentioned that for the series part of the displacement field of the classical strips both Y_m and Y_m' are present in the analysis of two-dimensional elasticity problems, where, in general, Y_m is used for u and Y_m' for v displacements (Figure 4.1). The above formulation is based on the theory of beams, in which the transverse deflection u is related to the longitudinal displacement v through

$$v = A\frac{du}{dy} \tag{4.1}$$

and therefore the general form for the displacements is given as

$$u = \sum_{m=1}^{r} f_m^{u}(x)\, Y_m \tag{4.2a}$$

$$v = \sum_{m=1}^{r} f_m^{v}(x)\, \frac{l}{\mu_m} Y_m' \tag{4.2b}$$

The above formulation can also be interpreted from the displacement functions for the simply supported case in which a sine series is used for u displacement and a cosine series for the v displacement.

For vibration problems, Y_m can take up any one of the seven functions (with the appropriate μ_m values) listed in Section 2.2.1.1, and accurate predictions of frequencies and mode shapes have been obtained for all cases. In static analysis, however, Eqs. 4.2a and 4.2b have been found to be valid for the simply supported case only.

For example, the function for the clamped-clamped case requires that $Y_m(0) = Y_m'(0) = Y_m(l) = Y_m'(l) = 0$, and this means that the shear stress which is given by

$$\tau_{xy} = G\left(\frac{\partial u}{\partial y} + \frac{\partial v}{\partial x}\right) = G\left(\sum_{m=1}^{r} f_m^u(x)\, Y_m' + \sum_{m=1}^{r} \frac{df_m^v(x)}{dx} \frac{l}{\mu_m} Y_m'\right) \tag{4.3}$$

will be equal to zero at the two ends. This is obviously an impossible situation in a beam problem because it means that no reaction will be provided to resist the vertical load at the supports.

From the above discussion it can be seen that for static analysis the assumed function should satisfy the natural boundary conditions[1] as well as the displacement boundary conditions. An alternative set of functions for the clamped-clamped case was developed and applied successfully to several problems.

The three sets of displacement functions which can be used in static analysis are listed below.

(i) Both ends simply supported [$u = 0$, $\sigma_y = E\,((\partial v/\partial y + \upsilon_x \partial u/\partial x)) = 0$ at $y = 0$ and $y = l$]

$$u = \sum_{m=1}^{r} f_m^u(x)\, Y_m = \sum_{m=1}^{r} f_m^u(x) \sin k_m y \tag{4.4a}$$

$$v = \sum_{m=1}^{r} f_m^u(x) \frac{l}{\mu_m} Y_m' = \sum_{m=1}^{r} f_m^v(x) \cos k_m y \tag{4.4b}$$

in which $k_m = \dfrac{m\pi}{l}$.

(ii) Both ends clamped [$u = v = 0$ at $y = 0$ and $y = l$]

$$u = \sum_{m=1}^{r} f_m^u(x) \sin k_m y \tag{4.5a}$$

$$v = \sum_{m=1}^{r} f_m^v(x) \sin k_{m+1} y \tag{4.5b}$$

in which k_{m+1} refers to $\dfrac{(m+1)\pi}{l}$.

(iii) One end clamped and the other end simply supported

$$u = \sum_{m=1}^{r} f_m^u(x)\, U_m \tag{4.6a}$$

$$v = \sum_{m=1}^{r} f_m^v(x)\, U_m' \tag{4.6b}$$

$$\text{where } U_m = \begin{cases} \dfrac{y}{l} \\ \sin(\dfrac{\mu_m y}{l}) - \sinh(\dfrac{\mu_m y}{l}) - \alpha_m[\cos(\dfrac{\mu_m y}{l}) - \cosh(\dfrac{\mu_m y}{l})] \end{cases}$$

$$\alpha_m = \frac{\sin\mu_m + \sinh\mu_m}{\cos\mu_m + \cosh\mu_m} \quad ;\mu_m = 1.875, 4.694, \ldots \frac{(2m-1)\pi}{2}$$

In the case of spline strips, special treatments on the spline functions of the end knots are also required to satisfy the boundary conditions.[2] For example, the modified spline functions for a simply supported strip are:

a) spline function associated with displacement u

$$\tilde{\phi}_{-1} = \text{eliminated} \; ; \; \tilde{\phi}_o = \phi_o - 4\phi_{-1} \; ; \; \tilde{\phi}_1 = \phi_1 - \phi_{-1}$$

b) spline function associated with displacement v

$$\tilde{\phi}_{-1} = \text{eliminated} \; ; \; \tilde{\phi}_o = \phi_o \quad ; \; \tilde{\phi}_1 = \phi_1 + \phi_{-1}$$

The modifications for other boundary conditions are tabulated in Table 4.1.

TABLE 4.1
Boundary conditions of strip defined by a combination of the boundary conditions of its B_3-spline representations.

Spline Representation	Simple Support	Sliding Support	Fixed Support	Free
ϕ_u	Simple Support $\phi_u = 0$ $\frac{\partial\phi_u}{\partial y} \neq 0$	Sliding clamped $\phi_u \neq 0$ $\frac{\partial\phi_u}{\partial y} = 0$	Simple Support $\phi_u = 0$ $\frac{\partial\phi_u}{\partial y} \neq 0$	Free $\phi_u \neq 0$ $\frac{\partial\phi_u}{\partial y} \neq 0$
ϕ_v	Sliding Clamped $\phi_v \neq 0$ $\frac{\partial\phi_v}{\partial y} = 0$	Simple Support $\phi_v = 0$ $\frac{\partial\phi_v}{\partial y} \neq 0$	Simple Support $\phi_v = 0$ $\frac{\partial\phi_v}{\partial y} \neq 0$	Free $\phi_v \neq 0$ $\frac{\partial\phi_v}{\partial y} \neq 0$

4.2 CLASSICAL PLANE STRESS STRIPS

The first rectangular plane stress strip was developed by Cheung[3] using a linear polynomial of x [Section 2.2.2] for both u and v displacements. This lower order strip with constant strain ε_x in the transverse direction tends to approximate the true σ_x and ε_x curves in a stepwise fashion if the results are output at the nodal lines and plotted accordingly. In order to avoid such abrupt stress and strain jumps at the interface of the strips, all σ_x and ε_x values should only be plotted at the centre line (x=b/2) of the strip.

A higher order strip having one additional internal nodal line was later presented by Loo and Cusens.[4] In this strip a parabolic variation of displacements across the section is assumed and it is found, as expected, that stress jumps at the interface of adjoining strips are much smaller than those found in the lower order strip analysis. In order to facilitate computation and to reduce the bandwidth of the overall matrix, the parameters associated with the internal nodal line are usually eliminated through static condensation before the assembly stage.

In a paper by Yoshida and Oka[5] several plane stress strips using $(u,\partial u/\partial x,v)$, $(u,v,\ \partial v/\partial x)$ and $(u,\ \partial u/\partial x,v,\ \partial v/\partial x)$ as nodal parameters respectively were suggested. After making some numerical comparisons it was concluded that the displacement functions with $(u,\ \partial u/\partial x,v,\ \partial v/\partial x)$ as nodal parameters should be adopted because of its superior accuracy. It should be pointed out, however, that such a strip cannot be used directly for problems in which an abrupt change in plate thickness or property occurs, simply because at such location the strains are always discontinuous.

The higher order strips have been formulated for the simply supported case only. Although no difficulty should be encountered in the extension to cover other boundary conditions, the materials presented in the subsequent sections on higher order strips will nevertheless be limited to that of a simply supported strip only.

4.2.1 LOWER ORDER STRIP WITH TWO NODAL LINES (LO2)

(a) Displacement functions

A typical strip is shown in Figure 4.1a in which only u and v displacements are used as nodal parameters. A suitable shape function is therefore found in [Section 2.2.2], and it is possible to write

$$u = \sum_{m=1}^{r} [(1-\bar{x})\ \ \bar{x}] \begin{Bmatrix} u_1 \\ u_2 \end{Bmatrix} Y_m \tag{4.7a}$$

$$v = \sum_{m=1}^{r} [(1-\bar{x})\ \ \bar{x}] \begin{Bmatrix} v_1 \\ v_2 \end{Bmatrix} \frac{l}{\mu_m} Y_m' \tag{4.7b}$$

in which $\bar{x} = x/b$.

The above equation can be written in matrix form as

$$\begin{Bmatrix} u \\ v \end{Bmatrix} = \sum_{m=1}^{r} \begin{bmatrix} (1-\bar{x})Y_m & 0 & \bar{x}Y_m & 0 \\ 0 & (1-\bar{x})\dfrac{l}{\mu_m}Y_m' & 0 & \bar{x}\dfrac{l}{\mu_m}Y_m' \end{bmatrix} \begin{Bmatrix} u_1 \\ v_1 \\ u_2 \\ v_2 \end{Bmatrix}$$

$$= \sum_{m=1}^{r} [N]_m \{\delta\}_m \tag{4.8}$$

Note that the constant l/μ_m before Y_m' is included simply for convenience because once the differentiation for Y_m is actually carried out, this quantity would be cancelled out.

(b) Strains

The strains for a plane stress problem are the two direct strains and a shear strain, and are given as

$$\{\varepsilon\} = \begin{Bmatrix} \varepsilon_x \\ \varepsilon_y \\ \gamma_{xy} \end{Bmatrix} = \begin{Bmatrix} \dfrac{\partial u}{\partial x} \\ \dfrac{\partial v}{\partial y} \\ \dfrac{\partial u}{\partial y} + \dfrac{\partial v}{\partial x} \end{Bmatrix} = \sum_{m=1}^{r} [B]_m \{\delta\}_m \tag{4.9}$$

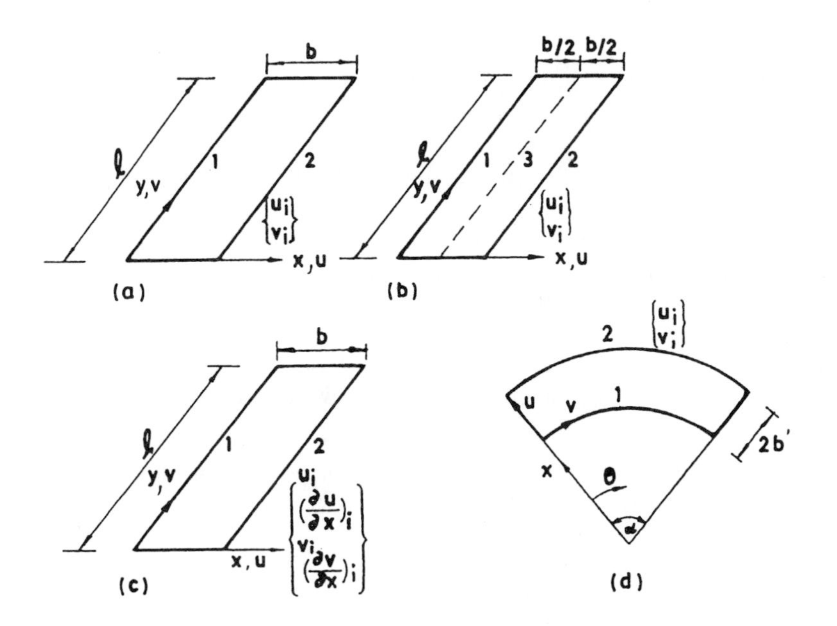

Figure 4.1 (a) LO2 strip. (b) HO3 strip. (c) HO2 strip. (d) Curved LO2 strip.

The strain matrix $[B]_m$ is obtained by performing the appropriate differentiation on the displacements in Eq. 4.6, and its explicit form is as follows:

$$[B]_m = \begin{bmatrix} \dfrac{-1}{b} Y_m & 0 & \dfrac{1}{b} Y_m & 0 \\ 0 & (1-\overline{x})\dfrac{l}{\mu_m} Y_m'' & 0 & \overline{x}\dfrac{l}{\mu_m} Y_m'' \\ (1-\overline{x}) Y_m' & \dfrac{-1}{b}\dfrac{l}{\mu_m} Y_m' & \overline{x} Y_m' & \dfrac{-1}{b}\dfrac{l}{\mu_m} Y_m' \end{bmatrix} \tag{4.10}$$

(c) Stresses

The stresses corresponding to ε_x , ε_y and γ_{xy} for orthotropic materials are σ_x , σ_y and τ_{xy} and the two sets are connected to each other through the following relationship:

$$\{\sigma\} = [D]\,\{\delta\} = [D] \sum_{m=1}^{r} [B]_m \{\delta\}_m \tag{4.11}$$

in which the property matrix for orthotropic materials is

$$[D] = \begin{bmatrix} \dfrac{E_x}{(1-\upsilon_x\upsilon_y)} & \dfrac{\upsilon_x E_y}{(1-\upsilon_x\upsilon_y)} & 0 \\ \dfrac{\nu_x E_y}{(1-\upsilon_x\upsilon_y)} & \dfrac{E_y}{(1-\upsilon_x\upsilon_y)} & 0 \\ 0 & 0 & G_{xy} \end{bmatrix}$$

For isotropic plate, the property matrix can be obtained by setting $E_x = E_y = E$, $\upsilon_x = \upsilon_y = \upsilon$, and $G_{xy} = E/2(1+\upsilon)$.

(d) Strain energy

Using Eqs. 4.9 and 4.10, one can show readily that the strain energy is

$$U = \sum_{m=1}^{r} \sum_{n=1}^{r} \int \{\delta\}_m^T [B]_m^T [D] [B]_n \{\delta\}_n \, d(\text{vol}) \tag{4.12}$$

(e) Stiffness matrix and consistent load vector

The stiffness matrix can be obtained by differentiating Eq. 4.12 with respect to $\{\delta\}_m$. Since the thickness 't' of a strip is assumed to be constant, the stiffness matrices can be simplified into the following form:

$$[K]_{mn} = t \int [B]_m^T [D] [B]_n \, d(\text{area}) \tag{4.13}$$

Eq. 4.13 has been used to work out the explicit form of the stiffness matrix for the lower order strip (Table 4.2). Note that the matrix coefficient

TABLE 4.2

In-plane stiffness matrix $[K]_{mn}$ of a rectangular strip.

$$t \left[\begin{array}{c:c:c:c} K_1(\frac{1}{b})I_1 & K_2(\frac{-1}{2C_2})I_3 & K_1(\frac{-1}{b})I_1 & K_2(\frac{-1}{2C_2})I_3 \\ +K_4(\frac{b}{3})I_2 & +K_4(\frac{-1}{2C_2})I_2 & +K_4(\frac{b}{6})I_2 & +K_4(\frac{1}{2C_2})I_2 \\ \hdashline K_2(\frac{-1}{2C_1})I_5 & K_3(\frac{b}{3C_1C_2})I_4 & K_2(\frac{1}{2C_1})I_5 & K_3(\frac{b}{6C_1C_2})I_4 \\ +K_4(\frac{-1}{2C_1})I_2 & +K_4(\frac{1}{bC_1C_2})I_2 & +K_4(\frac{-1}{2C_1})I_2 & +K_4(\frac{-1}{bC_1C_2})I_2 \\ \hdashline K_1(\frac{-1}{b})I_1 & K_2(\frac{1}{2C_2})I_3 & K_1(\frac{1}{b})I_1 & K_2(\frac{1}{2C_2})I_3 \\ +K_4(\frac{b}{6})I_2 & +K_4(\frac{-1}{2C_2})I_2 & +K_4(\frac{b}{3})I_2 & +K_4(\frac{1}{2C_2})I_2 \\ \hdashline K_2(\frac{-1}{2C_1})I_5 & K_3(\frac{b}{6C_1C_2})I_4 & K_2(\frac{1}{2C_1})I_5 & K_3(\frac{b}{3C_1C_2})I_4 \\ +K_4(\frac{1}{2C_1})I_2 & +K_4(\frac{-1}{bC_1C_2})I_2 & +K_4(\frac{1}{2C_1})I_2 & +K_4(\frac{1}{bC_1C_2})I_2 \end{array}\right]$$

$I_1 = \int_0^l Y_m Y_n dy$; $I_2 = \int_0^l Y_m' Y_n' dy$; $I_3 = \int_0^l Y_m Y_n'' dy$; $I_4 = \int_0^l Y_m'' Y_n'' dy$; $I_5 = \int_0^l Y_m'' Y_n dy$

$K_1 = \frac{E_x}{1-\upsilon_x\upsilon_y}$; $K_2 = \frac{\upsilon_x E_y}{1-\upsilon_x\upsilon_y}$; $K_3 = \frac{E_y}{1-\upsilon_x\upsilon_y}$; $K_4 = G_{xy}$; $C_1 = \frac{\mu_m}{l}$; $C_2 = \frac{\mu_n}{l}$

expressions listed there are really only valid for the particular formulation in which Y_m is used for u and Y_m' for v. For the clamped-clamped case given by Eq. 4.5, some modifications in the expressions will be necessary.

The stiffness matrix listed in Table 4.2 is valid for all seven types of support conditions and is widely used in vibration analysis. For the special case of two opposite ends simply supported, $[K]_{mn}$ becomes zero for m≠n, and the very simple stiffness matrix $[K]_{mm}$ is presented in Table 4.3.

The potential energy due to the applied load {q} can be readily shown to be

$$W = \sum_{m=1}^{r} \int \{\delta\}_m^T [N]_m \{q\}\, d(\text{area}) \tag{4.14}$$

Due to the linear displacement assumption in the transverse direction the consistent load vector for the uniform load case is particularly simple and can be written as

$$\{F\}_m = \frac{b}{2} \begin{Bmatrix} q_x \int_0^l Y_m dy \\ q_y \frac{l}{\mu_m} \int_0^l Y_m' dy \\ q_x \int_0^l Y_m dy \\ q_y \frac{l}{\mu_m} \int_0^l Y_m' dy \end{Bmatrix} \tag{4.15}$$

where q_x and q_y are the load components in the x and y directions respectively.

TABLE 4.3
In-plane stiffness matrix $[K]_{mm}$ of simply supported LO2 rectangular strip.

$$t\begin{bmatrix} \frac{lE_1}{2b}+\frac{lbk_m^2G}{6} & & & \text{SYM} \\ \frac{lk_m\upsilon_xE_2}{4}-\frac{lk_mG}{4} & \frac{lbk_m^2E_2}{6}+\frac{lG}{2b} & & \\ -\frac{lE_1}{2b}+\frac{lbk_m^2G}{12} & -\frac{lk_m\upsilon_xE_2}{4}-\frac{lk_mG}{4} & \frac{lE_1}{2b}+\frac{lbk_m^2G}{6} & \\ \frac{lk_m\upsilon_xE_2}{4}+\frac{lk_mG}{4} & \frac{lbk_m^2E_2}{6}-\frac{lG}{2b} & -\frac{lk_m\upsilon_xE_2}{4}+\frac{lk_mG}{4} & \frac{lbk_m^2E_2}{6}+\frac{lG}{2b} \end{bmatrix}$$

$$E_1 = \frac{E_x}{1-\upsilon_x\upsilon_y}\ ;\ E_2 = \frac{E_y}{1-\upsilon_x\upsilon_y}$$

4.2.2 HIGHER ORDER STRIP WITH ONE INTERNAL NODAL LINE (HO3)

Referring to Figure 4.1b it is seen that only displacement values exist at the three nodal lines, and therefore the situation concurs with that of case (d) in Section 2.2.2. A suitable set of displacement functions for the strip (with two opposite ends simply supported) is given by Eq. 4.16, in which C_1, C_2, and C_3 are the shape functions listed in [Section 2.2.2]

$$\{f\} = \begin{Bmatrix} u \\ v \end{Bmatrix} = \sum_{m=1}^{r} \begin{bmatrix} C_1 \sin k_m y & 0 & C_2 \sin k_m y & 0 & C_3 \sin k_m y & 0 \\ 0 & C_1 \cos k_m y & 0 & C_2 \cos k_m y & 0 & C_3 \cos k_m y \end{bmatrix} \begin{Bmatrix} u_1 \\ v_1 \\ u_2 \\ v_2 \\ u_3 \\ v_3 \end{Bmatrix}_m$$

$$= \sum_{m=1}^{r} [N]_m \{\delta\}_m \tag{4.16}$$

By applying Eqs. 4.9 to 4.16, the strain matrix $[B]_m$ is obtained such that

$$[B]_m = \begin{bmatrix} \frac{(-3+4\bar{x})}{b} S & 0 & \frac{(4-8\bar{x})}{b} S & 0 & \frac{(-1+4\bar{x})}{b} S & 0 \\ 0 & -(1-3\bar{x}+2\bar{x}^2) k_m S & 0 & -(4\bar{x}-4\bar{x}^2) k_m S & 0 & -(-\bar{x}+2\bar{x}^2) k_m S \\ (1-3\bar{x}+2\bar{x}^2) k_m C & \frac{(-3+4\bar{x})}{b} C & (4\bar{x}-4\bar{x}^2) k_m C & \frac{(4-8\bar{x})}{b} C & (-\bar{x}+2\bar{x}^2) k_m C & \frac{(-1+2\bar{x})}{b} C \end{bmatrix} \tag{4.17}$$

where $S = \sin k_m y$; $C = \cos k_m y$.

The stiffness matrix $[K]_{mm}$ is given in Table 4.4, while the consistent load vectors for several types of loading have also been worked out and are listed in Table 4.5. As discussed previously, from a computation point of view it is best to eliminate the parameters associated with the internal nodal line through static condensation, and the resulting reduced matrices have also been worked out by Loo.[6]

TABLE 4.4

Stiffness matrix for a HO3 strip for in-plane stress analysis.[6,7]

$$[K]_{mm} = \begin{bmatrix} K1 & & & & & \\ K2 & K7 & & & \text{SYM} & \\ K3 & -K4 & K11 & & & \\ K4 & K8 & 0 & K12 & & \\ K5 & -K6 & K3 & -K4 & K1 & \\ K6 & K10 & K4 & K8 & -K2 & K7 \end{bmatrix}$$

where

$$K1 = \frac{7l}{6b}A + \frac{lbk_m^2}{15}D \qquad K7 = \frac{lbk_m^2}{15}B + \frac{7l}{6b}D$$

$$K2 = \frac{lk_m}{4}C - \frac{lk_m}{4}D \qquad K8 = \frac{lbk_m^2}{30}B - \frac{4l}{3b}D$$

$$K3 = -\frac{4l}{3b}A + \frac{lbk_m^2}{30}D \qquad K10 = -\frac{lbk_m^2}{60}B + \frac{l}{6b}D$$

$$K4 = \frac{lk_m}{3}C + \frac{lk_m}{3}D \qquad K11 = -\frac{8l}{3b}A + \frac{4lbk_m^2}{15}D$$

$$K5 = \frac{l}{6b}A - \frac{lbk_m^2}{60}D \qquad K12 = \frac{4lbk_m^2}{15}B + \frac{8l}{3b}D$$

$$K6 = -\frac{lk_m}{12}C - \frac{lk_m}{12}D$$

in which $A = \frac{E_x}{1-\upsilon_x\upsilon_y}$; $B = \frac{E_y}{1-\upsilon_x\upsilon_y}$; $C = \upsilon_y A = \upsilon_x B$; $D = G_{xy}$

TABLE 4.5

Load vectors due to concentrated in-plane loads for a HO3 strip.[6,7]

a) Concentrated load U_o at (x_o, y_o)

$$\{F\}_m = \begin{Bmatrix} 1-3\bar{x}_o+2\bar{x}_o^2 \\ 0 \\ 4\bar{x}_o-4\bar{x}_o^2 \\ 0 \\ 2\bar{x}_o^2-\bar{x}_o \\ 0 \end{Bmatrix} U_o \sin k_m y_o$$

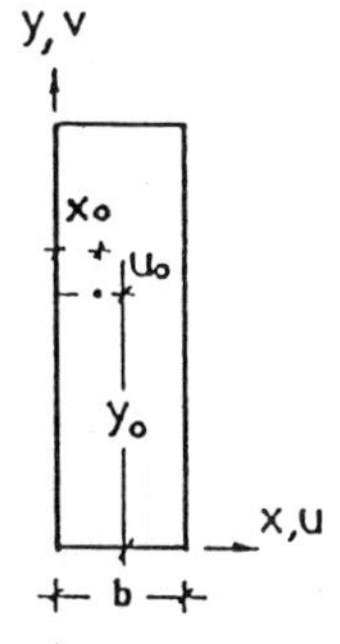

b) Concentrated load V_o at (x_o, y_o)

$$\{F\}_m = \begin{Bmatrix} 0 \\ 1-3\bar{x}_o+2\bar{x}_o^2 \\ 0 \\ 4\bar{x}_o-4\bar{x}_o^2 \\ 0 \\ 2\bar{x}_o^2-\bar{x}_o \end{Bmatrix} V_o \cos k_m y_o$$

$(\bar{x}_o = x_o / b)$

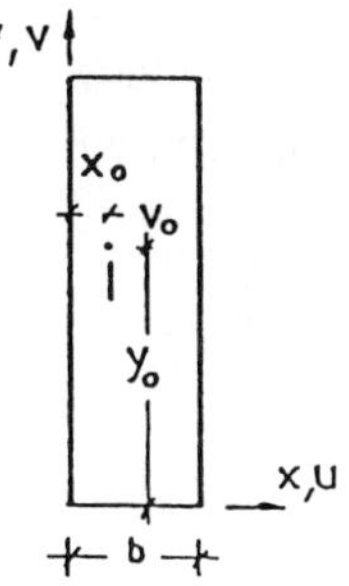

4.2.3 HIGHER ORDER STRIP WITH TWO NODAL LINES (HO2)

In this strip (Figure 4.1c), a higher order displacement function is arrived at through the use of four (instead of two for the strips presented previously) parameters at each of the two nodal lines. The parameters involved are u, $\partial u/\partial x$, v and $\partial v/\partial x$, and therefore the relevant shape functions are supplied through Section 2.2.2. The displacement function for the strip can now be written as in Eq. 4.18.

$$\{f\} = \begin{Bmatrix} u \\ v \end{Bmatrix} = \sum_{m=1}^{r} \begin{bmatrix} [C_1]\sin k_m y & & [C_2]\sin k_m y & \\ & [C_1]\cos k_m y & & [C_2]\cos k_m y \end{bmatrix} \begin{Bmatrix} \begin{Bmatrix} u_1 \\ (\frac{\partial u}{\partial x})_1 \end{Bmatrix} \\ \begin{Bmatrix} v_1 \\ (\frac{\partial v}{\partial x})_1 \end{Bmatrix} \\ \begin{Bmatrix} u_2 \\ (\frac{\partial u}{\partial x})_2 \end{Bmatrix} \\ \begin{Bmatrix} v_2 \\ (\frac{\partial v}{\partial x})_2 \end{Bmatrix} \end{Bmatrix}$$

$$= \sum_{m=1}^{r} [N]_m \{\delta\}_m \qquad (4.18)$$

The standard formulation procedure should now be followed to work out the stiffness matrix and load vector.

4.3 SPLINE PLANE STRESS STRIP[2]

4.3.1 FORMATION OF STRIP STIFFNESS MATRIX

The displacements of the lower order spline strip is

$$\{f\} = \begin{Bmatrix} u \\ v \end{Bmatrix} = \begin{bmatrix} C_1 & 0 & C_2 & 0 \\ 0 & C_1 & 0 & C_2 \end{bmatrix} \begin{bmatrix} [\phi]_1 & & & \\ & [\phi]_2 & & \\ & & [\phi]_3 & \\ & & & [\phi]_4 \end{bmatrix} \begin{Bmatrix} u_1 \\ v_1 \\ u_2 \\ v_2 \end{Bmatrix} \qquad (4.19)$$

where C_1 and C_2 are defined by Eq. 2.30 in Section 2.2.2.
u_i, v_i, u_j and v_j are the interpolation parameters of the i-th and j-th nodes, and $[\phi]_k$ (k=1, 2, 3, and 4) are the corresponding spline functions.

In short form,

$$\{f\} = [C]\,[\Phi]\,\{\delta\} = [N]\,\{\delta\} \qquad (4.20)$$

Following the variational method, the stiffness matrix can be derived:

$$[K^e] = t\int [B]^T\,[D]\,[B]\,d(\text{area}) \qquad (4.21)$$

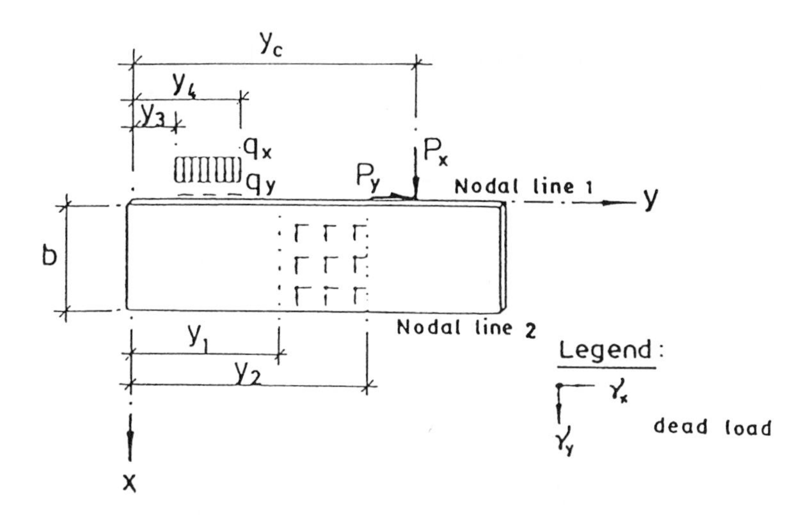

Figure 4.2 Uniformly distributed line load, dead load and concentrated load on rectangular plane strip.

The explicit forms of the strain matrix [B] and stiffness matrix [K^e] are given in Table 4.6.

4.3.2 LOAD VECTORS

The load vectors for several loading cases are as follows[(2)] (Figure 4.2):

a) uniformly distributed body load from y_1 to y_2

$$\{F^e\} = \frac{bt}{2} \int_{y_1}^{y_2} [\Phi]^T dy \begin{Bmatrix} \gamma_x \\ \gamma_y \\ \gamma_x \\ \gamma_y \end{Bmatrix} \tag{4.22}$$

b) uniformly distributed line load q_x and q_y on nodal line '1' from y_3 to y_4

$$\{F^e\} = t \int_{y_3}^{y_4} [\Phi]^T dy \begin{Bmatrix} q_x \\ q_y \\ 0 \\ 0 \end{Bmatrix} \tag{4.23}$$

c) concentrated loads P_x and P_y on nodal line '1' at $y = y_c$

$$\{F^e\} = t [\Phi(y_c)]^T \begin{Bmatrix} P_x \\ P_y \\ 0 \\ 0 \end{Bmatrix} \tag{4.24}$$

TABLE 4.6
Explicit form of strain and stiffness matrices for spline plane strip.[2]
a) strain matrix

$$\begin{bmatrix} \frac{-[\phi]_1}{b} & 0 & \frac{[\phi]_3}{b} & 0 \\ 0 & (1-\bar{x})\frac{\partial[\phi]_2}{\partial y} & 0 & \bar{x}\frac{\partial[\phi]_4}{\partial y} \\ (1-\bar{x})\frac{\partial[\phi]_1}{\partial y} & \frac{-[\phi]_2}{b} & \bar{x}\frac{\partial[\phi]_3}{\partial y} & \frac{[\phi]_4}{b} \end{bmatrix}$$

b) stiffness matrix

$$\begin{bmatrix} D_x I1_{11} + \frac{bD_{xy}I_{411}}{3} & -\frac{D_1 I_{222}}{2} - \frac{D_{xy}I_{312}}{2} & -\frac{D_x I_{113}}{b} + \frac{bD_{xy}I_{412}}{6} & -\frac{D_1 I_{214}}{2} + \frac{D_{xy}I_{314}}{6} \\ & \frac{bD_y I_{422}}{3} + \frac{D_{xy}I_{122}}{b} & \frac{D_1 I_{323}}{2} - \frac{D_{xy}I_{223}}{2} & \frac{bD_y I_{424}}{6} - \frac{D_{xy}I_{124}}{b} \\ \text{SYM} & & \frac{D_x I_{133}}{b} + \frac{bD_{xy}I_{433}}{3} & \frac{D_1 I_{234}}{2} + \frac{D_{xy}I_{334}}{2} \\ & & & \frac{bD_y I_{444}}{3} + \frac{D_{xy}I_{144}}{6} \end{bmatrix}$$

in which $I1_{ij} = \int_0^l t[\phi]_i^T[\phi]_j\,dy$; $I2_{ij} = \int_0^l t[\phi]_i^T[\phi]_j'dy$; $I3_{ij} = \int_0^l t[\phi]_i'^T[\phi]_j\,dy$;

$I4_{ij} = \int_0^l t[\phi]_i'^T[\phi]_j'dy$

4.4 COMPUTED SHAPE FUNCTION STRIP (COMSFUN)

A two-dimensional finite strip is shown in Figure 4.3a. In Figure 4.3b, a unit width of the strip is taken out as a fictitious beam with constant bending rigidity. A number of joints, say, r, are assigned as nodes. Since only C^o continuity is required, no restraint is applied to the rotational degree of freedom at each end of the fictitious beam; that is, the boundary condition at each end of the fictitious beam is either, respectively, simply supported or free when computing the shape functions for the internal nodes and the corresponding end node. A total of r computed shape functions are used in formulating each of the two in-plane displacement components. The displacements can be written as:

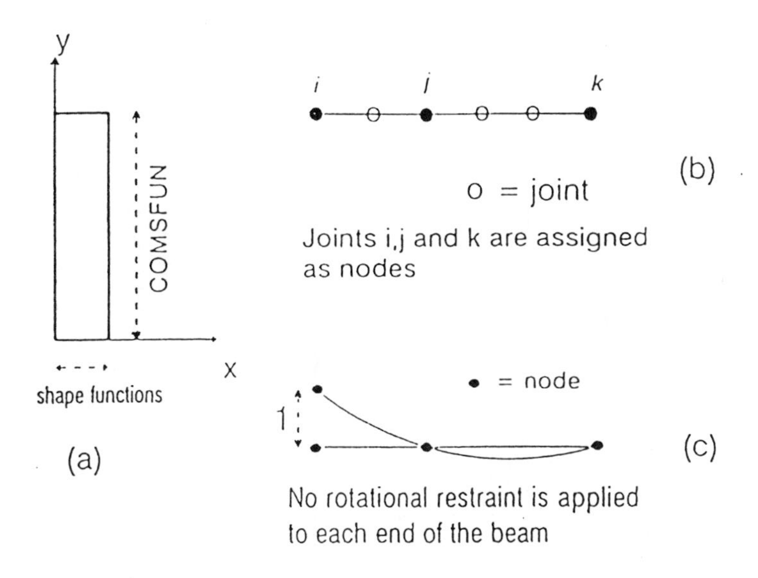

Figure 4.3 (a) A 2-D plane finite strip. (b) The fictitious beam and its discretisation. (c) COMSFUN ϕ_i of node i.

$$u(x,y) = \sum_{m=1}^{r} [N(x)] \phi_m(y) \{\alpha\}_m \tag{4.25}$$

$$v(x,y) = \sum_{m=1}^{r} [N(x)] \phi_m(y) \{\beta\}_m \tag{4.26}$$

It is also important to recognize that the interpolation parameters ($\{\alpha\}_m$ and $\{\beta\}_m$) represent the actual in-plane displacements at the nodes, that is

$$\{\alpha\}_m = \{u_1 \quad u_2\}_m \qquad \text{and} \qquad \{\beta\}_m = \{v_1 \quad v_2\}_m$$

As a consequence, the implementation of support conditions and the assembly of finite strips can be done in the same fashion as the finite element method.

It can be shown easily that C^o continuity is satisfied along the boundaries between adjacent strips. Referring to Figure 4.4, C^o continuity is maintained across the interface between strip '1' and strip '3' because displacements for both strips along the interface are uniquely defined by displacements along the same nodal line pk. The same principle can be applied to verify the C^o continuity across the interface between strip '1' and strip '2', as the displacements are defined uniquely by displacements at their common nodes j and k. Each strip is, therefore, C^o continuous around all edges.

The stiffness matrices and load vectors for each strip can be formed by the procedures as described in the previous section.

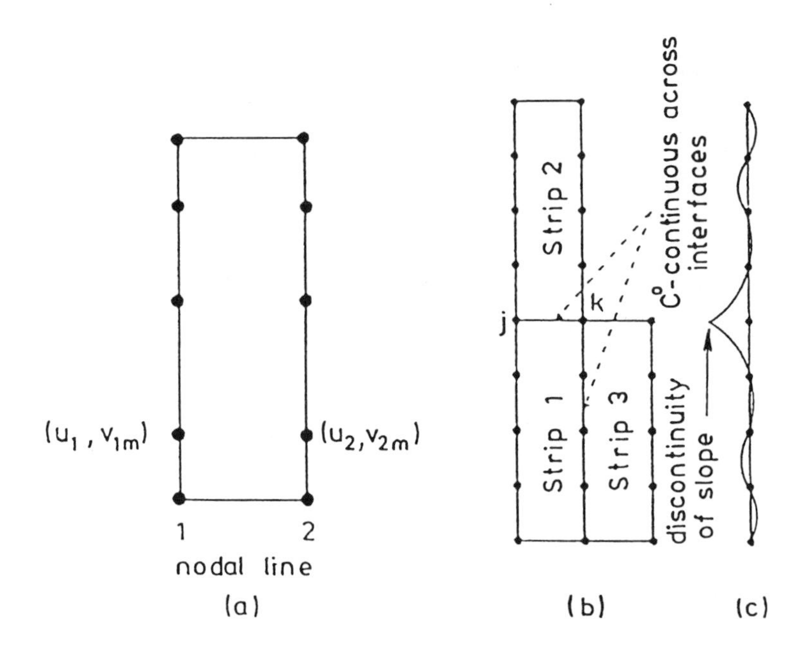

Figure 4.4 (a) A typical strip with 5 nodes on each nodal line. (b) An assemblage of three finite strips. (c) Connection of COMSFUN at node j.

4.5 CURVED PLANE STRESS STRIPS

In Section 3.5, it was shown how a curved bending strip can be evolved from a rectangular bending strip by transforming all the quantities associated with the Cartesian coordinate system into corresponding quantities associated with the polar coordinate system, and the same technique can be applied here in developing a curved plane stress strip. Such a curved strip has probably no direct application in actual two-dimensional engineering problems, but is mainly used in combination with a curved bending strip for analyzing curved box beams or stiffened plates, thus still serving a very important purpose.

A lower order classical strip (Figure 4.1d) was proposed by Cheung and Cheung[8,9] in which, again, a linear variation of displacements across the section was assumed. The displacement functions are simple transformations of Eq. 4.8 and are written as in Eq. 4.27, in which R, α, and Θ_m have already been defined in Section 2.2.4.

$$\{f\} = \begin{Bmatrix} u \\ v \end{Bmatrix} = \sum_{m=1}^{r} \begin{bmatrix} (1-\frac{R}{2})\Theta_m & 0 & \frac{R}{2}\Theta_m & 0 \\ 0 & (1-\frac{R}{2})\frac{\alpha}{\mu_m}\Theta_m' & 0 & \frac{R}{2}\frac{\alpha}{\mu_m}\Theta_m' \end{bmatrix} \begin{Bmatrix} u_1 \\ v_1 \\ u_2 \\ v_2 \end{Bmatrix} \quad (4.27)$$

The strain-displacement relationship[10] is defined by

$$\{\varepsilon\} = \begin{Bmatrix} \varepsilon_r \\ \varepsilon_\theta \\ \gamma_{r\theta} \end{Bmatrix} = \begin{Bmatrix} \frac{\partial u}{\partial r} \\ \frac{1}{r}\frac{\partial v}{\partial \theta} + \frac{u}{r} \\ \frac{1}{r}\frac{\partial u}{\partial \theta} + \frac{\partial v}{\partial r} \end{Bmatrix} = \sum_{m=1}^{r} [B]_m \{\delta\}_m \quad (4.28)$$

and the property matrix by

$$[D] = \begin{bmatrix} \frac{E_r}{(1-\upsilon_r\upsilon_\theta)} & \frac{\upsilon_\theta E_r}{(1-\upsilon_r\upsilon_\theta)} & 0 \\ \frac{\nu_\theta E_r}{(1-\upsilon_r\upsilon_\theta)} & \frac{E_\theta}{(1-\upsilon_r\upsilon_\theta)} & 0 \\ 0 & 0 & G_{r\theta} \end{bmatrix} \quad (4.29)$$

where E_r , E_θ , υ_r , υ_θ, and $G_{r\theta}$ are the elastic constants for the polar coordinate system.

The stiffness matrix for a constant thickness curved strip is computed through an equation similar to Eq. 4.12, that is

$$[K]_{mn} = t \int_0^\alpha \int_{r_1}^{r_2} [B]_m^T [D] [B]_n \, r dr d\theta \quad (4.30)$$

The expressions to be integrated in Eq. 4.30 are fairly complex, and, therefore, it is time consuming to work out all the integrals in closed form. Furthermore, due to the presence of 1/r terms, such closed form integrals are numerically sensitive to very small radii values, and a numerical integration scheme such as Gaussian quadrature should be used instead. In Table 4.7, $[K]_{mn}$ is shown in an explicit form, but with the integration step yet to be completed.

The characteristic matrices for a curved plane spline strip can be similarly formed. The displacements (Figure 4.5) are given as:

$$\{f\} = \begin{Bmatrix} u \\ v \end{Bmatrix} = \begin{bmatrix} N_1 & 0 & N_2 & 0 \\ 0 & N_1 & 0 & N_2 \end{bmatrix} \begin{bmatrix} [\phi]_1 & & & \\ & [\phi]_2 & & \\ & & [\phi]_3 & \\ & & & [\phi]_4 \end{bmatrix} \begin{Bmatrix} u_1 \\ v_1 \\ u_2 \\ v_2 \end{Bmatrix} \quad (4.31)$$

where $\begin{cases} N_1 = 1-\bar{r} \\ N_2 = \bar{r} \end{cases}$; and $\bar{r} = \frac{r-r_1}{b}$. In short form,

$$\{f\} = [N]\,[\Phi]\{\delta\} \quad (4.32)$$

TABLE 4.7
In-plane stiffness matrix $[K]_{mn}$ of a curved LO2 strip

$$[K]_{mn} = \int_{r_1}^{r_2} \begin{bmatrix} [S_{11}]_{mn} & [S_{12}]_{mn} \\ [S_{12}]_{mn}^T & [S_{22}]_{mn} \end{bmatrix} rdr$$

$$[S_{11}]_{mn} = \left[\begin{array}{c|c}
\begin{array}{c}
K_r \frac{1}{4b'^2} Y_m Y_n \\
-K_1 \frac{1}{rb'}(1-\frac{R}{2}) Y_m Y_n \\
+K_\theta \frac{1}{r^2}(1-\frac{R}{2})^2 Y_m Y_n \\
+K_{r\theta} \frac{1}{r^2}(1-\frac{R}{2})^2 Y'_m Y'_n
\end{array} &
\begin{array}{c}
-K_1 \frac{1}{2rb'} C_n (1-\frac{R}{2}) Y_m Y''_n \\
+K_\theta \frac{1}{r^2} C_n (1-\frac{R}{2})^2 Y_m Y''_n \\
-K_{r\theta} \frac{1}{2rb'} C_n (1-\frac{R}{2}) Y'_m Y'_n \\
-K_{r\theta} \frac{1}{r^2} C_n (1-\frac{R}{2}) Y'_m Y'_n
\end{array} \\ \hline
\begin{array}{c}
-K_1 \frac{1}{2rb'} C_m (1-\frac{R}{2}) Y''_m Y_n \\
+K_\theta \frac{1}{r^2} C_m (1-\frac{R}{2})^2 Y''_m Y_n \\
-K_{r\theta} \frac{1}{2rb'} C_m (1-\frac{R}{2}) Y'_m Y'_n \\
-K_{r\theta} \frac{1}{r^2} C_m (1-\frac{R}{2}) Y'_m Y'_n
\end{array} &
\begin{array}{c}
K_\theta \frac{1}{r^2} C_m C_n (1-\frac{R}{2})^2 Y''_m Y''_n \\
+K_{r\theta} \frac{1}{4b'^2} C_m C_n Y'_m Y'_n \\
+K_{r\theta} \frac{1}{rb'} C_m C_n (1-\frac{R}{2}) Y'_m Y'_n \\
+K_{r\theta} \frac{1}{r^2} C_m C_n (1-\frac{R}{2})^2 Y'_m Y'_n
\end{array}
\end{array}\right]$$

$$[S_{12}]_{mn} = \left[\begin{array}{c|c}
\begin{array}{c}
-K_r \frac{1}{4b'^2} Y_m Y_n \\
+K_1 \frac{1}{2rb'}(1-R) Y_m Y_n \\
+K_\theta \frac{1}{r^2}\frac{R}{2}(1-\frac{R}{2}) Y_m Y_n \\
+K_{r\theta} \frac{1}{r^2}\frac{R}{2}(1-\frac{R}{2}) Y'_m Y'_n
\end{array} &
\begin{array}{c}
-K_1 \frac{1}{2rb'} C_n \frac{R}{2} Y_m Y''_n \\
+K_\theta \frac{1}{r^2} C_n \frac{R}{2}(1-\frac{R}{2}) Y_m Y''_n \\
+K_{r\theta} \frac{1}{2rb'} C_n (1-\frac{R}{2}) Y'_m Y'_n \\
-K_{r\theta} \frac{1}{r^2} C_n \frac{R}{2}(1-\frac{R}{2}) Y'_m Y'_n
\end{array} \\ \hline
\begin{array}{c}
K_1 \frac{1}{2rb'} C_m (1-\frac{R}{2}) Y''_m Y_n \\
+K_\theta \frac{1}{r^2} C_m \frac{R}{2}(1-\frac{R}{2}) Y''_m Y_n \\
-K_{r\theta} \frac{1}{2rb'} C_m \frac{R}{2} Y'_m Y'_n \\
-K_{r\theta} \frac{1}{r^2} C_m \frac{R}{2}(1-\frac{R}{2}) Y'_m Y'_n
\end{array} &
\begin{array}{c}
K_\theta \frac{1}{r^2} C_m C_n \frac{R}{2}(1-\frac{R}{2}) Y''_m Y''_n \\
-K_{r\theta} \frac{1}{4b'^2} C_m C_n Y'_m Y'_n \\
+K_{r\theta} \frac{1}{2rb'} C_m C_n (R-1) Y'_m Y'_n \\
+K_{r\theta} \frac{1}{r^2} C_m C_n \frac{R}{2}(1-\frac{R}{2}) Y'_m Y'_n
\end{array}
\end{array}\right]$$

$$
[S_{22}]_{mn} = \left[\begin{array}{c|c}
\begin{array}{c}
K_r \dfrac{1}{4b'^2} Y_m Y_n \\
+K_1 \dfrac{1}{rb'} \dfrac{R}{2} Y_m Y_n \\
+K_\theta \dfrac{1}{r^2} \dfrac{R}{2} Y_m Y_n \\
+K_{r\theta} \dfrac{1}{r^2} (\dfrac{R}{2})^2 Y_m' Y_n'
\end{array} &
\begin{array}{c}
K_1 \dfrac{1}{2rb'} C_n \dfrac{R}{2} Y_m Y_n'' \\
+K_\theta \dfrac{1}{r^2} C_n (\dfrac{R}{2})^2 Y_m Y_n'' \\
+K_{r\theta} \dfrac{1}{2rb'} C_n \dfrac{R}{2} Y_m' Y_n' \\
-K_{r\theta} \dfrac{1}{r^2} C_n (\dfrac{R}{2})^2 Y_m' Y_n'
\end{array} \\
\hline
\begin{array}{c}
K_1 \dfrac{1}{2rb'} C_m \dfrac{R}{2} Y_m'' Y_n \\
+K_\theta \dfrac{1}{r^2} C_m (\dfrac{R}{2})^2 Y_m'' Y_n \\
+K_{r\theta} \dfrac{1}{2rb'} C_m \dfrac{R}{2} Y_m' Y_n' \\
-K_{r\theta} \dfrac{1}{r^2} C_m (\dfrac{R}{2})^2 Y_m' Y_n'
\end{array} &
\begin{array}{c}
K_\theta \dfrac{1}{r^2} C_m C_n (\dfrac{R}{2})^2 Y_m'' Y_n'' \\
+K_{r\theta} \dfrac{1}{4b'^2} C_m C_n Y_m' Y_n' \\
-K_{r\theta} \dfrac{1}{rb'} C_m C_n \dfrac{R}{2} Y_m' Y_n' \\
+K_{r\theta} \dfrac{1}{r^2} C_m C_n (\dfrac{R}{2})^2 Y_m' Y_n'
\end{array}
\end{array} \right]
$$

$K_r = \dfrac{E_r}{1-\upsilon_r\upsilon_\theta}$; $K_1 = \dfrac{\upsilon_\theta E_r}{1-\upsilon_r\upsilon_\theta}$; $K_\theta = \dfrac{E_\theta}{1-\upsilon_r\upsilon_\theta}$; $K_{r\theta} = G_{r\theta}$; $R = \dfrac{r-r_1}{b'}$; $b' = \dfrac{r-r_1}{2}$;

$Y_m Y_n = \int_o^\alpha \Theta_m \Theta_n d\theta$; $Y_m Y_n'' = \int_o^\alpha \Theta_m \Theta_n'' d\theta$; $Y_m'' Y_n'' = \int_o^\alpha \Theta_m'' \Theta_n'' d\theta$

$C_m = \dfrac{\alpha}{\mu_m}$; $C_n = \dfrac{\alpha}{\mu_n}$

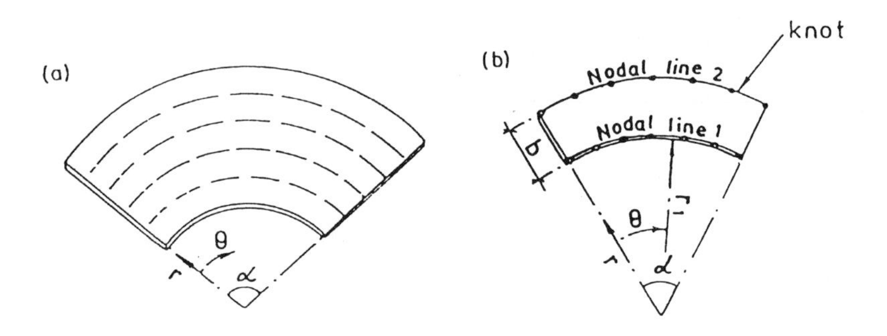

Figure 4.5 Circular plane strip analysis: (a) Diaphragm as an assembly of strips. (b) Typical strip.

The strain-displacement relationship is:

$$\begin{Bmatrix} \dfrac{\partial u}{\partial r} \\ \dfrac{1}{r}\dfrac{\partial v}{\partial \theta} + \dfrac{u}{r} \\ \dfrac{\partial u}{r\partial \theta} + \dfrac{\partial v}{\partial r} \end{Bmatrix} = \begin{bmatrix} -\frac{1}{b}[\phi]_1 & 0 & \frac{1}{b}[\phi]_3 & 0 \\ \frac{1}{r}(1-\bar{r})[\phi]_1 & \frac{1}{r}(1-\bar{r})[\phi]_2' & \frac{\bar{r}}{r}[\phi]_3 & \frac{\bar{r}}{r}[\phi]_4' \\ \frac{1}{r}(1-\bar{r})[\phi]_1' & -\frac{1}{b}[\phi]_2 & \frac{\bar{r}}{r}[\phi]_3' & \frac{\bar{r}}{r}[\phi]_4 \end{bmatrix} \{\delta\} \tag{4.33}$$

where $[\phi]_1' = \dfrac{\partial[\phi]_1}{\partial\theta}$; $[\phi]_2' = \dfrac{\partial[\phi]_2}{\partial\theta}$; $[\phi]_3' = \dfrac{\partial[\phi]_3}{\partial\theta}$; $[\phi]_4' = \dfrac{\partial[\phi]_4}{\partial\theta}$.

The stiffness matrix is tabulated in Table 4.8.

TABLE 4.8
Stiffness matrix of a circular plane spline strip.

$(dD_r - D_2 + gD_\theta)I1_{11}$ $+gD_{r\theta}I4_{11}$	$(-\frac{1}{2}D_2 + gD_\theta)I2_{12}$ $+\frac{1}{2}D_{r\theta}I4_{22}$	$(-dD_r + fD_\theta)I1_{13}$ $+fD_{r\theta}I4_{13}$	$(-\frac{1}{2}D_2 + fD_\theta)I2_{14}$ $+\frac{1}{2}D_{r\theta}I3_{14}$
	$dD_{r\theta}I1_{22}$ $+gD_{r\theta}I4_{22}$	$-\frac{1}{2}D_{r\theta}I2_{23}$ $+(\frac{1}{2}D_2 + fD_\theta)I3_{23}$	$-dD_{r\theta}I1_{24}$ $+fD_\theta I4_{24}$
	SYM	$(dD_r + D_2 + eD_\theta)I1_{33}$ $+eD_{r\theta}I4_{33}$	$(\frac{1}{2}D_2 + eD_\theta)I2_{34}$ $+\frac{1}{2}D_{r\theta}I3_{34}$
			$dD_{r\theta}I1_{44}$ $eD_\theta I4_{44}$

in which $I1_{ij} = \int_0^\alpha t[\phi]_i^T[\phi]_j d\theta$; $I2_{ij} = \int_0^\alpha t[\phi]_i^T[\phi]_j' d\theta$; $I3_{ij} = \int_0^\alpha t[\phi]_i'^T[\phi]_j d\theta$; $I4_{ij} = \int_0^\alpha t[\phi]_i'^T[\phi]_j' d\theta$; $d = \frac{1}{2} + \frac{r_1}{b}$; $e = \frac{1}{2} - \frac{r_1}{b} + (\frac{r_1}{b})^2 \ln\frac{r_1+b}{r_1}$; $f = 1 - e - \frac{r_1}{b}\ln\frac{r_1+b}{r_1}$; $g = -2 + e + 2d\ln\frac{r_1+b}{r_1}$

4.6 EXAMPLES

In order to demonstrate the accuracy of the finite strip method in plane stress analysis, a number of deep beam and ordinary beam problems have been analyzed using a LO2 rectangular strip.

The first example deals with the analysis of an isotropic, simply supported square deep beam under a uniform line load acting at the top, and the results are presented in Figure 4.6. It is found that both the longitudinal stress σ_y and transverse stress σ_x agree very well with the values given by finite element analysis (16x16 mesh of constant strain triangular elements). The stress σ_y is also compared with the values given in a book by Kalmanok[11] and again very good agreement is demonstrated.

The transverse stress σ_x has been plotted at the middle of each strip in order to avoid a stepwise type of representation. For the longitudinal stress σ_y the plotting location is immaterial because of the very small jump at the nodal line.

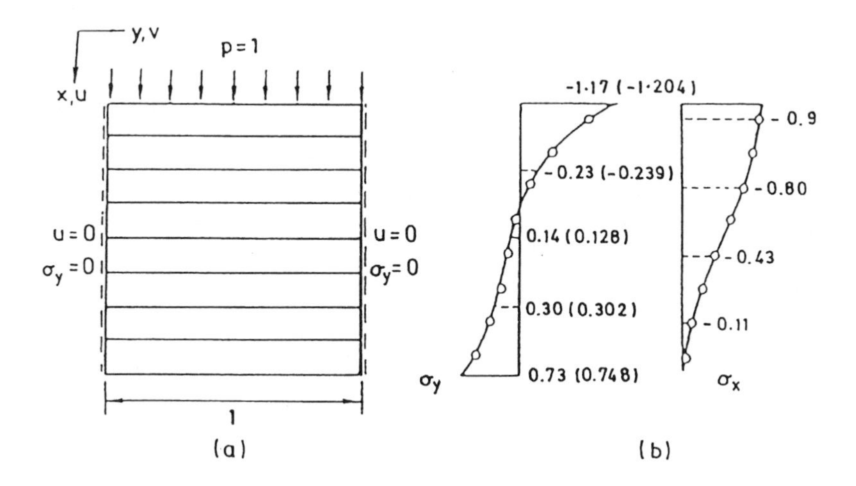

Figure 4.6 (a) Simply supported deep beam and its strip idealization. (b) Stress distributions at central section of the deep beam.
16x16 finite element solution; o o o finite strip solution; ( ) Kalmanok[11]

The second example demonstrates the use of another type of series other than the particular type used for simply supported conditions. The same deep beam with clamped ends is now analyzed using the displacement function

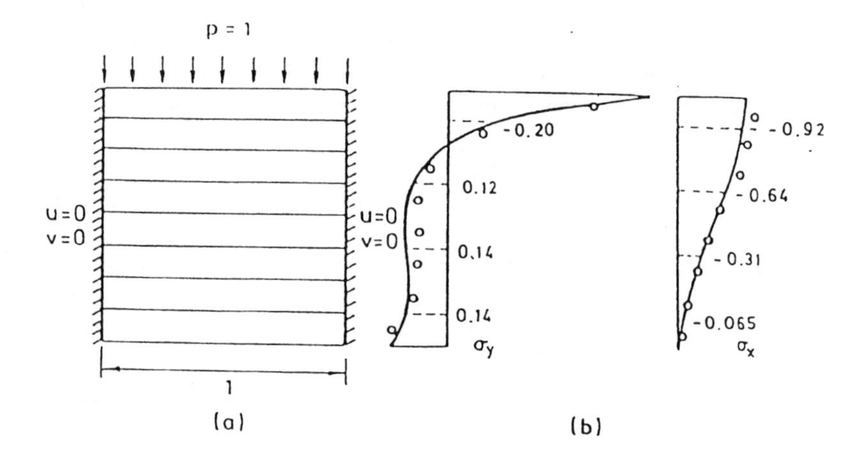

Figure 4.7 (a) A clamped deep beam and its strip idealization. (b) Stress distributions at central section of deep beam.
16x16 finite element solution; o o o finite strip solution.

given in Eq. 4.5. The results are presented in Figure 4.7, and once more the comparisons with the 16x16 finite element analysis results are excellent.

Kong[12] demonstrated the applications of the Computed Shape Function (COMSFUN) strip to two-dimensional multiply-connected domains. They analyzed a deep cantilever with step change in depth as shown in Figure 4.8. The beam was subjected to a point load at the free end and discretised into 15 strips. The number of nodes per nodal line was varied to study the convergence of the scheme. Figure 4.9 shows that the results converge rapidly except in the vicinity of geometrical discontinuity.

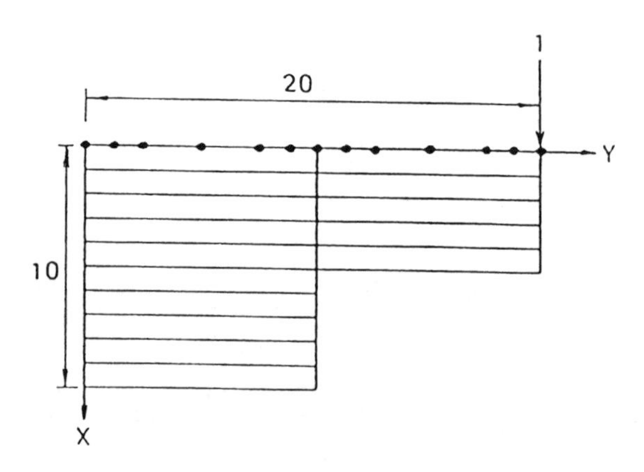

Figure 4.8 Deep cantilever with step change in depth at y=10.[12]

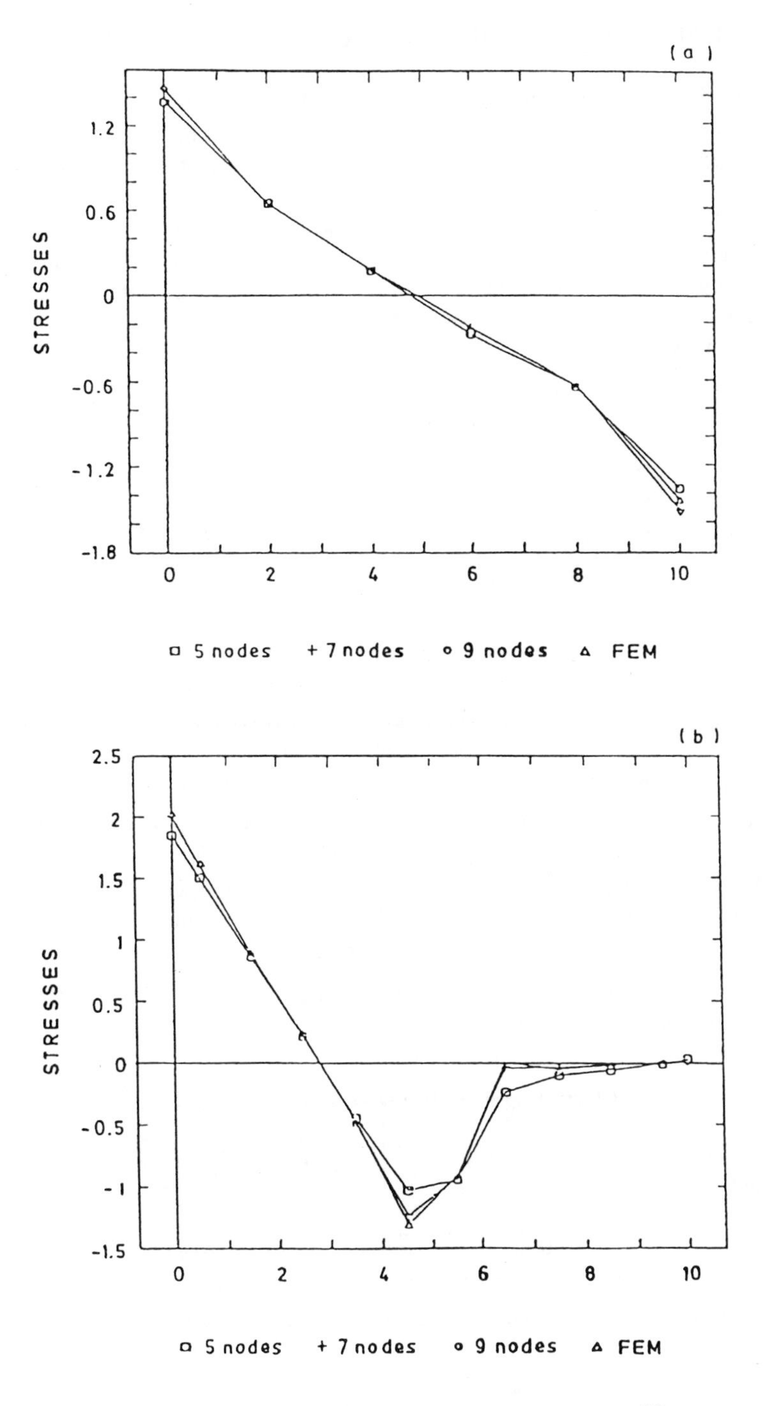

Figure 4.9 Stress distribution across the depth. (a) σ_y at y=0. (b) σ_y at y=9.5.[(12)]

4.7 MIXED FORMULATION OF PLANE STRESS ANALYSIS[13]

In the displacement approach, very accurate displacements can be obtained for most cases, but it is unfortunate that the stresses and/or moments derived from the displacement field are not necessarily continuous across the boundaries. In order to have stress continuity across the strip boundary, one may have to include the stresses and/or moments as independent variables in the formulation. Such an approach is referred to as Mixed Formulation, and it can be applied to both bending as well as plane problems. In this section, its application to the plane problem is used to illustrate the formulation principle.

The equilibrium equations for a plane problem can be defined as:

$$\frac{\partial \sigma_x}{\partial x}+\frac{\partial \tau_{xy}}{\partial y}=-X \quad ; \text{ or } \quad l_1^1(\phi)=f^1 \tag{4.34}$$

$$\frac{\partial \sigma_y}{\partial y}+\frac{\partial \tau_{xy}}{\partial x}=-Y \quad ; \text{ or } \quad l_1^2(\phi)=f^2 \tag{4.35}$$

The stress-strain relationships are:

$$\frac{\partial u}{\partial x} - C_1\sigma_x - C_2\sigma_y = \varepsilon_{xx}^{o} \quad ; \text{ or } \boldsymbol{l}_1^3(\phi)=f^3 \tag{4.36}$$

$$\frac{\partial v}{\partial y} - C_3\sigma_x - C_4\sigma_y = \varepsilon_{yy}^{o} \quad ; \text{ or } \boldsymbol{l}_1^4(\phi)=f^4 \tag{4.37}$$

$$\frac{\partial u}{\partial y} + \frac{\partial v}{\partial x} - C_5\tau_{xy} = \gamma_{xy}^{o} \quad ; \text{ or } \boldsymbol{l}_1^5(\phi)=f^5 \tag{4.38}$$

in which u and v are the displacement components in the x- and y-direction respectively. X and Y are the body force components; and $(\sigma_x, \sigma_y, \tau_{xy})$ and $(\varepsilon_{xx}^{o}, \varepsilon_{yy}^{o}, \gamma_{xy}^{o})$ are the stress and initial strain respectively. $(f^1, f^2, f^3, f^4, f^5) = (-X, -Y, \varepsilon_{xx}^{o}, \varepsilon_{yy}^{o}, \gamma_{xy}^{o})$. C_1 to C_5 are the material properties, which can be readily found in standard texts on elasticity. Thus the unknown parameters are

$$\phi = (u, v, \sigma_x, \sigma_y, \tau_{xy}) \tag{4.39}$$

The boundary conditions can be defined in terms of the differential operators along the boundary (τ) as

$$B(\phi) = g \tag{4.40}$$

where $B(\phi) = [b^1(\phi), b^2(\phi), \ldots, b^M(\phi)]$
$g = [g^1, g^2, \ldots, g^M]$

The problem can be restated as the minimization of the functional, ψ, which is defined as

$$\psi = \sum_{i=1}^{N} \int_{\Omega} t\alpha_i\, [\boldsymbol{l}_m^{i}(\phi) - f^{i}]^2\, d\Omega + \sum_{i=1}^{M} \int_{\tau} t\beta_i\, [b^{i}(\phi) - g^{i}]^2\, d\tau \tag{4.41}$$

The parameters α_i and β_i are chosen weighing functions. The variational of Eq. 4.28 is given as

$$\delta\psi = \sum_{i=1}^{N} \int_{\Omega} (t[\boldsymbol{l}_m^{i}(\phi)]^T \alpha_i\, [\boldsymbol{l}_m^{i}(\phi)])\, d\Omega\, \phi - \sum_{i=1}^{N} \int_{\Omega} (t[\boldsymbol{l}_m^{i}(\phi)]^T \alpha_i\, [f^{i}])\, d\Omega +$$

$$\sum_{i=1}^{M}\int_{\tau} (t[b^i(\phi)]^T\beta_i\,[b^i(\phi)])d\tau\,\phi - \sum_{i=1}^{M}\int_{\tau} (t[b^i(\phi)]^T\beta_i\,[g^i])d\tau \tag{4.42}$$

As in the standard energy approach for the finite element method, the 'stiffness' matrix and the 'load' vector can now be defined

$$[K^e] = \sum_{i=1}^{N}\int_{\Omega} (t[l_m^i(\phi)]^T\alpha_i\,[l_m^i(\phi)])\,d\Omega + \sum_{i=1}^{M}\int_{\tau} (t[b^i(\phi)]^T\beta_i\,[b^i(\phi)])d\tau \tag{4.43}$$

$$\{F^e\} = -\sum_{i=1}^{N}\int_{\Omega} (t[l_m^i(\phi)]^T\alpha_i\,[f^i\,])\,d\Omega - \sum_{i=1}^{M}\int_{\tau} (t[b^i(\phi)]^T\beta_i\,[g^i])d\tau \tag{4.44}$$

and the stiffness relationship as

$$[K^e]\,\{\phi\} + \{F^e\} = 0 \tag{4.45}$$

In the finite strip method, the domain Ω is discretised into subdomains with the variables inside each subdomain Ω' described by interpolating functions which comprise a continuous function in the x-direction and a polynomial in the y-direction, that is

$$u = \sum_{j=1}^{J}\sum_{i=1}^{I} N_i^u(y)\,U_j\,(x)\,u_{ij} \quad ; \quad v = \sum_{j=1}^{J}\sum_{i=1}^{I} N_i^v\,(y)\,V_j\,(x)\,v_{ij}$$

$$\sigma_x = \sum_{j=1}^{J}\sum_{i=1}^{I} N_i^x\,(y)\,S_j^x\,(x)\,s_{ij}^x \quad ; \sigma_y = \sum_{j=1}^{J}\sum_{i=1}^{I} N_i^y\,(y)\,S_j^y\,(x)\,s_{ij}^y$$

$$\tau = \sum_{j=1}^{J}\sum_{i=1}^{I} N_i^{xy}\,(y)\,S_j^{xy}\,(x)\,s_{ij}^{xy} \tag{4.46}$$

where S_j^x, S_j^y and S_j^{xy} are the series components of the interpolation shape functions. s_{ij}^x, s_{ij}^y and s_{ij}^{xy} are the interpolating parameters for the corresponding stresses.

Functions (Section 2.2.1) which satisfy the end boundary conditions can be chosen for the series part in the x-direction while polynomial interpolation functions described in Section 2.2.2 are to be used for the y-direction. By substituting the set of Eq. 4.46 into Eq. 4.45, the governing equations in each subdomain Ω' can be expressed in terms of the parameters (u_{ij} , v_{ij} , s_{ij}^x , s_{ij}^y , s_{ij}^{xy}), and the 'stiffness' matrix and the 'load' vector for each subdomain Ω' can be obtained from Eqs. 4.43 to 4.44. Summing up the matrices for all the subdomains and applying the appropriate boundary conditions in the y-direction, the problem can be solved in the standard manner.

The previous example (Section 4.4) of a simply supported square deep beam under a uniform line load acting at the top face was re-analyzed by this approach[13]. Parabolic polynomials (Eq. 2.36) were used in the y-direction, whereas the series parts were chosen as follows:

$$U_j\,(x) = \cos(\frac{j\pi x}{L});$$

$$V_j(x) = \sin(\frac{j\pi x}{L});$$

$$S_j^x(x) = \sin(\frac{j\pi x}{L});$$

$$S_j^y(x) = \sin(\frac{j\pi x}{L}); \text{ and}$$

$$S_j^{xy}(x) = \cos(\frac{j\pi x}{L})$$

The results are presented in Table 4.9, and they are compared with those published by Kalmanok[11] and those obtained by the displacement finite strips.

TABLE 4.9
Comparison of horizontal stress at mid-span.[13]

Depth (m)	Finite strip (Mixed form)	Kalmanok[11]	Classical strip (Displacement form)
1.0	-1.224	-1.204	-1.206
0.75	-0.245	-0.239	-0.243
0.50	0.130	0.128	0.132
0.25	0.302	0.302	0.302
.00	0.735	0.748	0.733

REFERENCES

1. Fung, Y. C., *Foundation of solid mechanics*, Prentice-Hall, 1965.
2. Fan, S. C., *Spline finite strip in structural analysis*, PhD thesis, Department of Civil and Structural Engineering, University of Hong Kong, 1982.
3. Cheung, Y. K., Analysis of box girder bridges by finite strip method, *Second International Symposium on Concrete Bridge Design*, Chicago, March 1969. Also Concrete bridge design, ACI Publication SP-26, 357-378, 1971.
4. Loo, Y. C. and Cusens, A. R., The auxiliary nodal line technique for the analysis of box girder bridge decks by the finite strip method, *Proc of Specialty Conf on the Finite Element Method in Civil Engineering*, McGill University, Montreal, June 1972.
5. Yoshida, K. and Oka, N., *A note on the in-plane displacement functions of strip elements*, Research Report, Department of Naval Architecture, Tokyo University, 1972.
6. Loo, Y. C., *Developments and applications of the finite strip method in the analysis of right bridge decks*, PhD thesis, University of Dundee, 1972.
7. Cheung, Y. K., *Finite strip method in structural analysis*, Pergamon Press, 1976.

8. Cheung, M. S. and Cheung, Y. K., Analysis of curved box girder bridges by finite strip method, *Publications, International Association for Bridges and Structural Engineering*, 31-I, 1971.
9. Cheung, M. S., *Finite strip analysis of structures*, PhD thesis, Department of Civil Engineering, University of Calgary, 1971.
10. Timoshenko, S. P. and Woinowsky-Krieger, S., *Theory of elasticity*, 2nd Edition, McGraw-Hill, 1951.
11. Kalmanok, A. S., *Calculation of plates*, National Press, Moscow, 1959 (in Russian).
12. Kong, J., *Analysis of plate type structures by finite strip, finite prism and finite layer method*, PhD thesis, Department of Civil and Structural Engineering, University of Hong Kong, 1994.
13. Cheung, Y. K. and Tham, L. G., Mixed formulation of finite strip method, *J of Eng Mech*, ASCE, 108(2), 452-456, 1982.

CHAPTER FIVE

VIBRATION AND STABILITY

5.1 MATRIX THEORY OF FREE VIBRATION

The analyses of a variety of elastic structures subject to static loads have been discussed in the previous chapters. It can be observed that all types of formulations will eventually lead to the same matrix equation of

$$[K]\{\delta\} = \{F\} \tag{5.1}$$

in which [K] is the stiffness matrix of a structure, $\{\delta\}$ the vector containing all nodal displacement parameters, and {F} the vector containing all nodal forces.

If the structure is moving then it is also possible to reduce the dynamic problem to a static one by applying D'Alembert's principle of dynamic equilibrium in which an inertia force equal to the product of the mass and the acceleration is assumed to act on the structure in the direction of negative acceleration. Thus at any instant of time the equilibrium equation for a structure in which both damping and external excitation forces are assumed to be non-existent is

$$[K]\{\delta(\tau)\} = -[\,[M]^c + [M]^e\,]\{\frac{\partial^2\delta(\tau)}{\partial\tau^2}\} = -[M]\{\frac{\partial^2\delta(\tau)}{\partial\tau^2}\} \tag{5.2}$$

where $\delta(\tau)$ is now a function of time.

In the above equation $[M]^c$ is a diagonal matrix of concentrated or line masses at the nodal lines, and is simply equal to zero when no such concentrated or line masses are acting on the structure, and $[M]^e$ is an overall mass matrix of the structure assembled from individual element consistent mass matrices $[M^e]$.(1)

For free vibration, the system is vibrating in a normal mode, and it is possible to make the substitutions

$$\{\delta(\tau)\} = \{\delta\}\sin\omega\tau \quad ; \quad \{\frac{\partial^2\delta(\tau)}{\partial\tau^2}\} = -\omega^2\{\delta\}\sin\omega\tau \tag{5.3}$$

into Eq. 5.2 to obtain

$$([K] - \omega^2[M])\{\delta\} = 0 \tag{5.4}$$

where ω is the natural frequencies of the modes and the common term $\sin\omega\tau$ has been cancelled out.

It is possible to transform Eq. 5.4 into

$$[K]^{-1}[M]\{\delta\} = \frac{1}{\omega^2}\{\delta\} \tag{5.5}$$

which becomes a standard eigenvalue problem with the form

$$[A]\{x\} = \lambda\{x\} \tag{5.6}$$

Various schemes have been developed for solving eigenvalue equations, and some of the standard techniques are described in texts such as the one by Bishop et al.[2] For more advanced techniques the mass condensation method[3], the subspace iteration method[4] and Lanczos method[5] should be used.

Note that a direct solution Eq. 5.5 is uneconomical because although both $[K]^{-1}$ and $[M]$ are symmetrical, the product $[K]^{-1}[M]$, however, is in general not symmetrical, and in practice some form of transformation similar to the process to be described in Chapter 8 of Reference 8 should be applied first.

For more detailed discussions on structural dynamics, readers should refer to other texts such as the one by Cheung and Leung.[5]

5.2 DERIVATION OF CONSISTENT MASS MATRIX OF CLASSICAL STRIPS

The displacement function for any strip has the general form of

$$\{f\} = [N]\{\delta\} = \sum_{m=1}^{r} [N]_m \{\delta\}_m \tag{5.7}$$

in which both $\{f\}$ and $\{\delta\}$ are time dependent.

If the mass is distributed throughout the strip, any acceleration will give rise to distributed inertia forces of the magnitude

$$\{q\} = -\rho t \left(\frac{d^2\{f\}}{d\tau^2}\right) \tag{5.8}$$

with ρ being the mass per unit volume and t the thickness of strip.

Substituting Eq. 5.7 into Eq. 5.8 and taking note of Eq. 5.3,

$$\{q\} = -\rho t [N] \left\{\frac{\partial^2\delta(\tau)}{\partial\tau^2}\right\} = \rho t\omega^2 [N]\{\delta\} \sin\omega\tau \tag{5.9}$$

From Eq. 5.9 it is seen that the term $\sin\omega\tau$ will be cancelled out and therefore duly dropped in all subsequent derivations. Eq. 5.9 is now simply

$$\{q\} = \rho t\omega^2 [N]\{\delta\} \tag{5.10}$$

Based on Eq. 5.10 it is possible at this stage to obtain equivalent nodal forces for the distributed inertia loadings. Thus

$$\begin{aligned}\{F^e\} &= \int [N]^T \{q\}\, d(\text{area}) = \int \rho t\omega^2 [N]^T [N]\{\delta\}\, d(\text{area}) \\ &= \int \rho t\omega^2 [[N]_1\ [N]_2 \ldots [N]_r]^T [[N]_1\ [N]_2 \ldots [N]_r]\{\delta\}\, d(\text{area}) \\ &= \int \rho t\omega^2 \begin{bmatrix} [N]_1^T[N]_1 & [N]_1^T[N]_2 & . & [N]_1^T[N]_r \\ [N]_2^T[N]_1 & [N]_2^T[N]_2 & . & [N]_2^T[N]_r \\ . & . & . & . \\ [N]_r^T[N]_1 & [N]_r^T[N]_2 & . & [N]_r^T[N]_r \end{bmatrix} \{\delta\}\, d(\text{area}) \\ &= \omega^2 [M^e]\{\delta\}\end{aligned} \tag{5.11}$$

Thus the basic unit submatrix in a consistent mass matrix is

$$[M]_{mn} = \int \rho t \; [N]_m^T \, [N]_n \, d(\text{area}) \tag{5.12}$$

5.2.1 CONSISTENT MASS MATRICES OF BENDING STRIPS

5.2.1.1 *Lower order classical rectangular strip (LO2)*

Using the rectangular bending strip discussed in Section 2.2.2 the matrix $[N]_m$ is defined as

$$\begin{aligned}[N]_m &= [(1-3\bar{x}^2+2\bar{x}^3) \;\; x(1-2\bar{x}+\bar{x}^2) \;\; (3\bar{x}^2-2\bar{x}^3) \;\; x(\bar{x}^2-\bar{x})]\, Y_m \\ &= [C]\, Y_m \end{aligned} \tag{5.13}$$

The product ρt in Eq. 5.13 is usually assumed to be constant in the transverse direction, and we have for the mass matrix

$$[M]_{mn} = \int_0^b [C]^T[C]\, dx \int_0^l \rho t Y_m Y_n \, dy \tag{5.14}$$

The integrals with respect to dx are quite simple and have been worked out by hand, and the submatrix $[M]_{mn}$ in a semi-complete form is given in Table 5.1.

TABLE 5.1
Consistent mass matrix of a L02 rectangular classical strip with any end conditions.

$$[M]_{mn} = \int_0^l \begin{bmatrix} \frac{13b}{35} & & & \\ \frac{11b^2}{210} & \frac{b^3}{105} & & \\ \frac{9b}{70} & \frac{13b^2}{420} & \frac{13b}{35} & \\ -\frac{13b^2}{420} & -\frac{3b^3}{420} & -\frac{11b^2}{210} & \frac{b^3}{105} \end{bmatrix} \rho t(y)\, Y_m Y_n dy$$

If the product $\rho t(y)$ is further assumed to be constant within the strip, then for all the six basic functions listed in Section 2.2.2 the integral $\int \rho t Y_m \, Y_n \, dy$ in Eq. 5.14 will be zero for $m \neq n$ because of their orthogonal properties. This will mean that all off-diagonal submatrices in the mass matrix will become zero, and as a result the consistent mass matrix can be greatly simplified and the integral in Table 5.1 is reduced to the standard form of

$$\int_0^l \rho t \, Y_m \, Y_n \, dy = \begin{cases} \frac{\rho t}{2} & \text{for } m = n \\ 0 & \text{for } m \neq n \end{cases} \tag{5.15}$$

5.2.1.2 *Higher order classical rectangular strip (HO2)*

For this strip a quintic polynomial is used in the transverse direction and the displacement function is described in Section 2.2.2. The corresponding matrix $[N]_m$ is given as

$$\begin{aligned}[N]_m = [&(1-10\bar{x}^3+15\bar{x}^4-6\bar{x}^5) \;\; x(1-6\bar{x}^2+8\bar{x}^3-3\bar{x}^4) \;\; x^2(0.5-1.5\bar{x}+1.5\bar{x}^2-1.5\bar{x}^3) \\ &(10\bar{x}^3-15\bar{x}^4+6\bar{x}^5) \;\; x(-4\bar{x}^2+7\bar{x}^3-3\bar{x}^4) \;\; x^2(0.5\bar{x}-\bar{x}^2+0.5\bar{x}^3)]Y_m\end{aligned}$$

$$= [C]Y_m \tag{5.16}$$

Once more it is possible to apply Eq. 5.14 for the computation of the mass matrix, which has also been worked out in a semi-complete form in Table 5.2.

TABLE 5.2
Consistent mass matrix of a higher order classical strip HO2.

$$[M]_{mn} = \left[\begin{array}{c|c|c|c|c|c} \frac{181}{462}b & \frac{311}{4620}b^2 & \frac{34931}{55440}b^3 & \frac{25}{231}b & -\frac{151}{4620}b^2 & \frac{181}{55440}b^3 \\ \hline \frac{311}{4620}b^2 & \frac{52}{3465}b^3 & \frac{23}{18480}b^4 & \frac{151}{4620}b^2 & -\frac{19}{1980}b^3 & \frac{13}{13860}b^4 \\ \hline \frac{34931}{55440}b^3 & \frac{23}{18480}b^4 & \frac{1}{9240}b^5 & \frac{181}{55440}b^3 & -\frac{13}{13860}b^4 & \frac{1}{11088}b^5 \\ \hline \frac{25}{231}b & \frac{151}{4620}b^2 & \frac{181}{55440}b^3 & \frac{181}{462}b & -\frac{311}{4620}b^2 & \frac{281}{55440}b^3 \\ \hline -\frac{151}{4620}b^2 & -\frac{19}{1980}b^3 & -\frac{13}{13860}b^4 & -\frac{311}{4620}b^2 & \frac{11}{693}b^3 & -\frac{23}{18480}b^4 \\ \hline \frac{181}{55440}b^3 & \frac{13}{13860}b^4 & \frac{1}{11088}b^5 & \frac{281}{55440}b^3 & -\frac{23}{18480}b^4 & -\frac{61}{5544}b^5 \end{array}\right] \int_o^l \rho t Y_m Y_n dy$$

5.2.1.3 *Mixed strip*[6]

Employing the same principle as outlined in Section 3.5.1, classical mixed strips can also be developed for vibration analyses.

A partially clamped rectangular plate is shown in Figure 5.1 and it can be seen that both simply supported and clamped conditions exist along the same boundary. Theoretically speaking both the finite element method and finite difference method can be used to deal with this type of mixed boundary problems, although in practice they are very often uneconomical in terms of computation cost.

The finite strip method provides a very convenient tool for the accurate analysis of rectangular or sector plates with an opposite pair of partially clamped sides. The solution procedure is started in the same way as in a standard finite strip analysis in which, first of all, a plate is divided into strips with preset end conditions which may differ from strip to strip (Figure 5.1). However, at the point where an abrupt change occurs it is not possible to satisfy the boundary condition exactly, and a so-called 'mixed' strip has to be introduced. This special strip is formulated by using two basic functions to satisfy the two different boundary conditions at the two nodal lines of the strip, and within these two sides the boundary conditions undergo a gradual change from one type to the other. In order to approximate the sudden change closely, such 'mixed' strips are usually made as narrow as possible.

Consider a typical 'mixed' strip shown in Figure 5.1. A suitable displacement function for the rectangular strip is

$$w = \sum_{m=1}^{r} [[C_1](Y_1)_m \quad [C_2](Y_2)_m] \{\delta\}_m \tag{5.17}$$

in which $(Y_i)_m$ represents the Y_m at the i-th edge and

$$\{\delta\}_m = [w_{1m} \ \theta_{1m} \ w_{2m} \ \theta_{2m}]^T \tag{5.18}$$

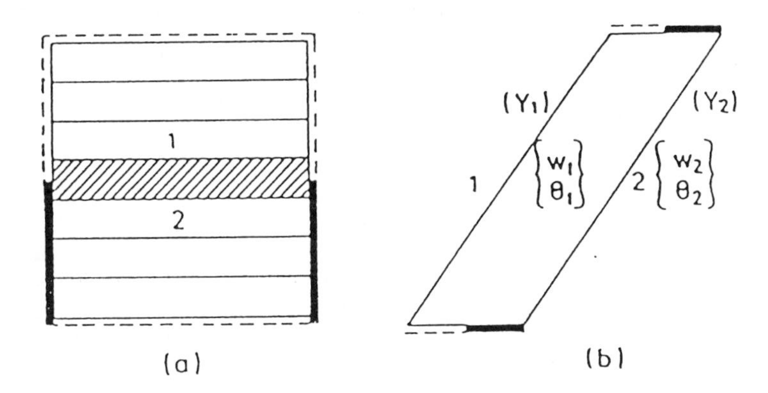

Figure 5.1 (a) A partially clamped plate and its strip idealization. (b) A mixed strip.

The shape functions $[C_1]$ and $[C_2]$ are given by Eq. 2.29. By definition (see Section 2.2.2) a shape function takes a value of unity at its own nodal line and a value of zero at the other nodal lines. Therefore at side 1 (x=0) we have

$$w = \sum_{m=1}^{r} [[1\ 0](Y_1)_m \quad [0\ 0](Y_2)_m]\,\{\delta\}_m \tag{5.19}$$

and at side 2 (x=b) we have

$$w = \sum_{m=1}^{r} [[0\ 0](Y_1)_m \quad [1\ 0](Y_2)_m]\,\{\delta\}_m \tag{5.20}$$

Eq. 5.19 simply states that at side 1, only $(Y_1)_m$ which satisfy the end conditions there is present. A similar statement applies to Eq. 5.20.

Within the range $0 < x < b$, both $[C_1]$ and $[C_2]$ will not be equal to zero, and the end conditions of either simply supported or clamped are never satisfied exactly.

With the displacement functions established, the formulation of the stiffness and mass matrices follows the standard procedure. For a rectangular strip, the two matrices are listed in Tables 5.3 and 5.4 respectively.

5.2.1.4 *Lower order classical curved strip (LO2)*

This lower order curved strip has been discussed in Section 3.5 and the matrix $[N]_m$ is reproduced below for convenience,

$$[N]_m = [(1-3\frac{R^2}{4}+\frac{R^3}{4}) \quad b(R-R^2+\frac{R^3}{4}) \quad (3\frac{R^2}{4}-\frac{R^3}{4}) \quad b(\frac{R^3}{4}-\frac{R^2}{2})]\,\Theta_m \tag{5.21}$$

Because the polar coordinate system is used here, it is necessary to change Eq. 5.14 to a slightly different form,

$$[M]_{mn} = \int_{r_2}^{r_1} [C]^T [C] r\, dr \int_0^{\alpha} \rho t \Theta_m \Theta_n \, d\theta \qquad (5.22)$$

TABLE 5.3
Stiffness matrix of a rectangular mixed classical strip.

$$[K]_{mn} = \left[\begin{array}{c:c:c:c}
\begin{array}{c}\frac{12D_x}{b^3}I_1 - \frac{6D_1}{5b}I_3\\ -\frac{6D_1}{5b}I_2 + \frac{13bD_y}{35}I_4\\ +\frac{24D_{xy}}{5b}I_5\end{array} &
\begin{array}{c}\frac{6D_x}{b^2}I_1 - \frac{11D_1}{10}I_3\\ -\frac{D_1}{10}I_2 + \frac{11b^2D_y}{210}I_4\\ +\frac{2D_{xy}}{5}I_5\end{array} &
\begin{array}{c}-\frac{12D_x}{b^3}I_6 + \frac{6D_1}{5b}I_{10}\\ +\frac{6D_1}{5b}I_7 + \frac{9bD_y}{35}I_8\\ -\frac{24D_{xy}}{5b}I_9\end{array} &
\begin{array}{c}\frac{6D_x}{b^2}I_6 - \frac{D_1}{10}I_{10}\\ -\frac{D_1}{10}I_7 - \frac{13b^2D_y}{420}I_8\\ +\frac{2D_{xy}}{5}I_9\end{array} \\ \hdashline
\begin{array}{c}\frac{6D_x}{b^2}I_1 - \frac{11D_1}{10}I_2\\ -\frac{D_1}{10}I_3 + \frac{11b^2D_y}{210}I_4\\ +\frac{2D_{xy}}{5}I_5\end{array} &
\begin{array}{c}\frac{4D_x}{b}I_1 - \frac{2bD_1}{15}I_3\\ -\frac{2bD_1}{15}I_2 + \frac{2b^3D_y}{210}I_4\\ +\frac{8bD_{xy}}{15}I_5\end{array} &
\begin{array}{c}-\frac{6D_x}{b^2}I_6 + \frac{D_1}{10}I_{10}\\ +\frac{D_1}{10}I_7 + \frac{26b^2D_y}{840}I_8\\ +\frac{2D_{xy}}{5}I_9\end{array} &
\begin{array}{c}\frac{2D_x}{b}I_6 + \frac{bD_1}{30}I_{10}\\ +\frac{bD_1}{30}I_7 - \frac{2b^3D_y}{280}I_8\\ -\frac{2bD_{xy}}{15}I_9\end{array} \\ \hdashline
\begin{array}{c}-\frac{12D_x}{b^3}I_6 + \frac{6D_1}{5b}I_7\\ +\frac{6D_1}{5b}I_{10} + \frac{9bD_y}{35}I_8\\ -\frac{24D_{xy}}{5b}I_9\end{array} &
\begin{array}{c}-\frac{6D_x}{b^2}I_6 + \frac{D_1}{10}I_7\\ +\frac{D_1}{10}I_{10} + \frac{26b^2D_y}{840}I_8\\ -\frac{2D_{xy}}{5}I_9\end{array} &
\begin{array}{c}\frac{12D_x}{b^3}I_{11} - \frac{6D_1}{5b}I_{13}\\ -\frac{6D_1}{5b}I_{12} + \frac{13bD_y}{35}I_{14}\\ +\frac{24D_{xy}}{5b}I_{15}\end{array} &
\begin{array}{c}-\frac{6D_x}{b^2}I_{11} + \frac{22D_1}{20}I_{13}\\ +\frac{2D_1}{20}I_{12} - \frac{11b^2D_y}{210}I_{14}\\ -\frac{2D_{xy}}{5}I_{15}\end{array} \\ \hdashline
\begin{array}{c}\frac{6D_x}{b^2}I_6 - \frac{D_1}{10}I_7\\ -\frac{D_1}{10}I_{10} - \frac{13b^2D_y}{420}I_8\\ +\frac{2D_{xy}}{5}I_9\end{array} &
\begin{array}{c}\frac{2D_x}{b}I_6 + \frac{bD_1}{30}I_7\\ +\frac{bD_1}{30}I_{10} - \frac{2b^3D_y}{280}I_8\\ -\frac{2bD_{xy}}{15}I_9\end{array} &
\begin{array}{c}-\frac{6D_x}{b^2}I_{11} + \frac{22D_1}{20}I_{12}\\ +\frac{2D_1}{20}I_{13} - \frac{11b^2D_y}{210}I_{14}\\ -\frac{2D_{xy}}{5}I_{15}\end{array} &
\begin{array}{c}\frac{4D_x}{b}I_{11} - \frac{2bD_1}{15}I_{13}\\ -\frac{2bD_1}{15}I_{12} + \frac{2b^3D_y}{210}I_{14}\\ +\frac{8bD_{xy}}{15}I_{15}\end{array}
\end{array}\right]$$

$I_1 = \int_0^l (Y_1)_m (Y_1)_n\, dy$; $I_2 = \int_0^l (Y_1)_m (Y_1'')_n\, dy$; $I_3 = \int_0^l (Y_1'')_m (Y_1)_n\, dy$; $I_4 = \int_0^l (Y_1'')_m (Y_1'')_n\, dy$;

$I_5 = \int_0^l (Y_1')_m (Y_1')_n\, dy$;

$I_6 = \int_0^l (Y_1)_m (Y_2)_n\, dy$; $I_7 = \int_0^l (Y_1)_m (Y_2'')_n\, dy$; $I_8 = \int_0^l (Y_1'')_m (Y_2'')_n\, dy$; $I_9 = \int_0^l (Y_1')_m (Y_2')_n\, dy$;

$I_{10} = \int_0^l (Y_1'')_m (Y_2)_n\, dy$;

$I_{11} = \int_0^l (Y_2)_m (Y_2)_n\, dy$; $I_{12} = \int_0^l (Y_2)_m (Y_2'')_n\, dy$; $I_{13} = \int_0^l (Y_2'')_m (Y_2)_n\, dy$; $I_{14} = \int_0^l (Y_2'')_m (Y_2'')_n\, dy$;

$I_{15} = \int_0^l (Y_2')_m (Y_2')_n\, dy$

TABLE 5.4
Consistent mass matrix of a rectangular mixed classical strip.

$$[M]_{mn} = \rho t \begin{bmatrix}
\frac{78b}{210}I_1 & \frac{11b^2}{210}I_1 & \frac{9b}{70}I_2 & -\frac{13b^2}{420}I_2 \\
\frac{11b^2}{210}I_1 & \frac{b^3}{105}I_1 & \frac{13b^2}{420}I_2 & -\frac{3b^3}{420}I_2 \\
\frac{9b}{70}I_4 & \frac{13b^2}{420}I_4 & \frac{13b}{35}I_3 & -\frac{11b^2}{420}I_3 \\
-\frac{13b^2}{420}I_4 & -\frac{3b^3}{420}I_4 & \frac{11b^2}{210}I_3 & \frac{b^3}{105}I_3
\end{bmatrix}$$

$I_1 = \int_0^l (Y_1)_m (Y_1)_n\, dy$; $I_2 = \int_0^l (Y_1)_m (Y_2)_n\, dy$; $I_3 = \int_0^l (Y_2)_m (Y_2)_n\, dy$; $I_4 = \int_0^l (Y_2)_m (Y_1)_n\, dy$

The algebraic expressions of the matrix coefficients for $[M]_{mn}$ are presented in Table 5.5. Again for the special simply supported case with constant ρt only the diagonal submatrix $[M]_{mm}$ is non-zero and it can be obtained from Table 5.5 by substituting $\rho t\alpha/2$ in place of the integral.

For a 'mixed' curved strip, a similar displacement function is written as

$$w = \sum_{m=1}^{r} \left[\; [(1-3\frac{R^2}{4} + \frac{R^3}{4}) \;\; b(R-R^2+\frac{R^3}{4})] \; (\Theta_1)_m \; [(3\frac{R^2}{4} - \frac{R^3}{4}) \;\; b(\frac{R^3}{4} - \frac{R^2}{2})] \; (\Theta_2)_m \; \right] \{\delta\}_m \tag{5.23}$$

in which $\{\delta\}_m = \{w_1 \;\; \psi_1 \;\; w_2 \;\; \psi_2\}_m$. The variables have the same meaning as was given in Section 3.5, and $(\Theta_1)_m$ is analogous to $(Y_1)_m$.

TABLE 5.5

Consistent mass matrix of a classical curved strip with any end conditions.

$$\begin{bmatrix} \frac{24}{70}b^2 + \frac{26}{35}br_1 & & & \\ \frac{2}{15}b^3 + \frac{8}{105}b^2r_1 & \frac{2}{35}b^4 + \frac{8}{105}b^3r_1 & \text{SYM} & \\ \frac{9}{35}b^2 + \frac{9}{35}br_1 & \frac{2}{15}b^3 + \frac{13}{105}b^2r_1 & \frac{24}{21}b^2 + \frac{26}{35}br_1 & \\ -\frac{4}{35}b^3 - \frac{13}{105}b^2r_1 & -\frac{2}{35}b^4 - \frac{2}{35}b^3r_1 & -\frac{2}{7}b^3 - \frac{22}{105}b^2r_1 & \frac{2}{21}b^4 + \frac{8}{105}b^3r_1 \end{bmatrix} \int_0^{\alpha} \rho t(\theta)\Theta_m\Theta_n d\theta$$

5.2.1.5 *Lower order classical skew strip (LO2)*

In Section 3.6 it was mentioned that the lower order skew plate strip was found to be unsatisfactory for static analysis of skew plates. This, however, has not been the case for vibration analysis, and the lower order skew strip has been successfully applied to the frequency analysis of skew orthotropic plates by Babu and Reddy.[7] Although only the simply supported case has been dealt with in the reference, there should not be any difficulty in extending the analysis to include the other types of support conditions.

For skew coordinates ξ and η, the displacement function for a skew plate strip similar to the one shown in Figure 3.1(f) is given as

$$f = w = \sum_{m=1}^{r} [(1-3\bar{\xi}^2+2\bar{\xi}^3) \;\; \xi(1-2\bar{\xi}+\bar{\xi}^2) \;\; (3\bar{\xi}^2-2\bar{\xi}^3) \;\; \xi(\bar{\xi}^2-\bar{\xi})]\sin k_m\eta \begin{Bmatrix} w_{1m} \\ \theta_{1m} \\ w_{2m} \\ \theta_{2m} \end{Bmatrix} \tag{5.24}$$

in which $\bar{\xi} = \xi/b$ and $\theta_{im} = (\partial w/\partial\xi)_{im}$.

Therefore

$$[N]_m = [(1-3\bar{\xi}^2+2\bar{\xi}^3) \;\; \xi(1-2\bar{\xi}+\bar{\xi}^2) \;\; (3\bar{\xi}^2-2\bar{\xi}^3) \;\; \xi(\bar{\xi}^2-\bar{\xi})]\sin k_m\eta \tag{5.25}$$

The stiffness matrix of this strip can be developed according to the presentation given in Section 3.7 and the mass matrix according to Eq. 5.14.

The terms of the series do not couple together in the case of a simply supported strip, and the stiffness and mass matrices for a typical m-th term are listed in Tables 5.6 and 5.7 respectively.

TABLE 5.6
Elements of the stiffness matrix $[K]_{mm}$ for the classical skew strip.[7,8]

$$
\left[\begin{array}{c|c|c|c}
\begin{array}{c}\frac{6Gl}{b^3}+\frac{6}{5}Hk_m^2\frac{l}{b}\\ +\frac{13}{70}Pk_m^4lb\\ +\frac{12}{5}Fk_m^2\frac{l}{b}\end{array} &
\begin{array}{c}\frac{3Gl}{b^2}+\frac{3}{5}Hk_m^2l\\ +\frac{11}{420}Pk_m^4lb^2\\ +\frac{1}{5}Fk_m^2l\end{array} &
\begin{array}{c}-\frac{6Gl}{b^3}-\frac{6}{5}Hk_m^2\frac{l}{b}\\ +\frac{9}{140}Pk_m^4lb\\ -\frac{12}{5}Fk_m^2\frac{l}{b}\end{array} &
\begin{array}{c}\frac{3Gl}{b^2}+\frac{1}{10}Hk_m^2l\\ -\frac{13}{840}Pk_m^4lb^2\\ +\frac{1}{5}Fk_m^2l\end{array} \\ \hline
\begin{array}{c}\frac{3Gl}{b^2}+\frac{3}{5}Hk_m^2l\\ +\frac{11}{420}Pk_m^4lb^2\\ +\frac{1}{5}Fk_m^2l\end{array} &
\begin{array}{c}\frac{2Gl}{b}+\frac{2}{15}Hk_m^2lb\\ +\frac{1}{210}Pk_m^4lb^3\\ +\frac{4}{15}Fk_m^2lb\end{array} &
\begin{array}{c}-\frac{3Gl}{b^2}-\frac{1}{10}Hk_m^2l\\ +\frac{13}{840}Pk_m^4lb^2\\ -\frac{1}{5}Fk_m^2l\end{array} &
\begin{array}{c}\frac{2Gl}{b}-\frac{1}{30}Hk_m^2lb\\ -\frac{1}{280}Pk_m^4lb^3\\ -\frac{1}{15}Fk_m^2lb\end{array} \\ \hline
\begin{array}{c}-\frac{6Gl}{b^3}-\frac{6}{5}Hk_m^2\frac{l}{b}\\ +\frac{9}{140}Pk_m^4lb\\ -\frac{12}{5}Fk_m^2\frac{l}{b}\end{array} &
\begin{array}{c}-\frac{3Gl}{b^2}-\frac{1}{10}Hk_m^2l\\ +\frac{13}{840}Pk_m^4lb^2\\ -\frac{1}{5}Fk_m^2l\end{array} &
\begin{array}{c}\frac{6Gl}{b^3}+\frac{6}{5}Hk_m^2\frac{l}{b}\\ +\frac{13}{70}Pk_m^4lb\\ +\frac{12}{5}Fk_m^2\frac{l}{b}\end{array} &
\begin{array}{c}-\frac{3Gl}{b^2}-\frac{3}{5}Hk_m^2l\\ -\frac{11}{420}Pk_m^4lb^2\\ -\frac{1}{5}Fk_m^2l\end{array} \\ \hline
\begin{array}{c}\frac{3Gl}{b^2}+\frac{1}{10}Hk_m^2l\\ -\frac{13}{840}Pk_m^4lb^2\\ +\frac{1}{5}Fk_m^2l\end{array} &
\begin{array}{c}\frac{2Gl}{b}-\frac{1}{30}Hk_m^2lb\\ -\frac{1}{280}Pk_m^4lb^3\\ -\frac{1}{15}Fk_m^2lb\end{array} &
\begin{array}{c}-\frac{3Gl}{b^2}-\frac{3}{5}Hk_m^2l\\ -\frac{11}{420}Pk_m^4lb^2\\ -\frac{1}{5}Fk_m^2l\end{array} &
\begin{array}{c}\frac{2Gl}{b}+\frac{2}{15}Hk_m^2lb\\ +\frac{1}{210}Pk_m^4lb^3\\ +\frac{4}{15}Fk_m^2lb\end{array}
\end{array}\right]
$$

where $G = [D_x + \frac{2D_1}{\tan^2\beta} + \frac{D_y}{\tan^4\beta} + \frac{4D_{xy}}{\tan^2\beta}]\sin\beta$; $P = \left[\frac{D_y}{\sin^4\beta}\right]\sin\beta$

$$H = [\frac{D_1}{\sin^2\beta} + \frac{D_y}{\sin^2\beta\tan^2\beta}]\sin\beta \ ; \ E = \left[\frac{D_y}{\sin^3\beta\tan\beta}\right]$$

$$C = [\frac{D_1}{\sin\beta\tan\beta} + \frac{D_y}{\sin\beta\tan^3\beta} + \frac{2D_{xy}}{\sin\beta\tan\beta}]\sin\beta \ ;$$

$$F = \left[\frac{D_{xy}}{\sin^2\beta}\right]\sin\beta + \left[\frac{D_y\cos^2\beta}{\sin^4\beta}\right]\sin\beta$$

$$k_m = \frac{m\pi}{l}$$

TABLE 5.7
Mass matrix $[M]_{mm}$ for skew strip.[7,8]

$$[M]_{mm} = t\rho l b \sin\beta \begin{bmatrix} \frac{13}{70} & \frac{11b}{420} & \frac{9}{140} & -\frac{13b}{840} \\ \frac{11b}{420} & \frac{b^2}{210} & \frac{13b}{840} & -\frac{b^2}{210} \\ \frac{9}{140} & \frac{13b}{840} & \frac{13}{70} & -\frac{11b}{420} \\ -\frac{13b}{840} & -\frac{b^2}{210} & -\frac{11b}{420} & \frac{b^2}{210} \end{bmatrix}$$

5.2.2 CONSISTENT MASS MATRICES FOR CLASSICAL PLANE STRIPS

A lower order rectangular plate strip was discussed in Section 3.3.1 and the reader might recall that the stiffness formulation was based on the assumption of Y_m for u displacement and $(l/\mu_m)Y_m'$ for v displacement. The same strip will be used here for vibration analysis.

In the previous discussions on flexural vibration problems, the submatrix $[M]_{mn}$ is always equal to zero for m≠n provided of course that ρt is constant within a strip. However, this is no longer true for plane stress strips because of the existence of Y_m' in the displacement function and hence the non-orthogonal integral $\int \rho t Y'Y' dy$ in the mass matrix. As a result, $[M]_{mn}$ will normally remain unequal to zero even if ρt is constant.

From Eq. 5.12,

$$[N]_m = \begin{bmatrix} (1-\bar{x})Y_m & 0 & \bar{x}Y_m & 0 \\ 0 & (1-\bar{x})\frac{l}{\mu_m}Y_m' & 0 & \bar{x}\frac{l}{\mu_m}Y_m' \end{bmatrix} \quad (5.26)$$

The mass matrix is fairly easy to work out because of the large number of zeros present. The explicit form is given in Table 5.8.

The mass matrix for the curved plane stress strip (Table 5.9) and the mixed strip can be derived along the same line and will not be elaborated here.

TABLE 5.8
In-plane mass matrix of a L02 rectangular classical strip.

$$[M]_{mn} = \rho t \begin{bmatrix} \frac{bI_1}{3} & 0 & \frac{bI_1}{6} & 0 \\ 0 & \frac{b}{3C_1C_2}I_2 & 0 & \frac{b}{6C_1C_2}I_2 \\ \frac{bI_1}{6} & 0 & \frac{bI_1}{3} & 0 \\ 0 & \frac{b}{6C_1C_2}I_2 & 0 & \frac{b}{3C_1C_2}I_2 \end{bmatrix}$$

$$I_1 = \int_0^l Y_m Y_n dy \;;\; I_2 = \int_0^l Y_m' Y_n' dy \;;\; C_1 = \frac{\mu_m}{l} \;;\; C_2 = \frac{\mu_n}{l}$$

TABLE 5.9
In-plane mass matrix of a LO2 classical curved slab strip.

$$[M]_{mn} = \rho t \begin{bmatrix} b(b+2r_1)\frac{I_1}{3} & 0 & b(b+r_1)\frac{I_1}{6} & 0 \\ 0 & \frac{b(b+2r_1)}{3C_1C_2}I_2 & 0 & \frac{b(b+r_1)}{6C_1C_2}I_2 \\ b(b+r_1)\frac{I_1}{6} & 0 & b\frac{I_1}{3} & 0 \\ 0 & \frac{b(b+r_1)}{6C_1C_2}I_2 & 0 & \frac{b(b+2r_1)}{3C_1C_2}I_2 \end{bmatrix}$$

$I_1 = \int_0^\alpha \Theta_m \Theta_n dy$; $I_2 = \int_0^\alpha \Theta'_m \Theta'_n dy$; $C_1 = \frac{\mu_m}{\alpha}$; $C_2 = \frac{\mu_n}{\alpha}$; r_1 = inner radius

5.3 DERIVATION OF CONSISTENT MASS MATRIX OF SPLINE STRIP[9]

Following the principle outlined in the previous section, one can show readily that the mass matrix for a rectangular bending strip is

$$[M^e] = \int [\Phi]^T [C]^T \rho [C] [\Phi] \, d(vol) \tag{5.27}$$

In the case of a lower order strip, the explicit form of $[M^e]$ is

$$[M^e] = \frac{b\rho}{420} \begin{bmatrix} 156I_{11} & 22bI_{12} & 54I_{13} & -13bI_{14} \\ & 4b^2I_{22} & 13bI_{23} & -3b^2I_{24} \\ & & 156I_{33} & -22bI_{34} \\ & & & 4b^2I_{44} \end{bmatrix} \tag{5.28}$$

where $I_{ij} = \int_0^l t[\phi]_i^T [\phi]_j \, dy$

and $[\phi]_i$, $[\phi]_j$ are B-3 spline representation for nodal line i and j as defined in Section 2.3.

In the case of a concentrated mass m attached to nodal line i at $y=y_c$, an additional mass matrix is obtained , that is

$$[M^c] = [\Phi(y_c)]^T \begin{bmatrix} m_c & 0 & 0 & 0 \\ 0 & 0 & 0 & 0 \\ 0 & 0 & 0 & 0 \\ 0 & 0 & 0 & 0 \end{bmatrix} [\Phi(y_c)] \tag{5.29}$$

The mass matrix of a strip is thus given by

$$[M] = [M^c] + [M^e] \tag{5.30}$$

Similarly, the mass matrix of a plane strip can be shown to be

$$[M^e] = \rho b \begin{bmatrix} \frac{bI_{11}}{3} & 0 & \frac{bI_{13}}{6} & 0 \\ & \frac{bI_{22}}{3} & 0 & \frac{bI_{24}}{6} \\ & & \frac{bI_{33}}{3} & 0 \\ & & & \frac{bI_{44}}{3} \end{bmatrix} \tag{5.31}$$

where $I_{ij} = \int_0^l t[\phi]_i^T [\phi]_j \, dy$

and $[\phi]_i$, $[\phi]_j$ are B-3 spline representation for nodal line i and j as defined in Section 2.3.

Readers can easily derive the mass matrices of the skew plates and curved plates along the same line.

5.4 EXAMPLES

(i) The natural frequencies of several square isotropic plates with different boundary conditions were obtained by using:
(a) eight LO2 rectangular bending strips for the whole plate and four terms of the series
(b) two HO2 strips and two terms,[10] and
(c) spline strips.[9]

The natural frequencies f_i ($= \frac{\omega_i}{a^2}\sqrt{\frac{D}{\rho t}}$, where a is the width of the plate) of the three analyses were compared in Table 5.10 with those due to Warburton[11] and very little difference can be found between the three sets of results. Note that four terms of the series were used in (a) because eight frequencies of high accuracy were originally required and only four frequencies were quoted here.

(ii) Examples of vibration of partially clamped square plates are given to demonstrate the accuracy achievable by the mixed classical strip. The length of the central clamped portion is equal to a/4, a/3, and a/2 respectively. The three lowest frequencies are given in Table 5.11 and they are compared with other solutions.[12,13]

The natural frequencies of spline strips[9] are also given in the table. In the analysis, relatively narrow strips with different end conditions on two nodal lines were used when there were sudden changes in the boundary conditions.

It can be noted that the agreement among the results is excellent. Furthermore, the frequencies also become higher with increased clamped length, as is expected.

TABLE 5.10
Natural frequencies of slabs with various boundary conditions.

		ω_1	ω_2	ω_3	ω_4	ω_5
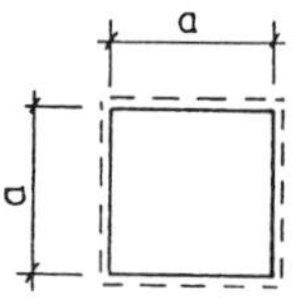 a, a	H02 L02 Spline strip Reference 11	19.74 19.74 19.74 19.74	49.35 49.32 49.36 49.35	49.36 49.34 49.35 49.35	78.97 78.91 78.98 78.95	98.64 98.64 98.80 98.69
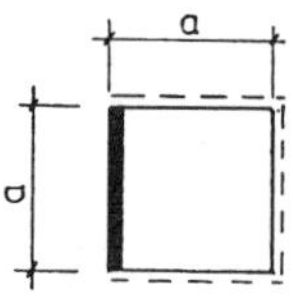 a, a	H02 L02 Spline strip Reference 11	23.66 23.62 23.65 23.77	51.79 51.62 51.71 52.00	58.66 58.65 58.67 58.65	86.41 86.16 86.17 86.26	100.35 100.35 100.80 100.84
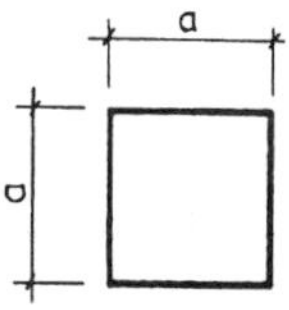 a, a	H02 L02 Spline strip Reference 11	36.05 36.01 36.00 36.00	73.44 73.48 73.46 73.41	73.75 73.96 73.55 73.41	108.53 108.91 108.40 108.24	133.53 132.09 132.20 131.90
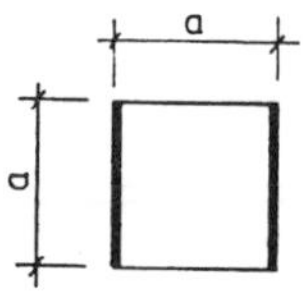 a, a	H02 L02 Spline strip Reference 11	22.33 22.29 22.20 22.38	27.19 27.08 26.45 27.33	45.19 44.76 43.65 54.49	61.55 61.53 61.39 61.68	68.33 68.29 67.40 68.73

TABLE 5.11
Natural frequencies of partially clamped plates.

Frequencies	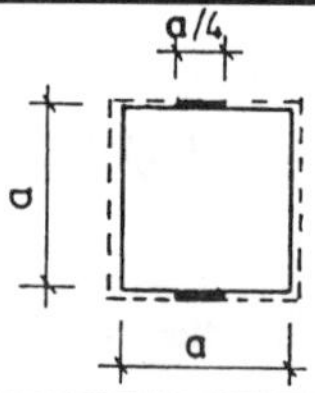 a/4, a, a	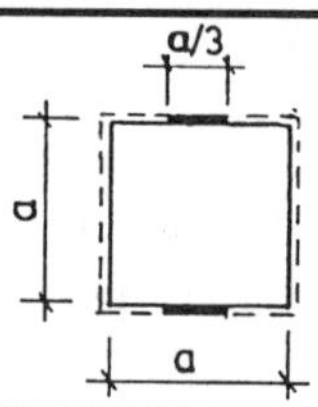 a/3, a, a	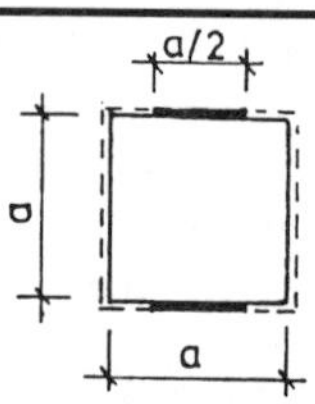 a/2, a, a
ω_1	26.46 26.27[9]	27.63 27.31[12] 27.83[9]	28.94 28.3[13] 28.65[9]
ω_2	50.83 51.63[9]	52.39 52.41[9]	54.26 54.00[9]
ω_3	62.11 64.72[9]	66.24 66.25[9]	68.07 68.58[9]

(iii) The LO2 classical rectangular bending strip was applied to a variable thickness plate problem, and the accuracy attainable for this type of structure was tested against the solution of Appl and Byers[14] who only calculated the fundamental frequency of simply supported plates with linearly varying thickness in one direction. The finite strip method, of course, can give as many frequencies as required for a variety of boundary conditions. A square plate with a thickness variation of 1.0-1.8 (Figure 5.2) was analyzed by using two

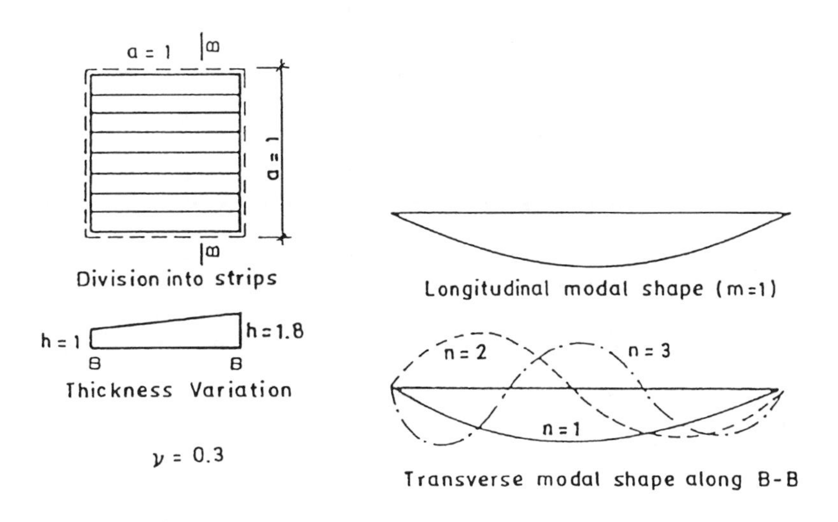

Figure 5.2 Simply supported variable thickness plates with some modal shape.

different types of strips. The first type was a standard constant thickness strip and the plate was approximated by a series of strips with different thicknesses, while the second type was a strip with linear variable thickness in the longitudinal direction and constant thickness in the transverse direction, so that the plate was made up of identical strips. It can be seen from Table 5.12 that both sets of results compared favourably with each other. The fundamental frequencies from both analyses also agree quite well with the result given in Reference 14.

Note that in the second case, due to the fact that the bending rigidities and ρt of the strip are now functions of y, the stiffness matrix $[K]_{mn}$ and the mass matrix $[M]_{mn}$ for $m \neq n$ will no longer be equal to zero although the strip is simply supported at the two ends.

TABLE 5.12
Frequencies of variable thickness plate.

Longitudinal modal shape	Transverse modal shape	Variable thickness strip	Constant thickness strip	Reference 14
(m = 1)	(n = 1)	27.45	27.35	27.35
(m = 1)	(n = 2)	68.20	67.99	
(m = 1)	(n = 3)	136.39	135.61	

(iv) The LO2 classical curved bending strip was used to analyze clamped sector plates and the results are compared with those computed by Ben-Amoz[15] (Figure 5.3). The agreement is very good on the whole, and the

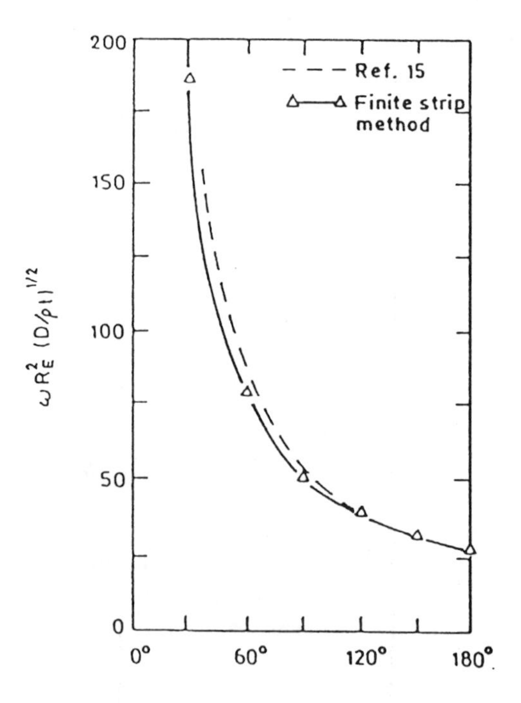

Figure 5.3 Variation of lowest frequency with sectorial angle.

frequencies, which in this case is defined as $\frac{\omega_i}{R_E{}^2}\sqrt{\frac{D}{\rho t}}$ (where R_E is the external radius) obtained by the two methods, are in fact identical for the range of subtended angles from 120° to 180°. Eight strips and four terms of the series are used in this analysis. It was mentioned in Chapter 3 that no numerical instability was ever experienced in the bending analysis of a clamped sector plate although the input data for the inner radius of the first strip is equal to zero. The same statement can be made for the vibration analysis of sector plates.

(v) Free vibrations of a variety of point-supported square plates of width 'a' were studied by Fan[9] using spline strips. Results are summarized in Table 5.13 and compared with the solutions obtained by Reed;[16] Cox;[17] Johns and Nataroja;[18] Rao[19] as well as Petyt and Mirza.[20]

Good agreement is observed in all cases except for the case of the edge-centre supported plate in which Johns' value is considerably lower than those obtained by Cox and the spline strips.

TABLE 5.13
Natural frequencies of rectangular plates with point supports.[8]

Support conditions	Reference	ω_1
	Reed[16] Petyt and Mirza[20] Spline strip:[9] (8 strips x 6 sections)	7.12 7.14 7.11
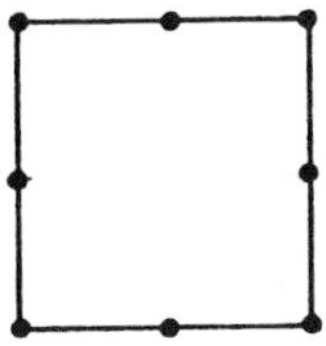	Rao[19] Spline strip:[9] (4 strips x 6 sections) (8 strips x 6 sections)	18.20 17.867 17.852
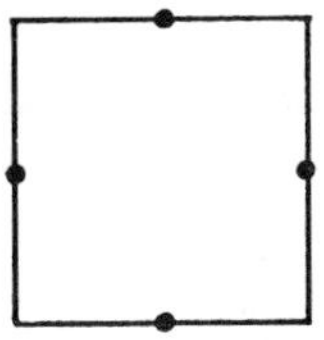	Cox[17] Johns and Nataroja[18] Spline strip:[9] (4 strips x 6 sections) (8 strips x 6 sections)	18.7 13.5 17.865 17.850

(vi) Continuous plates (Figure 5.4) with three panels in each direction were analyzed by the classical[21] and computed shape function strips.[22] The dimensionless frequencies of 2x2 and 3x3 simply supported plates were tabulated in Table 5.14 and 5.15 respectively. In the case of 2x2 plates, the results are also compared to those by Leissa,[23] Kim[24] and Liew.[25] It is obvious that the results are in very good agreement.

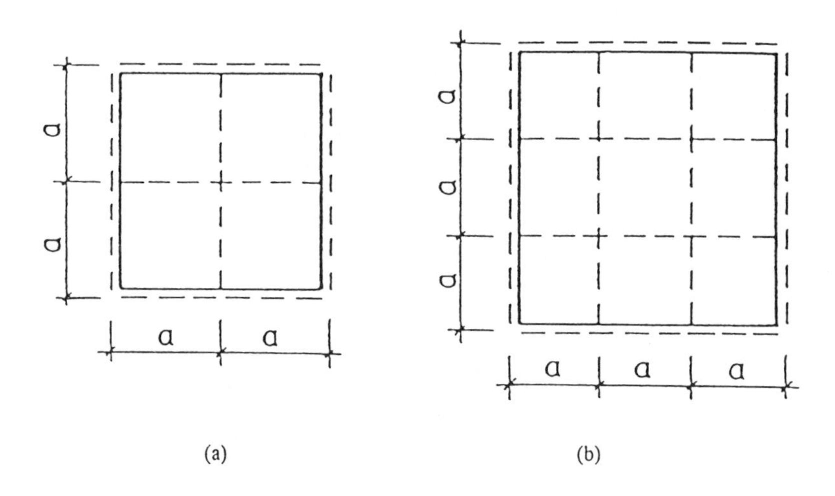

Figure 5.4 Continuous plates: (a) 2 x 2 simply supported plates. (b) 3 x 3 simply supported plates.

TABLE 5.14
Natural frequencies of 2x2 continuous plates.

Mode No	Dimensionless frequencies				
	Classical strip[21]	COMSFUN strip[22]	Leissa[23]	Kim[24]	Liew[25]
1	19.74	19.74	19.74	19.74	19.74
2	23.67	23.65	23.63	23.98	23.71
3	23.68	23.66	23.63	23.98	23.71
4	27.11	27.07	27.06	27.70	27.10

TABLE 5.15
Natural frequencies of 3x3 continuous plates.

Mode No	Dimensionless frequencies	
	Classical strip[21]	COMSFUN strip[22]
1	19.74	19.74
2	21.64	21.61
3	21.62	21.61
4	23.39	23.34

5.5 LINEAR INSTABILITY AND GEOMETRIC (STRESS) STIFFNESS MATRIX

In linear stability analysis, the critical loads are determined by solving the eigenvalue problem given by:

$$[K - \lambda K_g] = 0 \tag{5.32}$$

K_g is the geometric stiffness matrix which represents the energy loss of the membrane stresses during the buckling process. λ is the load factor. The buckling load is given in terms of the load factor λ_{cr} which is defined as

$$\lambda_{cr} = \sigma_{cr} \frac{a^2 t}{\pi^2 D}$$

where σ_{cr} is the critical reference stress, a is the reference dimension (length) and t is the thickness.

If classical finite strip is used in the analysis, it can be shown readily that the geometric stiffness matrix of a rectangular thin plate simply supported along two opposite edges is:

$$[K_g^e] = \frac{t}{2} \int\int [G]^T [\sigma_o] [G] \, d(\text{area}) \tag{5.33}$$

where $[\sigma_o] = \begin{bmatrix} \sigma_x^o & \tau_{xy}^o \\ \tau_{xy}^o & \sigma_y^o \end{bmatrix}$ (σ_x^o; σ_y^o and τ_{xy}^o are the applied initial stresses.)

$$[G] = \begin{bmatrix} (-\frac{6x}{b^2}+\frac{6x^2}{b^3})s & (1-\frac{4x}{b}+\frac{3x^2}{b^2})s & (\frac{6x}{b^2}-\frac{6x^2}{b^3})s & (\frac{3x^2}{b^2}-\frac{2x}{b})s \\ (1-\frac{3x^2}{b^2}+\frac{3x^3}{b^3})c & (x-\frac{2x^2}{b}+\frac{x^3}{b^2})c & (\frac{3x^2}{b^2}-\frac{2x^3}{b^3})c & (\frac{x^3}{b^2}-\frac{x^2}{b})c \end{bmatrix}$$

$(s = \sin\frac{m\pi}{l}\ ;\ c = \cos\frac{m\pi}{l})$

Readers may refer to Chapter 11 for the detailed derivations of the geometric stiffness matrix.

If the shear stress component is zero, the geometric stiffness matrix is uncoupled and its explicit form is listed in Table 5.16. The explicit forms of geometric matrices of other strips can be derived in the usual manner.

TABLE 5.16

Geometric stiffness matrix for classical rectangular plate under axial loads (l = length of the strip).

$$[K_g^e] = \sigma_x^o [K_g^x] + (\frac{m\pi}{l})^2 \sigma_y^o [K_g^y]$$

$$[K_g^x] = tl \begin{bmatrix} \frac{3}{5b} & & & \\ \frac{1}{20} & \frac{b}{15} & & \\ -\frac{3}{5b} & -\frac{1}{20} & \frac{3}{5b} & \\ \frac{1}{20} & -\frac{b}{60} & -\frac{1}{20} & \frac{b}{15} \end{bmatrix} \qquad [K_g^y] = \begin{bmatrix} \frac{13b}{70} & & & \\ \frac{11b^2}{420} & \frac{b^3}{210} & & \\ \frac{9b}{140} & \frac{13b^2}{840} & \frac{13b}{70} & \\ -\frac{13b^2}{840} & -\frac{b^3}{280} & -\frac{11b^2}{420} & \frac{b^3}{210} \end{bmatrix}$$

Babu et al.[26] carried out the analysis for rectangular plates of different aspect ratios and boundary conditions by dividing the plates into fifteen strips. The results are tabulated in Table 5.17 and they indicate that very accurate results can be achieved in all cases.

Szeto[28] as well as Tham and Szeto[29] carried out the formulation of the linear stability analysis using the spline strips. The explicit form of the geometric stiffness matrix is given in Table 5.18.

TABLE 5.17
The dimensionless buckling load factor for rectangular plates.[26]
$D_x = D_y = 1.0$; $\upsilon_x = \upsilon_y = 0.3$; $D_1 = 0.3$; $D_{xy} = 0.3$; $t = 0.1$

Cases	b/a	Finite Strip[26]	Bulson[27]
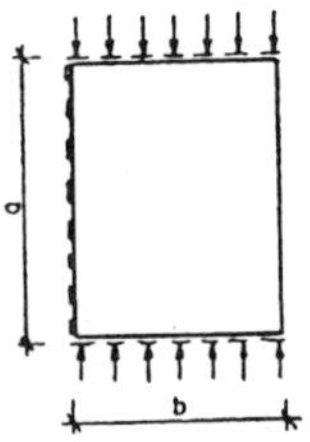	1.0 1.5 2.0	1.4018 0.8583 0.6653	1.37 0.83 0.65
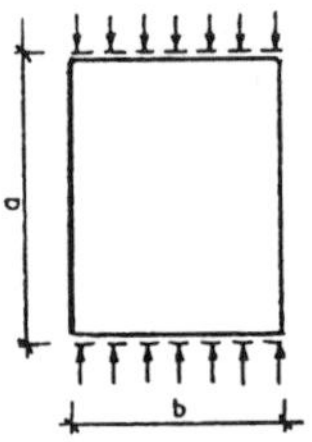	1.0 1.5 2.0	1.6528 1.2905 1.8436	1.62 1.29 1.82
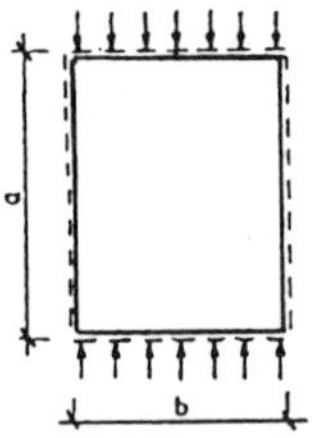	1.0 1.5 2.0	2.0030 1.4448 1.2502	2.00 1.44 1.23

TABLE 5.18
Explicit form of the geometric stiffness matrix of spline finite strip.[28]

$$
\begin{bmatrix}
\begin{array}{l} S_{11}\sigma_x^o I_{11}^1 \\ +S_{12}\gamma_{xy}^o I_{11}^2 \\ +S_{21}\gamma_{xy}^o I_{11}^3 \\ +S_{22}\sigma_y^o I_{11}^4 \end{array} &
\begin{array}{l} S_{11}\sigma_x^o I_{12}^1 \\ +S_{12}\gamma_{xy}^o I_{12}^2 \\ +S_{21}\gamma_{xy}^o I_{12}^3 \\ +S_{22}\sigma_y^o I_{12}^4 \end{array} &
\begin{array}{l} S_{11}\sigma_x^o I_{13}^1 \\ +S_{12}\gamma_{xy}^o I_{13}^2 \\ +S_{21}\gamma_{xy}^o I_{13}^3 \\ +S_{22}\sigma_y^o I_{13}^4 \end{array} &
\begin{array}{l} S_{11}\sigma_x^o I_{14}^1 \\ +S_{12}\gamma_{xy}^o I_{14}^2 \\ +S_{21}\gamma_{xy}^o I_{14}^3 \\ +S_{22}\sigma_y^o I_{14}^4 \end{array} \\
 &
\begin{array}{l} S_{11}\sigma_x^o I_{22}^1 \\ +S_{12}\gamma_{xy}^o I_{22}^2 \\ +S_{21}\gamma_{xy}^o I_{22}^3 \\ +S_{22}\sigma_y^o I_{22}^4 \end{array} &
\begin{array}{l} S_{11}\sigma_x^o I_{23}^1 \\ +S_{12}\gamma_{xy}^o I_{23}^2 \\ +S_{21}\gamma_{xy}^o I_{23}^3 \\ +S_{22}\sigma_y^o I_{23}^4 \end{array} &
\begin{array}{l} S_{11}\sigma_x^o I_{24}^1 \\ +S_{12}\gamma_{xy}^o I_{24}^2 \\ +S_{21}\gamma_{xy}^o I_{24}^3 \\ +S_{22}\sigma_y^o I_{24}^4 \end{array} \\
 & \text{SYM} &
\begin{array}{l} S_{11}\sigma_x^o I_{33}^1 \\ +S_{12}\gamma_{xy}^o I_{33}^2 \\ +S_{21}\gamma_{xy}^o I_{33}^3 \\ +S_{22}\sigma_y^o I_{33}^4 \end{array} &
\begin{array}{l} S_{11}\sigma_x^o I_{34}^1 \\ +S_{12}\gamma_{xy}^o I_{34}^2 \\ +S_{21}\gamma_{xy}^o I_{34}^3 \\ +S_{22}\sigma_y^o I_{34}^4 \end{array} \\
 & & &
\begin{array}{l} S_{11}\sigma_x^o I_{44}^1 \\ +S_{12}\gamma_{xy}^o I_{44}^2 \\ +S_{21}\gamma_{xy}^o I_{44}^3 \\ +S_{22}\sigma_y^o I_{44}^4 \end{array}
\end{bmatrix}
$$

$S_{11} = t\int[N']^T[N']d\xi$; $S_{12} = t\int[N]^T[N']d\xi$; $S_{21} = t[N']^T[N]d\xi$; $S_{22} = t\int[N]^T[N]d\xi$

$I_{ij}^1 = \int[\phi]_i^T[\phi]_j d\eta$; $I_{ij}^2 = \int[\phi]_i'^T[\phi]_j d\eta$; $I_{ij}^3 = \int[\phi]_i^T[\phi]_j' d\eta$; $I_{ij}^4 = \int[\phi]_i'^T[\phi]_j' d\eta$

The buckling load factors for a square plate of thickness 't' under biaxial loads were computed by the spline strip model.[28] The plate (Figure 5.5) was under compression force in one direction (x-direction) and tension in the other direction (y-direction). Figure 5.5 depicts the variation of the critical load factor with the stress ratio $\frac{\sigma_y^o}{\sigma_x^o}$. The load factor is defined in terms of the compression load. Comparison with results of Reference 30 shows that good agreement was achieved

Buckling of plate under combined compression and shear loads was also studied.[28] The interaction curve is shown in Figure 5.6 and is compared with that of Reference 30. Note that the stresses have been normalized against the respective critical buckling stresses.

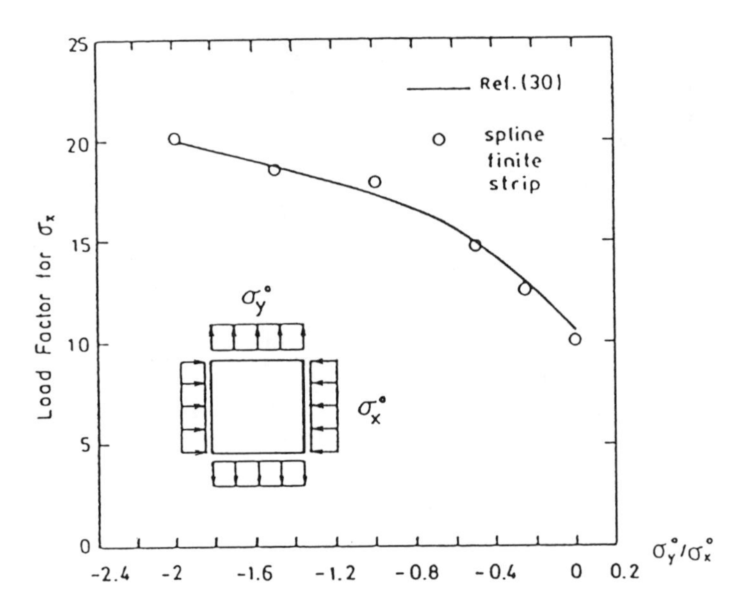

Figure 5.5 Clamped square plate under biaxial loading (tensile load in y-direction).[28]

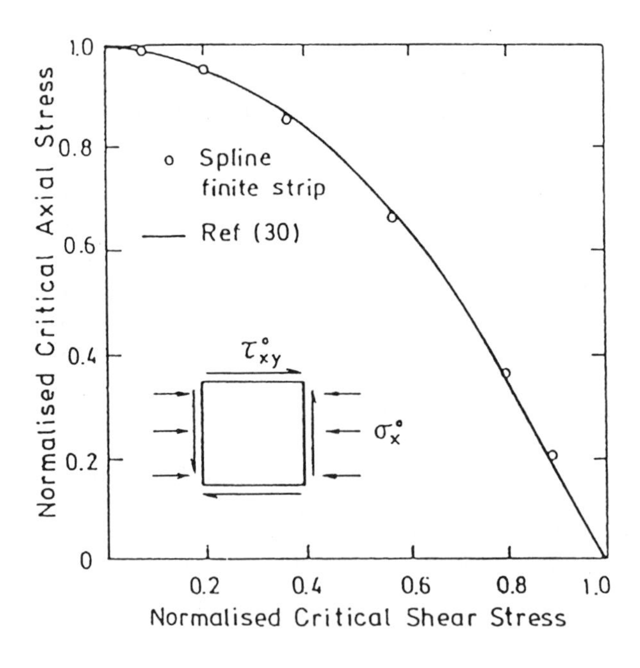

Figure 5.6 Interaction curve for buckling of square plate under combined shear and uniaxial load.[28]

To demonstrate the versatility of the spline strip, plates of variable thickness and with various end conditions were analyzed. The results are tabulated in Table 5.19 and they are compared with those published in Reference 30. Good agreement with other published results is again noted.

TABLE 5.19
Load factor for plates of variable thickness.[28]
(support conditions are specified from the top edge in an anti-clockwise direction)

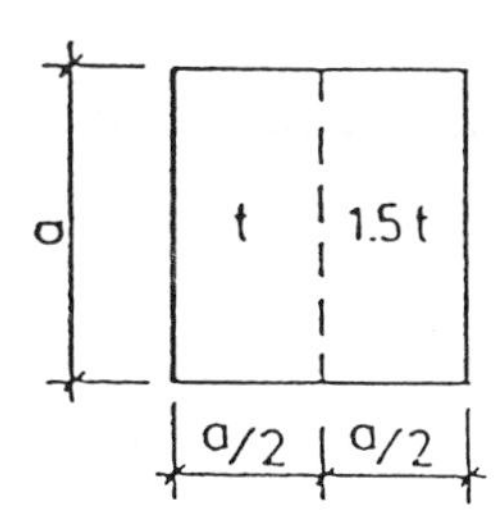

Support conditions	Load directions	Spline finite strip[28]	Reference 30
S-S-S-S	y x	5.988 1.512	5.988 1.511
S-C-C-S	y x	8.512 2.257	8.512 2.255
C-C-C-C	y x	12.683 2.584	12.681 2.582
S-S-S-S support at centre	x	2.275	2.272

Using lower order spline strips, the buckling analysis of parallelogram plates of skew angles (θ) varying from 45° to 75° was also carried out. The buckling coefficients for various boundary conditions under biaxial loads are given in Table 5.20. The relation between the buckling loads and skew angles for uniaxial loading utilizing higher order strips with interpolation functions given by Eq. 2.32 is shown graphically in Figure 5.7. The results are in good agreement with those obtained by Yoshimura.[31]

TABLE 5.20
Buckling load factors for parallelogram plates.

Support conditions	Method	Skew Angle		
		45°	60°	75°
Uniaxial load (simply supported)	spline strip[28]	10.36	5.93	4.40
	Reference 31	9.08	5.62	4.38
Biaxial load (clamped)	spline strip[28]	9.96	6.87	5.64
	Reference 31	9.88	6.85	5.64

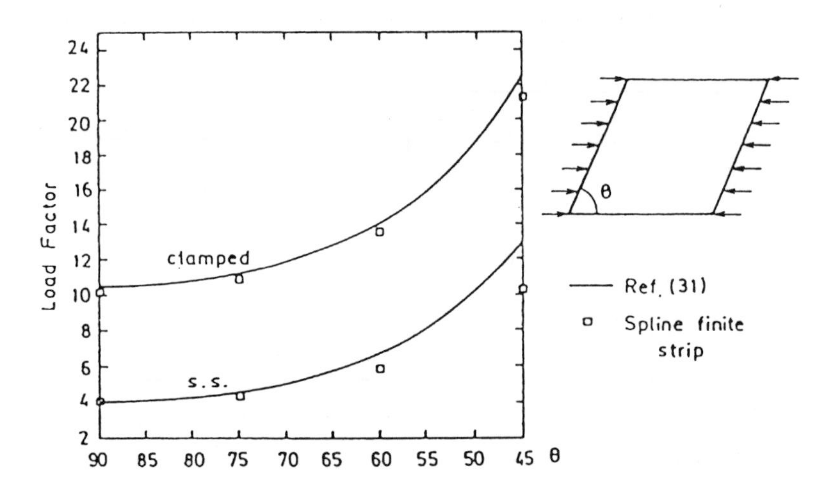

Figure 5.7 Buckling load for parallelogram plate under uniaxial load.[28]

5.6 VIBRATION AND STABILITY ANALYSIS OF MODERATELY THICK PLATES

Vibration and stability of moderately thick plates were studied by Hinton[32,33] and Dawe.[34,35] Based on the Mindlin theory, one can show readily that the kinetic energy per unit surface area of a vibrating plate is

$$\frac{\rho t^3}{24}\left[\left(\frac{\partial \theta_x}{\partial \tau}\right)^2 + \left(\frac{\partial \theta_y}{\partial \tau}\right)^2\right] + \frac{\rho t}{2}\left(\frac{\partial w}{\partial \tau}\right)^2 \tag{5.34}$$

Therefore, the mass matrix for vibration analysis is

$$[M^e] = \int [N]^T [P][N]\, dxdy \tag{5.35}$$

where [N] is the shape function matrix (Eq. 5.7), and

$$[P] = \begin{bmatrix} \frac{\rho t}{2} & 0 & 0 \\ 0 & \frac{\rho t^3}{24} & 0 \\ 0 & 0 & \frac{\rho t^3}{24} \end{bmatrix}$$

Recalling Eq. 5.33, it can be shown readily that the potential energy due to the buckling process is the integrated sum of products of the membrane stresses and the corresponding second order terms. The mathematical expression for such energy is:[35]

$$E = \iint \frac{t}{2}\begin{Bmatrix} \frac{\partial w}{\partial x} & \frac{\partial w}{\partial y} \end{Bmatrix} [\sigma_o] \begin{Bmatrix} \frac{\partial w}{\partial x} \\ \frac{\partial w}{\partial y} \end{Bmatrix} dxdy + \iint \frac{t^3}{24}\begin{Bmatrix} \frac{\partial \theta_x}{\partial x} & \frac{\partial \theta_y}{\partial y} \end{Bmatrix} [\sigma_o] \begin{Bmatrix} \frac{\partial \theta_x}{\partial x} \\ \frac{\partial \theta_y}{\partial y} \end{Bmatrix} dxdy$$

$$+ \iint \frac{t^3}{24}\begin{Bmatrix} \frac{\partial \theta_y}{\partial x} & \frac{\partial \theta_x}{\partial y} \end{Bmatrix} [\sigma_o] \begin{Bmatrix} \frac{\partial \theta_y}{\partial x} \\ \frac{\partial \theta_x}{\partial y} \end{Bmatrix} dxdy \tag{5.36}$$

where $[\sigma_o] = \begin{bmatrix} \sigma_x^o & \tau_{xy}^o \\ \tau_{xy}^o & \sigma_y^o \end{bmatrix}$

The geometric stiffness matrix for a plate simply supported along opposite edges is:

$$[K_g]_{mm} = l\, t \int [S]_m^T [\sigma_o] [S]_m\, dx + \frac{l t^3}{12} \int [Q]_m^T [\sigma_o] [Q]_m\, dx + \frac{l t^3}{12} \int [T]_m^T [\sigma_o] [T]_m\, dx \tag{5.37}$$

where

$$[S]_m = \begin{bmatrix} \frac{\partial N_1}{\partial x} & 0 & 0 & \frac{\partial N_2}{\partial x} & 0 & 0 & \cdots & \cdots \\ \frac{\partial N_1}{\partial y} & 0 & 0 & \frac{\partial N_2}{\partial y} & 0 & 0 & \cdots & \cdots \end{bmatrix}_m$$

$$[Q]_m = \begin{bmatrix} 0 & \frac{\partial N_1}{\partial x} & 0 & 0 & \frac{\partial N_2}{\partial x} & 0 & \cdots & \cdots \\ 0 & 0 & \frac{m\pi N_1}{l} & 0 & 0 & \frac{m\pi N_2}{l} & \cdots & \cdots \end{bmatrix}_m$$

$$[T]_m = \begin{bmatrix} 0 & 0 & \frac{\partial N_1}{\partial x} & 0 & 0 & \frac{\partial N_2}{\partial x} & \cdots & \cdots \\ 0 & \frac{m\pi N_1}{l} & 0 & 0 & \frac{m\pi N_2}{l} & 0 & \cdots & \cdots \end{bmatrix}_m$$

and l is the length of the strip.

Geometric stiffness matrices for plates with other boundary conditions can be formed accordingly by substituting the shape functions of the displacements into Eq. 5.39.

The fundamental frequencies of square moderately thick plates (a x a) clamped on four sides were computed by Roufaeil and Dawe.[35] The frequency parameter Ω ($= \omega\sqrt{\frac{2(1+\upsilon)\rho a}{E}}$) obtained by using different number of strips and terms in the analysis is tabulated in Table 5.21. Also in the table are solutions obtained by Rayleigh-Ritz[36] and classical approaches.[22]

Though the presence of shear stress will lead to a coupled geometric stiffness matrix, accurate solutions can still be obtained by the finite strip method, and Dawe[34] studied the buckling of plates under the action of

TABLE 5.21

Frequency parameters Ω for clamped plate, t/a = 0.1 (υ = 0.3).[35]

No of strip	No of series terms	Mode					
		1,1	2,1	1,2	2,2	3,1	1,3
1	1	1.593		4.161			8.458
	2	1.593	3.032	4.161	5.032		8.458
	3	1.589	3.032	4.156	5.032	5.021	8.457
2	1	1.593		3.043			5.100
	2	1.593	3.030	3.043	4.260		5.100
	3	1.588	3.030	3.031	4.260	5.010	5.090
3	1	1.592		3.041			5.041
	2	1.592	3.029	3.041	4.258		5.041
	3	1.588	3.029	3.029	4.258	4.997	5.040
4	1	1.592		3.041			5.040
	2	1.592	3.029	3.041	4.258		5.040
	3	1.588	3.029	3.029	4.258	4.997	5.038
Raleigh-Ritz[36]		1.588	3.029	3.029	4.256	4.997	5.040
Classical[22]		1.756	3.581	3.581	5.280	6.421	6.451

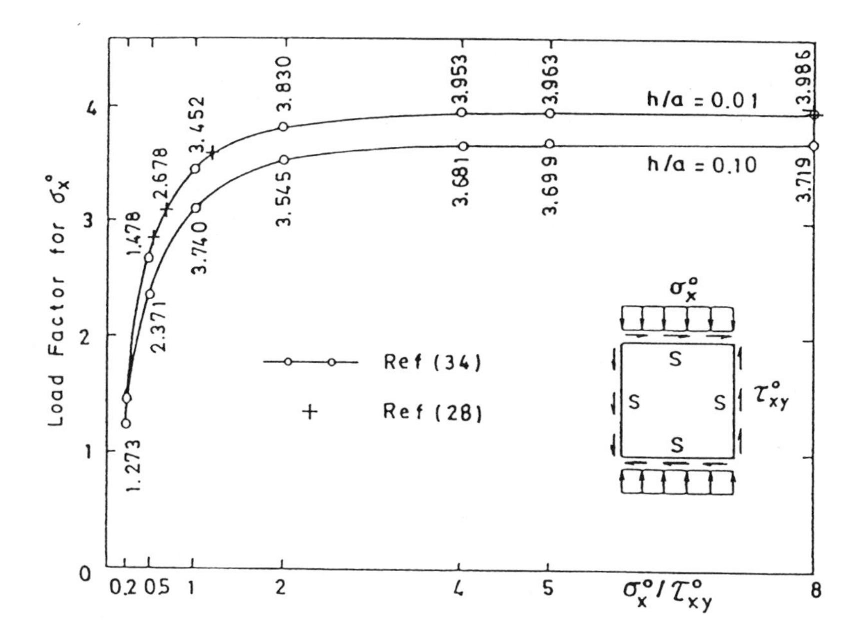

Figure 5.8 Variation of buckling load with stress ratio.[34]

uniaxial direct stress and shear stress. Figure 5.8 shows the variation of the buckling load for a square plate with different stress ratios. The stress ratio is defined as σ_x^o / τ_{xy}^o . In the figure, the results obtained by the spline strips are also shown. Comparison shows that they are in good agreement with other published results.

REFERENCES

1. Zienkiewicz, O. C. and CHEUNG, Y. K., The finite element method for analysis of elastic isotropic and orthotropic slabs, *Proc Inst Civil Eng*, 28, 471, 1964.
2. Bishop, R. E. D., Gladwell, G. M. L. and Michaelson, S., *The matrix analysis of vibration*, Cambridge University Press, 1965.
3. Irons, B. M., Eigenvalue economisers in vibration problems, *J Aeronaut Soc*, 67, 526, 1963.
4. Bathe, K. J., *Solution methods for large generalised eigenvalue problems in structural engineering*, SESM Report, 71-20, Department of Civil Engineering, University of California, Berkeley, 1971.
5. Cheung, Y. K. and Leung, A. Y. T., *Finite element methods in dynamics*, Kluwer Academic Publ, Dordrecht, the Netherlands & Science Press, Beijing, China, 1991.
6. Cheung, M. S., *Finite strip analysis of structures*, PhD thesis, Department of Civil Engineering, University of Calgary, 1971.

7. Babu, P. V. T., and Reddy, D. V., Frequency analysis of skew orthotropic plates by the finite strip method, *J of Sound and Vib*, 18, 465-474, 1971.
8. Cheung, Y. K., *Finite strip method in structural analysis*, Pergamon Press, 1976.
9. Fan, S. C., *Spline finite strip in structural analysis*, PhD thesis, Department of Civil Engineering, University of Hong Kong, 1982.
10. Cheung, M. S. and Cheung, Y. K., Static and dynamic behaviour of rectangular plates using higher order finite strips, *Building Sci*, 7(3), 415-419, 1972.
11. Warburton, G. B., The vibration of rectangular plates, *Proc Instn Mech Engrs*, 168(12), 371-384, 1954.
12. Leissa, A. W., Free vibration of elastic plates, *AIAA, 7th Aerospace Science Meeting*, New York, 1969.
13. Kurata, M. and Okamura, H., Natural vibration of partially clamped plates, *J of Eng Mech, ASCE*, 89, June, 1963.
14. Appl, F. C. and Byers, N. R., Fundamental frequency of simply supported rectangular plates with linearly varying thickness, *J Appl Mech, Trans ASME*, 32, 163-168, 1965.
15. Ben-Amoz, M., Note on deflections and flexural vibration of clamped sectorial plates, *J Appl Mech, Tran ASME*, 26, 136, 1959.
16. Reed Jr, R. E., *Comparison of methods in calculating frequencies of corner supported rectangular plates*, NASA Report NASA TN D-3030, 1965.
17. Cox, H. L., Vibration of certain square plates having similar adjacent edges, *Quart J Mech Appl Math*, 8, 454-456, 1955.
18. Johns, D. J. and Nataroja, R., Vibration of a square plate symmetrically supported at four points, *J of Sound and Vib*, 25, 75-82, 1972.
19. Rao, G. V., Fundamental frequency of a square panel with multiple point supports on edges, *J of Sound and Vib*, 38, 271, 1975.
20. Petyt, M. and Mirza, W. H., Vibration of column-supported floor slabs, *J of Sound and Vib*, 21, 355-364, 1972.
21. Wu, C. I. and Cheung, Y. K., Frequency analysis of rectangular plates continuous in one or two directions, *Earthquake Eng and Struct Dyn*, 3, 3-14, 1974.
22. Kong, J., *Analysis of plate-type structures by finite strip, finite prism and finite layer methods*, PhD thesis, Department of Civil and Structural Engineering, University of Hong Kong, 1994.
23. Leissa, A. W., The free vibration of rectangular plates, *J of Sound and Vib*, 31, 257-293, 1973.
24. Kim, C. S. and Dickinson, S. M., The flexural vibration of line supported rectangular plate systems, *J of Sound and Vib*, 114, 129-142, 1987.
25. Liew, K.M. and Lam, K. Y., Vibration analysis of multi-span plates having orthogonal straight edges, *J of Sound and Vib*, 147, 255-264, 1991.

26. Babu, T. P. V. and Reddy, D. V., Stability analysis of skew orthotropic plates by the finite strip method, *Comp and Struct*, 8, 599-607, 1978.
27. Bulson, P. S., *The stability of flat plates*, American Elsevier, New York, 1969.
28. Szeto, H. Y., *Application of spline finite strip method in stability analysis of arbitrarily shaped plates*, MPhil thesis, Department of Civil and Structural Engineering, University of Hong Kong, 1992.
29. Tham, L. G. and Szeto, H. Y., Buckling analysis of arbitrarily shaped plates by spline finite strip method, *Comp and Struct*, 36(4), 729-735, 1990.
30. Column Research Committee of Japan, *Handbook of structural stability*, Corona, Lund, 1971.
31. Yoshimura, Y. and Iwata, K., Buckling of simply supported oblique plates, *J of Appl Mech*, 30, 363-366, 1963.
32. Hinton, E., Buckling of initially stressed Mindlin plates using a finite strip method, *Comp and Struct*, 8, 99-105, 1978.
33. Benson, P. R. and Hinton, E. A., Thick finite strip solution for static, free vibration and stability problems, *Int J for Num Methods in Eng*, 10, 665-678, 1976.
34. Dawe, D. J. and Roufaeil, O. L., Buckling of rectangular Mindlin plates, *Comp and Struct*, 15, 461-471, 1982.
35. Roufaeil, O. L. and Dawe, D. J., Vibration analysis of rectangular Mindlin plates by the finite strip method, *Comp and Struct*, 12, 833-842, 1980.
36. Dawe, D. J. and Roufaeil, O. L., Rayleigh-Ritz vibration analysis of Mindlin plates, *J of Sound and Vib*, 69, 345-359, 1980.

CHAPTER SIX

MODELLING OF THREE-DIMENSIONAL SOLIDS : FINITE PRISM AND FINITE LAYER METHODS

6.1 INTRODUCTION

In the previous chapters, the development of the finite strip method and application to two-dimensional problems have been discussed in detail. Here attention is paid to the modelling of three-dimensional solids.

It was mentioned briefly in Chapter 2 that it is possible to reduce a three-dimensional problem to a two-dimensional one by writing the displacement function as

$$\delta = \sum_{m=1}^{r} f_m(x,z)Y_m(y) \tag{6.1a}$$

The above displacement function can then be used to formulate the stiffness matrix of a finite prism. This method is best suited to structures with variable thicknesses or with holes in the cross-section, such as voided slabs or thick-walled box girder bridges. The investigation of local stresses at web and flange junctions, which normally would not be justified because of the high cost of a full three-dimensional finite element analysis, can now be carried out at moderate cost. In conjunction with finite strips, the method can also be applied to the analysis of formed sandwich plates, which are made up of weak cores sandwiched between stiff thin layers. The cores and thin layers are modelled by finite prisms and finite strips respectively. A detailed discussion of this approach will be made in Chapter 10.

Similarly, a three-dimensional problem can be reduced to a one-dimensional one by assuming the shape functions in the form of a double series:

$$\delta = \sum_{n=1}^{r} \sum_{m=1}^{r} f_{mn}(x)Y_m(y)Z_n(z) \tag{6.1b}$$

thus forming the basis of the finite layer method. In this method, a thick medium is imagined to be divided into a number of horizontal layers with each being supported along four vertical sides (Figure 6.1a), while a cylinder is divided into a number of concentric cylindrical layers (Figure 6.1b). Since each layer can be assigned individual material properties and thickness, the method is ideally suited for the analysis of thick layered plates and shells. The method is also widely applied to the analysis of geotechnical problems (Chapter 10) in which one can assume that the soil is layered in nature and that the displacements as well as stresses take up zero value at sufficiently large distance away from the loading zone.

This chapter will focus on the prisms and layers based on semi-analytical functions, and it has to be pointed out that other types of prisms and layers can be easily generated.

6.2 FINITE PRISM METHOD

6.2.1 INTRODUCTION

In the finite element approach it is often convenient to define the coordinates of the more complex element shapes by the nodal coordinates ξ_k ,

$$\xi = \sum_{k=1}^{s} \phi_k \xi_k \tag{6.2a}$$

in which s refers to the number of nodes of the element.

Similarly, the displacements δ of a point within the element can also be expressed in terms of the nodal displacements δ_k ,

$$\delta = \sum_{k=1}^{s} \psi_k \delta_k \tag{6.2b}$$

where δ_k and ψ_k are functions of a coordinate system (Cartesian, skew, curvilinear, etc.). For the special case in which ϕ_k and ψ_k are identical, the element is termed isoparametric. This concept was first introduced, in the finite element formulation, by Taig(1) and by Irons(2) and later extended by Ergatoudis et al.(3,4) as well as Irons and Zienkiewicz.(5) In this section, the concept is further extended to the formulation of straight and curved finite prisms.(6) All Serendipity as well as Lagrange elements can be used and examples of such elements have already been given in Chapter 2. Further developments have demonstrated that the isoparametric restriction can be lifted, that is, one can use different elements for interpolation and domain transformation. For example, one may use higher order elements which include derivative terms as their interpolation parameters for approximation of the displacement functions while retaining the Serendipity element functions for the geometric transformation. This approach is referred to as sub-parametric.(7) (Detailed discussion on domain transformation is made in Chapter 7.)

6.2.2 SERENDIPITY ELEMENTS: 8-NODED SERENDIPITY ELEMENT

Without much loss of generality, the following discussion will be restricted to the 8-noded Serendipity elements.

Referring to Figure 6.1 and Eq. 2.1b it can be seen that in the plane of the cross-section the shape functions given in Eq. 2.38 should be used, while for the axial direction Y_m there are again some Fourier series which satisfy the simple support end conditions u=w=∂v/∂y=0. A suitable set of displacement functions for a straight prism is

$$u = \sum_{m=1}^{r} \sum_{k=1}^{s} C_k u_{km} \sin k_m y \tag{6.3a}$$

$$v = \sum_{m=1}^{r} \sum_{k=1}^{s} C_k v_{km} \cos k_m y \tag{6.3b}$$

$$w = \sum_{m=1}^{r} \sum_{k=1}^{s} C_k w_{km} \sin k_m y \tag{6.3c}$$

The coordinates of the eight-node isoparametric section can thus be similarly defined as

$$x = \sum_{k=1}^{s} C_k x_k \tag{6.4a}$$

$$z = \sum_{k=1}^{s} C_k z_k \tag{6.4b}$$

In Eqs. 6.3 and 6.4, u_{km} , v_{km} and w_{km} are the m-th term displacements at the k-th node, x_k and z_k are the x and z coordinates of the k-th node, while the shape function C_k has been defined by Eq. 2.38 for corner node, midside node with $\xi = 0$ and midside node with $\eta = 0$ respectively. As can be seen from Eq. 2.38, C_k is given in terms of a set of curvilinear coordinates ξ and η.

The strain matrix, in the Cartesian coordinates, of a three-dimensional solid is

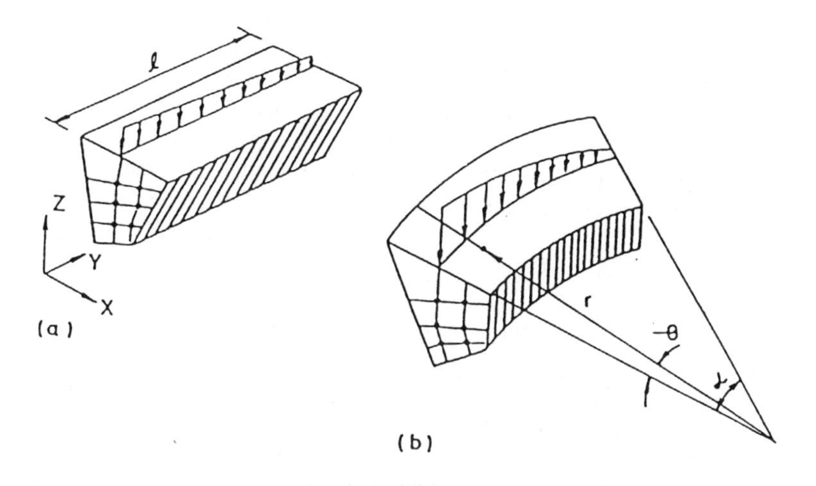

Figure 6.1 Typical prismatic situations: (a) straight prism. (b) curved prism.

$$\{\varepsilon\} = \begin{Bmatrix} \varepsilon_x \\ \varepsilon_y \\ \varepsilon_z \\ \gamma_{xy} \\ \gamma_{yz} \\ \gamma_{zx} \end{Bmatrix} = \begin{Bmatrix} \frac{\partial u}{\partial x} \\ \frac{\partial v}{\partial y} \\ \frac{\partial w}{\partial z} \\ \frac{\partial u}{\partial y} + \frac{\partial v}{\partial x} \\ \frac{\partial v}{\partial z} + \frac{\partial w}{\partial y} \\ \frac{\partial w}{\partial x} + \frac{\partial u}{\partial z} \end{Bmatrix} \tag{6.5}$$

The stress-strain relationships are

$$\{\sigma\} = [D]\,\{\varepsilon\} = [D] \sum_{m=1}^{r} [B]_m \{\delta\}_m = [D]\,[B]\,\{\delta\} \tag{6.6}$$

in which $\{\sigma\}$ represents the stresses corresponding to the strains $\{\varepsilon\}$ and is given by

$$\{\sigma\} = \begin{Bmatrix} \sigma_x \\ \sigma_y \\ \sigma_z \\ \tau_{xy} \\ \tau_{yz} \\ \tau_{zx} \end{Bmatrix} \tag{6.7}$$

[D] is the property matrix of the material for the particular prism under consideration which can be isotropic, orthotropic, or general anisotropic.

$[B_i]_m$ is obtained from the strain displacement relationship given by Eq. 6.5, that is

$$[B_i]_m = \begin{bmatrix} \frac{\partial C_i}{\partial x} s_m & 0 & 0 \\ 0 & -C_i k_m s_m & 0 \\ 0 & 0 & \frac{\partial C_i}{\partial z} s_m \\ C_i k_m c_m & \frac{\partial C_i}{\partial x} c_m & 0 \\ 0 & \frac{\partial C_i}{\partial z} c_m & C_i k_m c_m \\ \frac{\partial C_i}{\partial z} s_m & 0 & \frac{\partial C_i}{\partial x} s_m \end{bmatrix} \tag{6.8}$$

where $s_m = \sin k_m y$ and $c_m = \cos k_m y$.

Having defined the stresses and strains, one can show readily that the basic unit of the stiffness matrix is

$$[K_{ij}]_{mn} = \int [B_i]_m^T \, [D] \, [B_j]_n \, d(\text{vol}) \tag{6.9}$$

In the above equation, the property matrix for orthotropic materials can be written as

$$[D] = \begin{bmatrix} D_{11} & D_{12} & D_{13} & 0 & 0 & 0 \\ D_{21} & D_{22} & D_{23} & 0 & 0 & 0 \\ D_{31} & D_{32} & D_{33} & 0 & 0 & 0 \\ 0 & 0 & 0 & D_{44} & 0 & 0 \\ 0 & 0 & 0 & 0 & D_{55} & 0 \\ 0 & 0 & 0 & 0 & 0 & D_{66} \end{bmatrix} \tag{6.10}$$

If $[B_i]_m$ and [D] are now substituted into Eq. 6.9, we have

$$[K_{ij}]_{mn} = \int_A \int_0^l \left[\begin{array}{c|c|c} \begin{array}{l} D_{11}\frac{\partial C_i}{\partial x}\frac{\partial C_j}{\partial x}s_m s_n \\ +D_{44}C_iC_j k_m k_n c_m c_n \\ +D_{66}\frac{\partial C_i}{\partial z}\frac{\partial C_j}{\partial z}s_m s_n \end{array} & \begin{array}{l} -D_{12}C_j\frac{\partial C_i}{\partial x}k_m s_m s_n \\ +D_{44}C_i\frac{\partial C_j}{\partial x}k_m s_m s_n \end{array} & \begin{array}{l} D_{13}\frac{\partial C_i}{\partial x}\frac{\partial C_j}{\partial z}s_m s_n \\ +D_{66}\frac{\partial C_i}{\partial x}\frac{\partial C_j}{\partial z}s_m s_n \end{array} \\ \hline \begin{array}{l} -D_{21}C_i\frac{\partial C_j}{\partial x}k_m s_m s_n \\ +D_{44}C_j\frac{\partial C_i}{\partial x}k_n c_m c_n \end{array} & \begin{array}{l} D_{22}C_iC_j k_m k_n s_m s_n \\ +D_{44}\frac{\partial C_i}{\partial x}\frac{\partial C_j}{\partial x}c_m c_n \\ +D_{55}\frac{\partial C_i}{\partial z}\frac{\partial C_j}{\partial z}c_m c_n \end{array} & \begin{array}{l} -D_{23}C_i\frac{\partial C_j}{\partial z}k_m s_m s_n \\ +D_{55}C_j\frac{\partial C_i}{\partial z}k_m c_m c_n \end{array} \\ \hline \begin{array}{l} D_{31}\frac{\partial C_i}{\partial z}\frac{\partial C_j}{\partial x}s_m s_n \\ +D_{66}\frac{\partial C_i}{\partial x}\frac{\partial C_j}{\partial z}s_m s_n \end{array} & \begin{array}{l} -D_{32}C_j\frac{\partial C_i}{\partial z}k_m s_m s_n \\ +D_{55}C_i\frac{\partial C_j}{\partial z}k_m c_m c_n \end{array} & \begin{array}{l} D_{33}\frac{\partial C_i}{\partial z}\frac{\partial C_j}{\partial z}s_m s_n \\ +D_{55}C_iC_j k_m k_n c_m c_n \\ +D_{66}\frac{\partial C_i}{\partial x}\frac{\partial C_j}{\partial x}s_m s_n \end{array} \end{array}\right] dxdydz \tag{6.11}$$

All the coefficients inside the submatrix $[K_{ij}]_{mn}$ contain integrals of the form $\int_0^l \sin k_m y \sin k_n y\, dy$ or $\int_0^l \cos k_m y \cos k_n y\, dy$, which will vanish for $m \neq n$; consequently, the different terms of the series are uncoupled.

It is interesting to note that decoupling will not occur for general anisotropic materials with twenty-one constants[8] because of the presence of the integral $\int_0^l \sin k_m y \cos k_n y dy$ which does not exhibit orthogonal properties.

A closer examination of Eq. 6.11 shows that while the shape function C_i is given in terms of local coordinates ξ and η, all differentiation and integration are in terms of the global coordinate system since all the strains are written as derivatives of x, y and z. Hence it is necessary to (i) rewrite the derivatives with respect to local coordinate system, that is, transform $\partial C_i/\partial x$, $\partial C_i/\partial z$ to

$\partial C_i/\partial \xi$, $\partial C_i/\partial \eta$, (ii) change the area dxdz to $d\xi d\eta$, and (iii) change the integration limits to suit the local coordinate system. In short, domain transformation has to be carried out to map the domain from the Cartesian coordinate plane into a natural coordinate plane. (A general discussion on the domain transformation will be made in Chapter 7.)

By applying the chain rule in partial differentiation it is possible to write

$$\frac{\partial C_i}{\partial \xi} = \frac{\partial C_i}{\partial x}\frac{\partial x}{\partial \xi} + \frac{\partial C_i}{\partial z}\frac{\partial z}{\partial \xi} \tag{6.12a}$$

$$\frac{\partial C_i}{\partial \eta} = \frac{\partial C_i}{\partial x}\frac{\partial x}{\partial \eta} + \frac{\partial C_i}{\partial z}\frac{\partial z}{\partial \eta} \tag{6.12b}$$

Thus we have in matrix form

$$\begin{Bmatrix} \dfrac{\partial C_i}{\partial \xi} \\ \dfrac{\partial C_i}{\partial \eta} \end{Bmatrix} = \begin{bmatrix} \dfrac{\partial x}{\partial \xi} & \dfrac{\partial z}{\partial \xi} \\ \dfrac{\partial x}{\partial \eta} & \dfrac{\partial z}{\partial \eta} \end{bmatrix} \begin{Bmatrix} \dfrac{\partial C_i}{\partial x} \\ \dfrac{\partial C_i}{\partial z} \end{Bmatrix} = [J] \begin{Bmatrix} \dfrac{\partial C_i}{\partial x} \\ \dfrac{\partial C_i}{\partial z} \end{Bmatrix} \tag{6.13}$$

The matrix [J], referred to as the Jacobian matrix, can be written explicitly for isoparametric elements by virtue of Eq. 6.4. The expanded form of the Jacobian matrix is

$$[J] = \begin{bmatrix} \dfrac{\partial x}{\partial \xi} & \dfrac{\partial z}{\partial \xi} \\ \dfrac{\partial x}{\partial \eta} & \dfrac{\partial z}{\partial \eta} \end{bmatrix} = \sum_{k=1}^{8} \begin{bmatrix} \dfrac{\partial C_k}{\partial \xi} x_k & \dfrac{\partial C_k}{\partial \xi} z_k \\ \dfrac{\partial C_k}{\partial \eta} x_k & \dfrac{\partial C_k}{\partial \eta} z_k \end{bmatrix}$$

$$= \begin{bmatrix} \dfrac{\partial C_1}{\partial \xi} & \dfrac{\partial C_2}{\partial \xi} & \dfrac{\partial C_3}{\partial \xi} & \dfrac{\partial C_4}{\partial \xi} & \dfrac{\partial C_5}{\partial \xi} & \dfrac{\partial C_6}{\partial \xi} & \dfrac{\partial C_7}{\partial \xi} & \dfrac{\partial C_8}{\partial \xi} \\ \dfrac{\partial C_1}{\partial \eta} & \dfrac{\partial C_2}{\partial \eta} & \dfrac{\partial C_3}{\partial \eta} & \dfrac{\partial C_4}{\partial \eta} & \dfrac{\partial C_5}{\partial \eta} & \dfrac{\partial C_6}{\partial \eta} & \dfrac{\partial C_7}{\partial \eta} & \dfrac{\partial C_8}{\partial \eta} \end{bmatrix} \begin{bmatrix} x_1 & z_1 \\ x_2 & z_2 \\ x_3 & z_3 \\ x_4 & z_4 \\ x_5 & z_5 \\ x_6 & z_6 \\ x_7 & z_7 \\ x_8 & z_8 \end{bmatrix} \tag{6.14}$$

With [J] computed it is possible to write the global derivatives as

$$\begin{Bmatrix} \dfrac{\partial C_i}{\partial x} \\ \dfrac{\partial C_i}{\partial z} \end{Bmatrix} = [J]^{-1} \begin{Bmatrix} \dfrac{\partial C_i}{\partial \xi} \\ \dfrac{\partial C_i}{\partial \eta} \end{Bmatrix} \tag{6.15}$$

The elemental area dxdz, if expressed in the curvilinear coordinate system, becomes

$$dxdz = \left| \, J \, \right| d\xi d\eta \tag{6.16}$$

Finally, the integration limits are changed to +1 and -1 since ξ and η take up such values at the four sides of the quadrilateral section. This particular property is quite valuable because Gaussian quadrature can be applied directly.

The extension to circular prisms is obvious, and the displacement functions given in Eq. 6.3 require only a small modification which consists of replacing the variable 'y' by the angle θ and the span l by the subtended angle α. Thus

$$u = \sum_{m=1}^{r} \sum_{k=1}^{8} C_k u_{km} \sin\left(\frac{m\pi\theta}{\alpha}\right) \tag{6.17a}$$

$$v = \sum_{m=1}^{r} \sum_{k=1}^{8} C_k v_{km} \cos\left(\frac{m\pi\theta}{\alpha}\right) \tag{6.17b}$$

$$w = \sum_{m=1}^{r} \sum_{k=1}^{8} C_k w_{km} \sin\left(\frac{m\pi\theta}{\alpha}\right) \tag{6.17c}$$

with C_k being given by Eq. 2.38 as before.

The strain displacement relationship has been given as

$$\{\varepsilon\} = \begin{Bmatrix} \varepsilon_x \\ \varepsilon_\theta \\ \varepsilon_z \\ \gamma_{\theta z} \\ \gamma_{zx} \\ \gamma_{x\theta} \end{Bmatrix} = \begin{Bmatrix} \dfrac{\partial u}{\partial x} \\ (1+\frac{z}{r})^{-1}(\frac{1}{r_1}\frac{\partial v}{\partial \theta} + \frac{w}{r}) \\ \dfrac{\partial w}{\partial z} \\ \frac{\partial v}{\partial z} + (1+\frac{z}{r})^{-1}(\frac{1}{r}\frac{\partial w}{\partial \theta} - \frac{v}{r}) \\ \frac{\partial u}{\partial z} + \frac{\partial w}{\partial x} \\ \frac{\partial v}{\partial z} + (1+\frac{z}{r})^{-1}(\frac{1}{r}\frac{\partial u}{\partial \theta}) \end{Bmatrix} \tag{6.18}$$

or

$$\{\varepsilon\} = [B]\{\delta\} = \sum_{m=1}^{r} [B]_m \{\delta\}_m \tag{6.19}$$

The formulation of the stiffness matrix can now proceed along the same lines as for the straight prism.

6.2.3 HIGHER ORDER QUADRILATERAL FINITE PRISM[11]

Another method to improve the accuracy of the prism is by including rotations as unknowns in the displacement interpolations. Any higher order element (Section 2.2.2) developed for stress analysis in finite elements can be used for this purpose. Examples of these elements are given in Figure 6.2. Though displacements are interpolated by higher order element shape functions, the domain transformations are still to be carried out by employing Serendipity and Lagrange element shape functions. Therefore, the Jacobian matrix for the

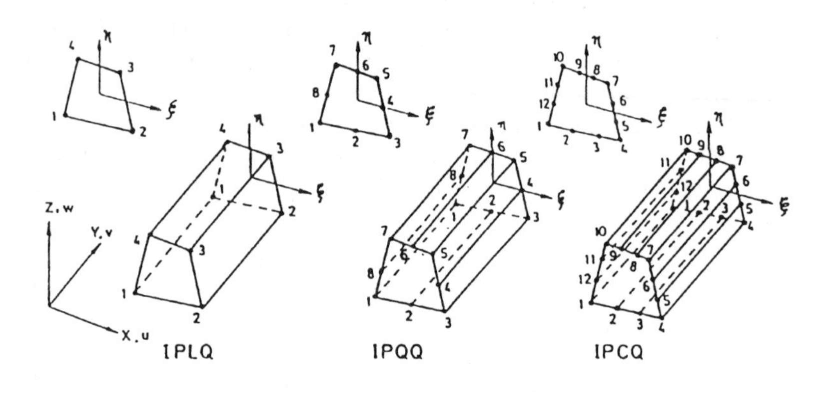

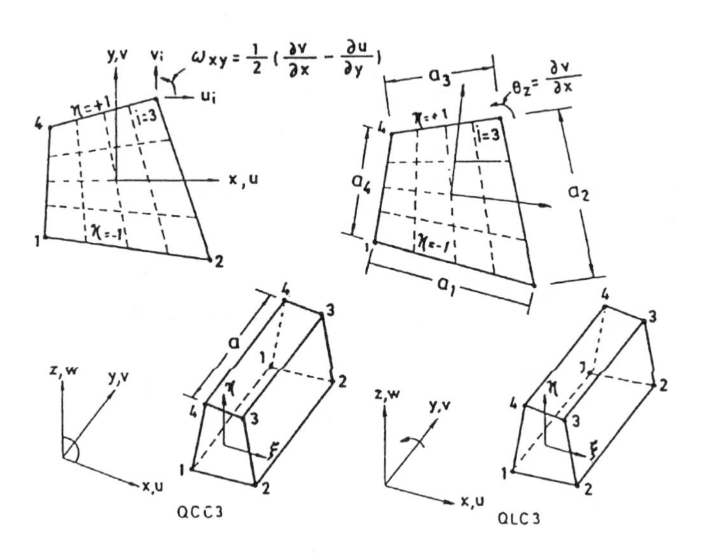

Figure 6.2 Higher order finite prisms.[11]

geometric transformation remains unchanged, and the pertinent characteristic matrices can be formed in the usual manner.

6.2.4 EXAMPLES

In order to assess the accuracy of the finite prism method, a simply supported straight beam subject to a central concentrated load (Figure 6.3) was investigated[6,9] using different mesh size and eleven odd harmonics. The displacements and stresses for meshes (a), (b), (c) of Figure 6.3 are listed in Table 6.1 and compared with an exact solution.[10] The comparison can be regarded as satisfactory for coarse mesh (a) and good for the fine mesh (c). It is interesting to note that local effects can be predicted with reasonable accuracy for a small cost.

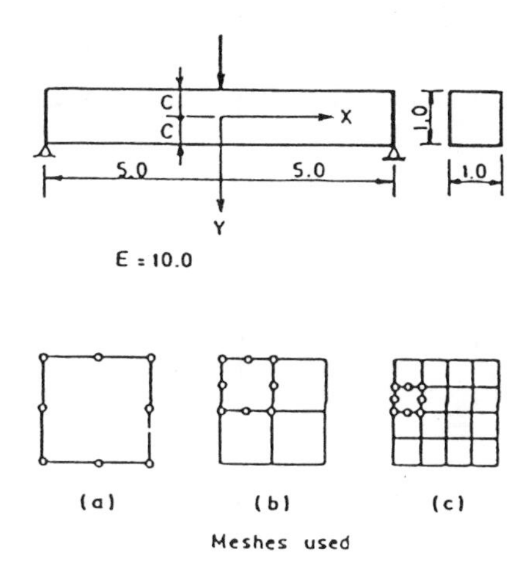

Figure 6.3 Simply supported beam under point load.[6,9]

TABLE 6.1
Comparison of stresses of a simply supported beam under point load.[6,9]

Stress	y	Mesh			Exact
		(a)	(b)	(c)	
σ_x at mid-span	-c	-1571.03	-1907.69	-2288.86	∞
	-c/2		-770.75	-749.02	-664.4
	0	71.69	3.63	18.89	24.2
	c/2		730.31	727.11	722.8
	c	1395.45	1465.30	1475.55	1473.4
σ_y at mid-span	-c	-296.56	-985.72	-1847.60	∞
	-c/2		-558.94	-579.23	-246.0
	0	56.63	-184.15	-103.04	-91.2
	c/2		-23.45	-29.60	-29.0
	c	-83.42	43.16	9.80	0.0
τ_{xy} at 1/4-span	-c	40.81	10.01	1.20	0
	-c/2		47.62	55.24	56.25
	0	60.98	80.76	73.40	75.00
	c/2		47.67	55.10	56.20
	c	40.88	10.98	2.63	0
w at mid-span		2545.15	2562.42	2563.90	2567.28

To demonstrate the application of the higher order prisms, Cheung et al.[11] carried out a benchmark test for the simply supported beam problem using the five different elements shown in Figure 6.2. Their results are summarized in Tables 6.2. Comparing the results with those obtained by beam

TABLE 6.2
Comparison of stress for a simply supported beam under uniformly distributed load.[11]

Strip type & mesh size	σ_y at mid-span				
	z = 2.0	z = 1.5	z =1.0	z = 0.5	z = 0.0
IPLQ 1x1	-18.752				18.751
IPLQ 2x2	-19.153		0.003		19.157
IPLQ 4x4	-19.310	-9.255	.000	9.248	19.319
IPQQ 1x1	-18.708				18.733
IPQQ 2x2	-18.832	-9.212	0.003	9.207	18.846
IPQQ 4x4	-18.271	-8.979	0.001	8.979	18.298
IPCQ 1x1	-18.830				18.830
IPCQ 2x2	-18.759		0.002		18.785
IPCQ 4x4	-18.102	-8.920	0.001	8.920	18.101
QCC3 1x1	-18.491				18.275
QCC3 2x2	-18.930		0.003		19.032
QCC3 4x4	-19.167	-9.259	0.000	9.259	19.117
QLC3 1x1	-18.755				18.755
QLC3 2x2	-19.146		0.003		19.153
QLC3 4x4	-19.255	-9.253	0.000	9.246	19.256
Beam Theory[10]	-18.950	-9.288	0.000	9.288	18.950

theory, one can conclude that the two higher order elements show better convergence over the lower order ones.

The second example is of a more practical nature. A thick-walled straight box girder bridge was analyzed under a sinusoidal load which is uniformly distributed in the transverse direction. The dimensions of the bridge and the mesh used are shown in Figure 6.4. Since the load distribution is proportional to $\sin(\pi y/l)$, in the longitudinal direction, only the stiffness matrix corresponding to the first term of the series needs to be solved. The stresses obtained are presented as stress contours in Figure 6.5a, and 6.5b. Figure 6.5a

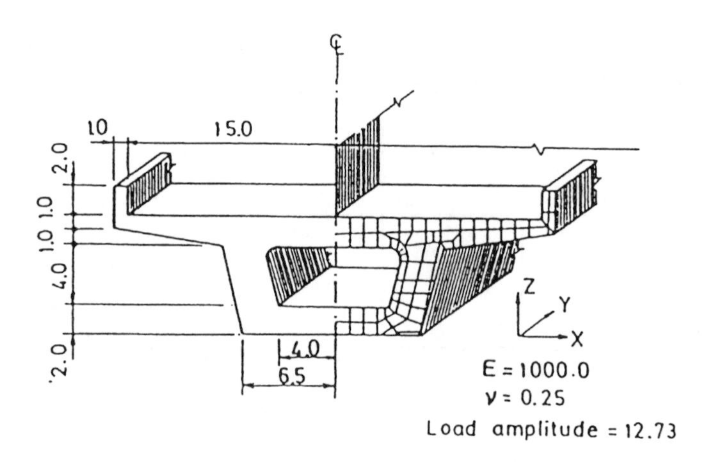

Figure 6.4 Dimension and meshes of a simply supported thick bridge box.[6,9]

shows the σ_y stress normal to the plane of section, and it can be seen that the assumption of linear stress distribution in simple beam theory holds quite well here. Figure 6.5b shows the transverse stress σ_x, and here it is possible to see the transverse bending of the upper flange and also the high stress concentration at the bottom root of the cantilever slab.

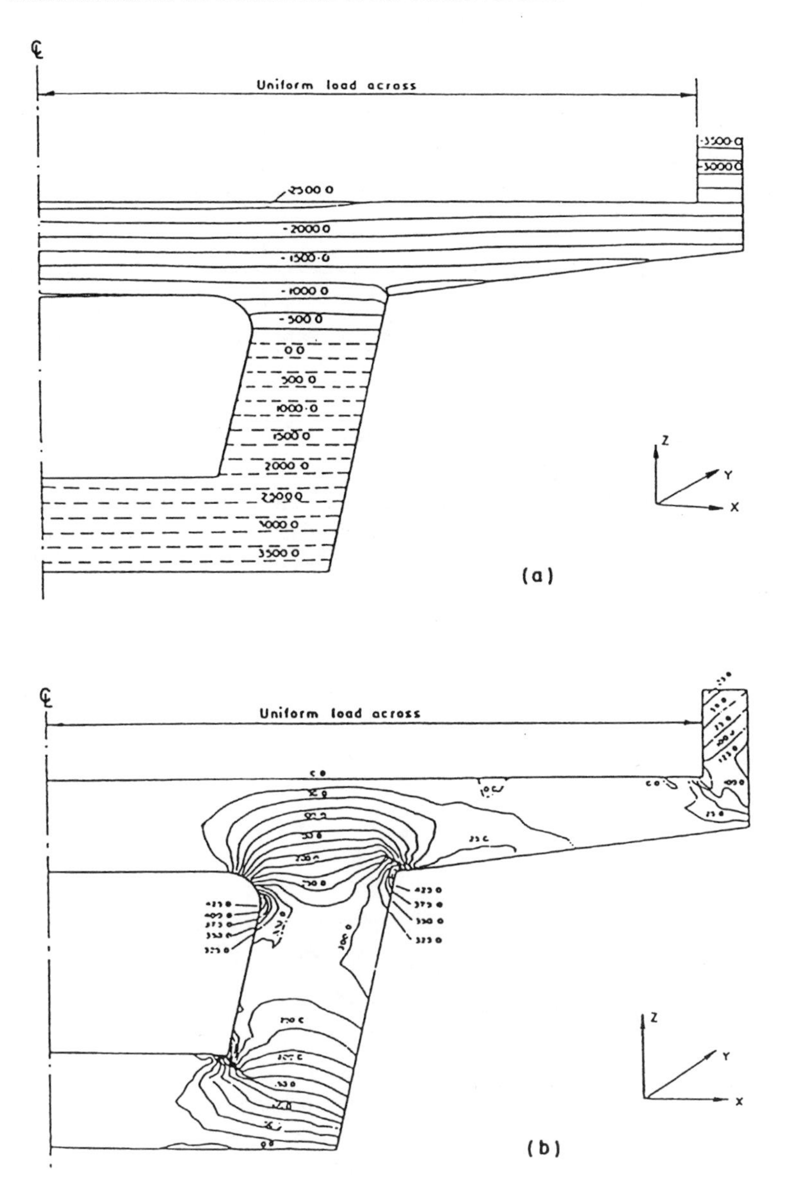

Figure 6.5 Simply supported straight thick bridge box under sinusoidal load: (a) Y-stress at mid-span. (b) X-stress at mid-span.[6,9]

6.3 FINITE LAYER METHOD

6.3.1 RECTANGULAR LAYER

The rectangular layer shown in Figure 6.6a has two nodal surfaces and a suitable set of displacement functions is selected as

$$u = \sum_{m=1}^{r} \sum_{n=1}^{t} [(1-\bar{z})u_{1mn} + \bar{z}\, u_{2mn}]\, X_m'(x)\, Y_n(y) \qquad (6.20a)$$

$$v = \sum_{m=1}^{r} \sum_{n=1}^{t} [(1-\bar{z})v_{1mn} + \bar{z}\, v_{2mn}]\, X_m(x)\, Y_n'(y) \qquad (6.20b)$$

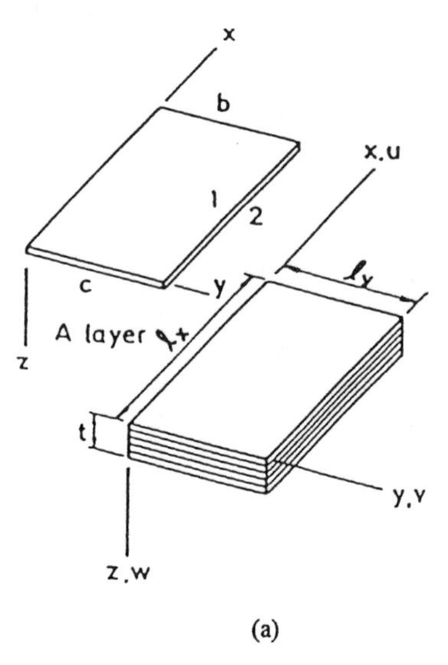

(a)

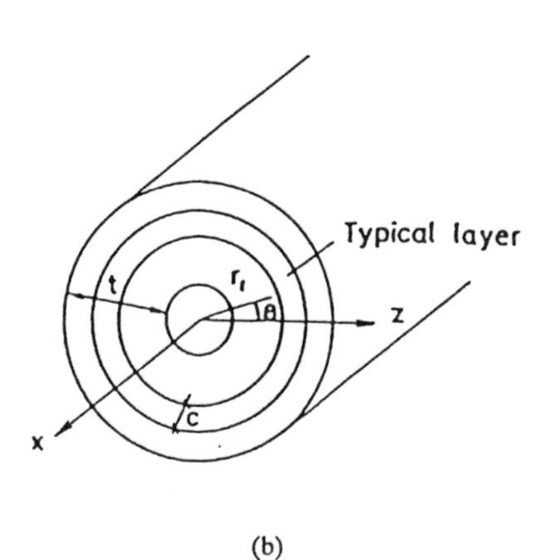

(b)

Figure 6.6 (a) Finite layer idealization of a thick plate. (b) Layer of a thick cylinder.

$$w = \sum_{m=1}^{r} \sum_{n=1}^{t} [(1-\bar{z})w_{1mn} + \bar{z}\, w_{2mn}]\, X_m(x)\, Y_n(y) \tag{6.20c}$$

in which u_{1mn}, v_{1mn}, w_{1mn}, ..., etc., refer to the displacement parameters, and $X_m(x)$, $X_m'(x)$, $Y_m(y)$, $Y_m'(y)$ are the basic functions and their first derivatives. Note that $X'(x)$ and $Y'(y)$ are incorporated in the expressions for u and v respectively because of the relationship $u=A(\partial w/\partial x)$ and $v=B(\partial w/\partial y)$ in linear plate theory, and they are used here as an approximation for the individual layer. For two-dimensional problems the validity of such an assumption has been amply demonstrated in the examples of Chapter 4. The displacement in matrix form is

$$\{f\} = \begin{Bmatrix} u \\ v \\ w \end{Bmatrix} = \sum_{m=1}^{r} \sum_{n=1}^{t} [N]_{mn}\, \{\delta\}_{mn} = [N]\, \{\delta\} \tag{6.21}$$

in which $\{\delta\}_{mn} = [u_{1mn} \;\; v_{1mn} \;\; w_{1mn} \;\; u_{2mn} \;\; v_{2mn} \;\; w_{2mn}]^T$.

In the finite layer model, the strain matrix of a three-dimensional solid can be obtained by substituting the displacement functions into Eq. 6.5 and carrying out the appropriate differentiation, that is

$$\{\varepsilon\} = \sum_{m=1}^{r} \sum_{n=1}^{t} [B]_{mn}\, \{\delta\}_{mn} = [B]\, \{\delta\} \tag{6.22}$$

The stress-strain relationships are:

$$\{\sigma\} = [D]\, \{\varepsilon\} = [D] \sum_{m=1}^{r} \sum_{n=1}^{t} [B]_{mn}\, \{\delta\}_{mn} = [D]\, [B]\, \{\delta\} \tag{6.23}$$

The standard formula for the stiffness is given as

$$[K^e] = \int [B]^T\, [D][B]\, d(vol) \tag{6.24}$$

The expanded form of the strain matrix is

$$[B] = \left[\; [B]_{11} \;\; [B]_{12} \; ...[B]_{1t} \;\; [B]_{21} \; ...[B]_{2t} \quad [B]_{r1} \; [B]_{r2} \; ...[B]_{rt} \;\right] \tag{6.25}$$

Therefore, the stiffness matrix for a layer is

$$[K^e] = \int_0^{l_x}\int_0^{l_y}\int_0^{c} \begin{bmatrix} [B]_{11}^T[D][B]_{11} & [B]_{11}^T[D][B]_{12} & \cdots & [B]_{11}^T[D][B]_{1t} & [B]_{11}^T[D][B]_{21} & [B]_{11}^T[D][B]_{22} & \cdots & [B]_{11}^T[D][B]_{2t} & \cdots & [B]_{11}^T[D][B]_{rt} \\ [B]_{12}^T[D][B]_{11} & [B]_{12}^T[D][B]_{12} & \cdots & [B]_{12}^T[D][B]_{1t} & [B]_{12}^T[D][B]_{21} & [B]_{12}^T[D][B]_{22} & \cdots & [B]_{12}^T[D][B]_{2t} & \cdots & [B]_{12}^T[D][B]_{rt} \\ \cdots & \cdots & \cdots & \cdots & \cdots & \cdots & \cdots & \cdots & \cdots & \cdots \\ [B]_{1t}^T[D][B]_{11} & [B]_{1t}^T[D][B]_{12} & \cdots & [B]_{1t}^T[D][B]_{1t} & [B]_{1t}^T[D][B]_{21} & [B]_{1t}^T[D][B]_{22} & \cdots & [B]_{1t}^T[D][B]_{2t} & \cdots & [B]_{1t}^T[D][B]_{rt} \\ [B]_{21}^T[D][B]_{11} & [B]_{21}^T[D][B]_{12} & \cdots & [B]_{21}^T[D][B]_{1t} & [B]_{21}^T[D][B]_{21} & [B]_{21}^T[D][B]_{22} & \cdots & [B]_{21}^T[D][B]_{2t} & \cdots & [B]_{21}^T[D][B]_{rt} \\ [B]_{22}^T[D][B]_{11} & [B]_{22}^T[D][B]_{12} & \cdots & [B]_{22}^T[D][B]_{1t} & [B]_{22}^T[D][B]_{21} & [B]_{22}^T[D][B]_{22} & \cdots & [B]_{22}^T[D][B]_{2t} & \cdots & [B]_{22}^T[D][B]_{rt} \\ \cdots & \cdots & \cdots & \cdots & \cdots & \cdots & \cdots & \cdots & \cdots & \cdots \\ [B]_{rt}^T[D][B]_{11} & [B]_{rt}^T[D][B]_{12} & \cdots & [B]_{rt}^T[D][B]_{1t} & [B]_{rt}^T[D][B]_{21} & [B]_{rt}^T[D][B]_{22} & \cdots & [B]_{rt}^T[D][B]_{2t} & \cdots & [B]_{rt}^T[D][B]_{rt} \end{bmatrix} dxdydz \tag{6.26}$$

The consistent mass matrix for a layer is developed in a similar way. We have

$$[M^e] = \int_0^{l_x}\int_0^{l_y}\int_0^{c} \begin{bmatrix} [N]_{11}^T[N]_{11} & [N]_{11}^T[N]_{12} & \cdots & [N]_{11}^T[N]_{1t} & [N]_{11}^T[N]_{21} & [N]_{11}^T[N]_{22} & \cdots & [N]_{11}^T[N]_{2t} & \cdots & [N]_{11}^T[N]_{rt} \\ [N]_{12}^T[N]_{11} & [N]_{12}^T[N]_{12} & \cdots & [N]_{12}^T[N]_{1t} & [N]_{12}^T[N]_{21} & [N]_{12}^T[N]_{22} & \cdots & [N]_{12}^T[N]_{2t} & \cdots & [N]_{12}^T[N]_{rt} \\ \cdots & \cdots & \cdots & \cdots & \cdots & \cdots & \cdots & \cdots & \cdots & \cdots \\ [N]_{1t}^T[N]_{11} & [N]_{1t}^T[N]_{12} & \cdots & [N]_{1t}^T[N]_{1t} & [N]_{1t}^T[N]_{21} & [N]_{1t}^T[N]_{22} & \cdots & [N]_{1t}^T[N]_{2t} & \cdots & [N]_{1t}^T[N]_{rt} \\ [N]_{21}^T[N]_{11} & [N]_{21}^T[N]_{12} & \cdots & [N]_{21}^T[N]_{1t} & [N]_{21}^T[N]_{21} & [N]_{21}^T[N]_{22} & \cdots & [N]_{21}^T[N]_{2t} & \cdots & [N]_{21}^T[N]_{rt} \\ [N]_{22}^T[N]_{11} & [N]_{22}^T[N]_{12} & \cdots & [N]_{22}^T[N]_{1t} & [N]_{22}^T[N]_{21} & [N]_{22}^T[N]_{22} & \cdots & [N]_{22}^T[N]_{2t} & \cdots & [N]_{22}^T[N]_{rt} \\ \cdots & \cdots & \cdots & \cdots & \cdots & \cdots & \cdots & \cdots & \cdots & \cdots \\ [N]_{rt}^T[N]_{11} & [N]_{rt}^T[N]_{12} & \cdots & [N]_{rt}^T[N]_{1t} & [N]_{rt}^T[N]_{21} & [N]_{rt}^T[N]_{22} & \cdots & [N]_{rt}^T[N]_{2t} & \cdots & [N]_{rt}^T[N]_{rt} \end{bmatrix} \rho dxdydz \tag{6.27}$$

Note that the final forms of the stiffness matrix and mass matrix [Eqs. 6.26 and 6.27] are somewhat different from the ones given by Eq. 3.14 and Eq. 5.12 because of the double summation used in the displacement functions.

For the static analysis of a simply supported plate we have

$$\int_0^{l_x} \int_0^{l_y} \int_0^{c} [B]_{mn}^T [D] [B]_{rs} \, dxdydz = 0 \qquad \text{for } mn \neq rs \qquad (6.28)$$

that is, all the off-diagonal submatrices are zero and the terms of the series are now decoupled. For vibration problems, however, all combinations of end conditions are allowable.

To demonstrate the application of the finite layer method to the solution of thick plate problems, the following two examples have been considered.

The first example(9,12) is on the static analysis of a simply supported plate (a x b) with uniformly distributed load q acting at the top of the surface. The results for various ratios of b/a and t/a are shown in Table 6.3, using twenty-five terms of the series (m= 1,3,5,6,7 and n=1,3,5,7,9). In the analysis, the plate was divided into three equal layers for t/a ratios of 0.10 and 0.25, and into five equal layers when t/a is 0.50. Note that for t/a=0.10 the results are close to those given by the thin-plate theory, and the deflections and stresses are almost equal at the upper and lower extreme fibres, although in the finite layer method the thin plate is treated as a three-dimensional solid. For higher t/a ratios, the thin plate theory is no longer applicable, and both the stresses and deflections at the extreme fibres show marked differences in magnitudes.

TABLE 6.3
Stresses and deflections at centre of rectangular plates under uniformly distributed load.(9,12)

b/a	t/a	0.1		0.25		0.50	
		Top	Bottom	Top	Bottom	Top	Bottom
1.0	w	491.9639	491.4200	37.7139	36.4321	9.0208	6.0001
	σ_z	-26.0287	25.9095	-4.2806	4.1699	-1.2427	1.0938
	σ_y	-26.087	25.9095	-4.2806	4.1699	-1.2427	1.0938
1.5	w	924.2100	923.6643	67.4901	66.1891	13.0846	11.0972
	σ_z	-46.7853	46.7066	-7.6061	7.5305	-2.0725	1.9684
	σ_y	-24.1373	23.9917	-4.0081	3.8476	-1.1594	0.9990
2.0	w	1207.200	1206.653	86.6447	85.3310	16.5533	13.9514
	σ_z	-60.0407	59.9828	-9.7299	9.6724	-2.6056	2.5186
	σ_y	-20.0641	19.9060	-3.3642	3.1857	-1.0032	0.8134

(a=10; E=1; υ=0.15; q=1)

Note that ε_z is constant through the thickness of each layer and consequently for non-zero Poisson's ratios, stress jumps can be expected at the interfaces of adjoining layers for the normal stress component σ_z. Therefore the stresses should be averaged at the interface or plotted at the mid-height of layers to obtain a smooth stress distribution.

The second problem(9,13) concerns the free vibration of isotropic plates with various types of boundary conditions for three different thickness-span ratios. Each plate was analyzed with five identical layers and three terms of double series. Thus, in each case, 162 equations were involved in the computation of the frequencies. Only results for the first five lowest flexural frequencies are shown in Table 6.4. It is observed that the frequencies for the smallest thickness-span ratio are close to those given by Warburton[14] for thin plates, while the frequencies for higher thickness-span ratios tend to be of lower values.

A point worth mentioning is that in comparison with the exact values the thin plate theory tends to overestimate the frequencies and the error increases with higher modes of vibration.

TABLE 6.4
Natural frequencies f (Hz) for various isotropic homogeneous plates.[9,13]

Cases	Frequency number	Thin Plate Theory[13] f/t	h = 0.05	h = 0.10	h = 0.20
	1	1.7402	0.0887	0.1662	0.2744
	2	3.5518	0.1777	0.3220	0.4981
	3	3.5518	0.1777	0.3220	0.4981
	4	5.2402	0.2576	0.4547	0.6741
	5	6.3813	0.3043	0.5166	0.7245
	1	1.1671	0.0597	0.1134	0.1930
	2	1.9525	0.0975	0.1819	0.2981
	3	3.0812	0.1550	0.2834	0.4451
	4	3.7245	0.1833	0.3334	0.5207
	5	3.9550	0.1937	0.3491	0.5346
	1	1.3100		0.1268	0.2222
	2	2.9377	0.1454	0.2688	0.4326
	3	2.9377	0.1454	0.2688	0.4326
	4	4.4853	0.2190	0.3937	0.6024
	5	5.5475	0.2671	0.4702	0.6949
	1	1.3972	0.0709	0.1344	0.2278
	2	2.6520	0.1305	0.2430	0.3984
	3	3.3402	0.1677	0.3056	0.4775
	4	4.5675	0.2239	0.4013	0.6110
	5	4.9538	0.2374	0.4260	0.6513
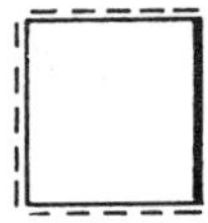	1	1.1418	0.0574	0.1109	0.1977
	2	2.5044	0.1228	0.2309	0.3859
	3	2.8250	0.1403	0.2602	0.4214
	4	4.1548	0.2025	0.3675	0.5730
	5	4.8570	0.2327	0.4193	0.6452

Data : 5 identical layers; m=1,2,3; n=1,2,3; a=b=1.0; E=1.0; υ=0.3; ρ=1.0
Simply supported ━ ━ ━ ; free ——— ; clamped ━━━

6.3.2 CYLINDRICAL LAYER

Two types of cylindrical layers have been developed. The first is a higher order layer (HO3) with an additional internal nodal surface developed by Nelson et al.[15] for the study of free vibration of infinite cylinders. The second is a lower order layer (LO2) used by Cheung and Wu[9,16] for the determination of the frequencies of thick, layered cylinders with any combinations of boundary conditions at the axial ends of the cylinder.

(a) Lower order cylindrical layer (LO2)

For a typical cylindrical layer with nodal surfaces 1 and 2, the thickness is designated by c. The layer is referred to a right-hand screw coordinate system (x,θ,z) with x being taken along the axial direction, θ the circumferential direction, and z the radial direction, which is positive when measured outwards from the origin of the coordinate system. The displacements u, v and w for the layer in the x, θ and z directions respectively are selected as follows:
(1) Linear polynomial in the thickness direction (z), using displacement values only at the two nodal surfaces.
(2) Fourier series in the circumferential direction (θ). This is a well-known technique for axisymmetric analysis.
(3) Basic functions and their first derivatives in the axial direction that satisfy the end conditions.

The final displacement functions are written as

$$u = \sum_{m=1}^{r} \sum_{n=1}^{t} [(1-\bar{z})u_{1mn} + \bar{z}\, u_{2mn}]\, X_m'(x) \cos(n\theta+\theta_o) \qquad (6.29a)$$

$$v = \sum_{m=1}^{r} \sum_{n=1}^{t} [(1-\bar{z})v_{1mn} + \bar{z}\, v_{2mn}]\, X_m(x) \sin(n\theta+\theta_o) \qquad (6.29b)$$

$$w = \sum_{m=1}^{r} \sum_{n=1}^{t} [(1-\bar{z})w_{1mn} + \bar{z}\, w_{2mn}]\, X_m(x) \cos(n\theta+\theta_o) \qquad (6.29c)$$

or, written in matrix form,

$$\{f\} = \begin{Bmatrix} u \\ v \\ w \end{Bmatrix} = \sum_{m=1}^{r} \sum_{n=1}^{t} [N]_{mn} \{\delta\}_{mn} = [N]\{\delta\} \qquad (6.30)$$

in which $\bar{z} = (z/c)$, $\{\delta\}_{mn} = [u_{1mn} \;\; v_{1mn} \;\; w_{1mn} \;\; u_{2mn} \;\; v_{2mn} \;\; w_{2mn}]^T$, n is an integer representing the circumferential wave number and θ_o is a phase angle which is introduced for the study of axisymmetric cases (when n=0), in which $\theta_o = 0^o$ indicates a motion primarily radial and longitudinal and $\theta_o = 90^o$ indicates a torsional motion. The strain-displacement relationships are again given by Eq. 6.18, that is

$$\{\varepsilon\} = [B]\{\delta\} = \sum_{m=1}^{r} \sum_{n=1}^{t} [B]_{mn} \{\delta\}_{mn} \qquad (6.31)$$

The stresses corresponding to $\{\varepsilon\}$ are

$$\{\sigma\} = \begin{Bmatrix} \sigma_x \\ \sigma_y \\ \sigma_z \\ \tau_{xy} \\ \tau_{yz} \\ \tau_{zx} \end{Bmatrix} = [D]\{\varepsilon\} = [D] \sum_{m=1}^{r} \sum_{n=1}^{t} [B]_{mn} \{\delta\}_{mn} \qquad (6.32)$$

with the non-zero elastic constants equal to

$D_{11} = \eta E_x (1-\upsilon_{z\theta}\, \upsilon_{\theta z})$; $D_{12} = \eta E_x (\upsilon_{\theta z}+\upsilon_{\theta z}\, \upsilon_{zx}) = D_{21}$
$D_{22} = \eta E_\theta (1-\upsilon_{xz}\, \upsilon_{zx})$; $D_{23} = \eta E_\theta (\upsilon_{z\theta}+\upsilon_{zx}\, \upsilon_{x\theta}) = D_{32}$
$D_{33} = \eta E_z (1-\upsilon_{\theta x}\, \upsilon_{x\theta})$; $D_{13} = \eta E_x (\upsilon_{zx}+\upsilon_{\theta x}\, \upsilon_{z\theta}) = D_{31}$
$D_{44} = G_{z\theta}$; $D_{55} = G_{zx}$; $D_{66} = G_{x\theta}$
$\eta = 1/(1-\upsilon_{zx}\upsilon_{xz}-\upsilon_{\theta x}\upsilon_{x\theta}-\upsilon_{\theta z}\upsilon_{z\theta}-\upsilon_{\theta z}\upsilon_{zx}\upsilon_{x\theta}-\upsilon_{z\theta}\upsilon_{\theta x}\upsilon_{zx})$
with $\upsilon_{ij}\, E_j = \upsilon_{ji}\, E_i$ $\quad (i,j = x,\theta,z)$

Both the stiffness matrix and mass matrix are integrated numerically by Gaussian quadrature.

(b) Higher order cylindrical layer (HO3)

For such a layer a shape function given by Eq. 2.33 should be used. The displacement given by Nelson et al.[15] is of the following form:

$$u = \sum_{m=1}^{r} \sum_{n=1}^{t} [(1-3\bar{r}+2\bar{r}^2)u_1+(2\bar{r}^2-\bar{r})u_2+(4\bar{r}-4\bar{r}^2)u_3]\cos(n\theta)\sin(\tfrac{\pi z}{\lambda}) \qquad (6.33a)$$

$$v = \sum_{m=1}^{r} \sum_{n=1}^{t} [(1-3\bar{r}+2\bar{r}^2)v_1+(2\bar{r}^2-\bar{r})v_2+(4\bar{r}-4\bar{r}^2)v_3]\sin(n\theta)\cos(\tfrac{\pi z}{\lambda}) \qquad (6.33b)$$

$$w = \sum_{m=1}^{r} \sum_{n=1}^{t} [(1-3\bar{r}+2\bar{r}^2)w_1+(2\bar{r}^2-\bar{r})w_2+(4\bar{r}-4\bar{r}^2)w_3]\cos(n\theta)\cos(\tfrac{\pi z}{\lambda}) \qquad (6.33c)$$

in which λ is the axial wave length, and

$$\bar{r} = \frac{r-r_1}{c} = \frac{r-r_1}{r_2-r_1}$$

For the case of axisymmetric torsion (n=0), it is only necessary to interchange the $\sin(n\theta)$ and $\cos(n\theta)$ terms in the v, u and w expressions respectively of Eq. 6.33.

The wavelength λ is usually not specified, and the natural frequencies are usually given in terms of the axial wave number $\xi=t/\lambda$, where t is the thickness of the cylinder.

The stiffness matrix and mass matrices are given in Reference 16.

6.3.3 EXAMPLES

To demonstrate the accuracy and effectiveness of the present analysis, the numerical results presented are compared with previous analytical and experimental investigations for the free vibration of cylinders over a wide range of thickness-radius ratios and with various boundary conditions.

In Table 6.5 it is seen that the frequencies of isotropic freely supported solid cylinders obtained by Armenakas et al.[17] and Nelson et al.[15] using fifty HO3 layers, and, finally, by Cheung and Wu[9,16] using twenty LO2 layers, are all in excellent agreement. This shows that while both types of elements produce accurate results, the higher order finite layer analysis leads to much more complicated algebra and a significantly larger matrix without demonstrating a corresponding improvement in accuracy. Hence the choice of a simple linear variation of displacements through the thickness of a layer is entirely justified.

The LO2 layer was also applied to the analysis of a cylinder with layered materials, and the frequencies obtained (see Table 6.6) are again in very good agreement with those presented by Weingarten[18] and by Bert et al.[19]

TABLE 6.5

Dimensionless frequencies ($\Omega = \frac{\omega t}{\pi}\sqrt{\frac{\rho}{G}}$) for freely supported isotropic solid cylinders. (n=1, t/R=2.0, No of elements = 20, Poisson's ratio=0.3)

H/L	Method	Ω_1	Ω_3	Ω_5
0.01	Reference 17	0.000253	0.89677	2.04326
	H03	0.000253	0.89668	2.04305
	L02	0.000253	0.89705	2.04821
0.1	Reference 17	0.02423	0.90402	2.04855
	H03	0.02424	0.90393	2.04834
	L02	0.02423	0.90429	2.05356
0.2	Reference 17	0.08703	0.92654	2.06364
	H03	0.08703	0.92645	2.06343
	L02	0.08703	0.92681	2.06881
0.4	Reference 17	0.26658	1.02614	2.11383
	H03	0.26659	1.02606	2.11362
	L02	0.26660	1.02654	2.11941
0.6	Reference 17	0.46827	1.20161	2.17858
	H03	0.46828	1.20154	2.17837
	L02	0.46832	1.20199	2.18440
0.8	Reference 17	0.67316	1.39598	2.25398
	H03	0.67317	1.39590	2.25378
	L02	0.67327	1.39666	2.25985

TABLE 6.6

Comparison of natural frequencies (Hz) for a layered freely supported cylinder.[9]

No of axial half wave	No of circumferential wave	Reference 18	Reference 19	L02
1	3	1148	1149	1148.5
	5	502	502	502.2
	11	561	562	561.9
2	4	2146	2147	2146.3
	8	807	807	807.2
	12	769	770	770.2
3	4	3398	3399	3398.4
	5	2719	2720	2718.9
	7	1793	1794	1793.7

Data :

Layer	Thickness (m)	No of elements	E (kN/m^2)	Poisson's ratio	Density ($kN/s^2/m^4$)
Inside	0.000254	10	4.8265×10^6	0.30	1.31425
Middle	0.000127	5	206.85×10^6	0.30	8.52666
Outside	0.000254	10	4.8265×10^6	0.30	1.31425

TABLE 6.7

Comparison of natural frequencies (Hz) between experimental results and present analysis.[9]

No of circum-ferential wave	No of axial half wave	Frequencies					
		freely supported[a]		clamped-clamped[a]		clamped-free[b]	
		Reference 20	LO2	Reference 20	LO2	Reference 20	LO2
2	1		824.5		1071.0		329.2
	2		1892.9		2005.0		1217.2
	3		2509.4		2563.7		2188.1
3	1		461.5		708.8	155.0	174.8
	2		1292.8		1462.2		770.1
	3		1961.6		2038.3		1578.1
4	1	287.0	287.6		496.5	107.0	111.8
	2		900.9		1092.7		519.0
	3		1511.9		1609.4		1149.0
5	1	203.0	201.8	321.0	366.0	89.0	94.6
	2		651.6		838.6	341.0	372.5
	3		1172.1		1284.5		860.2
6	1	175.0	166.6	263.0	289.1	102.0	106.3
	2		493.6		662.2	276.0	288.5
	3		924.3		1043.3		666.3
7	1	163.0	166.2	233.0	249.9	130.0	134.7
	2		396.7	479.0	541.3	240.0	248.1
	3		747.4		865.4		537.4
8	1	188.0	189.3	227.0	242.1	166.0	172.6
	2	345.0	345.0	418.0	463.1	227.0	241.5
	3		625.2		736.4		456.8

Data : five identical layers ;
$E = 6.895 \times 10^7 kN/m^2$; Poisson's ratio = 0.30 ; density = $2.714 kN\ s^2/m^4$
$R = 0.24227m$; $t = 0.0006477m$; $L = 0.6096m$[a] ($0.6255m$[b])

For cylinders with various boundary conditions, the results obtained from LO2 finite layer analysis are compared with experimental measurements[20] in Table 6.7. Reasonable agreements can be observed. Only thin shells have been analyzed in this example because no previous investigations on free vibration of thick cylinders having any conditions other than simply supported are available for comparison.

REFERENCES

1. Taig, I. C., *Structural analysis by displacement method*, English Electric Aviation Ltd, Report SO 17 (1961) (unpublished).
2. Irons, B. M., *Stress analysis of stiffnesses using numerical integration*, Rolls Royce Ltd, ASM 622 June 1963 (internal report).
3. Ergatoudis, J., Irons, B. M. and Zienkiewicz, O. C., Curved isoparametric quadrilateral elements for finite element analysis. *Int J Solids Structures*, 7, 33-42, 1968.
4. Ergatoudis, J., Irons, B. M. and Zienkiewicz, O. C., Three dimensional analysis of arch dams and their foundations, *Symposium on Arch Dams, Institution of Civil Engineers, London*, March, 1968.
5. Irons, B. M. and Zienkiewicz, O. C., The isoparametric system - a new concept in finite element analysis. Conf - *Recent advances in stress analysis*, JBCSA, Royal Aeronautical Society London, March 1968.
6. Too, J. J. M., *Two-dimension plate, shell and finite prism isoparametric elements and their applications*, PhD thesis, Department of Civil Engineering, University of Wales, Swansea, 1971.
7. Zienkiewicz, O. C., *Finite element method*, 4th Ed., Mc-Graw Hill, London, 1989.
8. Bisplinghoff, R. L., Mar, J. W. and Pian, T. H. H., *Static of deformable solids*, Addison-Wesley, 1965.
9. Cheung, Y. K., *Finite strip method in structural analysis*, Pergamon Press, 1976.
10. Timoshenko, S. P. and Goodier, J. N., *Theory of elasticity*, 2nd Ed., McGraw-Hill, 1951.
11. Cheung, M. S. and Chan, M. Y. T., Three-dimensional finite strip analysis of elastic solids, *Comp and Struct*, 9, 629-638, 1978.
12. Cheung, Y. K. and Chakrabarti, S., Analysis of simply supported thick, layered plates, *J of Eng Mech, ASCE*, 97(3), 1039-1044, 1971.
13. Cheung, Y. K. and Chakrabarti, S., Free vibration of thick, layered rectangualr plates by finite layer method, *J of Sound and Vib*, 21(3), 277-284, 1972.

14. Warburton, G. B., The vibration of rectangular plates, *Proceedings of the Institution of Mechanical Engineers*, 168(12), 371-384, 1954.
15. Nelson, R. B., Dong, S. B. and Kalra, R. D., Vibrations and waves in laminated orthotropic circular cylinders, *J of Sound and Vib*, 18(3), 429-444, 1971.
16. Cheung, Y. K. and Wu, C. I., Free vibrations of thick, layered cylinders having finite length with various boundary conditions, *J of Sound and Vib*, 24(2), 189-200, 1972.
17. Armenakas, A. E., Gazis, D. C. and Herrmann, G., *Free vibration of cylindrical shells*, Pergamon Press, 1969.
18. Weingarten, V. I., Free vibration of multilayered cylindrical shells, *Exp Mech*, 4, 200-205, 1964.
19. Bert, C. W., Baker, J. L. and Egle, D. M., Free vibrations of multilayer anisotropic cylindrical shells, *J Composite Materials*, 3, 480-499, 1969.
20. Sewall, J. L. and Naumann, E. C., *An experimental and analytical vibration study of thin cylindrical shells with and without longitudinal stiffeners*, NASA TN D-4705, 1968.

CHAPTER SEVEN

DOMAIN TRANSFORMATION: TREATMENT FOR ARBITRARILY SHAPED STRUCTURES

7.1 INTRODUCTION

For some problems with complex geometry it is possible to obtain analytical solutions by applying appropriate transformations which map the complex domains into simpler ones such that analyses can be carried out more easily. Examples include the conformal transformation as well as the adoption of a curvilinear coordinate system in favour of the Cartesian coordinates.

In finite element analyses of problems with complex geometrical shapes, simple rectangular and triangular elements can still be used but a large number of elements will be required to minimize the geometrical error. To reduce the number of elements used, Taig[1] suggested the mapping of the geometry into simpler shapes by coordinate transformation. Such transformation functions usually take the same form as the displacement shape functions. Irons[2,3] generalized this concept which is now commonly referred to as isoparametric mapping. Some practical methods of generating curved surfaces for engineering design led to the establishment of a similar definition by Coons[4] and Forrest[5] independently. To further minimize the geometrical error, blending function can be used as the mapping function and thus another form of mapping has been introduced.[6,7] This method was originally developed for representation of complex motor-car body shapes.

The same domain transformation concept was introduced by Li et al.[8] into the finite strip method. Since the finite strip method requires the shape of the structures to be rectangular, the aim of the domain transformation is to transfer a non-regular structure into a rectangular shape mathematically. Various forms of transformation have been proposed and the essential points will be discussed in this chapter. Different ways for the representation of arbitrarily shaped plane curves are first discussed, and this is followed by examples as illustrations of their applications in describing the geometry of plates, shells and box-girder bridges.

7.2 PARAMETRIC REPRESENTATION OF ARBITRARY PLANE CURVES

The boundaries of a structure can adopt different forms of curves, such as straight lines (degenerated curves), circles, parabolas, spirals, etc. To define the geometry of a structure, one has to describe these boundary curves accurately and conveniently. Parametric representations, which permit description of various forms of curves by using the same set of algebraic equations by varying the coefficients, are definitely very attractive. Depending on the accuracy one tries to achieve in the approximation of a curve, various parametric representations can be chosen. In the following sub-sections, some of the most common parametric representations are discussed.

7.2.1 CONIC CURVES

Conic curves (Figure 7.1), including the circular curve, are often used in engineering practice and can be easily expressed by the parametric equations in terms of η as

$$x_o(\eta) = A_1\cos T + A_2\sin^2 T + \frac{A_3}{\cos T} - A_1\cos A_s - A_2\sin^2 A_s - \frac{A_3}{\cos A} \qquad (7.1a)$$

$$y_o(\eta) = B_1\sin T + B_2\tan T - B_1\sin A_s - B_2\tan A_s \qquad (7.1b)$$

where $T = \frac{A_e - A_s}{2}(1 + \eta) + A_s$, where $1 \geq \eta \geq -1$. The parameters A_1, A_2, A_3, A_e, A_s, B_1 and B_2 depend on the geometry of the curve and are listed in Table 7.1 and Figure 7.1.

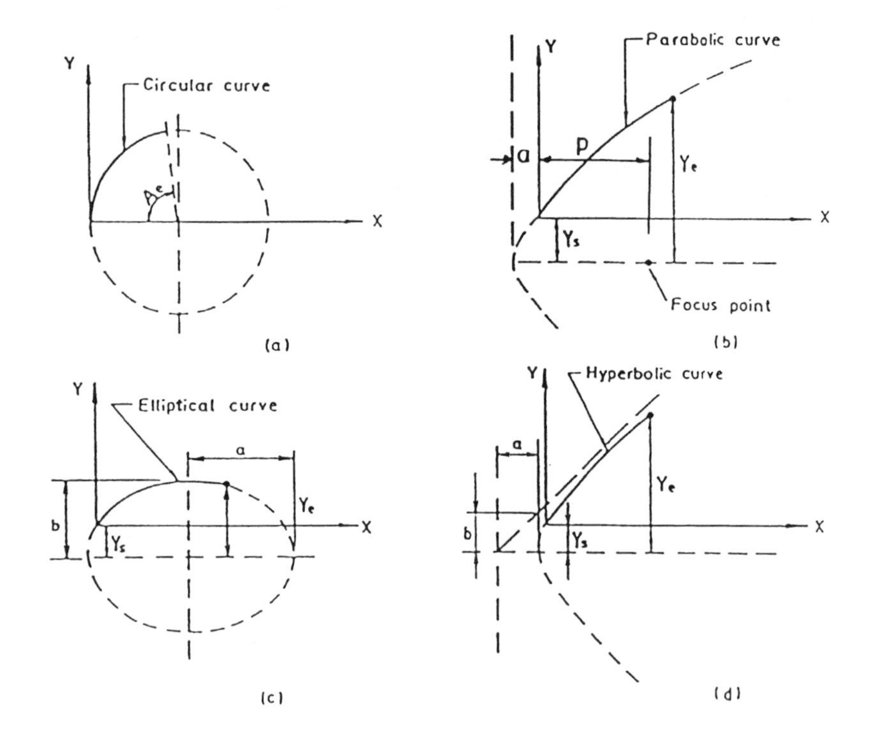

Figure 7.1 Conic curves: (a) circular curve. (b) parabolic curve. (c) elliptical curve. (d) hyperbolic curve.

TABLE 7.1
Parameters for conic curve.

Curve type	A_1	A_2	A_3	A_s	A_e	B_1	B_2
Circular	-R	0	0	0	0	R	0
Elliptical	a	0	0	$\sin^{-1} y_s/b$	$\sin^{-1} y_e/b$	b	0
Parabolical	0	$a^2/2p$	0	$\sin^{-1} y_s/a$	$\sin^{-1} y_e/a$	a	0
Hyperbolical	0	0	a	$\tan^{-1} y_s/b$	$\tan^{-1} y_e/b$	o	b

7.2.2 CUBIC SERENDIPITY SHAPE FUNCTION

It can be shown readily that any given arbitrary curve can be approximated by a Serendipity shape function. For example, the relations for an approximation by the cubic member of the family are

$$x_o(\eta) = \frac{1}{16}[x_3(9\eta^2 - 1)(\eta + 1) - 9x_2(3\eta + 1)(\eta^2 - 1) + 9x_1(3\eta - 1)(\eta^2 - 1)] \tag{7.2a}$$

$$y_o(\eta) = \frac{1}{16}[y_3(9\eta^2 - 1)(\eta + 1) - 9y_2(3\eta + 1)(\eta^2 - 1) + 9y_1(3\eta - 1)(\eta^2 - 1)] \tag{7.2b}$$

where x_i and y_i ($i = 1,2,3$) are the coordinates of three control points of the curve and the curve is assumed to be passing through the origin (0,0) of the coordinate system (Figure 7.2).

Some geometrical errors are inevitable for complex curves but it has been proven that acceptable accuracy can still be achieved.

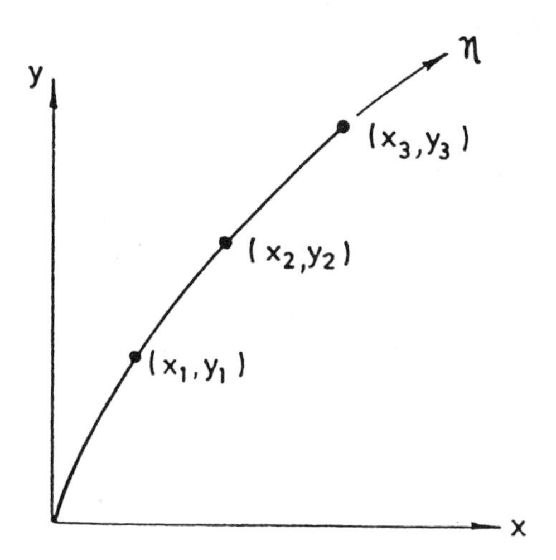

Figure 7.2 Cubic Serendipity shape functions.

7.2.3 SPLINE FUNCTIONS

For curves with complex geometry, it is more suitable to adopt the B-3 spline function for their representations. Thus we have

$$x_o(\eta) = T \tag{7.3a}$$

$$y_o(\eta) = \sum_{i=-1}^{m} a_i \, \phi_i(T) \tag{7.3b}$$

where $T = \frac{X_e}{2}(\eta+1)$; ϕ_i is the B-3 spline function with m-sections (Eqs. 2.39a and 2.39b). X_e is the coordinates of the end point of the curve.

The coefficients a_i depend on the coordinates of the control points and the slope of the curve at its two ends (Figure 7.3). The advantage of using such a function for complex curves is that it is possible to select more control points to make a better approximation while the polynomial is kept at the same order.

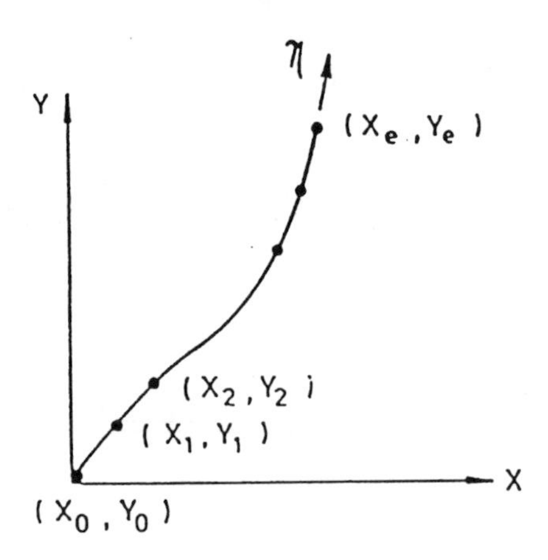

Figure 7.3 Spline functions.

7.2.4 ISOMETRIC FAMILY OF CURVES

Irrespective of the form of representation for a given curve, it is always possible to generate an isometric family of curves (Figure 7.4) from such a curve, which is referred to as the generating curve. An arbitrary generating curve can be defined by one of the parametric equations defined in the previous section:

$$x = x_o(\eta) \quad ; \quad y = y_o(\eta) \tag{7.4}$$

where $x_o(\eta)$ and $y_o(\eta)$ are arbitrary functions of the parameter η and $-1 \leq \eta \leq 1$.

An isometric family of curves generated on a plane can be defined in terms of the parameter d_o as:

$$x = x_o(\eta) + d_o \sin\varphi \quad ; \quad y = y_o(\eta) - d_o \cos\varphi \tag{7.5}$$

where $\sin\varphi = \frac{dy_o(\eta)}{d\eta} R(\eta)$; $\cos\varphi = \frac{dx_o(\eta)}{d\eta} R(\eta)$

$$R(\eta) = [(\frac{dy_o(\eta)}{d\eta})^2 + (\frac{dx_o(\eta)}{d\eta})^2]^{1/2}$$

Every value of d_o defines a new curve, which is parallel to the generating curve and at a distance d_o from it. This means that for every value of η, the straight line segment joining the two corresponding points on the two curves is the common normal of the curves and its length is equal to d_o. φ is the angle that the normal of the generating curve at point η makes with the x-axis as shown in Figure 7.4.

This family of curves can be easily extended to the spatial case, such that

$$x = x_o(\eta) + d_o(z_o) \sin\varphi \quad (7.6a)$$
$$y = y_o(\eta) - d_o(z_o) \cos\varphi \quad (7.6b)$$
$$z = z_o \quad (7.6c)$$

Note that the curves are in the plane $z = z_o$. $d_o(z_o)$ is defined in Figure 7.4 . When the curve is projected onto the x-y plane, $d_o(z_o)$ and φ have the same meaning as given in Eq. 7.5.

It will be shown at a later stage that such isometric families of curves play an important role in defining the geometry of shells and box girder bridges.

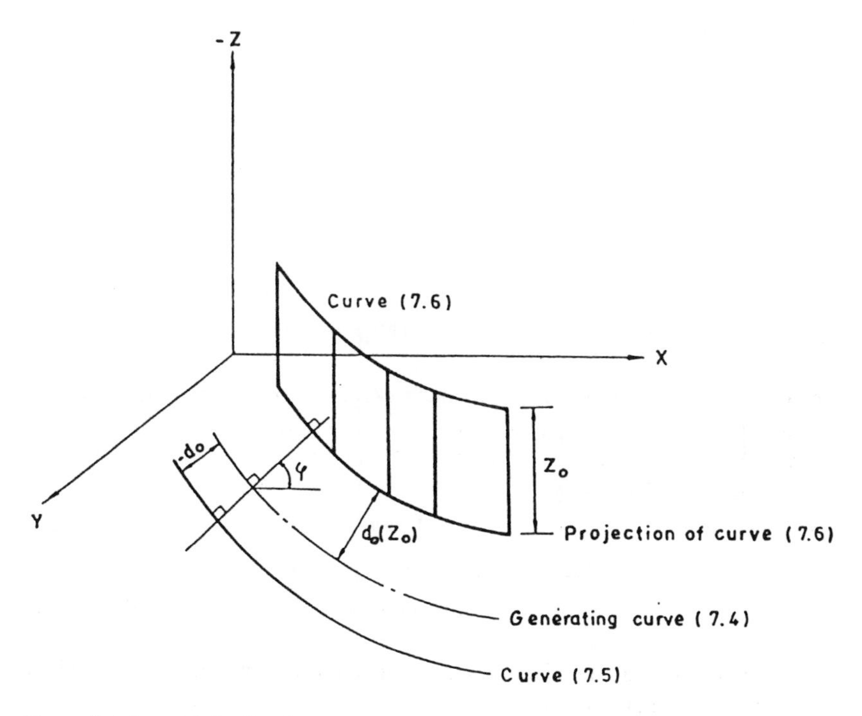

Figure 7.4 Isometric family of curve.

7.3 TRANSFORMATION FOR ARBITRARILY SHAPED QUADRILATERAL PLATES

In many plate problems, quadrilateral plates with curved or straight edges may be encountered. As it is a prerequisite that the domain must be mapped into a rectangular domain, simple and effective transformation methods which can transform an arbitrarily shaped quadrilateral plate to a rectangular domain will be presented in this section.

7.3.1 CUBIC SERENDIPITY SHAPE FUNCTION

If the shape of the quadrilateral plate is not very complicated, the Serendipity shape functions will be adequate for the transformation. The method has already been highlighted in the previous chapter by using the eight noded element as an example.

Here a transformation based on the cubic Serendipity shape functions will be discussed in detail. For an arbitrarily shaped quadrilateral plate shown in Figure 7.5, four points on each curved side are selected as the control points for the transformation. The plate can then be mapped approximately into the region [-1,1]x[-1,1] in the ξ-η plane (Figure 7.5) by using the following cubic Serendipity shape functions, that is

$$x = \sum_{i=1}^{12} N_i(\xi,\eta)\, x_i \quad ; \quad y = \sum_{i=1}^{12} N_i(\xi,\eta)\, y_i \tag{7.7}$$

where (x_i, y_i) are the coordinates of the i-th boundary node of the plate and $N_i(\xi,\eta)$ are the shape functions as given in Table 2.2.

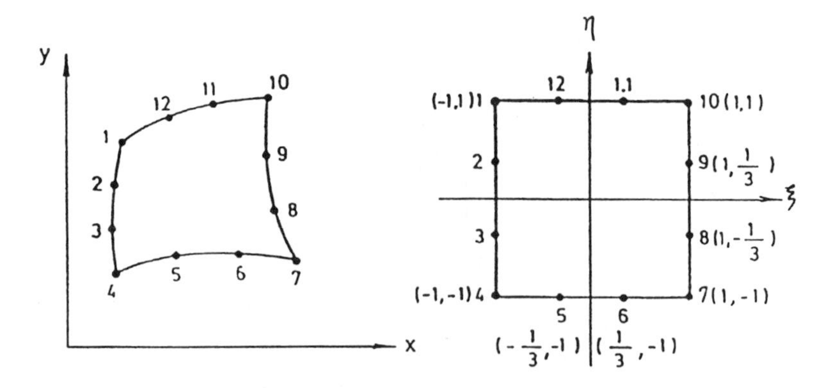

Figure 7.5 Transformation by cubic Serendipity function.

7.3.2 BLENDING FUNCTIONS[9,10]

For plates with complex geometries, it is necessary to develop a generalized mapping technique to ensure a close fitting of the boundary curves to minimize the geometrical error. Unlike the cubic Serendipity shape function method, this technique employs the blending functions for the transformation, by which the exact transformation of the arbitrary quadrilateral plate can be achieved. The main points are illustrated as follows:

Expressing the boundaries of the domain (Figure 7.6) in terms of a parameter t, we have similar to Eq. 7.4:

$$x = X_i(t) \; ; \; y = Y_i(t) \qquad \text{for } -1 \leq t \leq 1 \tag{7.8}$$

where i=1,2,3,4 represent the four sides of the domain. $X_i(\eta)$ and $Y_i(\eta)$ can be defined by the parametric representations given in the previous section. The transformation for the general case then becomes:

$$x = \Phi_{1x} + \Phi_{2x} - \Phi_{3x} \tag{7.9a}$$

$$y = \Phi_{1y} + \Phi_{2y} - \Phi_{3y} \tag{7.9b}$$

where

$$\Phi_{1x} = N(\eta)X_4(\xi) + M(\eta)X_2(\xi)$$

$$\Phi_{2x} = N(\xi)X_3(\eta) + M(\xi)X_1(\eta)$$

$$\Phi_{3x} = N(\xi)M(\eta)X_3(1) + N(\xi)M(\eta)X_2(1) + N(\xi)M(\eta)X_4(-1) + N(\xi)M(\eta)X_1(-1)$$

$$N(r) = \frac{1+r}{2} \; ; \; M(r) = \frac{1-r}{2} \qquad (r = \xi \text{ or } \eta)$$

Similarly, Φ_{1y}, Φ_{2y} and Φ_{3y} can be obtained by interchanging x with y. In this way, the given domain is transformed into [-1,1]x[-1,1] in the ξ-η plane (Figure 7.6).

If the parametric equations of the curved sides cannot be expressed exactly, the transformation is , of course, still an approximate one. Fortunately, most of the curves encountered in practical cases can be expressed by exact parametric equations.

7.3.3 AN EXACT TRANSFORMATION METHOD FOR PLATES WITH A PAIR OF STRAIGHT PARALLEL EDGES

For a plate with a pair of straight opposite edges (Figure 7.7), one can express the curved boundaries exactly by the following parametric equations:

$$x = X_i(\eta) \; ; \; y = Y_i(\eta) \qquad \text{for } -1 \leq \eta \leq 1 \tag{7.10}$$

with i= 1 or 2 representing one of the two curved sides respectively.

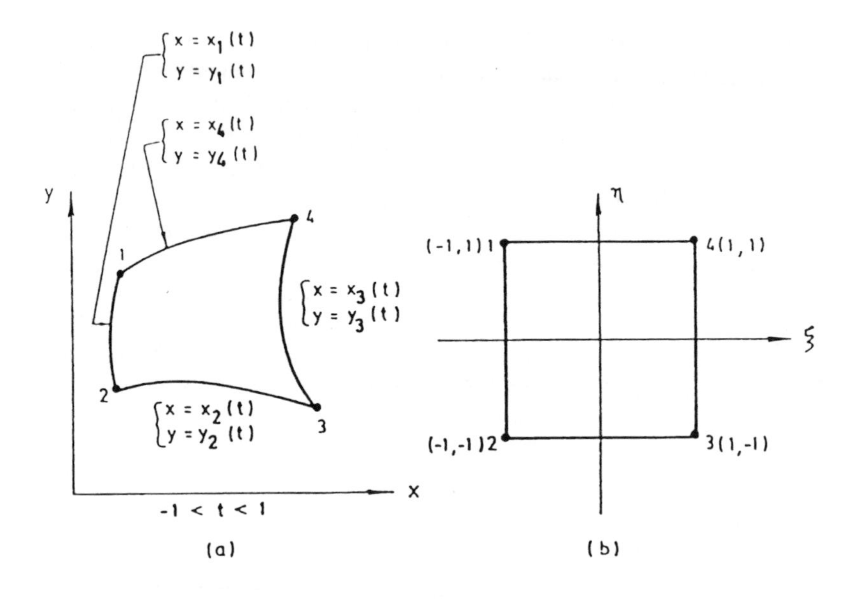

Figure 7.6 Transformation by blending functions.

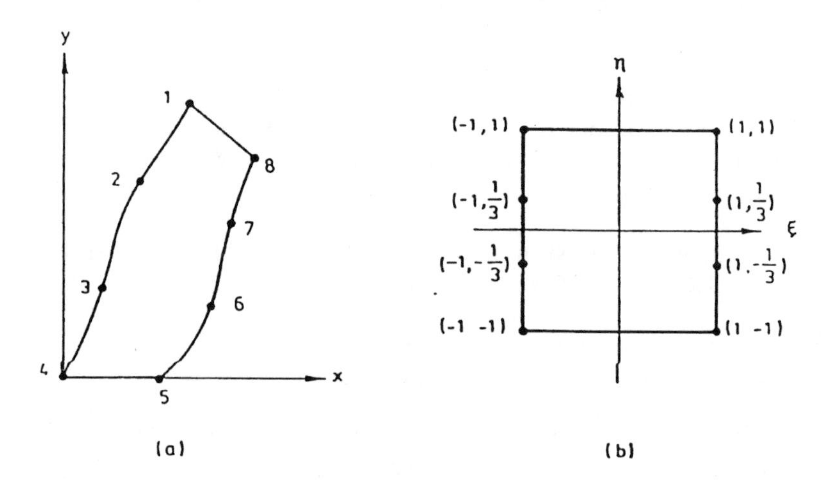

Figure 7.7 Transformation for plates with a pair of straight opposite edges.

In this case, a suitable transformation function for the domain can be shown readily to be:

$$x = \frac{1}{2}\,[X_1(\eta)\,(1 - \xi) + X_2(\eta)\,(1 + \xi)] \tag{7.11a}$$

$$y = \frac{1}{2}\,[Y_1(\eta)\,(1 - \xi) + Y_2(\eta)\,(1 + \xi)] \tag{7.11b}$$

7.3.4 RELATED TRANSFORMATION MATRICES

In our discussion, all the analyses of arbitrarily shaped plates are carried out in the ξ-η plane. For convenience of description, let the transformation be

$$x = X(\xi,\eta)\ ;\ y = Y(\xi,\eta) \tag{7.12}$$

The first and second derivatives of any variables, say, displacement w, related to x and y should be then replaced by the derivatives related to ξ and η in the analysis. These two types of derivatives are related by

a) first derivatives

$$\begin{Bmatrix} \frac{\partial w}{\partial x} \\ \frac{\partial w}{\partial y} \end{Bmatrix} = \begin{bmatrix} \frac{\partial \xi}{\partial x} & \frac{\partial \eta}{\partial x} \\ \frac{\partial \xi}{\partial y} & \frac{\partial \eta}{\partial y} \end{bmatrix} \begin{Bmatrix} \frac{\partial w}{\partial \xi} \\ \frac{\partial w}{\partial \eta} \end{Bmatrix} \tag{7.13}$$

b) second derivatives

$$\begin{Bmatrix} \frac{\partial^2 w}{\partial y^2} \\ \frac{\partial^2 w}{\partial x^2} \\ 2\frac{\partial^2 w}{\partial x \partial y} \end{Bmatrix} = [T_B] \begin{Bmatrix} \frac{\partial^2 w}{\partial \eta^2} \\ \frac{\partial w}{\partial \eta} \\ \frac{\partial^2 w}{\partial \xi^2} \\ \frac{\partial w}{\partial \xi} \\ 2\frac{\partial^2 w}{\partial \xi \partial \eta} \end{Bmatrix} \tag{7.14}$$

where $[T_B] =$

$$\left[\begin{array}{c:c:c:c:c} (\frac{\partial \eta}{\partial y})^2 & \frac{\partial \xi}{\partial y}\frac{\partial}{\partial \xi}(\frac{\partial \eta}{\partial y}) + \frac{\partial \eta}{\partial y}\frac{\partial}{\partial \eta}(\frac{\partial \eta}{\partial y}) & (\frac{\partial \xi}{\partial y})^2 & \frac{\partial \xi}{\partial y}\frac{\partial}{\partial \xi}(\frac{\partial \xi}{\partial y}) + \frac{\partial \eta}{\partial y}\frac{\partial}{\partial \eta}(\frac{\partial \xi}{\partial y}) & \frac{\partial \eta}{\partial y}\frac{\partial \xi}{\partial y} \\ \hdashline (\frac{\partial \eta}{\partial x})^2 & \frac{\partial \xi}{\partial x}\frac{\partial}{\partial \xi}(\frac{\partial \eta}{\partial x}) + \frac{\partial \eta}{\partial x}\frac{\partial}{\partial \eta}(\frac{\partial \eta}{\partial x}) & (\frac{\partial \xi}{\partial x})^2 & \frac{\partial \xi}{\partial x}\frac{\partial}{\partial \xi}(\frac{\partial \xi}{\partial x}) + \frac{\partial \eta}{\partial x}\frac{\partial}{\partial \eta}(\frac{\partial \xi}{\partial x}) & \frac{\partial \eta}{\partial x}\frac{\partial \xi}{\partial x} \\ \hdashline 2\frac{\partial \eta}{\partial y}\frac{\partial \eta}{\partial x} & 2(\frac{\partial \xi}{\partial y}\frac{\partial}{\partial \xi}(\frac{\partial \eta}{\partial x}) + \frac{\partial \eta}{\partial y}\frac{\partial}{\partial \eta}(\frac{\partial \eta}{\partial x})) & 2\frac{\partial \xi}{\partial y}\frac{\partial \xi}{\partial x} & 2(\frac{\partial \xi}{\partial y}\frac{\partial}{\partial \xi}(\frac{\partial \xi}{\partial x}) + \frac{\partial \eta}{\partial y}\frac{\partial}{\partial \eta}(\frac{\partial \xi}{\partial x})) & \frac{\partial \eta}{\partial x}\frac{\partial \xi}{\partial y} + \frac{\partial \xi}{\partial x}\frac{\partial \eta}{\partial y} \end{array}\right]$$

and $\partial\xi/\partial x$, $\partial\xi/\partial y$, $\partial\eta/\partial x$ and $\partial\eta/\partial y$ can be obtained by differentiating Eq. 7.12:

$$1 = \frac{\partial X}{\partial \xi}\,\frac{\partial \xi}{\partial x} + \frac{\partial X}{\partial \eta}\,\frac{\partial \eta}{\partial x}\quad ;\quad 0 = \frac{\partial Y}{\partial \xi}\,\frac{\partial \xi}{\partial x} + \frac{\partial Y}{\partial \eta}\,\frac{\partial \eta}{\partial x} \tag{7.15a, b}$$

$$0 = \frac{\partial X}{\partial \xi}\frac{\partial \xi}{\partial y} + \frac{\partial X}{\partial \eta}\frac{\partial \eta}{\partial y} \quad ; \quad 1 = \frac{\partial Y}{\partial \xi}\frac{\partial \xi}{\partial y} + \frac{\partial Y}{\partial \eta}\frac{\partial \eta}{\partial y} \qquad (7.16a, b)$$

7.3.5 SOME APPLICATIONS OF DOMAIN TRANSFORMATION IN THE ANALYSIS OF PLATES

In this section, examples of circular and triangular plates are presented to demonstrate the applications of domain transformation. Since cubic Serendipity shape functions can map the shapes of these plates accurately into square domains, they have been used in these examples. In the case of circular plates, the control points are equally spaced around the circumference, whereas the control points for triangular plates (for example, 45°-90°-45°) are as shown in Figure 7.8.

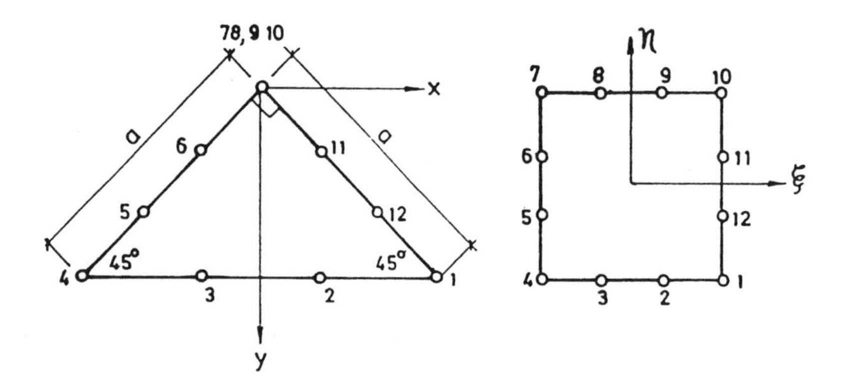

Figure 7.8 Mapping for triangular plates.

TABLE 7.2
Mid-point deflection for a circular plate. (radius R =1; D=1; υ = 0.3)[11]

Load Case	Method (Mesh)		Boundary condition: simply-supported	Boundary condition: clamped
uniformly distributed load (q=1)	spline strip	(12x12)	0.06383	0.01567
		(8x8)	0.06391	0.01568
	Analytical[12]		0.06370	0.01563
Point load at r =1 (P = 1)	spline strip	(12x12)	0.05046	0.01985
		(8x8)	0.05040	0.01975
	Analytical[12]		0.05050	0.01989

Both simply-supported and clamped circular plates with unit radius had been analyzed by Li[11] and the results are checked against the closed form solutions.[12] The central deflections of the plates under uniformly distributed

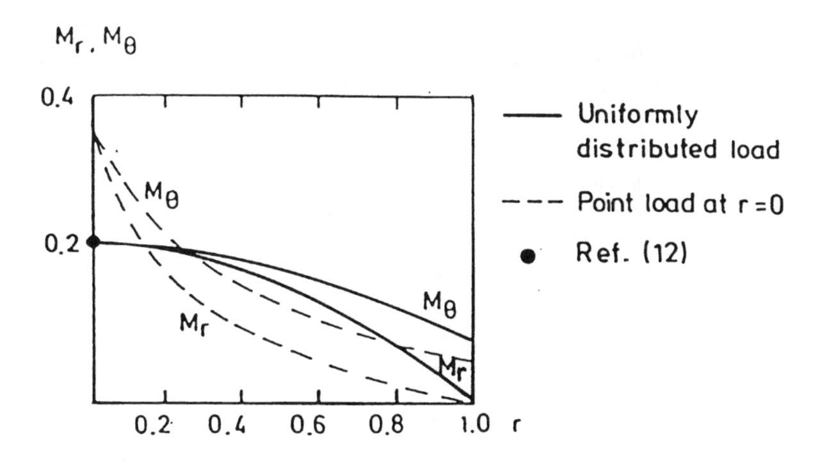

Figure 7.9 Distribution for M_r and M_θ for a simply supported circular plate.[11]

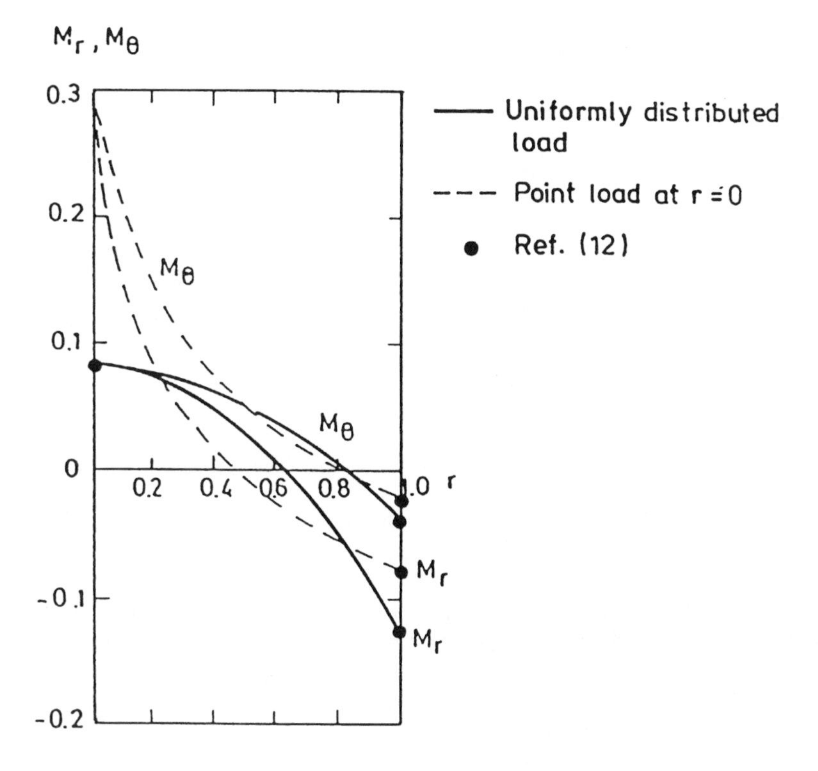

Figure 7.10 Distribution for M_r and M_θ for a clamped circular plate.[11]

load and a point load are tabulated in Table 7.2. It can be seen that the results of the present method converge rapidly. In addition, the moments (M_r and M_θ) also demonstrated good agreement (Figures 7.9 and 7.10) with the analytical results.[12]

The present method can also handle mixed support conditions easily.[11] A circular plate with half of the circumference clamped and the remaining half free had been analyzed and the distribution of moments is depicted in Figure 7.11.

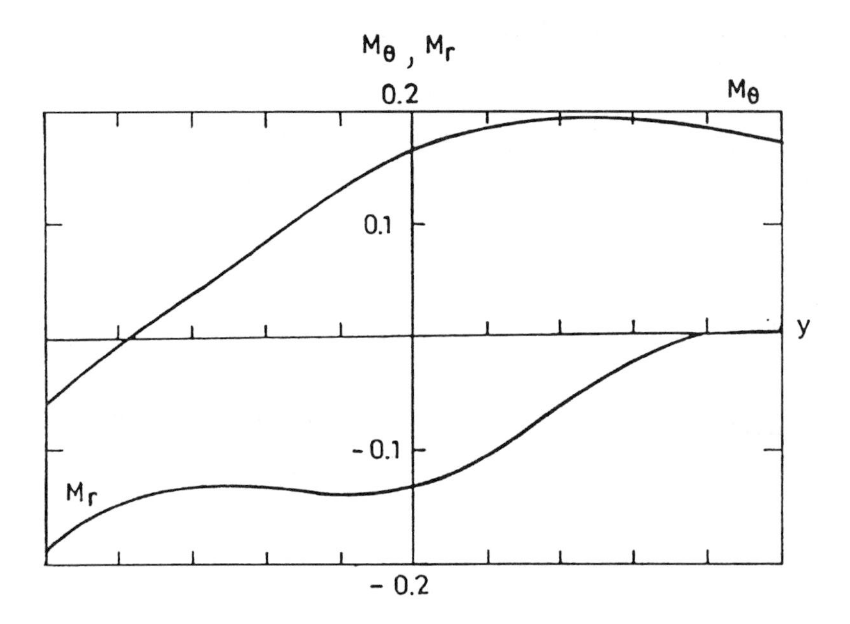

Figure 7.11 Distribution of moment for partially clamped circular plate under uniformly distributed load. ($\upsilon = 0.3$; $D = 1$; $R = 1$)[11]

The method can also be applied to free vibration analyses.[11] The analytical solution[13] for circular plates provides an excellent check on the accuracy of the present method. Both simply supported and clamped plates are analyzed. The lowest six eigenvalues obtained by a (6x6) mesh are given in Tables 7.3 and 7.4. There is no doubt whatsoever that very high accuracy can be achieved by the transformed spline strips.

A 45°-90°-45° plate with three edges simply supported under a uniformly distributed load was also studied. In this case, one side of the quadrilateral is degenerated to a point to form a triangle in the program. The displacements and moments along the symmetrical axis are again compared with Timoshenko's.[12] Note that the results listed in Table 7.5 are in good agreement.

TABLE 7.3
Eigenvalues of circular plate clamped at circumference.[11]

Mode order	$\omega R^2 \sqrt{\frac{\rho t}{D}}$	
	Analytical solution[13]	spline strip
1	10.2158	10.2062
2	21.26	21.27
3	34.88	34.94
4	39.77	40.21
5	51.04	52.05
6	60.82	61.97

TABLE 7.4
Eigenvalues of circular plate simply supported at circumference.[11]

Mode order	$\omega R^2 \sqrt{\frac{\rho t}{D}}$	
	Analytical solution[13]	spline strip
1	4.977	4.927
2	13.94	13.88
3	25.65	25.54
4	29.76	29.84
5		40.30
6	48.51	48.93

TABLE 7.5
45°-90°-45° triangular plate simply supported at all edges under uniformly distributed load.
($w=\frac{\alpha}{100}\frac{qa^4}{Et^2}$; $M_{x1} = \beta qa^2/10$; $M_{y1} = \gamma qa^2/10$; $\upsilon = 0.3$)[11]

	Method	Point position			
		$x_1 = 0$	$x_1 = b/4$	$x_1 = b/2$	$x_1 = 3b/4$
α	spline strip	0.000	0.267	0.662	0.616
	Timoshenko	0.000	0.267	0.662	0.615
β	spline strip	-0.144	0.000	0.182	0.210
	Timoshenko	-0.134	0.000	0.181	0.208
γ	spline strip	0.140	0.166	0.180	0.131
	Timoshenko	0.134	0.166	0.181	0.130

7.4 TRANSFORMATION OF ARBITRARILY SHAPED SHELLS

7.4.1 THE GENERAL DESCRIPTION OF SHELLS

In general, the mid-surface of arbitrarily shaped shells can be described by curvilinear coordinates. In thin shell analysis, the orthogonal curvilinear coordinate system is most commonly used. The relationship between the two coordinate systems is:

$$x = X(\xi,\eta,\zeta) \ ; \ y = Y(\xi,\eta,\zeta) \ ; \ z = Z(\xi,\eta,\zeta) \tag{7.17}$$

where x, y, z are the coordinates of the Cartesian system, while ξ,η,ζ are the coordinates of the orthogonal curvilinear coordinate system. To facilitate future discussions, it is worthwhile to define at this stage the Lame's coefficients which represent the characteristics of the transformation and are the most important parameters in the analysis. They are:

$$H_1^2 = (\frac{\partial x}{\partial \xi})^2 + (\frac{\partial y}{\partial \xi})^2 + (\frac{\partial z}{\partial \xi})^2 \tag{7.18a}$$

$$H_2^2 = (\frac{\partial x}{\partial \eta})^2 + (\frac{\partial y}{\partial \eta})^2 + (\frac{\partial z}{\partial \eta})^2 \tag{7.18b}$$

$$H_3^2 = (\frac{\partial x}{\partial \varsigma})^2 + (\frac{\partial y}{\partial \varsigma})^2 + (\frac{\partial z}{\partial \varsigma})^2 \tag{7.18c}$$

where $H_1 = dS_1/d\xi$, $H_2 = dS_2/d\eta$, $H_3 = dS_3/d\zeta$. S_1, S_2 and S_3 are the differential of arc length along the curvilinear coordinate lines (ξ-line, η-line and ζ-line) respectively. In general, ζ is the normal direction of the mid-surface of the shell in the thin shell theory and H_3 is always equal to 1.

According to differential geometry, there are two directions at every point of a spatial surface, perpendicular to each other, along which the curvatures of the surface are maximum and minimum. The curves on this surface, of which the tangent coincides with the maximum or minimum curvature direction, are called curvature lines. It is more convenient to select the curvature lines to be the coordinate lines of the curvilinear coordinate system in the analysis.

If an arbitrarily shaped shell is bounded by the maximum and minimum curvature lines, one can show readily that it is possible to carry out the analysis by spline finite strip model without difficulty. To perform the analysis, the first step is to find out a simple and effective method to describe the arbitrarily shaped shells.

7.4.2 ORTHOGONAL CURVILINEAR COORDINATE SYSTEM

In terms of the expression of the isometric family of curves (Section 7.2.4), the mid-surface of one arbitrarily shaped shell can be described as

$$x = x_o(\eta) + f(\xi) \sin\varphi \tag{7.19a}$$

$$y = y_o(\eta) + f(\xi) \cos\varphi \tag{7.19b}$$

$$z = H\frac{(1+\xi)}{2} \quad (\text{for } -1 \le \eta \le 1 ; -1 \le \xi \le 1) \qquad (7.19c)$$

where H is the vertical height of the shell. Obviously, this type of shell is formed by a series of parallel curves and $f(\xi)$ represents the profile of the shell along the z-direction.

Noting that the three functions $x_o(\eta)$, $y_o(\eta)$ and $f(\xi)$ can be selected arbitrarily, it is not surprising to find that the shape of this type of shell can be rather arbitrary. This means that not only is the generating curve in the x-y plane of the shell arbitrary but also the profile curve is arbitrary as shown in Figure 7.12.

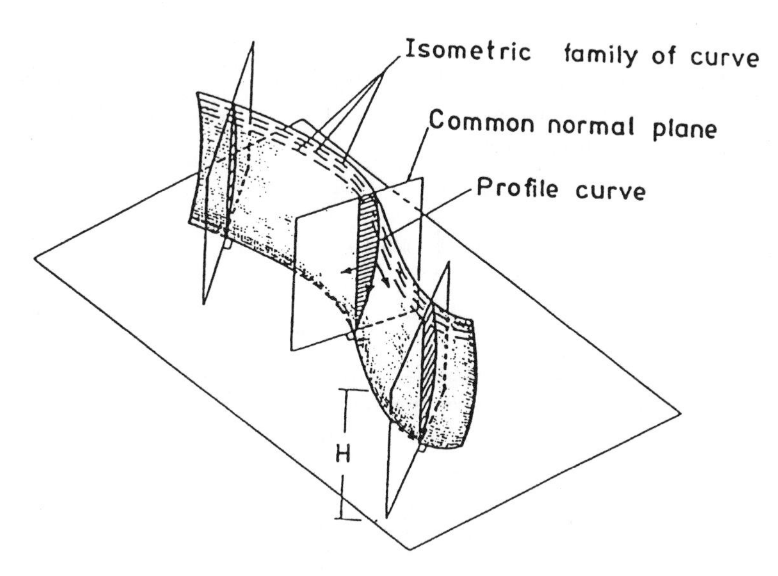

Figure 7.12 Isometric curve shell.

The relationship between the curvilinear coordinate system and the Cartesian system of this type of shell is:

$$x = x_o(\eta) + f(\xi)\sin\varphi + n_x\,\zeta \;\; ; \;\; y = y_o(\eta) + f(\xi)\cos\varphi + n_y\,\zeta \qquad (7.20a, b)$$

$$z = H\frac{(1+\xi)}{2} + n_z\,\zeta \qquad (7.20c)$$

where $\mathbf{n} = n_x\, i + n_y\, j + n_z\, k$ is the vector along the z-direction, in which **n** is the normal direction of the shell and $|n| = 1$.

Lame's coefficients on the mid-surface of the shell are

$$A = H_1 \Big|_{\varsigma=0} = [\frac{df(\xi)}{d\xi} + \frac{H^2}{4}]^{1/2} \qquad (7.21a)$$

$$B = H_2 \Big|_{\varsigma=0} = \frac{1}{R(\eta)}[1+K_o(\eta)f(\xi)] \qquad (7.21b)$$

$$C = H_3 \Big|_{\varsigma=0} = 1 \qquad (7.21c)$$

$$K_o(\eta) = (R(\eta))^3 \left[\frac{d^2y_o(\eta)}{d\eta^2} \frac{dx_o(\eta)}{d\eta} - \frac{d^2x_o(\eta)}{d\eta^2} \frac{dy_o(\eta)}{d\eta} \right] \quad (7.21d)$$

in which $K_o(\eta)$ is the curvature of the generating curve and

$$R(\eta) = \left[\left(\frac{dy_o(\eta)}{d\eta} \right)^2 + \left(\frac{dx_o(\eta)}{d\eta} \right)^2 \right]^{1/2}$$

Furthermore, it can be proved mathematically that the coordinate lines of the discussed system are also the curvature lines of the shell. The Lame's coefficients A, B and C will be further discussed in Section 8.1.1.

7.4.3 EXAMPLES

(1) A cylindrical panel with radius R is shown in Figure 7.13. The subtended angle for the panel is α and the height is H. It can be shown readily that the parametric form for such a panel is[11]

$$x = -R\cos\frac{1}{2}(1+\eta)\alpha + R \quad (7.22a)$$

$$y = R\sin\frac{1}{2}(1+\eta)\alpha \quad (7.22b)$$

$$z = \frac{1}{2}(1+\xi)H \quad (7.22c)$$

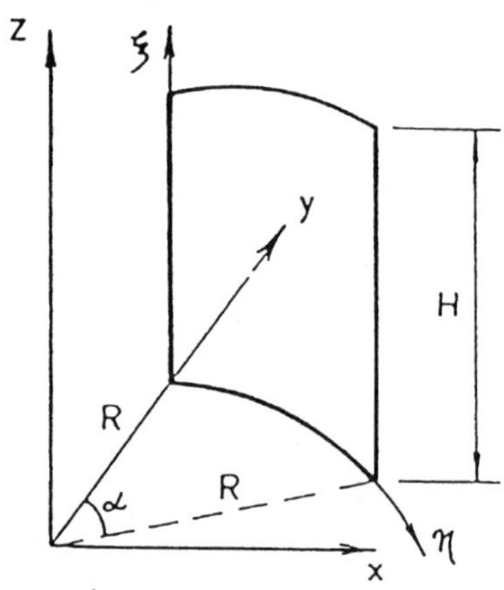

Figure 7.13 Cylindrical panel.

(2) The parametric form for the shallow spherical shell on square planform shown in Figure 7.14 can be shown to be:[11]

$$x = -R\cos\frac{1}{2}(1+\eta)\alpha + R + f(\xi)\sin\varphi \quad (7.23a)$$

$$y = R\sin\frac{1}{2}(1+\eta)\alpha - f(\xi)\cos\varphi \quad (7.23b)$$

$$z = \xi \quad (7.23c)$$

where $f(\xi) = R - R\cos[\sin^{-1}(\frac{\xi}{R})]$

$$\sin\varphi = (\frac{\alpha}{2})^2 \cos\frac{1}{2}(1+\eta)\alpha \;\; ; \;\; \cos\varphi = (\frac{\alpha}{2})^2 \sin\frac{1}{2}(1+\eta)\alpha$$

$$\varphi = 2\sin^{-1}\frac{a}{2r_o^2} \qquad (r_o = \sqrt{R^2 - 0.25a^2})$$

Adopting such representation, the boundaries will be defined as:

$$\eta = -1 \;\; ; \;\; \eta = 1$$

$$\xi = -\frac{a}{2} \;\; ; \;\; \xi = \frac{a}{2}$$

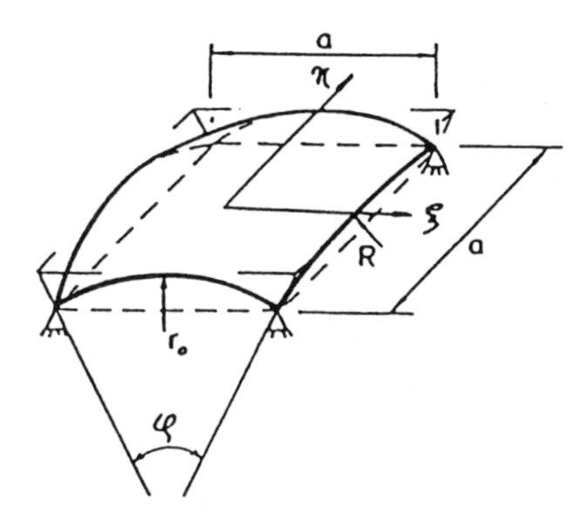

Figure 7.14 Spherical shell on square planform.

A hyperbolic paraboloidal shell on square planform (Figure 7.15) can also be defined by Eqs. 7.23a to 7.23c, except that $f(\xi)$ and r_o are defined respectively as:[(11)]

$$f(\xi) = R - (R^2 + \xi^2)^{1/2} \tag{7.24}$$

$$r_o = \sqrt{R^2 + 0.25a^2} \tag{7.25}$$

7.5 TRANSFORMATION FOR LONGITUDINALLY ARBITRARILY CURVED BOX GIRDERS

7.5.1 GEOMETRY OF STRUCTURES

A mixed plate-shell structure comprised of shells and curved plates (Figure 7.16) is considered herein. The structure is characterized by having longitudinal lines of its contour being parallel to the centre line as described by Eq. 7.17 and straight lines forming the transverse boundaries. Although the study is confined to box-girder structures, the geometry of open or closed section thin-walled curved beams and arches and curved stiffened panels can also be mapped into regular domain by this method.

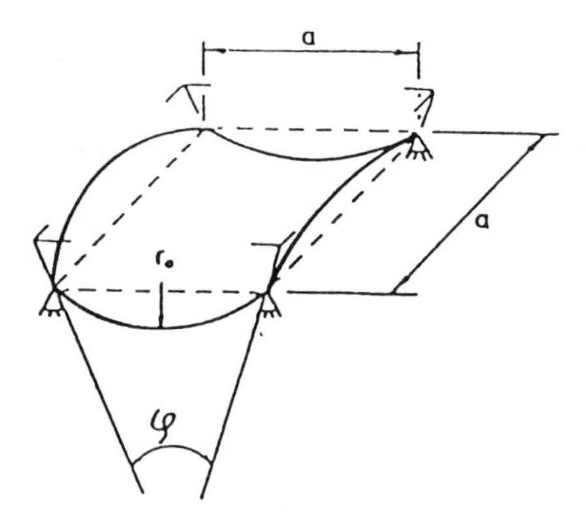

Figure 7.15 Hyperbolic paraboloidal shell on square planform.

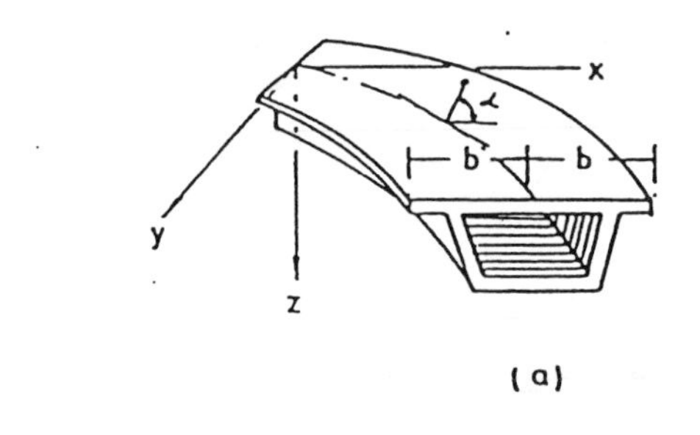

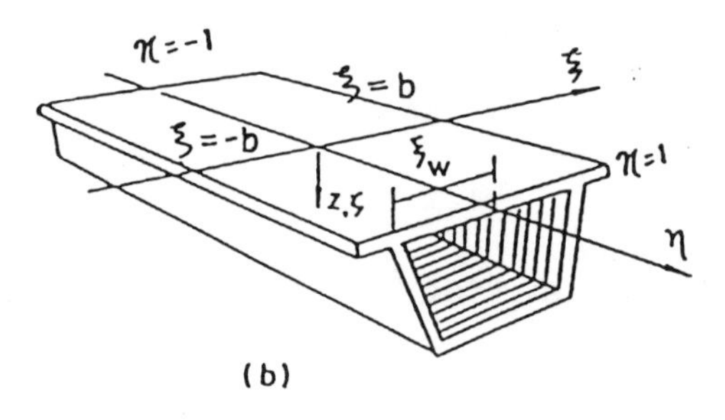

Figure 7.16 Horizontal curved box-girder bridge: (a) Cartesian coordinate. (b) Curvilinear coordinate system.

7.5.2 CURVILINEAR COORDINATE SYSTEM

Without any loss of generality, the top plane of the box-girder is selected as the x-y plane for analysis. The centre line is represented by the line $\xi = 0$ in the ξ-η plane, and it is related to the Cartesian coordinates by

$$x = x_o(\eta) \quad ; \quad y = y_o(\eta) \tag{7.26}$$

where $x_o(\eta)$ and $y_o(\eta)$ can be expressed by Eqs. 7.1, 7.2 and 7.3 or any other parametric expressions. For some simple cases, it is convenient to express $x_o(\eta)$ and $y_o(\eta)$ by Eq. 7.2 (the cubic Serendipity shape functions) and the accuracy is satisfactory for most cases.

In this type of structure, it is convenient to establish a separate coordinate system for every element of the structure.

The curvilinear coordinate system then can be expressed as

$$x = x_o(\eta) + (\xi \cos\phi + \zeta \sin\phi) \sin\varphi \tag{7.27a}$$

$$y = y_o(\eta) - (\xi \cos\phi + \zeta \sin\phi) \sin\varphi \tag{7.27b}$$

$$z = \zeta \cos\phi - \xi \sin\phi \tag{7.27c}$$

where φ is defined as Eqs. 7.5; the geometrical meaning is the same as in Section 7.2.4. For the curved plates, ϕ is equal to zero; while for the shell, ϕ can be found in Figure 7.10. In this way, every element of the structure is transformed into a rectangular domain mathematically (Figure 7.17).

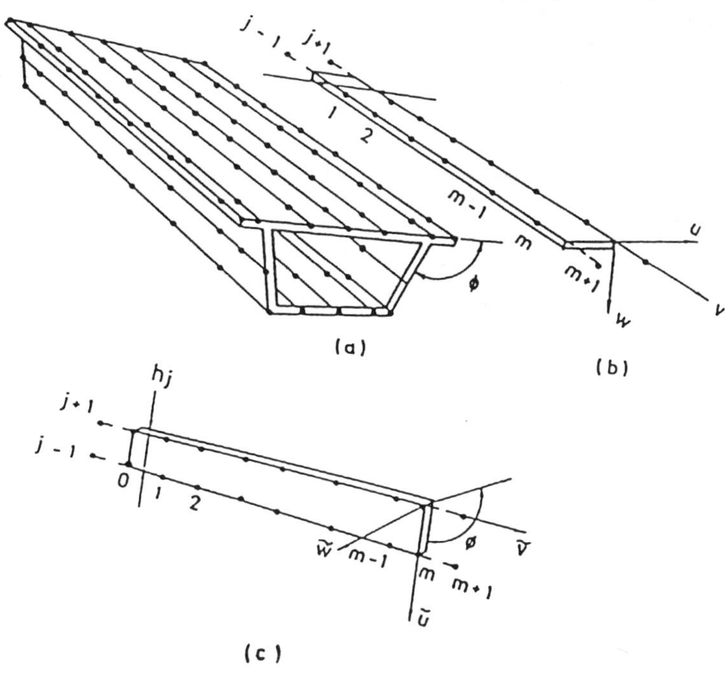

Figure 7.17 (a) Spline finite strip model. (b) Flange strip. (c) Web strip.

REFERENCES

1. Taig, I. C., *Structural analysis by the matrix displacement method*, English Electric Aviation Report No S017, 1961.

2. Irons, B. M., Numerical integration applied to finite element methods, *Conf use of digital computers in struct eng*, University of Newcastle-on-Tyne, 1966.
3. Irons, B. M., Engineering application of numerical integration in stiffness method, *J of AIAA*, 14, 2035-2037, 1966.
4. Coons, S. A., *Surfaces for computer aided design of space form*, MIT Project MAC, MAC-TR-41, 1967.
5. Forrest, A. R., *Curves and surfaces for computer aided design*, Computer Aided Design Group, Cambridge, England, 1968.
6. Gordon, W. J., Blending-function methods of bivariate and multivariate interpolation and approximation, *SIAM J Num Anal*, 8, 158-177, 1971.
7. Gordon, W. J. and Hall, C. A., Construction of curvilinear coordinate systems and application to mesh generation, *Int J for Num Meth of Eng*, 7, 461-477, 1973.
8. Li, W. Y., Cheung, Y. K. and Tham, L. G., Spline finite strip analysis of general plates, *J of Eng Mech, ASCE*, 112(1), 43-54, 1986.
9. Zienkiewicz, O. C., *The finite element method*, Third Edition, McGraw-Hill, UK.
10. Gordon, W. J. and Hall, C. A., Transfinite element methods blending-function interpolation over arbitrary curved element domains, *Num Math*, 21, 109-129, 1973.
11. Li, W. Y., *Spline finite strip analysis of arbitrarily shaped plates and shells*, PhD thesis, Department of Civil Engineering, University of Hong Kong, 1988.
12. Timoshenko, S. P. and Woinowsky-Krieger, S., *Theory of plates and shells*, 2nd Ed., McGraw-Hill, 1959.
13. Leissa, A. W., *Vibration of plates*, NASA SP-160, 1969.

CHAPTER EIGHT

APPLICATIONS TO SHELL STRUCTURES AND BRIDGES

8.1 SHELL STRUCTURES

The analysis of folded plate structures, a special type of shell structures, (Figure 8.1) has been the centre of attention for many researchers, and much information on this topic can be found in an ASCE report.[1] The elasticity method, which was developed by Goldberg and Leve[2] and subsequently programmed and applied by De Fries-Skene and Scordelis[3] as a stiffness approach, became quite popular because of its accuracy. The method however suffers from being fairly complex, and is difficult to apply to orthotropic folded plate structures and to dynamic analysis. The elasticity method has also been used in the study of simply supported box girder bridges by Scordelis[4] and by Chu and Dudnik.[5]

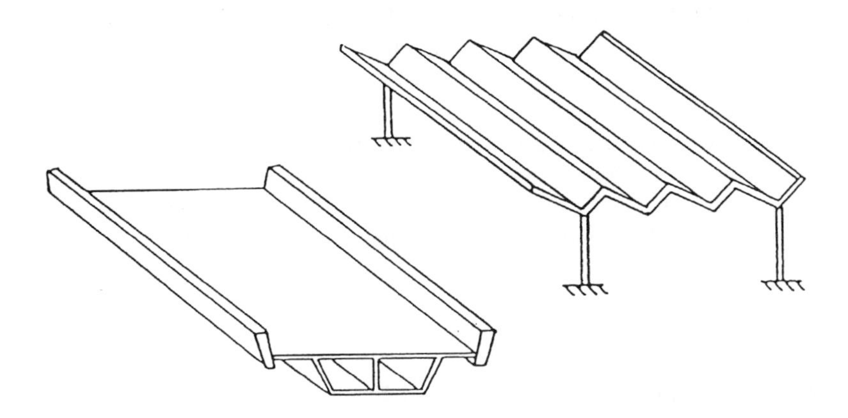

Figure 8.1 Folded plate structures.

Subsequently, the elasticity method has been extended to cover the analysis of curved box girder bridges,[6,7,8] but the disadvantages outlined above exist also for this case.

The finite strip analysis of prismatic folded plate structures and box girder bridge was introduced by Cheung[9,10] who combined together a third order bending strip and a linear plane stress strip to form a shell strip with both bending and membrane actions. The lower order shell was then applied to the study of orthotropic folded plates[11] and eccentrically stiffened folded plates.[12] A higher order flat shell strip with one internal nodal line was later developed by Loo and Cusens[13] and applied to the analysis of right box girder bridge decks. Solutions for curved folded plates and box girder bridges were presented by Meyer and Scordelis[14] and independently by Cheung and Cheung,[15] in which a conical frustrum shell strip was used. Furthermore, continuous folded plate structures were analyzed by Delcourt and Cheung.[16]

Recently, Fan[17] carried out the analyses of this class of structures by the spline strip model. Examples on prismatic folded structures, including box girders, were reported in Reference 18. Further extensions to circular curve bridges as well as shallow shells were made by Fan and Cheung.[19] Linear stability analysis of formed columns was a subject of study by Lau.[20] In addition, nonlinear analyses by the spline model were covered by Cheung and Zhu[21] as well as Kwon,[22] and a detailed exposition on this topic will be made in Chapter 11. Incorporating the concept of the domain transformation outlined in the last chapter, Li[23] presented solutions of the spline model for arbitrarily shaped shells and folded plate structures. He proposed a lower order strip model for singly curved shells and a higher order strip model, which can improve the rate of convergence, for doubly curved shells. On the other hand, the Mindlin plate theory was adopted in the analysis of folded plates by Tham[24] (spline model) and Kong[25] (computed shaped function model).

Puckett et al.[26] developed the compound strip method by modifying the finite strip method for the analysis of plates with transverse stiffeners or intermediate column supports. The stiffness of the stiffeners and columns is accounted for by the direct stiffness method. The interpolation functions of the compound strips are expressed as products of piecewise polynomials and trigonometrical series as in the case of classical strips. Maleki[27] also adopted this approach in his study of folded plates and box girders with intermediate supports. Spline compound strips were proposed by Chen et al.[28] to analyze folded plates.

8.1.1 GENERAL SHELL THEORIES

Based on different assumptions and approximations, different shell theories have been proposed. Delving into the details of such theories is beyond the scope of this book, but some of these theories are rewritten into forms more readily applicable to the strip models. In addition, this textbook focuses on several types of shells discussed in Chapter 7.

In the orthogonal curvilinear coordinate system, the following geometrical relations can be established:

$$k_1 = \frac{H}{2A^3}\frac{d^2 f(\xi)}{d\xi^2} \quad ; \quad k_2 = -\left(\frac{HK_0}{2ABR}\right) \tag{8.1}$$

$R = R(\eta)$; $A = A(\xi)$; $B = B(\xi,\eta)$; $K_o = K(\eta)$.

where A and B are the values of the Lame's coefficients on the mid-surface of the shell and their expressions have been given in Chapter 7 (Eqs. 7.21a and 7.21b); k_1 and k_2 are the curvatures along ξ and η respectively. K_o and $R(\eta)$ are also defined in Chapter 7.

Since A and k_1 are only functions of ξ (Eq. 8.1), Donnell[29] showed that the strain and displacement relationships for a shell can be written as:

$$\varepsilon_\xi{}^d = \frac{1}{A}\frac{\partial u}{\partial \xi} + k_1 w \tag{8.2a}$$

$$\varepsilon_\eta{}^d = \frac{1}{B}\frac{\partial v}{\partial \eta} + \frac{1}{AB}\frac{\partial B}{\partial \xi} u + k_2 w \tag{8.2b}$$

$$\varepsilon_{\xi\eta}{}^d = \frac{1}{B}\frac{\partial u}{\partial \eta} + \frac{B}{A}\frac{\partial}{\partial \xi}\left(\frac{v}{B}\right) \tag{8.2c}$$

$$\chi_\xi{}^d = -\frac{1}{A}\frac{\partial}{\partial \xi}\left(\frac{1}{A}\frac{\partial w}{\partial \xi}\right) \tag{8.2d}$$

$$\chi_\eta{}^d = -\frac{1}{B}\frac{\partial}{\partial \eta}\left(\frac{1}{B}\frac{\partial w}{\partial \eta}\right) - \frac{1}{A^2 B}\frac{\partial B}{\partial \xi}\frac{\partial w}{\partial \xi} \tag{8.2e}$$

$$2\chi_{\xi\eta}{}^d = -\frac{2}{AB}\left(\frac{\partial^2 w}{\partial \xi \partial \eta} - \frac{1}{B}\frac{\partial B}{\partial \xi}\frac{\partial w}{\partial \eta}\right) \tag{8.2f}$$

The superscript 'd' denotes Donnell's Theory. The first three strains represent the in-plane components whereas the other three are the bending components.

Reissner[30] included higher order terms in the bending strains and modified the strain displacement relationships to:

$$\varepsilon_\xi{}^r = \varepsilon_\xi{}^d \;;\; \varepsilon_\eta{}^r = \varepsilon_\eta{}^d \;;\; \varepsilon_{\xi\eta}{}^r = \varepsilon_{\xi\eta}{}^d \tag{8.3a, b, c}$$

$$\chi_\xi{}^r = \chi_\xi{}^d + \frac{1}{A}\frac{\partial}{\partial \xi}(k_1 u) \;;\; \chi_\eta{}^r = \chi_\eta{}^d + \frac{k_1 u}{AB}\frac{\partial B}{\partial \xi} + \frac{1}{B}\frac{\partial}{\partial \eta}(k_2 v) \tag{8.3d, e}$$

$$2\chi_{\xi\eta}{}^r = 2\chi_{\xi\eta}{}^d + \frac{k_1}{B}\frac{\partial u}{\partial \eta} + \frac{1}{A}\frac{\partial}{\partial \xi}(k_2 u) - \frac{k_2 v}{AB}\frac{\partial B}{\partial \xi} \tag{8.3f}$$

The superscript 'r' denotes Reissner's Theory.

In the case of the rectangular flat shell strip, the following relations can be established:

$$A = 1 \;;\; B = 1 \;;\; k_1 = 0 \text{ and } k_2 = 0.$$

Therefore, Eqs. 8.2a to 8.2f are reduced to:

$$\varepsilon_\xi = \frac{\partial u}{\partial \xi} \;;\; \varepsilon_\eta = \frac{\partial v}{\partial \eta} \;;\; \varepsilon_{\xi\eta} = \frac{\partial u}{\partial \eta} + \frac{\partial v}{\partial \xi} \tag{8.4a, b, c}$$

$$\chi_\xi = -\frac{\partial^2 w}{\partial \xi^2} \;;\; \chi_\eta = -\frac{\partial^2 w}{\partial \eta^2} \;;\; 2\chi_{\xi\eta} = -2\frac{\partial^2 w}{\partial \xi \partial \eta} \tag{8.4d, e, f}$$

Note that the relations of Reissner's Theory will also be reduced to the same forms as those of Donnell's Theory. In addition, the in-plane and bending components are uncoupled.

If the mid-surface of a shallow shell is given as z=G(x,y), one can express the strain-displacement relations as

$$\varepsilon_\xi = \frac{\partial u}{\partial \xi} - k_x w \;;\; \varepsilon_\eta = \frac{\partial v}{\partial \eta} - k_y w \;;\; \varepsilon_{\xi\eta} = \frac{\partial u}{\partial \eta} + \frac{\partial v}{\partial \xi} \qquad (8.5a, b, c)$$

$$\chi_\xi = -\frac{\partial^2 w}{\partial \xi^2} \;;\; \chi_\eta = -\frac{\partial^2 w}{\partial \eta^2} \;;\; 2\chi_{\xi\eta} = -2\frac{\partial^2 w}{\partial \xi \partial \eta} \qquad (8.5d, e, f)$$

where k_x ($= \partial^2 G/\partial x^2$) and k_y ($= \partial^2 G/\partial y^2$) are the curvatures of the shell surface. The values of curvatures for a variety of common shallow shells are tabulated in Table 8.1. The above theory was proposed by Vlazov.[(31)]

TABLE 8.1
k_x and k_y for shallow shells z=G(x,y).

Shell type	k_x	k_y
Flat shell	0	0
Cylindrical shell	$\frac{1}{r_x}$ 0	0 $\frac{1}{r_y}$
Elliptical paraboloidal shell	$\frac{1}{r_x}$	$\frac{1}{r_y}$
Hyperbolic paraboloidal shell	$\frac{1}{r_x}$ $-\frac{1}{r_x}$	$-\frac{1}{r_y}$ $\frac{1}{r_y}$

On the other hand, the strain displacement relationship for a conical shell, which is a common unit in the curved folded plate structures, can be obtained by substituting the following relations into Eqs. 8.2a to 8.2f:

$$A = 1 \;;\; B = r \;;\; \frac{\partial B}{\partial \xi} = \sin\phi \;;\; k_1 = 0 \;;\; k_2 = \frac{\cos\phi}{r}$$

The Reissner strain-displacement relations (ξ = x and $\eta = \theta$) are thus:

$$\varepsilon_x = \frac{\partial u}{\partial x} \qquad (8.6a)$$

$$\varepsilon_\theta = \frac{1}{r}\frac{\partial v}{\partial \theta} + \frac{w\cos\phi + u\sin\phi}{r} \qquad (8.6b)$$

$$\gamma_{r\theta} = \frac{1}{r}\frac{\partial u}{\partial \theta} + \frac{\partial v}{\partial x} - \frac{v\sin\phi}{r} \qquad (8.6c)$$

$$\chi_x = -\frac{\partial^2 w}{\partial x^2} \qquad (8.6d)$$

$$\chi_\theta = -\frac{1}{r^2}\frac{\partial^2 w}{\partial \theta^2} + \frac{\cos\phi}{r^2}\frac{\partial v}{\partial \theta} - \frac{\sin\phi}{r}\frac{\partial w}{\partial x} \qquad (8.6e)$$

$$\chi_{x\theta} = 2\left(-\frac{1}{r}\frac{\partial^2 w}{\partial x \partial \theta} + \frac{\sin\phi}{r^2}\frac{\partial w}{\partial \theta} + \frac{\cos\phi}{r}\frac{\partial v}{\partial x} - \frac{\sin\phi\cos\phi}{r^2}v\right) \qquad (8.6f)$$

The conical frustrum shell strip changes to a cylindrical strip in the vertical position (for $\phi = 0^o$), and to a flat sector strip when horizontal (for $\phi = 90^o$).

One must point out that the above theories are chosen on the ground of convenience, and that other theories are equally applicable.

8.1.2 CLASSICAL FINITE STRIPS FOR SHELLS

8.1.2.1 *Rectangular flat shell strips*

Referring to Figure 8.1, it can be seen readily that the displacements of the shell can be described in terms of u, v and w.

In the context of classical strips, the shape functions for the w-displacement of a rectangular strip (Figure 8.1a) can be given by Eq. 2.31, Eq. 2.32 or Eq. 2.34, depending on the choice of the order of the strips. On the other hand, the shape functions for u and v can be chosen from the family of plane strips (Figure 8.1b) described in Section 4.2.

As there is no interaction between the bending and in-plane actions (Eqs. 8.4a to 8.4f), a flat shell strip can be formed through the simple combination of a bending strip and a plane stress strip in the following manner.

Let both bending and in-plane nodal displacements act simultaneously; then, at each of the two nodal lines of a LO2 strip there will be four displacement components and four corresponding force components. The displacement components are related to the force components through a stiffness matrix. A typical submatrix $[K_{ij}]_{mn}$ of the stiffness matrix is made up of the bending and in-plane stiffness submatrices, that is

$$[K_{ij}]_{mn} = \begin{bmatrix} [K^{p}_{ij}]_{mn} & [0] \\ [0] & [K^{b}_{ij}]_{mn} \end{bmatrix} \tag{8.7}$$

The above formulation is generally valid for rectangular flat shell strips and applies to both lower order as well as higher order strips.

For a LO2 strip, $[K^{b}_{ij}]_{mm}$ and $[K^{p}_{ij}]_{mm}$ are given in Tables 3.2 and 4.2 respectively, while for a HO3 strip with one internal nodal line, the corresponding matrices can be found in Tables 3.3 and 4.4.

The comprehensive mass matrix can also be made up in a similar fashion. Let $[M^{p}_{ij}]_{mn}$ and $[M^{b}_{ij}]_{mn}$ refer to the in-plane mass matrix (Table 5.9) and bending mass matrix (Table 5.1) respectively, and let $[M]_{mn}$ represent the comprehensive mass matrix. Then

$$[M_{ij}]_{mn} = \begin{bmatrix} [M^{p}_{ij}]_{mn} & [0] \\ [0] & [M^{b}_{ij}]_{mn} \end{bmatrix} \tag{8.8}$$

In order to predict correctly the torsional or overall buckling of stiffened panels, it is necessary to take into account the second order terms of the strains as well. For example, in the case of long flat-walled structures mainly submitted to compression along the longitudinal direction (y-direction) only

σ_y^o needs to be considered because there is no conceivable in-plane instability that can arise from the action of σ_x^o and τ_{xy}^o. The relevant secondary strains are

$$\varepsilon_x^o = \frac{1}{2}\left(\frac{\partial w}{\partial x}\right)^2 \tag{8.9a}$$

$$\varepsilon_y^o = \frac{1}{2}\left[\left(\frac{\partial u}{\partial y}\right)^2 + \left(\frac{\partial v}{\partial y}\right)^2 + \left(\frac{\partial w}{\partial y}\right)^2\right] \tag{8.9b}$$

$$\gamma_{xy}^o = \frac{\partial w}{\partial x}\frac{\partial w}{\partial y} \tag{8.9c}$$

The geometric stiffness matrix is then

$$[K_G^e] = t\iint [G_u]^T [D_s][G_u]dxdy + t\iint [G_v]^T [D_s][G_v]dxdy + t\iint [G_w]^T [D_s][G_w]dxdy \tag{8.10}$$

Other non-linear terms may also be included in the secondary strains, but fortunately, the geometric stiffness matrix can be formulated in a similar way by simply including the additional terms in the integrals.

8.1.2.2 *Conical frustrum shell strip*

The displacement functions for the conical shell shown in Figure 8.2 can be expressed in terms of $\bar{x}$ and θ. The m-th term is written as:

$$u_m = [(1-\bar{x})\,u_{1m} + \bar{x}\,u_{2m}]\sin\left(\frac{m\pi\theta}{\alpha}\right) \tag{8.11a}$$

$$v_m = [(1-\bar{x})\,v_{1m} + \bar{x}\,v_{2m}]\cos\left(\frac{m\pi\theta}{\alpha}\right) \tag{8.11b}$$

$$w_m = [(1-3\bar{x}^2+2\bar{x}^3)w_{1m} + x(1-2\bar{x}+\bar{x}^2)\psi_{1m}+(3\bar{x}^2-2\bar{x}^3)w_{2m}+x(\bar{x}^2-\bar{x})\,\psi_{2m}]\sin\left(\frac{m\pi\theta}{\alpha}\right) \tag{8.11c}$$

in which $\bar{x}=x/2b'$. Using Eqs. 8.6a to 8.6f, the strains of the m-th term can be written as

$$\begin{Bmatrix} \varepsilon_x \\ \varepsilon_\theta \\ \gamma_{x\theta} \\ \chi_x \\ \chi_\theta \\ \chi_{x\theta} \end{Bmatrix}_m = \begin{Bmatrix} \dfrac{\partial u}{\partial x} \\ \dfrac{1}{r}\dfrac{\partial v}{\partial \theta} + \dfrac{w\cos\phi + u\sin\phi}{r} \\ \dfrac{1}{r}\dfrac{\partial u}{\partial \theta} + \dfrac{\partial v}{\partial x} - \dfrac{v\sin\phi}{r} \\ -\dfrac{\partial^2 w}{\partial x^2} \\ -\dfrac{1}{r^2}\dfrac{\partial^2 w}{\partial \theta^2} + \dfrac{\cos\phi}{r^2}\dfrac{\partial v}{\partial \theta} - \dfrac{\sin\phi}{r}\dfrac{\partial w}{\partial x} \\ 2\left(-\dfrac{1}{r}\dfrac{\partial^2 w}{\partial x\partial \theta} + \dfrac{\sin\phi}{r^2}\dfrac{\partial w}{\partial \theta} + \dfrac{\cos\phi}{r}\dfrac{\partial v}{\partial x} - \dfrac{\sin\phi\cos\phi}{r^2}v\right) \end{Bmatrix}_m \tag{8.12}$$

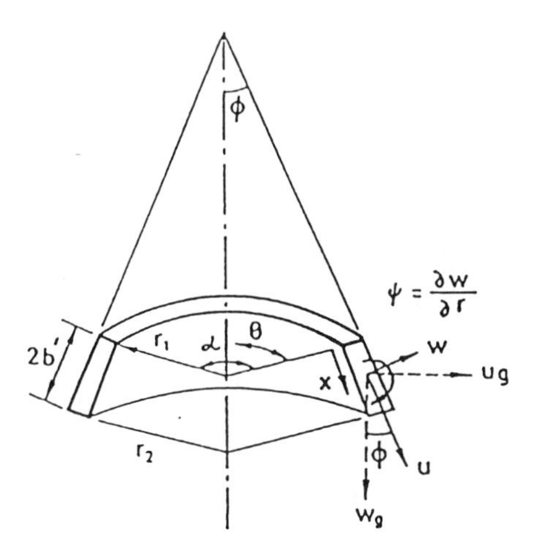

Figure 8.2 Conical frustrum shell.

In short form,

$$\{\varepsilon\}_m = [B]_m \{\delta\}_m \tag{8.13}$$

The elasticity matrix for a general orthotropic material reads as follows:

$$[D] = \begin{bmatrix} K_x & K_2 & 0 & 0 & 0 & 0 \\ K_2 & K_\theta & 0 & 0 & 0 & 0 \\ 0 & 0 & K_{x\theta} & 0 & 0 & 0 \\ 0 & 0 & 0 & D_x & D_2 & 0 \\ 0 & 0 & 0 & D_2 & D_\theta & 0 \\ 0 & 0 & 0 & 0 & 0 & D_{x\theta} \end{bmatrix} \tag{8.14}$$

in which $K_x = \dfrac{E_x t}{(1-\upsilon_x \upsilon_\theta)}$; $K_2 = \upsilon_\theta K_x$; $K_{x\theta} = G_{x\theta}\, t$; $K_\theta = \dfrac{E_\theta t}{(1-\upsilon_x \upsilon_\theta)}$

$D_x = \dfrac{E_x t^3}{12(1-\upsilon_x \upsilon_\theta)}$; $D_2 = \upsilon_\theta D_x$; $D_{x\theta} = G_{x\theta} \dfrac{t^3}{12}$; $D_\theta = \dfrac{E_\theta t^3}{12(1-\upsilon_x \upsilon_\theta)}$

The stiffness matrix is obtained in the usual way through the application of Eq. 3.14 but with the volume integral changed to an area integral. The strain matrix $[B]_m$ and the stiffness matrix $[K]_{mm}$ are listed in Tables 8.2 and 8.3 respectively. The terms B_{ij} in Table 8.3 refer to the corresponding coefficients of the matrix $[B]_m$ with ij indicating the location of a coefficient. Note that $[K]_{mn}$ is always equal to zero for m≠n because of the orthogonal Fourier series.

TABLE 8.2
Strain matrix of a conical web strip.

$-\frac{1}{b}S$	0	0	0
$\frac{1}{r}(1-\frac{x}{b})SS\phi$	$-\frac{1}{r}(1-\frac{x}{b})k_mS$	$\frac{1}{r}(1-\frac{3x^2}{b^2}+\frac{2x^3}{b^3})SC\phi$	$\frac{1}{r}(x-\frac{2x^2}{b}+\frac{x^3}{b^2})SC\phi$
$\frac{1}{r}(1-\frac{x}{b})k_mC$	$-\frac{1}{b}C+\frac{1}{r}(1-\frac{x}{b})CS\phi$	0	0
0	0	$(\frac{6}{b^2}-\frac{12x}{b^3})S$	$(\frac{4}{b}-\frac{6x}{b^2})S$
0	$-\frac{1}{r^2}(1-\frac{x}{b})k_mSC\phi$	$\frac{1}{r^2}(1-\frac{3x^2}{b^2}+\frac{2x^3}{b^3})k_m^2S-\frac{1}{r}(-\frac{6x}{b^2}+\frac{6x^2}{b^3})SS\phi$	$\frac{1}{r^2}(x-\frac{2x^2}{b}+\frac{x^3}{b^2})k_m^2S-\frac{1}{r}(1-\frac{4x}{b}+\frac{3x^2}{b^2})SS\phi$
0	$-\frac{2}{rb}CC\phi-\frac{2}{r^2}(1-\frac{x}{b})CS\phi C\phi$	$\frac{2}{r}(\frac{6x}{b^2}-\frac{6x^2}{b^3})k_mC+\frac{2}{r^2}(1-\frac{3x^2}{b^2}+\frac{2x^3}{b^3})k_mCS\phi$	$\frac{2}{r}(-1+\frac{4x}{b}-\frac{3x^2}{b^2})k_mC+\frac{2}{r^2}(x-\frac{2x^2}{b}+\frac{x^3}{b^2})k_mCS\phi$

$\frac{1}{b}S$	0	0	0
$\frac{1}{r}\frac{x}{b}SS\phi$	$-\frac{1}{r}\frac{x}{b}k_mS$	$\frac{1}{r}(\frac{3x^2}{b^2}-\frac{2x^3}{b^3})SC\phi$	$\frac{1}{r}(-\frac{x^2}{b}+\frac{x^3}{b^2})SC\phi$
$\frac{1}{r}\frac{x}{b}k_mC$	$\frac{1}{b}C-\frac{1}{r}\frac{x}{b}CS\phi$	0	0
0	0	$(-\frac{6}{b^2}+\frac{12x}{b^3})S$	$(\frac{2}{b}-\frac{6x}{b^2})S$
0	$-\frac{1}{r^2}\frac{x}{b}k_mSC\phi$	$\frac{1}{r^2}(\frac{3x^2}{b^2}-\frac{2x^3}{b^3})k_m^2S-\frac{1}{r}(\frac{6x}{b^2}-\frac{6x^2}{b^3})SS\phi$	$+\frac{1}{r^2}(\frac{x^3}{b^2}-\frac{x^2}{b})k_m^2S-\frac{1}{r}(-\frac{2x}{b}+\frac{3x^2}{b^2})SS\phi$
0	$+\frac{2}{rb}CC\phi-\frac{2}{r^2}\frac{x}{b}CS\phi C\phi$	$\frac{2}{r}(-\frac{6x}{b^2}+\frac{6x^2}{b^3})k_mC+\frac{2}{r^2}(\frac{3x^2}{b^2}-\frac{2x^3}{b^3})k_mCS\phi$	$\frac{2}{r}(\frac{2x}{b}-\frac{3x^2}{b^2})k_mC+\frac{2}{r^2}(-\frac{x^2}{b}+\frac{x^3}{b^2})k_mCS\phi$

$k_m = \frac{m\pi}{\alpha}$; $S = \sin k_m\theta$; $C = \cos k_m\theta$; $S\phi = \sin\phi$; $C\phi = \cos\phi$; $b = 2b'$

TABLE 8.3
Stiffness matrix of shell strip.

$$[K]_{mm} = \frac{\alpha}{2}\int_0^b [\overline{K}]_{mm}\, r\,dr \quad ; \quad [\overline{K}]_{mm} = \begin{bmatrix} [\overline{K}_{11}]_{mm} & [\overline{K}_{12}]_{mm} \\ [\overline{K}_{21}]_{mm} & [\overline{K}_{22}]_{mm} \end{bmatrix}$$

$[\overline{K}_{11}]_{mm} =$

$$\begin{bmatrix}
\begin{array}{c} K_x B_{11}^2 + 2K_2 B_{21} B_{11} \\ + K_\theta B_{21}^2 + K_{x\theta} B_{31}^2 \\ + D_x B_{41}^2 + 2D_2 B_{51} B_{41} \\ + D_\theta B_{51}^2 + D_{x\theta} B_{61}^2 \end{array} &
\begin{array}{c} K_x B_{12} B_{11} + K_2 B_{22} B_{11} \\ + K_2 B_{12} B_{21} + K_\theta B_{22} B_{21} \\ + K_{x\theta} B_{32} B_{31} + D_x B_{42} B_{41} \\ + D_2 B_{52} B_{41} + D_2 B_{42} B_{51} \\ + D_\theta B_{52} B_{51} + D_{x\theta} B_{62} B_{61} \end{array} &
\begin{array}{c} K_x B_{13} B_{11} + K_2 B_{23} B_{11} \\ + K_2 B_{13} B_{21} + K_\theta B_{23} B_{21} \\ + K_{x\theta} B_{33} B_{31} + D_x B_{43} B_{41} \\ + D_2 B_{53} B_{41} + D_2 B_{43} B_{51} \\ + D_\theta B_{53} B_{51} + D_{x\theta} B_{63} B_{61} \end{array} &
\begin{array}{c} K_x B_{14} B_{11} + K_2 B_{24} B_{11} \\ + K_2 B_{14} B_{21} + K_\theta B_{24} B_{21} \\ + K_{x\theta} B_{34} B_{31} + D_x B_{44} B_{41} \\ + D_2 B_{54} B_{41} + D_2 B_{44} B_{51} \\ + D_\theta B_{54} B_{51} + D_{x\theta} B_{64} B_{61} \end{array} \\
 &
\begin{array}{c} K_x B_{12}^2 + 2K_2 B_{22} B_{12} \\ + K_\theta B_{22}^2 + K_{x\theta} B_{32}^2 \\ + D_x B_{42}^2 + 2D_2 B_{52} B_{42} \\ + D_\theta B_{52}^2 + D_{x\theta} B_{62}^2 \end{array} &
\begin{array}{c} K_x B_{13} B_{12} + K_2 B_{23} B_{12} \\ + K_2 B_{13} B_{22} + K_\theta B_{23} B_{22} \\ + K_{x\theta} B_{33} B_{32} + D_x B_{43} B_{42} \\ + D_2 B_{53} B_{42} + D_2 B_{43} B_{52} \\ + D_\theta B_{53} B_{52} + D_{x\theta} B_{63} B_{62} \end{array} &
\begin{array}{c} K_x B_{14} B_{12} + K_2 B_{24} B_{12} \\ + K_2 B_{14} B_{22} + K_\theta B_{24} B_{22} \\ + K_{x\theta} B_{34} B_{32} + D_x B_{44} B_{42} \\ + D_2 B_{54} B_{42} + D_2 B_{44} B_{52} \\ + D_\theta B_{54} B_{52} + D_{x\theta} B_{64} B_{62} \end{array} \\
 & \text{SYM} &
\begin{array}{c} K_x B_{13}^2 + 2K_2 B_{23} B_{13} \\ + K_\theta B_{23}^2 + K_{x\theta} B_{33}^2 \\ + D_x B_{43}^2 + 2D_2 B_{53} B_{43} \\ + D_\theta B_{53}^2 + D_{x\theta} B_{63}^2 \end{array} &
\begin{array}{c} K_x B_{14} B_{13} + K_2 B_{24} B_{13} \\ + K_2 B_{14} B_{23} + K_\theta B_{24} B_{23} \\ + K_{x\theta} B_{34} B_{33} + D_x B_{44} B_{43} \\ + D_2 B_{54} B_{43} + D_2 B_{44} B_{53} \\ + D_\theta B_{54} B_{53} + D_{x\theta} B_{64} B_{63} \end{array} \\
 & & &
\begin{array}{c} K_x B_{14}^2 + 2K_2 B_{24} B_{14} \\ + K_\theta B_{24}^2 + K_{x\theta} B_{34}^2 \\ + D_x B_{44}^2 + 2D_2 B_{54} B_{44} \\ + D_\theta B_{54}^2 + D_{x\theta} B_{64}^2 \end{array}
\end{bmatrix}$$

$[\overline{K}_{21}]_{mm}^T = [\overline{K}_{12}]_{mm} =$

$$\begin{bmatrix}
\begin{array}{c} K_x B_{15} B_{11} + K_2 B_{25} B_{11} \\ + K_2 B_{15} B_{21} + K_\theta B_{25} B_{21} \\ + K_{x\theta} B_{35} B_{31} + D_x B_{45} B_{41} \\ + D_2 B_{35} B_{41} + D_2 B_{45} B_{51} \\ + D_\theta B_{55} B_{51} + D_{x\theta} B_{65} B_{61} \end{array} &
\begin{array}{c} K_x B_{16} B_{11} + K_2 B_{26} B_{11} \\ + K_2 B_{16} B_{21} + K_\theta B_{26} B_{21} \\ + K_{x\theta} B_{36} B_{31} + D_x B_{46} B_{41} \\ + D_2 B_{36} B_{41} + D_2 B_{46} B_{51} \\ + D_\theta B_{56} B_{51} + D_{x\theta} B_{66} B_{61} \end{array} &
\begin{array}{c} D_x B_{17} B_{11} + K_2 B_{27} B_{11} \\ + K_2 B_{17} B_{21} + K_\theta B_{27} B_{21} \\ + K_{x\theta} B_{37} B_{31} + D_x B_{47} B_{41} \\ + D_2 B_{37} B_{41} + D_2 B_{47} B_{51} \\ + D_\theta B_{57} B_{51} + D_{x\theta} B_{67} B_{61} \end{array} &
\begin{array}{c} D_x B_{18} B_{11} + K_2 B_{28} B_{11} \\ + K_2 B_{18} B_{21} + K_\theta B_{28} B_{21} \\ + K_{x\theta} B_{38} B_{31} + D_x B_{48} B_{41} \\ + D_2 B_{38} B_{41} + D_2 B_{48} B_{51} \\ + D_\theta B_{58} B_{51} + D_{x\theta} B_{68} B_{61} \end{array} \\
\begin{array}{c} K_x B_{15} B_{12} + K_2 B_{25} B_{12} \\ + K_2 B_{15} B_{22} + K_\theta B_{25} B_{22} \\ + K_{x\theta} B_{35} B_{32} + D_x B_{45} B_{42} \\ + D_2 B_{35} B_{42} + D_2 B_{45} B_{52} \\ + D_\theta B_{55} B_{52} + D_{x\theta} B_{65} B_{62} \end{array} &
\begin{array}{c} K_x B_{16} B_{12} + K_2 B_{26} B_{12} \\ + K_2 B_{16} B_{22} + K_\theta B_{26} B_{22} \\ + K_{x\theta} B_{36} B_{32} + D_x B_{46} B_{42} \\ + D_2 B_{36} B_{42} + D_2 B_{46} B_{52} \\ + D_\theta B_{56} B_{52} + D_{x\theta} B_{66} B_{62} \end{array} &
\begin{array}{c} D_x B_{17} B_{12} + K_2 B_{27} B_{12} \\ + K_2 B_{17} B_{22} + K_\theta B_{27} B_{22} \\ + K_{x\theta} B_{37} B_{32} + D_x B_{47} B_{42} \\ + D_2 B_{37} B_{42} + D_2 B_{47} B_{52} \\ + D_\theta B_{57} B_{52} + D_{x\theta} B_{67} B_{62} \end{array} &
\begin{array}{c} D_x B_{18} B_{12} + K_2 B_{28} B_{12} \\ + K_2 B_{18} B_{22} + K_\theta B_{28} B_{22} \\ + K_{x\theta} B_{38} B_{32} + D_x B_{48} B_{42} \\ + D_2 B_{38} B_{42} + D_2 B_{48} B_{52} \\ + D_\theta B_{58} B_{52} + D_{x\theta} B_{68} B_{62} \end{array} \\
\begin{array}{c} K_x B_{15} B_{13} + K_2 B_{25} B_{13} \\ + K_2 B_{15} B_{23} + K_\theta B_{25} B_{23} \\ + K_{x\theta} B_{35} B_{33} + D_x B_{45} B_{43} \\ + D_2 B_{35} B_{43} + D_2 B_{45} B_{53} \\ + D_\theta B_{55} B_{53} + D_{x\theta} B_{65} B_{63} \end{array} &
\begin{array}{c} K_x B_{16} B_{13} + K_2 B_{26} B_{13} \\ + K_2 B_{16} B_{23} + K_\theta B_{26} B_{23} \\ + K_{x\theta} B_{36} B_{33} + D_x B_{46} B_{43} \\ + D_2 B_{36} B_{43} + D_2 B_{46} B_{53} \\ + D_\theta B_{56} B_{53} + D_{x\theta} B_{66} B_{63} \end{array} &
\begin{array}{c} D_x B_{17} B_{13} + K_2 B_{27} B_{13} \\ + K_2 B_{17} B_{23} + K_\theta B_{27} B_{23} \\ + K_{x\theta} B_{37} B_{33} + D_x B_{47} B_{43} \\ + D_2 B_{37} B_{43} + D_2 B_{47} B_{53} \\ + D_\theta B_{57} B_{53} + D_{x\theta} B_{67} B_{63} \end{array} &
\begin{array}{c} D_x B_{18} B_{13} + K_2 B_{28} B_{13} \\ + K_2 B_{18} B_{23} + K_\theta B_{28} B_{23} \\ + K_{x\theta} B_{38} B_{33} + D_x B_{48} B_{43} \\ + D_2 B_{38} B_{43} + D_2 B_{48} B_{53} \\ + D_\theta B_{58} B_{53} + D_{x\theta} B_{68} B_{63} \end{array} \\
\begin{array}{c} K_x B_{15} B_{14} + K_2 B_{25} B_{14} \\ + K_2 B_{15} B_{24} + K_\theta B_{25} B_{24} \\ + K_{x\theta} B_{35} B_{34} + D_x B_{45} B_{44} \\ + D_2 B_{35} B_{44} + D_2 B_{45} B_{54} \\ + D_\theta B_{55} B_{54} + D_{x\theta} B_{65} B_{64} \end{array} &
\begin{array}{c} K_x B_{16} B_{14} + K_2 B_{26} B_{14} \\ + K_2 B_{16} B_{24} + K_\theta B_{26} B_{24} \\ + K_{x\theta} B_{36} B_{34} + D_x B_{46} B_{44} \\ + D_2 B_{36} B_{44} + D_2 B_{46} B_{54} \\ + D_\theta B_{56} B_{54} + D_{x\theta} B_{66} B_{64} \end{array} &
\begin{array}{c} D_x B_{17} B_{14} + K_2 B_{27} B_{14} \\ + K_2 B_{17} B_{24} + K_\theta B_{27} B_{24} \\ + K_{x\theta} B_{37} B_{34} + D_x B_{47} B_{44} \\ + D_2 B_{37} B_{44} + D_2 B_{47} B_{54} \\ + D_\theta B_{57} B_{54} + D_{x\theta} B_{67} B_{64} \end{array} &
\begin{array}{c} D_x B_{18} B_{14} + K_2 B_{28} B_{14} \\ + K_2 B_{18} B_{24} + K_\theta B_{28} B_{24} \\ + K_{x\theta} B_{38} B_{34} + D_x B_{48} B_{44} \\ + D_2 B_{38} B_{44} + D_2 B_{48} B_{54} \\ + D_\theta B_{58} B_{54} + D_{x\theta} B_{68} B_{64} \end{array}
\end{bmatrix}$$

Table 8. 3 (cont'd)

$[\overline{K}_{22}]_{mm} =$

$$
\left[\begin{array}{c:c:c:c}
\begin{array}{c} K_xB_{15}^2+2K_2B_{25}B_{15} \\ +K_\theta B_{25}^2+K_{x\theta}B_{35}^2 \\ +D_xB_{45}^2+2D_2B_{55}B_{45} \\ +D_\theta B_{55}^2+D_{x\theta}B_{65}^2 \end{array} &
\begin{array}{c} K_xB_{16}B_{15}+K_2B_{26}B_{15} \\ +K_2B_{16}B_{25}+K_\theta B_{26}B_{25} \\ +K_{x\theta}B_{36}B_{35}+D_xB_{46}B_{45} \\ +D_2B_{56}B_{45}+D_2B_{46}B_{55} \\ +D_\theta B_{56}B_{55}+D_{x\theta}B_{66}B_{65} \end{array} &
\begin{array}{c} K_xB_{17}B_{15}+K_2B_{27}B_{15} \\ +K_2B_{17}B_{25}+K_\theta B_{27}B_{25} \\ +K_{x\theta}B_{37}B_{35}+D_xB_{47}B_{45} \\ +D_2B_{57}B_{45}+D_2B_{47}B_{55} \\ +D_\theta B_{57}B_{55}+D_{x\theta}B_{67}B_{65} \end{array} &
\begin{array}{c} K_xB_{18}B_{15}+K_2B_{28}B_{15} \\ +K_2B_{18}B_{25}+K_\theta B_{28}B_{25} \\ +K_{x\theta}B_{38}B_{35}+D_xB_{48}B_{45} \\ +D_2B_{58}B_{45}+D_2B_{48}B_{55} \\ +D_\theta B_{58}B_{55}+D_{x\theta}B_{68}B_{65} \end{array} \\ \hdashline
 &
\begin{array}{c} K_xB_{16}^2+2K_2B_{26}B_{16} \\ +K_\theta B_{26}^2+K_{x\theta}B_{36}^2 \\ +D_xB_{46}^2+2D_2B_{56}B_{46} \\ +D_\theta B_{56}^2+D_{x\theta}B_{66}^2 \end{array} &
\begin{array}{c} K_xB_{17}B_{16}+K_2B_{27}B_{16} \\ +K_2B_{17}B_{26}+K_\theta B_{27}B_{26} \\ +K_{x\theta}B_{37}B_{36}+D_xB_{47}B_{46} \\ +D_2B_{57}B_{46}+D_2B_{47}B_{56} \\ +D_\theta B_{57}B_{56}+D_{x\theta}B_{67}B_{66} \end{array} &
\begin{array}{c} K_xB_{18}B_{16}+K_2B_{28}B_{16} \\ +K_2B_{18}B_{26}+K_\theta B_{28}B_{26} \\ +K_{x\theta}B_{38}B_{36}+D_xB_{48}B_{46} \\ +D_2B_{58}B_{46}+D_2B_{48}B_{56} \\ +D_\theta B_{58}B_{56}+D_{x\theta}B_{68}B_{66} \end{array} \\ \hdashline
 & \text{SYM} &
\begin{array}{c} K_xB_{17}^2+2K_2B_{27}B_{17} \\ +K_\theta B_{27}^2+K_{x\theta}B_{37}^2 \\ +D_xB_{47}^2+2D_2B_{53}B_{47} \\ +D_\theta B_{57}^2+D_{x\theta}B_{67}^2 \end{array} &
\begin{array}{c} K_xB_{18}B_{17}+K_2B_{28}B_{17} \\ +K_2B_{18}B_{27}+K_\theta B_{28}B_{27} \\ +K_{x\theta}B_{38}B_{37}+D_xB_{48}B_{47} \\ +D_2B_{58}B_{47}+D_2B_{48}B_{57} \\ +D_\theta B_{58}B_{57}+D_{x\theta}B_{68}B_{67} \end{array} \\ \hdashline
 & & &
\begin{array}{c} K_xB_{18}^2+2K_2B_{28}B_{18} \\ +K_\theta B_{28}^2+K_{x\theta}B_{38}^2 \\ +D_xB_{48}^2+2D_2B_{58}B_{48} \\ +D_\theta B_{58}^2+D_{x\theta}B_{68}^2 \end{array}
\end{array}\right]
$$

The mass matrix is made up of the in-plane and bending components, and it can be written as:

$$
[M_{ij}]_{mm} = \begin{bmatrix} [M^{p}_{ij}]_{mm} & [0] \\ [0] & [M^{b}_{ij}]_{mm} \end{bmatrix} \tag{8.15}
$$

where

$$
[M^{p}_{ij}]_{mm} = \begin{bmatrix}
b(\frac{b}{12}S\phi+\frac{r_1}{3})I_1 & & & \text{SYM} \\
0 & b(\frac{b}{12}S\phi+\frac{r_1}{3})I_2 & & \\
b(\frac{b}{12}S\phi+\frac{r_1}{6})I_1 & 0 & b(\frac{b}{4}S\phi+\frac{r_1}{3})I_1 & \\
0 & b(\frac{b}{12}S\phi+\frac{r_1}{6})I_2 & 0 & b(\frac{b}{4}S\phi+\frac{r_1}{3})I_2
\end{bmatrix}
$$

$$
[M^{b}_{ij}]_{mm} = \begin{bmatrix}
b(\frac{6bS\phi}{70}+\frac{13r_1}{35})I_1 & & & \\
b^2(\frac{bS\phi}{60}+\frac{11r_1}{210})I_1 & b^3(\frac{bS\phi}{280}+\frac{r_1}{105})I_1 & & \\
b(\frac{9bS\phi}{140}+\frac{9r_1}{70})I_1 & b^2(\frac{bS\phi}{60}+\frac{13r_1}{420})I_1 & b(\frac{6bS\phi}{21}+\frac{13r_1}{35})I_1 & \\
-b^2(\frac{bS\phi}{70}+\frac{13r_1}{420})I_1 & -b^3(\frac{bS\phi}{280}+\frac{r_1}{140})I_1 & -b^2(\frac{bS\phi}{28}+\frac{11r_1}{210})I_1 & b^3(\frac{bS\phi}{168}+\frac{r_1}{105})I_1
\end{bmatrix}
$$

$S\phi = \sin\phi$; $I_1 = \int_0^{\alpha} \Theta_m\Theta_n d\theta$; $I_2 = \frac{\alpha^2}{\mu_m\mu_n}\int_0^{\alpha} \Theta_m^{'}\Theta_n^{'} d\theta$

8.1.3 SPLINE SHELL STRIP

It is more convenient to express the displacement functions of a spline strip in terms of the natural coordinates ξ and η. The displacement field of a lower order strip is thus defined as:

$$\{f\} = \begin{Bmatrix} u \\ v \\ w \end{Bmatrix}$$

$$= \begin{bmatrix} C_1 & 0 & 0 & 0 & C_2 & 0 & 0 & 0 \\ 0 & C_1 & 0 & 0 & 0 & C_2 & 0 & 0 \\ 0 & 0 & C_3 & C_4 & 0 & 0 & C_5 & C_6 \end{bmatrix} \begin{bmatrix} \phi_1 & & & & & & & \\ & \phi_2 & & & & & & \\ & & \phi_3 & & & & & \\ & & & \phi_4 & & & & \\ & & & & \phi_5 & & & \\ & & & & & \phi_6 & & \\ & & & & & & \phi_7 & \\ & & & & & & & \phi_8 \end{bmatrix} \begin{Bmatrix} u_1 \\ v_1 \\ w_1 \\ \theta_1 \\ u_2 \\ v_2 \\ w_2 \\ \theta_2 \end{Bmatrix} \quad (8.16)$$

In short form,

$$\{f\} = [C]\,[\Phi]\,\{\delta\} \quad (8.17)$$

As in the case of classical strip, the in-plane and bending stiffnesses of rectangular shells are uncoupled and the total strip stiffness can again be given by Eq. 8.7, with the in-plane and bending stiffness matrices as derived in Sections 4.3 and 3.4. Similarly, the comprehensive mass matrices and geometric stiffness matrices can be formed by combining the in-plane and bending components.

Fan(18) had shown that the strain-displacement relationship for a conical frustrum shallow spline shell strip with curvature k_x in x'-z' plane is:

$$\{\varepsilon\} = \begin{Bmatrix} \dfrac{\partial \tilde{u}}{\partial x'} - k_x \tilde{w} \\ \dfrac{1}{r}\dfrac{\partial v}{\partial \psi} + \dfrac{\tilde{u}\cos\beta - \tilde{w}\sin\beta}{r} \\ \dfrac{1}{r}\dfrac{\partial \tilde{u}}{\partial \psi} + \dfrac{\partial v}{\partial x'} - \dfrac{v\cos\beta}{r} \\ -\dfrac{\partial^2 \tilde{w}}{\partial x'^2} \\ -\dfrac{1}{r^2}\dfrac{\partial^2 \tilde{w}}{\partial \psi^2} - \dfrac{\sin\beta}{r^2}\dfrac{\partial v}{\partial \psi} - \dfrac{\cos\beta}{r}\dfrac{\partial \tilde{w}}{\partial x'} \\ 2\left(-\dfrac{1}{r}\dfrac{\partial^2 \tilde{w}}{\partial x'\partial \psi} + \dfrac{\cos\beta}{r^2}\dfrac{\partial \tilde{w}}{\partial \psi} - \dfrac{\sin\beta}{r}\dfrac{\partial v}{\partial x'} + \dfrac{\sin\beta\cos\beta}{r^2}v\right) \end{Bmatrix} \quad (8.18)$$

The displacements are defined in Figure 8.3. The explicit form of the strain matrix [B] is given in Table 8.4. Similarly, the stiffness matrix of such strips can be obtained in the usual manner.

TABLE 8.4
Strain matrix of a conical frustrum shallow shell strip.

$$\begin{bmatrix}
-\frac{1}{b}\phi_1 & 0 & \begin{matrix}-k_x\phi_3\\(1-3\bar{x}^2+2\bar{x}^3)\end{matrix} & \begin{matrix}-k_x\phi_4\\x'(1-2\bar{x}+\bar{x}^2)\end{matrix} & \frac{1}{b}\phi_5 & 0 & \begin{matrix}-k_x\phi_7\\(3\bar{x}^2-2\bar{x}^3)\end{matrix} & \begin{matrix}-k_x\phi_8\\x'(\bar{x}^2-\bar{x})\end{matrix} \\
\frac{c}{r}\phi_1(1-\bar{x}) & \frac{1}{r}\phi_2'(1-\bar{x}) & \begin{matrix}-\frac{S}{r}\phi_3\\(1-3\bar{x}^2+2\bar{x}^3)\end{matrix} & -\frac{S}{r}\phi_4 x'(1-2\bar{x}+\bar{x}^2) & \frac{c}{r}\phi_5\bar{x} & \frac{1}{r}\phi_6'\bar{x} & \begin{matrix}-\frac{S}{r}\phi_7\\(3\bar{x}^2-2\bar{x}^3)\end{matrix} & -\frac{S}{r}\phi_4 x'(\bar{x}^2-\bar{x}) \\
\frac{1}{r}\phi_1'(1-\bar{x}) & \begin{matrix}-\frac{1}{b}\phi_2\\+\frac{c}{r}\phi_2(1-\bar{x})\end{matrix} & 0 & 0 & \frac{1}{r}\phi_5'\bar{x} & \begin{matrix}\frac{1}{b}\phi_6\\-\frac{c}{r}\phi_6\bar{x}\end{matrix} & 0 & 0 \\
0 & 0 & \frac{6}{b^2}\phi_3(1-2\bar{x}) & \frac{2}{b}\phi_4(2-3\bar{x}) & 0 & 0 & -\frac{6}{b^2}\phi_7(1-2\bar{x}) & \frac{2}{b}\phi_8(1-3\bar{x}) \\
0 & -\frac{S}{r^2}\phi_2'(1-\bar{x}) & \begin{matrix}-\frac{1}{r^2}\phi_3''(1-3\bar{x}^2+2\bar{x}^3)\\-\frac{C}{rb}\phi_3(-6\bar{x}+6\bar{x}^2)\end{matrix} & \begin{matrix}-\frac{1}{r^2}\phi_4''x'(1-2\bar{x}+\bar{x}^2)\\-\frac{C}{r}\phi_4(1-4\bar{x}+3\bar{x}^2)\end{matrix} & 0 & -\frac{S}{r^2}\phi_6'\bar{x} & \begin{matrix}-\frac{1}{r^2}\phi_7''(3\bar{x}^2-2\bar{x}^3)\\-\frac{C}{rb}\phi_7(6\bar{x}-6\bar{x}^2)\end{matrix} & \begin{matrix}-\frac{1}{r^2}\phi_8''x'(\bar{x}^2-\bar{x})\\-\frac{C}{r}\phi_8(3\bar{x}^2-2\bar{x})\end{matrix} \\
0 & \begin{matrix}\frac{2S}{rb}\phi_2\\+\frac{2CS}{r^2}\phi_2(1-\bar{x})\end{matrix} & \begin{matrix}-\frac{2}{rb}\phi_3'(-6\bar{x}+6\bar{x}^2)\\+\frac{2C}{r^2}\phi_3'(1-3\bar{x}^2+2\bar{x}^3)\end{matrix} & \begin{matrix}-\frac{2}{r}\phi_4'(1-4\bar{x}+3\bar{x}^2)\\+\frac{2C}{r^2}\phi_4'x'(1-2\bar{x}+\bar{x}^2)\end{matrix} & 0 & \begin{matrix}-\frac{2S}{rb}\phi_6\\+\frac{2CS}{r^2}\phi_6\bar{x}\end{matrix} & \begin{matrix}-\frac{2}{rb}\phi_7'(6\bar{x}-6\bar{x}^2)\\+\frac{2C}{r^2}\phi_7'(3\bar{x}^2-2\bar{x}^3)\end{matrix} & \begin{matrix}-\frac{2}{r}\phi_8'(3\bar{x}^2-2\bar{x})\\+\frac{2C}{r^2}\phi_8'x'(\bar{x}^2-\bar{x})\end{matrix}
\end{bmatrix}$$

where $\phi_i' = \frac{\partial\phi_i}{\partial\psi}$; $\phi_i'' = \frac{\partial^2\phi_i}{\partial\psi^2}$; $S = \sin\beta$; $C = \cos\beta$; $\bar{x} = \frac{x'}{b'}$

On the other hand, Li[(23)] defined the characteristic matrices of a singly curved shell in terms of the natural coordinate system (ξ, η) as follows:
1) Strain matrix (based on Donnell's theory)

$$
[B] = \left[\begin{array}{cc:c:c}
-\frac{[\Phi_s]}{\Delta\xi} & \frac{[\Phi_s]}{\Delta\xi} & [0] & [0] \\
\hat{B}(1-\bar{\xi})[\Phi_s] & \hat{B}\bar{\xi}[\Phi_s] & \frac{(1-\bar{\xi})[\Phi_s']}{B} & \frac{\bar{\xi}[\Phi_s']}{B} \\
\frac{(1-\bar{\xi})[\Phi_s']}{B} & \frac{\bar{\xi}[\Phi_s']}{B} & -(\hat{B}(1-\bar{\xi})+\frac{1}{\Delta\xi})[\Phi_s] & -(\hat{B}\bar{\xi}+\frac{1}{\Delta\xi})[\Phi_s] \\
[0] & [0] & [0] & [0] \\
[0] & [0] & [0] & [0] \\
[0] & [0] & [0] & [0]
\end{array}\right.
$$

$$
\left.\begin{array}{c:c:c:c}
[0] & [0] & [0] & [0] \\
N_1[\Phi_s]k_2 & \Delta\xi N_2[\Phi_s]k_2 & N_3[\Phi_s]k_2 & \Delta\xi N_4[\Phi_s]k_2 \\
[0] & [0] & [0] & [0] \\
-\frac{N_1''[\Phi_s]}{\Delta\xi^2} & -\frac{N_2''[\Phi_s]}{\Delta\xi} & -\frac{N_3''[\Phi_s]}{\Delta\xi^2} & -\frac{N_4''[\Phi_s]}{\Delta\xi} \\
\begin{array}{c} -\frac{1}{B^2}N_1[\Phi_s''] \\ +\frac{1}{B^3}\frac{\partial B}{\partial\eta}N_1[\Phi_s'] \\ -\hat{B}N_1'\frac{[\Phi_s]}{\Delta\xi} \end{array} & \begin{array}{c} -\frac{1}{B^2}N_2[\Phi_s''] \\ +\frac{1}{B^3}\frac{\partial B}{\partial\eta}N_2[\Phi_s'] \\ -\hat{B}N_2'[\Phi_s] \end{array} & \begin{array}{c} -\frac{1}{B^2}N_3[\Phi_s''] \\ +\frac{1}{B^3}\frac{\partial B}{\partial\eta}N_3[\Phi_s'] \\ -\hat{B}N_3'\frac{[\Phi_s]}{\Delta\xi} \end{array} & \begin{array}{c} -\frac{1}{B^2}N_4[\Phi_s''] \\ +\frac{1}{B^3}\frac{\partial B}{\partial\eta}N_4[\Phi_s'] \\ -\hat{B}N_4'[\Phi_s] \end{array} \\
\frac{2}{B}(\hat{B}N_1-\frac{N_1'}{\Delta\xi})[\Phi_s'] & \frac{2}{B}(\hat{B}N_2-N_2')[\Phi_s'] & \frac{2}{B}(\hat{B}N_3-\frac{N_3'}{\Delta\xi})[\Phi_s'] & \frac{2}{B}(\hat{B}N_4-N_4')[\Phi_s']
\end{array}\right] \quad (8.19)
$$

$$[\Phi_s'] = \frac{d[\Phi_s]}{d\eta} \;;\; [\Phi_s''] = \frac{d^2[\Phi_s]}{d\eta^2} \;;\; N_i' = \frac{dN_i(\bar{\xi})}{d\bar{\xi}} \;;\; N_i'' = \frac{d^2N_i(\bar{\xi})}{d\bar{\xi}^2} \;,\; \hat{B} = \frac{1}{B}\frac{\partial B}{\partial\xi} \;;$$

$\Delta\xi$ is the width of the strip. N_i are the beam function as given in Eq. 2.31 of Section 2.2.2.1. $\bar{\xi} = (\xi - \xi_i)/\Delta\xi$

It can be demonstrated readily that the strain matrix for the Reissner's theory can be formed similarly.
2) Stiffness matrix

$$[K^e] = \iint [B]^T [D] [B] \, |J| \, d\xi d\eta \quad (8.20)$$

where $|J|$ is the Jacobian determinant.
3) Mass matrix

$$[M^e] = \rho t \iint [\Phi]^T [N]^T [N] [\Phi] \, |J| \, d\xi d\eta \quad (8.21)$$

Note that the integrations have to be carried out by numerical integration.

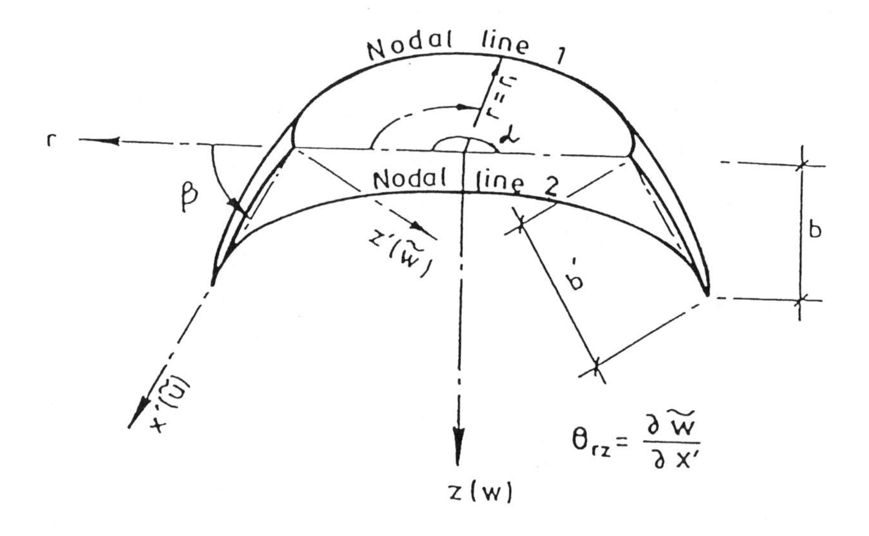

Figure 8.3 Spline strip for conical frustrum shell.

In the case of doubly curved shells, Li[23] concluded from his study that the rate of convergence of the lower order spline strip is much too slow and unsatisfactory. It was recommended that the higher order strips with all u, v and w expressed in terms of the corresponding variables and the first derivatives should be used. The shape functions are:

$$\begin{Bmatrix} u \\ v \\ w \end{Bmatrix} = [C]\,[\Phi]\,\{\delta\} \tag{8.22}$$

where

$$[C] = \begin{bmatrix} [C_1] & 0 & 0 & [C_2] & 0 & 0 \\ 0 & [C_1] & 0 & 0 & [C_2] & 0 \\ 0 & 0 & [C_1] & 0 & 0 & [C_2] \end{bmatrix}$$

$$[\Phi] = \begin{bmatrix} \phi_1 & & & & & & & & & & & \\ & \phi_2 & & & & & & & & & & \\ & & \phi_3 & & & & & & & & & \\ & & & \phi_4 & & & & & & & & \\ & & & & \phi_5 & & & & & & & \\ & & & & & \phi_6 & & & & & & \\ & & & & & & \phi_7 & & & & & \\ & & & & & & & \phi_8 & & & & \\ & & & & & & & & \phi_9 & & & \\ & & & & & & & & & \phi_{10} & & \\ & & & & & & & & & & \phi_{11} & \\ & & & & & & & & & & & \phi_{12} \end{bmatrix}$$

$\{\delta\}^T = \{(\alpha_u)_j \ (\alpha_{u,\xi})_j \ (\alpha_v)_j \ (\alpha_{v,\xi})_j \ (\alpha_w)_j \ (\alpha_{w,\xi})_j \ (\alpha_u)_{j+1} \ (\alpha_{u,\xi})_{j+1} \ (\alpha_v)_{j+1} \ (\alpha_{v,\xi})_{j+1} \ (\alpha_w)_{j+1} \ (\alpha_{w,\xi})_{j+1}\}^T$. $[C_1]$ and $[C_2]$ are given by Eq. 2.31 (Section 2.2.2.1).

The strain and stiffness matrices of such shells, although fairly complicated, can still be derived by using Eqs. 8.2 (Donnell's Theory) and 8.3 (Reissner's Theory).

8.1.4 COORDINATE TRANSFORMATION FOR FOLDED PLATE STRUCTURES

The stiffness matrices for various types of strips have so far been derived in terms of a set of local axes, two of which usually coincide with the mid-surface of a strip. Such stiffness matrices can be used directly in plate bending or plane stress problems because all the strips are coplanar. In folded plate structures, however, any two plates will in general meet at an angle, and in order to establish the equilibrium of nodal forces at nodal lines common to non-coplanar strips, a common coordinate system is obviously required.

In Figure 8.4 the individual coordinates of a strip are labelled as x', y', z' and the common coordinates as x, y, z. y and y' are coincident with each other and also with the intersection line of two adjoining strips. The transformation of forces and displacements between the two sets of coordinate systems is given by

$$\{F\}_m = [R]\,\{F'\}_m \qquad (8.23)$$

$$\{\delta'\}_m = [R]^T\,\{\delta\}_m \qquad (8.24)$$

in which [R] is the transformation matrix

$$[R] = \begin{bmatrix} [r] & [0] \\ [0] & [r] \end{bmatrix} \qquad (8.25)$$

$$\text{with } [r] = \begin{bmatrix} \cos\beta & 0 & -\sin\beta & 0 \\ 0 & 1 & 0 & 0 \\ \sin\beta & 0 & \cos\beta & 0 \\ 0 & 0 & 0 & 1 \end{bmatrix}$$

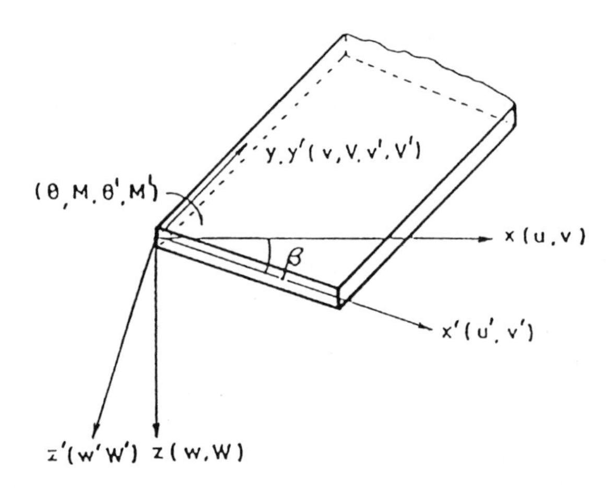

Figure 8.4 Coordinate transformation.

[0] is the null matrix and β is the angle between the x and x' axes (clockwise positive).

The final result of the transformation is

$$\begin{aligned}\{F\}_m &= [R]\ \{F'\}_m \\ &= [R]\ [K']_{mm}\ \{\delta'\}_m \\ &= [R]\ [K']_{mm}\ [R]^T\ \{\delta\}_m \\ &= [K]_{mm}\ \{\delta\}_m \qquad (8.26)\end{aligned}$$

Once the stiffness matrix of the strip has been transformed into the common coordinate system, it is ready to be assembled into the overall stiffness matrix of the structure in the conventional manner.

In the process of transformation, two points stand out in favour of the finite strip method when compared with the finite element method. The first concerns the compatibility of displacements after transformation. In the finite strip method, the displacement functions have been chosen in such a way that u and w, the two components that are involved in the transformation, have the same variation in the longitudinal (y) direction, and a rotation of coordinate axes x and z will consequently not affect the compatibility of displacements at the nodal lines. This is, however, not so for flat shell finite elements since the majority of such elements use different order polynomials for the u and w displacement components. Even if both u and w are compatible displacement functions when examined individually, after transformation they will be combined in a certain proportion depending on the direction cosines of the element concerned and compatibility of displacements will in general be lost. Secondly, for most flat shell finite elements, only five DOF (u', v', w', θ_x', θ_y') exist at the six DOF (u, v, w, θ_x , θ_y , θ_z) common coordinate system. As a result, in finite element analysis either an artificial sixth DOF or, indeed, some artificial stiffness coefficients have to be inserted, or else a set of surface coordinate axes, the orientation of which changes from node to node, has to be used in order to retain the five DOF system. Meanwhile the strips always retain their four DOF per nodal line, and only the standard transformation used in plane frame analysis is needed.

8.1.5 STATIC ANALYSIS

(i) *Prismatic folded plate under ridge loads*[32]

A typical folded plate cross-section (Figure 8.5) was analyzed by the classical[9] and spline strip methods[18] and compared with the results obtained by De Fries-Skene and Scordelis.[3] Four different span lengths of 100m, 70m, 30m and 10m were used in order to demonstrate the applicability of the method to long and short shells. Poisson's ratio was taken as zero in the examples. The structures were assumed to be under ridge loads, although the amount of work involved in analyzing problems with distributed load would be just the same. Note that the ridge loads were really equivalent loads obtained from conventional elementary (one-way slab) analysis, and they were used in the example of the Task Force report.[1] In practice, such a simplification should not be accepted for shorter folded plates, since a significant portion of the distributed load will be carried directly by longitudinal bending moments and torsional moments to end supports. This was confirmed in a comparative study carried out by Cheung.[9]

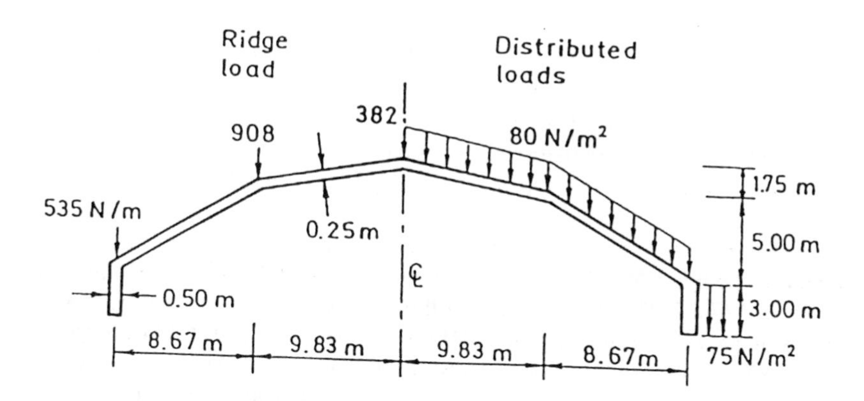

Figure 8.5 Folded plate sectional dimensions and loadings.

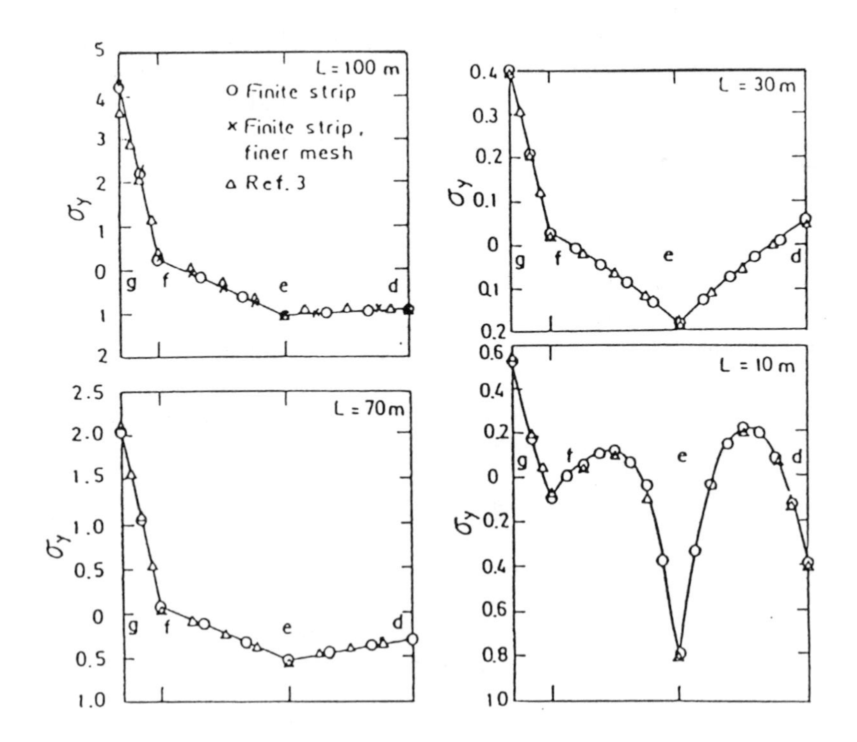

Figure 8.6 Longitudinal stress at midspan for various span lengths L.

Five non-zero harmonics with the strain-displacement relations given by Eqs. 8.4a to 8.4f were used in the analysis, and excellent agreement is observed in comparison with those obtained by De Fries-Skene and Scordelis[3] as shown in Figure 8.6.

(ii) *Curved beam*[14]

The accuracy of the curved finite strip was tested in this example in which a curved beam was analyzed by ordinary curved beam theory neglecting warping torsion and by the finite strip method for the four cross-sections depicted in Figure 8.7a. By maintaining a constant value for the length of the beam $R\alpha$ and denoting the ratio between the bending moments in the curved beam and in the corresponding straight beam of equal length $L=R\alpha$ by

$$\rho = \frac{M_{curved}}{M_{straight}}$$

it is possible to plot ρ against the included angle α. Figures 8.7 demonstrates clearly the good agreement between the finite strip and curved beam theory results. The moments are plotted 1 m away from the concentrated mid-span load because too many Fourier series terms would be needed to represent accurately the value of the bending moment directly underneath or in the immediate vicinity of the concentrated load.

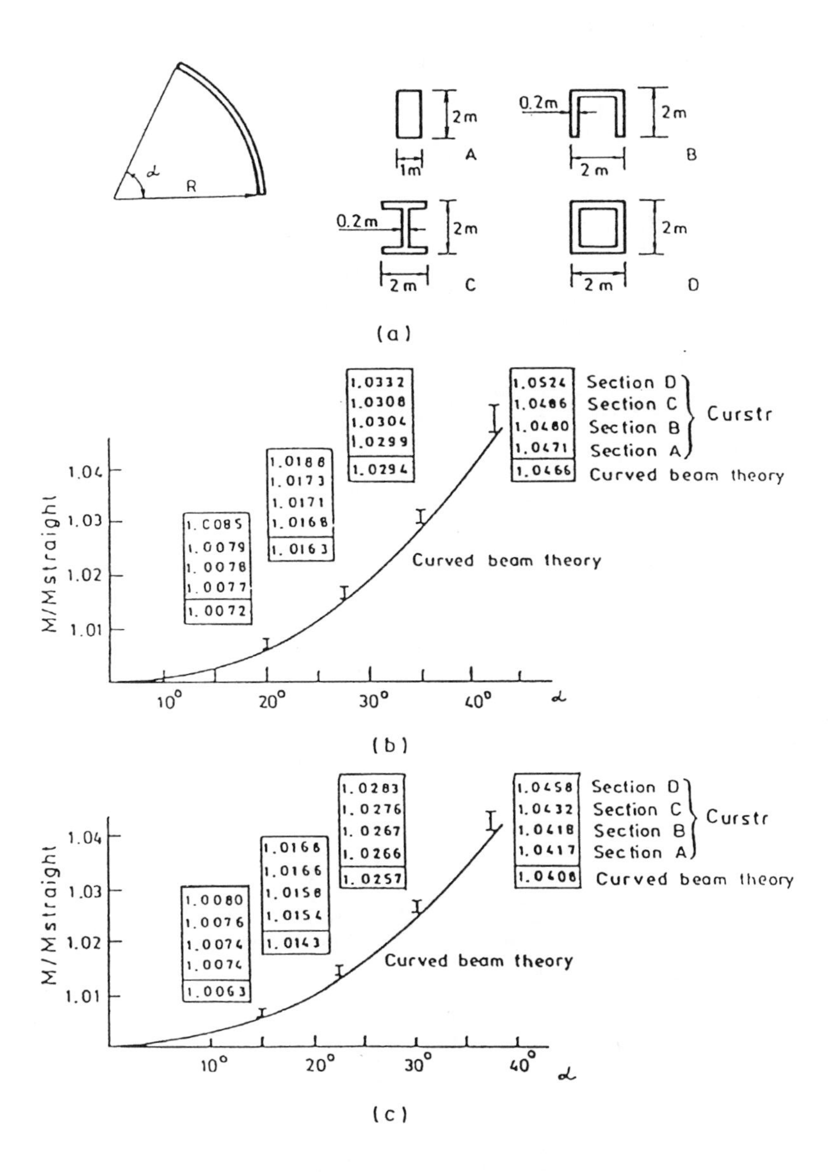

Figure 8.7 Curved beam problem (Meyer and Scordelis[14]): (a) Curved beam dimensions and cross-sectional types. (b) Moment variation for concentrated load. (c) Moment variation for concentrated load.

(iii) *Shallow shells*

To demonstrate the applications of spline strips, Fan[18,19] carried out analyses for the following shells:

1) Clamped saddle-shaped shallow shell (hyperbolic paraboloidal) with square base (Figure 8.8);

2) Simply supported cylindrical shell (Figure 8.9); and

3) Simply supported hyperbolic paraboloidal shallow shell with straight edges (Figure 8.10).

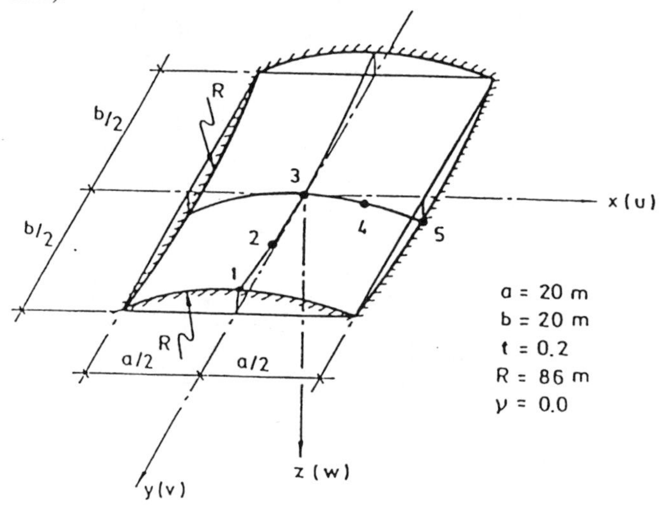

Figure 8.8 Clamped saddle-shaped shallow shell with square base subjected to uniformly distributed load.[18,19]

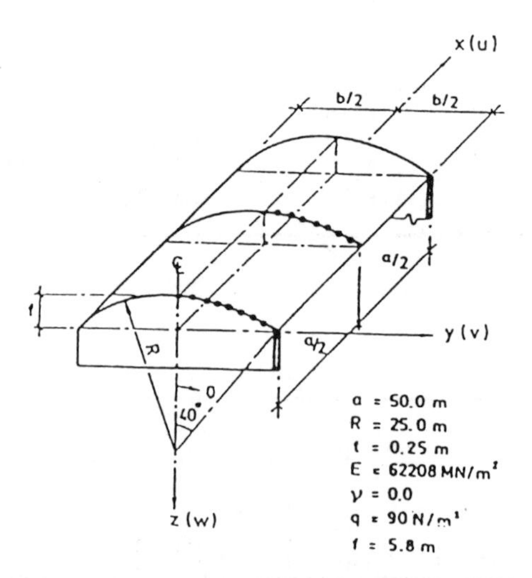

Figure 8.9 Cylindrical shell approximated by shallow shell subjected to uniformly distributed load.[18,19]

The strain-displacement relations for these shells are given by Eqs. 8.5a to 8.5f with k_x and k_y as defined in Table 8.1. Displacements for such shells subjected to uniformly distributed load are tabulated in Tables 8.5 to 8.7. The distributions of longitudinal, transverse as well as twisting moments of the cylindrical shell are depicted in Figures 8.11 and 8.12.

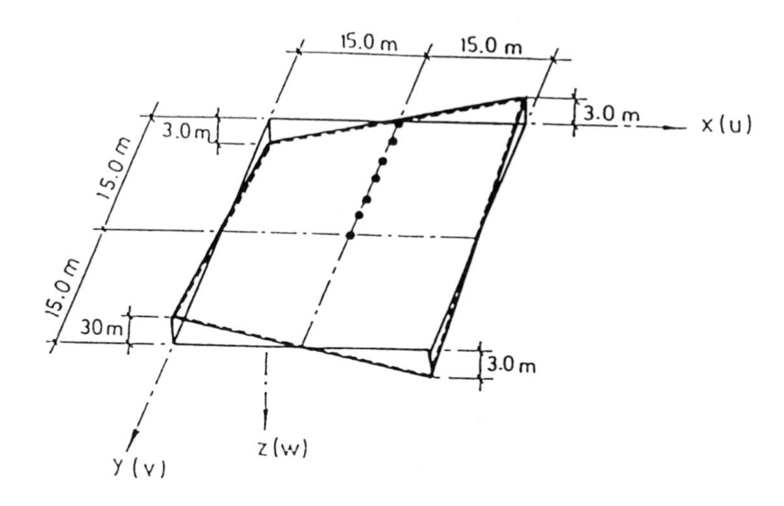

Figure 8.10 Simply supported hyperbolic paraboloidal shallow shell subjected to uniformly distributed load.[18,19]

TABLE 8.5
Displacement of clamped square saddle-shaped shallow shell.[18,19]

Discretisation strip x section	Vertical displacement			Horizontal displacement	
	w_3	w_2	w_4	v_2	v_4
6 x 6	21.10	14.57	14.60	113.78	114.15
8 x 8	21.25	14.64	14.66	114.61	114.80
10 x10	21.28	14.65	14.67	114.76	114.85
12 x 12	21.31	14.67	14.68	114.88	11.94
Series solution	20.0	15.0			
Multiplication factor	q/D			q/Et	

TABLE 8.6
Deflection of cylindrical shell.[18,19]

Discretisation strip x section	Vertical deflection				
	$\theta = 0^{\circ}$	$\theta = 10^{\circ}$	$\theta = 20^{\circ}$	$\theta = 30^{\circ}$	$\theta = 40^{\circ}$
6 x 6	-0.0377	0.0011	0.1005	0.2229	0.3378
8 x 8	-0.0382	0.0015	0.1033	0.2284	0.3456
10 x 10	-0.0383	0.0017	0.1045	0.2306	0.3488
12 x 12	-0.0385	0.0018	0.1051	0.2319	0.3507
Megard[33]	-0.045				0.285
Scordelis[34]	-0.046				0.308

TABLE 8.7
Deflection of hyperbolic paraboloidal shallow shell.[18,19]

Discretisation strip x section	Vertical deflection				
	y = 0	y = 0.25	y = 5	y = 10	y = 15
6 x 6	0	0.0151	0.0262	0.0347	0.0358
8 x 8	0	0.0164	0.0278	0.0361	0.0370
10 x 10	0	0.0162	0.0274	0.0358	0.0368
12 x 12	0	0.0164	0.0277	0.0361	0.0371

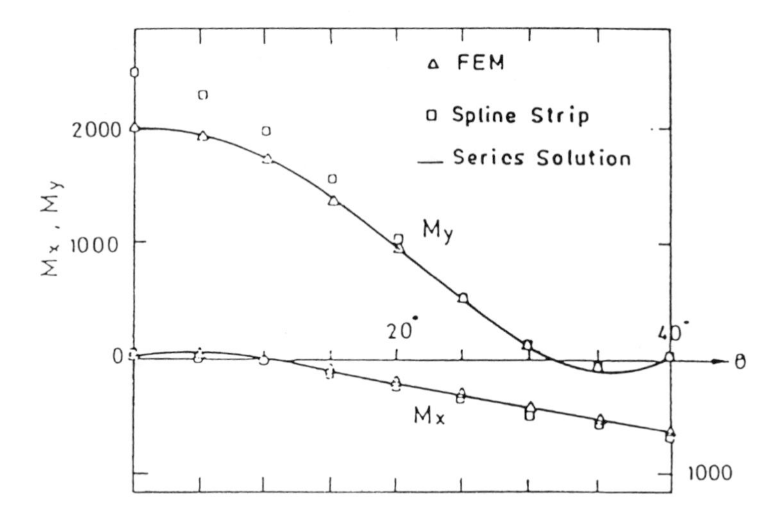

Figure 8.11 Cylindrical shell. Distribution of longitudinal and transverse moment at midspan.[18,19]

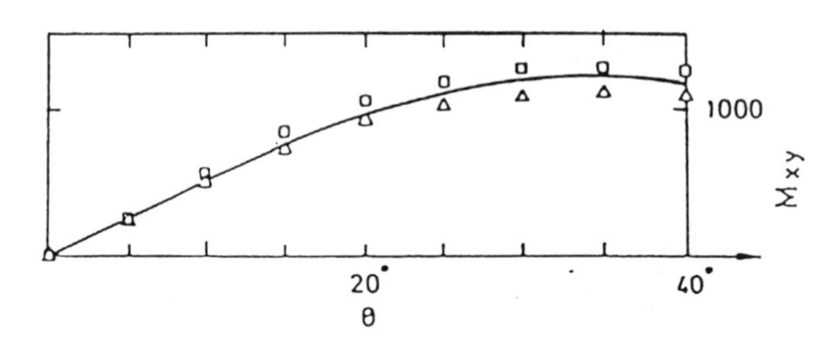

Figure 8.12 Cylindrical shell. Distribution of twisting moment at support.[18,19]

(iv) *Continuous folded plate structures by classical finite strips*

A folded plate structure[16] was analyzed using the functions developed in Section 2.2.1.1. The dimensions and loadings of the structure are shown in Figure 8.13. The modulus of elasticity was taken as $25.25 x 10^6 kN/m^2$ and Poisson's ratio as zero. The analysis was conducted by using nine non-zero terms, and the structure was sub-divided into nine strips in the longitudinal direction. The results were compared with those obtained by Lo et al.[35] using the elasticity method.

Tables 8.8 and 8.9 tabulate the longitudinal variations of the deflection and transverse moment along sections 'B' and 'D' respectively . The two sets of results are in fairly good agreement. In addition the variation of the longitudinal stress along the midspan is depicted in Figure 8.14.

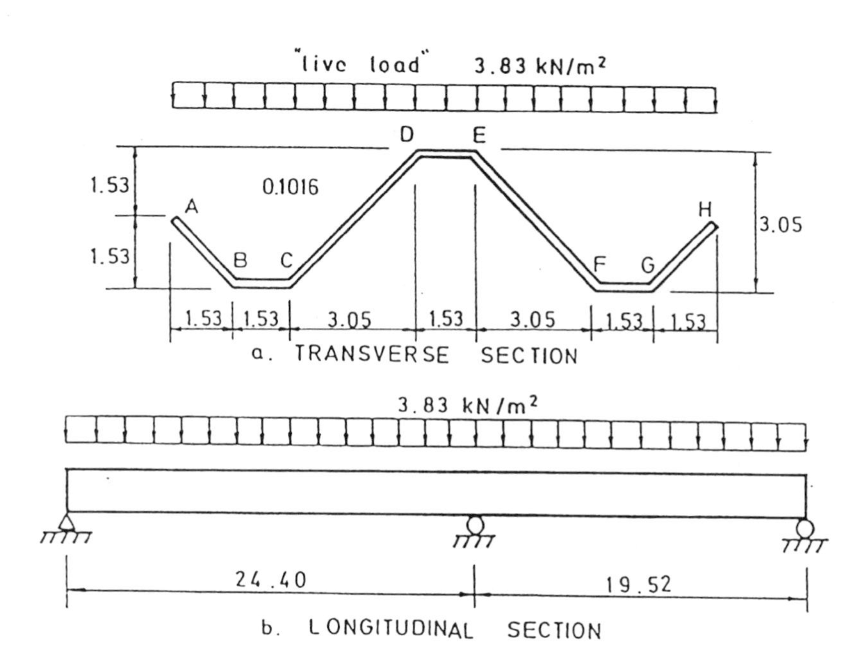

Figure 8.13 Continuous folded plate structures.[16]

TABLE 8.8

Comparison of vertical deflections, in metres time 10^4. [16]

x	Finite strip		Elasticity Theory[35]	
(metres)	B	D	B	D
3.66	42.40	17.39	42.70	18.00
10.98	80.52	33.55	81.74	34.47
18.30	44.23	18.91	54.45	19.83
30.50	15.56	3.36	16.47	3.97
37.82	27.15	7.32	27.76	7.93

TABLE 8.9
Comparison of transverse moments (kg-m/m).[16]

x (metres)	Finite strip		Elasticity Theory[35]	
	B	D	B	D
3.66	-334	-183	-336	-190
10.98	-378	-154	-397	-169
18.30	-381	-199	-398	-205
30.50	-381	-190	-391	-187
37.82	-358	-172	-372	-171

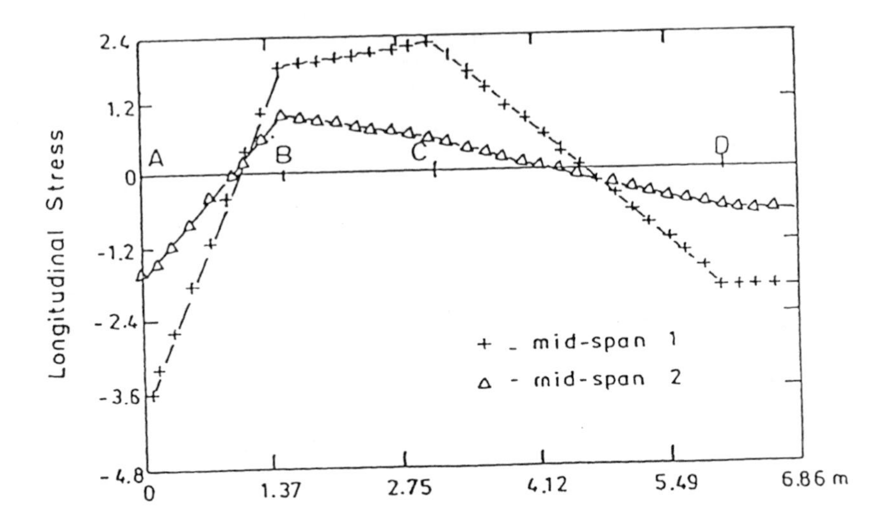

Figure 8.14 Longitudinal stress at midspan.[16]

(v) *Continuous folded structures by spline strips*

Another continuous folded plate structure (Figure 8.15) subjected to uniformly distributed load was analyzed by Fan[18] using the flat shell spline strips. By symmetry, analyses were performed only on one span and half the

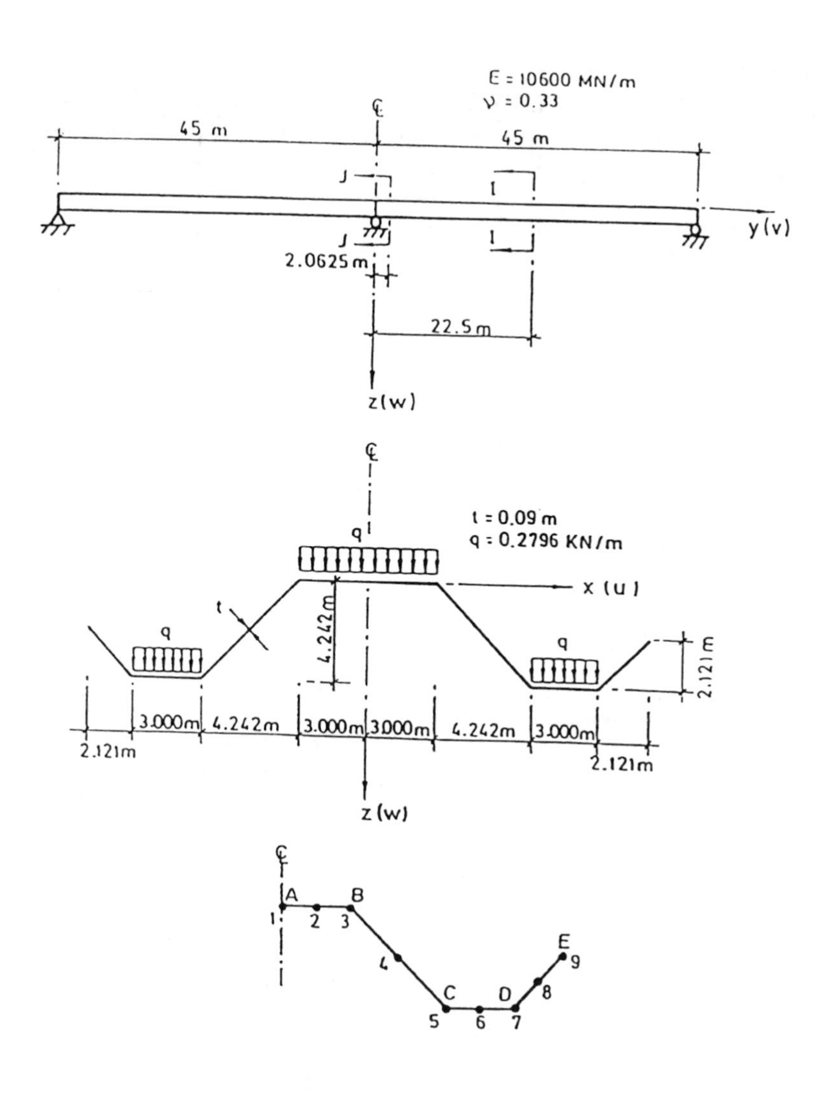

Figure 8.15 Two-equal span folded plate subjected to uniform vertical load.[18]

width of the folded plate by meshes consisting of 7 strips with subdivisions of 4, 6, 8, 10 or 12 sections.

The longitudinal membrane stresses at section K-K (Figure 8.15) are tabulated in Tables 8.10. It can be seen that rates of convergence are very fast and only slight differences can be found in the refined meshes analyses. In Figure 8.16, the longitudinal membrane stresses obtained by the 12-section mesh are compared to Beaufait's theoretical values.[36,37] Good agreement is observed.

TABLE 8.10
Longitudinal membrane stress at Section K-K.[18]

Discretisation	Node Point*							
	1	2	3	4	5	6	7	8
7 x 4	137.38	149.21	188.96	-21.19	-240.28	-229.81	-265.75	530.44
7 x 6	138.59	151.33	196.40	-21.43	-247.37	-233.14	-274.65	545.58
7 x 8	139.91	153.02	198.68	-21.60	-250.25	-235.97	-278.19	553.62
7 x 10	140.36	154.15	198.51	-21.75	-250.75	-238.50	-278.81	557.49
7 x 12	140.03	154.75	197.25	-21.96	-248.67	-240.44	-277.67	558.62

(* Refer to Figure 8.15)

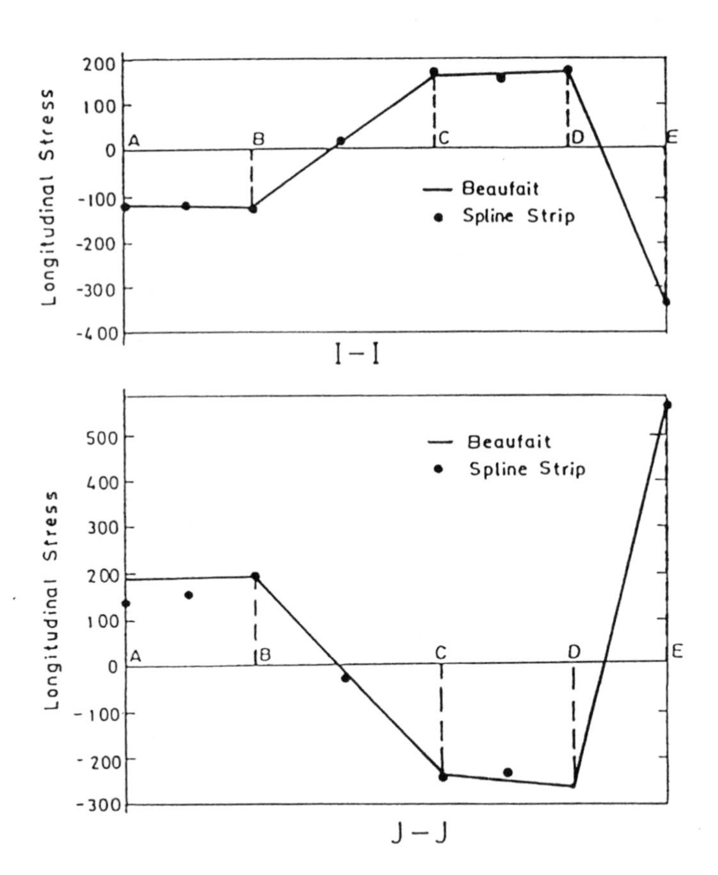

Figure 8.16 Distribution of longitudinal membrane stress.[18]

8.1.6 VIBRATION ANALYSIS

(i) *Stiffened panel*

A four-stiffener panel (Figure 8.17) with two ends simply supported and the other edges free was analyzed using the classical flat shell strip as well as the computed shape function strip.[38] The Poisson's ratio υ is assumed to be 0.3 throughout. The lowest ten natural frequencies of the panel are presented in Table 8.11, in terms of a dimensionless parameter $\bar{n}$ defined by

$$\bar{n} = \omega l \left(\frac{E}{\rho}\right)^{-1/2}$$

In the same table comparison is made with the results of Reference 39, and very good agreement is observed.

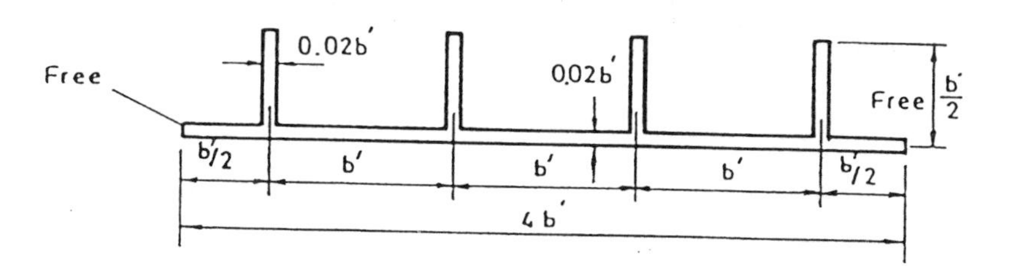

Figure 8.17 A simply supported four-stiffener panel (l = 6b').

The mode shapes corresponding to the lowest five natural frequencies are sketched in Figure 8.18. The data from which these sketches were prepared consist of displacements and rotations at the top and bottom of all four stiffeners. It should be pointed out that only half of the cross-section is shown because of symmetry or anti-symmetry conditions. Since the structure is simply supported at two opposite ends, all the longitudinal modes are uncoupled and only a small matrix of fifty-two equations has to be solved for each term of the series.

It is interesting to note that all the frequencies given by the finite strip method in Table 8.11 constitute an upper bound to the frequencies computed in Reference 39 through the use of exact stiffness and mass matrices for isotropic plates. In the table, the results obtained by Kong[38] using the COMSFUN strips were also given for comparison.

(ii) *Cylindrical panels*

Two cylindrical panels, one simply supported and the other clamped, were analyzed using both classical cylindrical shell strips and flat shell strips.

TABLE 8.11
Natural frequencies of a four-stiffener panel (12 strips).

Mode number	$\bar{n} = \omega l\,(E/\rho)^{-1/2}$			Wave number (m)	Type of symmetry
	Classical strip	COMSFUN[38]	Reference 39		
1	0.0287	0.0302	0.0286	1	A
2	0.0292	0.0307	0.0291	1	S
3	0.0365	0.0384	0.0359	1	S
4	0.0366	0.0385	0.0362	1	A
5	0.0394	0.0413	0.0391	1	S
6	0.0396	0.0414	0.0395	2	S
7	0.0411	0.0431	0.0410	2	A
8	0.0504	0.0530	0.0504	3	S
9	0.0521	0.0547	0.0519	3	A
10	0.0557	0.0584	0.0555	2	S

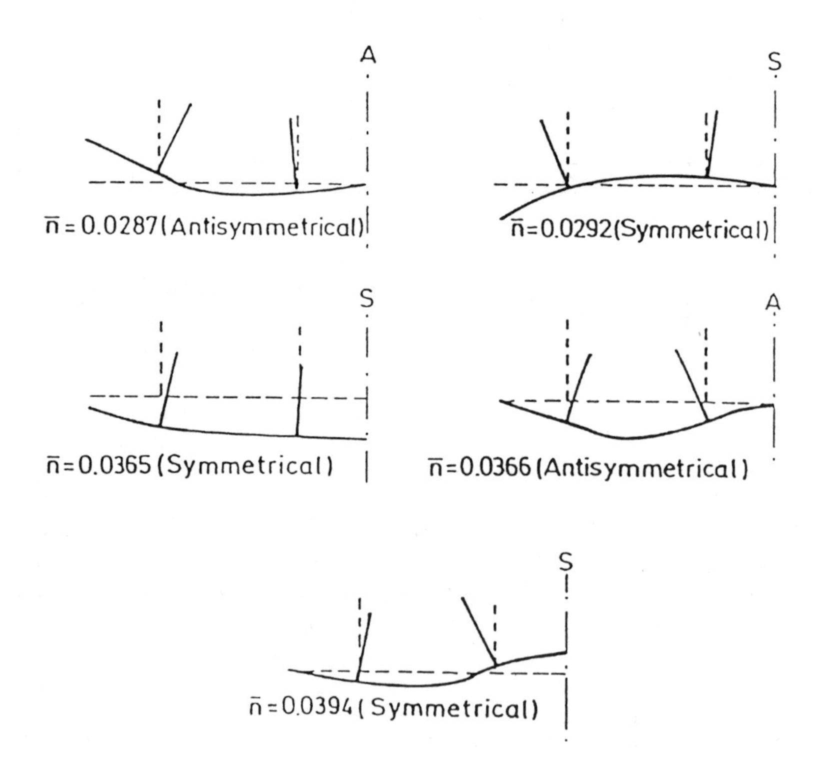

Figure 8.18 Modal shapes for the first five natural frequencies of a four-stiffener panel.

The circular frequencies computed for the two cases together with the results of References 40 and 42 are given in Table 8.12. For each case six strips and four terms of the series are used.

It can be seen that the agreement between the three sets of results is quite good. However, from experience it has been found that the idealization of a curved panel as an assembly of flat shell strips is more suitable for shallow shells than for deep shells, due to the fact that for the latter case a great number of strips is required to approximate the geometric shape before accurate frequencies can be obtained.

TABLE 8.12
Natural frequencies (radian/sec) of cylindrical panels.

Panel shape and boundary	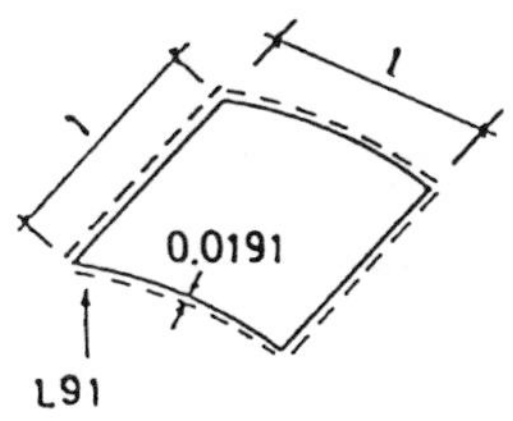 0.0191		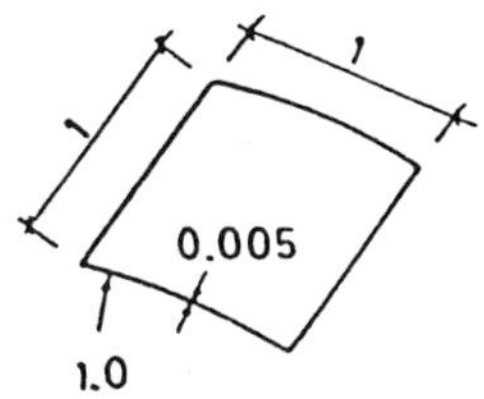 0.005; 1.0	
Frequencies	Curved strip	Flat strip	Curved strip	Flat strip
ω_1	0.277 (0.282)[42]	0.285	0.101 (0.108)[40]	0.119
ω_2	0.277	0.305	0.127	0.119
ω_3	0.466	0.512	0.139	0.142
ω_4	0.493	0.530	0.166	0.177
ω_5	0.542	0.573	0.196	0.206

(iii) *Continuous I-beam*

The free vibration of a continuous beam, for which analytical solutions are available, was reported by Cheung and Delcourt[43] to demonstrate the validity of the classical finite strip formulation. The cross-section of the beam was an I section made up of two equal flanges of width 0.5 and a web of 1.0 and a constant thickness of 0.1. A series of problems involving different end conditions, different number of spans (S) and span lengths (l_i, i = 1,2,..S) was attempted. Five strips (two for each flange and one for the web) and four terms of the series were used in the analysis. The lowest frequencies are compared with exact solutions (wherever possible) in Table 8.13.

In the table, the results obtained by computed shape function strips[38] were also tabulated for comparison.

TABLE 8.13
The lowest natural frequencies of continuous I-beams.
$E = 192$; $\upsilon = 0.3$; $m = 1$; $l = 5$; $\alpha = l_1/l_2$; $\beta = l_1/l_3$; $\gamma = l_1/l_4$

Lengths and boundary conditions	Circular frequency		
	Classical strip	Kong[38]	Biggs[44]
$S = 2$; $\alpha = 1$ simply supported ends	0.5696	0.5718	0.5692
$S = 2$; $\alpha = 0.75$ simply supported ends	0.3819		0.3809
$S = 2$; $\alpha = 1$ built in ends	0.8899		0.8896
$S = 2$; $\alpha = 1.272$ one end clamped, other end simply supported	0.902		0.903
$S = 3$; $\alpha = \beta = 1$ simply supported ends	0.5696		0.5692
$S = 3$; $\alpha = 0.5$; $\beta = 1$ simply supported ends	0.2270	0.2265	0.2224
$S = 3$; $\alpha = 1.285$; $\beta = 1$ built in end	0.9035		0.9039
$S = 4$; $\alpha = \beta = \gamma = 1$ simply supported end	0.5696		0.5692

(iv) *Doubly curved shells*[23,45]

Deep spherical shells with hinged and clamped support conditions along the circumference were employed to demonstrate the applications of the spline strip in shell vibration analysis. Analyses were based on both Donnell's and Reissner's theories. Transformation was carried out by using the equations given in Section 7.3. The results are tabulated and compared with other published ones[46,47,48] in Tables 8.14 and 8.15.

TABLE 8.14
Dimensionless frequencies of deep spherical hinged shell.[23,45]
$\lambda_I = \omega_i R(\rho/E)^{1/2}$; semi-angle = 60° ; $\upsilon = 0.3$; $R/t = 20$; R = radius of shell

λ_i	Reference 46	Reference 47	Reference 48	Spline strip[45]	
				Donnell	Reissner
λ_1	0.963	0.962		0.977	0.971
λ_2	1.338	1.334		1.354	1.342
λ_3	1.653			1.654	1.652
λ_4	2.131	2.128	2.124	2.165	2.149
λ_5	2.141			2.113	2.112
λ_6	3.185	3.176	3.178	3.235	3.216

TABLE 8.15
Dimensionless frequencies of deep spherical clamped shell.[23,45]
$\lambda_I = \omega_i R(\rho/E)^{1/2}$; semi-angle = 60° ; υ = 0.3 ; R/t = 20 ; R = radius of shell

λ_i	Reference 46	Reference 47	Reference 48	Spline strip[45]	
				Donnell	Reissner
λ_1	1.008	1.006		1.024	1.018
λ_2	1.395	1.391	1.335	1.410	1.402
λ_3	1.702			1.714	1.706
λ_4	2.126			2.133	2.133
λ_5	2.387	2.375	2.368	2.416	2.400
λ_6	3.506	3.486	3.478	3.553	3.537

(v) *Cooling tower*

Li et al.[23,45] also reported an example of the vibrational analysis of a cooling tower by spline strips. The geometrical details of the shell are given in Figure 8.19. Results based on both Donnell's and Reissner's theories are tabulated in Table 8.16, and they are compared to those obtained by Hashish[49] and Yang and Kapania[50] using 36DOF and 48DOF finite element models.

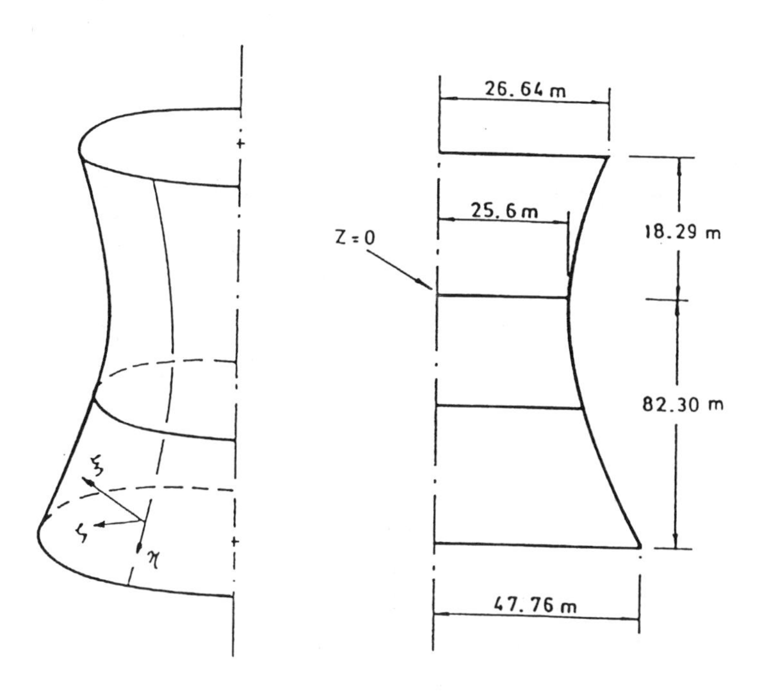

(E = 20.92GN/m^2; υ = 0.15; density = 2406.5kg/m^3)

Figure 8.19 Modelling of cooling tower by spline finite strip.[23,45]

TABLE 8.16
Natural frequencies (Hz) of a fixed base cooling tower.[23,45]
C = circumferential wave number ; L = longitudinal wave number

C	L	Hashish[49]	36DOF[50]	48DOF[50]	Donnell	Reissner
1	1	3.3345	3.3137	3.2901	3.3723	3.3723
	2	6.8816	7.0656	6.7958	6.9667	6.9447
	3	10.525		10.624	10.788	10.787
2	1	1.7848	1.7723	1.7683	1.8160	1.8149
	2	3.7234	3.7510	3.6953	3.7963	3.7958
	3	6.9553	7.4466	6.9813	7.1413	7.1410
3	1	1.3927	1.3731	1.3809	1.4146	1.4109
	2	2.0150	1.9982	1.9996	2.0614	2.0590
	3	4.3353	4.4536	4.3595	4.4380	4.4316
4	1	1.2003	1.1512	1.1885	1.2349	1.2280
	2	1.4597	1.4195	1.4615	1.4802	1.4731
	3	2.772	2.7946	2.7941	2.8926	2.8899

8.1.7 STABILITY ANALYSIS

(i) *Formed sections*

To demonstrate that the finite strip formulation is also valid for stability problems, Dawe[51] has investigated the buckling of a Z-section (Figure 8.20) under uniform axial compression. A direct comparison of the strip results is made to those obtained by Viswanathan et al.[52] in Table 8.17. They are in excellent agreement. The typical buckled mode shape in the plane of the cross-section is also shown in Figure 8.20.

TABLE 8.17
Buckling of Z-section.[51]

No. of longitudinal half wave	Values of $10^3 \times \sigma_1/E$	
	Dawe[51]	Viswanathan[52]
2	6.071	6.049
3	5.765	5.741
4	6.465	6.437

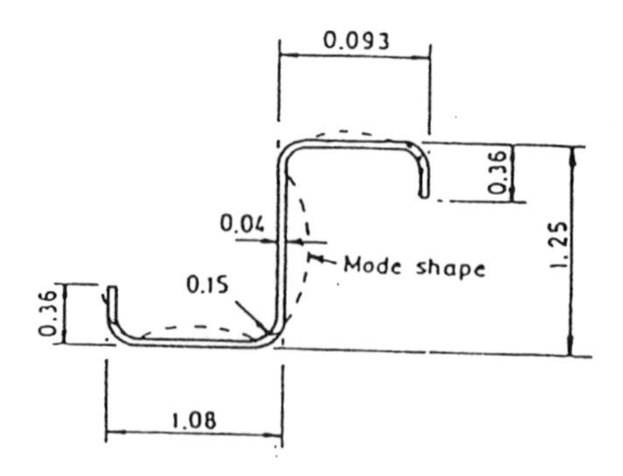

Figure 8.20 Details of formed Z-section.[51]

(ii) *Continuous I-beam*

Cheung et al.[43] analyzed a number of continuous beams with the same I section as described in the previous example (Section 8.2.6.3). The results are in good agreement with available exact solutions (Table 8.18).[53]

TABLE 8.18
Stability analysis ; Ω = cross-sectional area = 0.2; $P_{cr} = \Omega\sigma_{cr}$.[43]

Lengths and boundary conditions	Critical stress	
	Classical strip	Timoshenko[53]
$S = 2$; $\alpha = 1$ simply supported ends	$\sigma_{cr1} = 0.854$ $\sigma_{cr2} = 1.6913$	0.821 1.6773
$S = 2$; $\alpha = 0.75$ simply supported ends	0.5887	0.5831
$S = 2$; $\alpha = 1$ built in ends	1.6719	1.6773
$S = 3$; $\alpha = \beta = 1$ simply supported ends	0.822	0.821

(iii) *Square stiffened panel*

Another simply supported square stiffened panel (Figure 8.21) was analyzed by subdividing into two strips between stiffeners, one strip for each stiffener, and six spline sections longitudinally.[20] The details of the panel are given in Table 8.19. Kong[38] had also analyzed the panel using the COMSFUN strip. The results are tabulated in Table 8.20. Note that the results are compared with the finite element ones.[54] The three sets of results are in fairly good agreement with COMSFUN, which predicts slightly higher values.

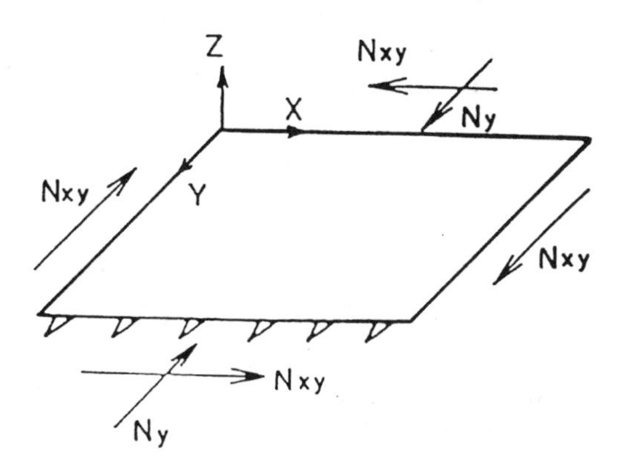

Figure 8.21 Square stiffened panel.[38]

TABLE 8.19
Loading, dimensions and boundary conditions of the stiffened panel.[38]

Length (L)	0.7620m	Skin thickness	2.1336m
Width (b)	0.7620m	Stiffener thickness	1.4732m
Stiffener spacing	0.1270m	Young's modulus	0.724×10^5 MPa
Stiffener depth	0.03434m	Poisson's ratio	0.32

Buckling boundary conditions are simply supported at all four edges :

TABLE 8.20
Buckling loads of square stiffened panel.[38]

Load		Load factor		
N_x (kN/m)	N_{xy} (kN/m)	Finite element[54]	Spline strip[20]	COMSFUN[38]
0	175.1	0.8138	0.8371	
70	175.1	0.7195	0.7371	0.7490
175.1	175.1	0.6061	0.6183	0.6304
350.3	175.1	0.4444	0.4542	0.4612
875.6	175.1	0.1929	0.1977	0.1990
175.1	0	0.9759	1.0003	1.0070

8.1.8 ANALYSIS OF FOLDED PLATE STRUCTURES USING MINDLIN PLATE THEORY

The advantages of the formulation of the plate bending problem by the Mindlin plate theory have already been reviewed in Section 3.7. Combining the stiffness matrix given by Eq. 3.58 with the in-plane stiffness matrix (Eq. 4.12), one can proceed to carry out the analysis for folded plate structures. Based on this approach, Tham[24] carried out an analysis of a 32mx36m game hall, which was covered by nine V-shaped thin wall reinforced concrete beams each 4m wide and 2m deep (Figure 8.22). The same folded form continued from the roof to the side walls, which were slightly inclined. This example was first presented by Lee et al.,[55] and as it was demonstrated that the edge effect diminishes rapidly, it is possible to carry out the analysis separately for two typical bays: an internal bay and an external bay. The results of the external bay are used to demonstrate the accuracy of the method. The loading details are given in Table 8.21.

TABLE 8.21
Plate thickness and loading.

Member	Thickness (m)	Loading (kN/m^2)
1	0.15	7.5
2	0.25	12.5
3	0.21	8.4
4	0.35	14.0

Figures 8.23 and 8.24 show the variations of deflection along Line 'B' and the transverse moment along Line 'C'. The figures show the results obtained by the spline strips and higher order finite elements.[55]

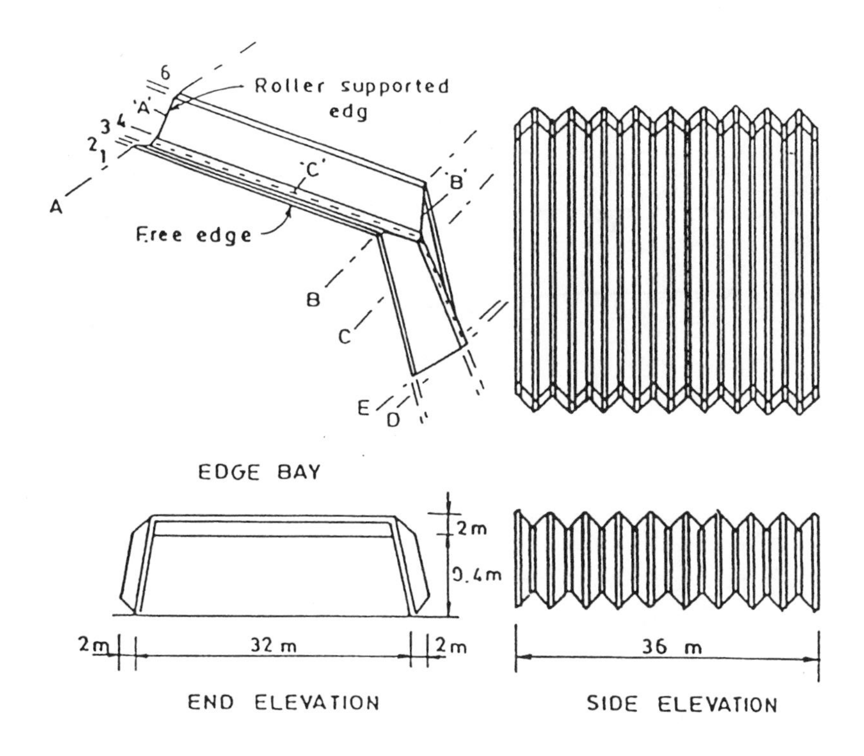

Figure 8.22 Details of the indoor game hall.[24]

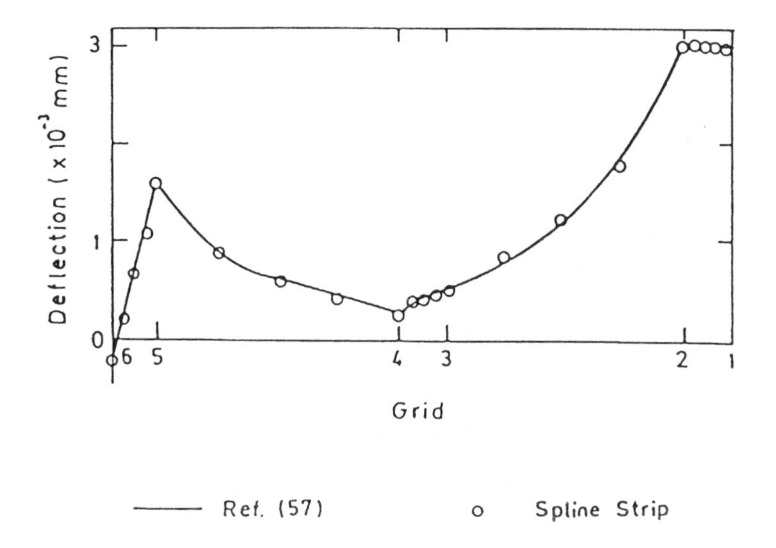

Figure 8.23 Deflection along Line 'B'.[24]

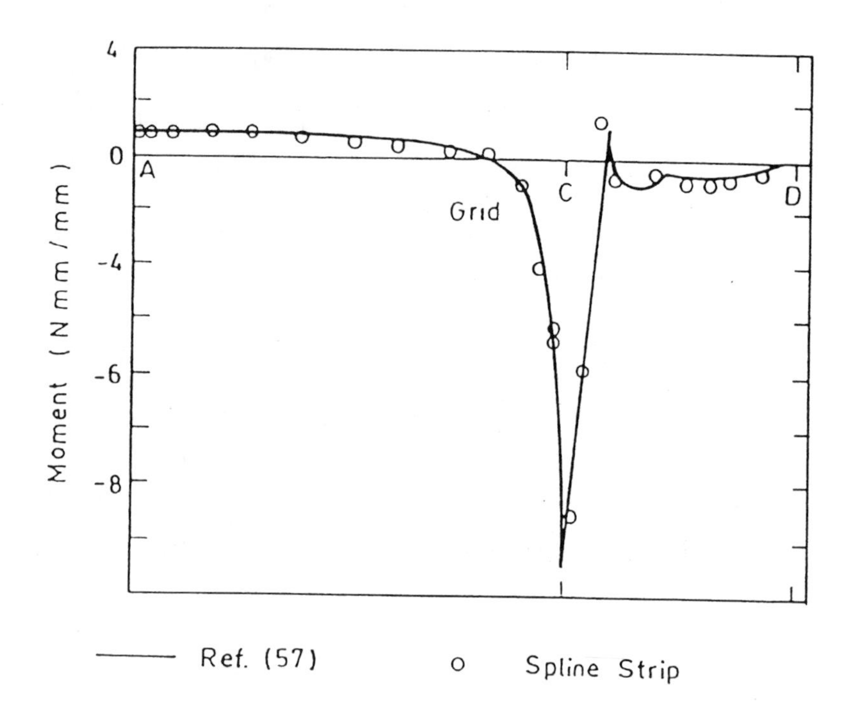

Figure 8.24 Transverse moment along Line C.[24]

8.1.9 COMPOUND STRIP METHOD

Many plates are stiffened by beams which are arranged transversely and longitudinally. The exact modelling of such systems, though achievable by the folded plate model, is not always necessary as reasonably accurate results can be attained more efficiently by the compound strips.

To demonstrate the formulation of the compound strip, a typical stiffened plate (Figure 8.25) with both transverse and longitudinal stiffeners is used.

Figure 8.25(a) show a transverse stiffener across a typical strip at a distance y_t from one end. The bending and torsional rigidities of the stiffener are EI_t and GJ_t respectively. Puckett[26] had shown that the stiffness of stiffeners can be written in terms of the strip displacement parameters as:

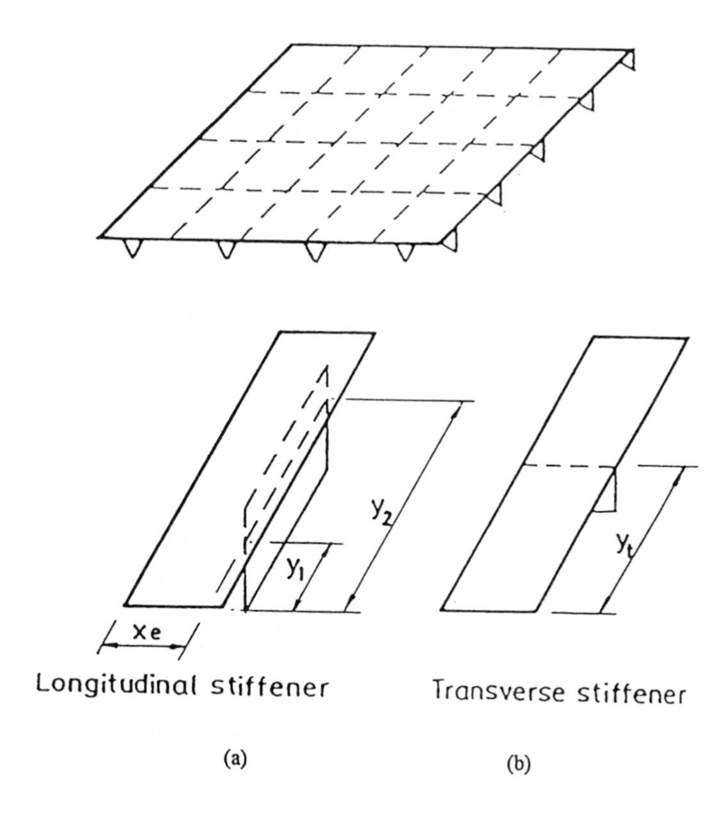

Figure 8.25 Stiffened plate: (a) Longitudinal stiffener. (b) Transverse stiffener.

a) bending stiffness:

$$EI_t\, Y_m(y_t)\, Y_n(y_t) \begin{bmatrix} \frac{12}{b^3} & \frac{6}{b^2} & -\frac{12}{b^3} & \frac{6}{b^2} \\ & \frac{4}{b} & -\frac{6}{b^2} & \frac{2}{b} \\ & & \frac{12}{b^3} & -\frac{6}{b^2} \\ \text{SYM} & & & \frac{4}{b} \end{bmatrix} \tag{8.27}$$

b) torsional stiffness:

$$\frac{1}{30b} GJ_t \frac{dY_m(y_t)}{dy} \frac{dY_n(y_t)}{dy} \begin{bmatrix} 36 & 3b & -36 & 3b \\ & 4b^2 & -3b & b^2 \\ & & 36 & 3b \\ \text{SYM} & & & 4b^2 \end{bmatrix} \tag{8.28}$$

Similarly, the stiffness of a longitudinal stiffener can be expressed as:
a) bending stiffness:

$$EI_l \int_{y_1}^{y_2} \frac{dY_m(y)}{dy}\frac{dY_n(y)}{dy}dy \begin{bmatrix} C_1C_1 & C_1C_2 & C_1C_3 & C_1C_4 \\ & C_2C_2 & C_2C_3 & C_2C_4 \\ & & C_3C_3 & C_3C_4 \\ SYM & & & C_4C_4 \end{bmatrix} \quad (8.29)$$

b) torsional stiffness:

$$GJ_l \int_{y_1}^{y_2} \frac{dY_m(y)}{dy}\frac{dY_n(y)}{dy}dy \begin{bmatrix} C_1'C_1' & C_1'C_2' & C_1'C_3' & C_1'C_4' \\ & C_2'C_2' & C_2'C_3' & C_2'C_4' \\ & & C_3'C_3' & C_3'C_4' \\ SYM & & & C_4'C_4' \end{bmatrix} \quad (8.30)$$

where $C_1 = (1\text{-}3\,\overline{x}_l^{\,2} + 2\,\overline{x}_l^{\,3})$; $C_1' = \frac{1}{b}(-6\overline{x}_l + 6\overline{x}_l^{\,2})$ (8.31a)

$C_2 = x(1\text{-}2\,\overline{x}_l + \overline{x}_l^{\,2})$; $C_2' = (1 - 4\overline{x}_l + 3\overline{x}_l^{\,2})$ (8.31b)

$C_3 = (3\,\overline{x}_l^{\,2}\text{-}2\,\overline{x}_l^{\,3})$; $C_3' = \frac{1}{b}(6\overline{x}_l - 6\overline{x}_l^{\,2})$ (8.31c)

$C_4 = x(\overline{x}_l^{\,2} - \overline{x}_l)$; $C_4' = (3\overline{x}_l^{\,2} - 2\overline{x}_l)$ (8.31d)

(x_l, y_1) and (x_l, y_2) are the end coordinates of the stiffener; and

$\overline{x}_l = \frac{x_l}{l}$.

One should have noted that the inclusion of transverse stiffeners will lead to the coupling of the terms of the series, and the stiffness matrices thus required to be solved will be much greater in size. A large number of terms will also be required to model transverse stiffeners, especially in the case of stiff ones. Furthermore, the sectional properties of the stiffeners are difficult to be estimated accurately. In spite of these limitations, the compound strips have been adopted by a number of researchers.

A simply supported folded plate roof with a gable at mid-span (Figure 8.26) was analyzed by Maleki[27] using classical strips. The gable was rectangular in shape with a cross-sectional area of 0.3 in x 1.0 in. Other pertinent geometric dimensions as well as loadings on the roof are given in Figure 8.26. The results are compared to those obtained by Pultar[56] and Mark[57] in Figure 8.27. It is obvious that they are in good agreement.

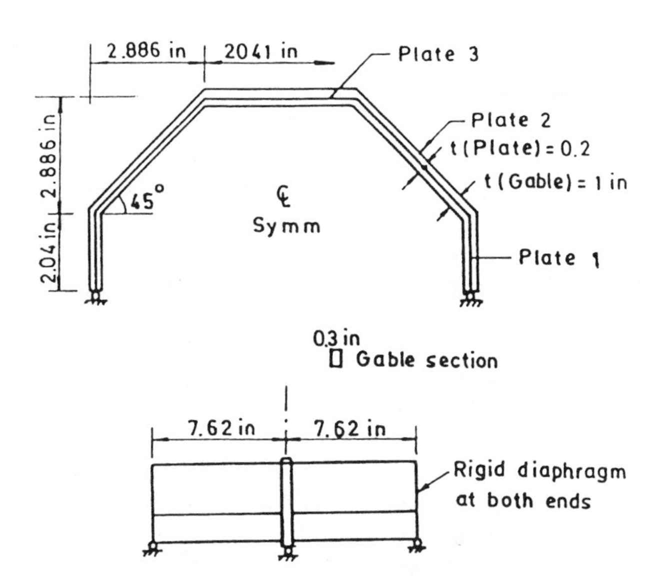

	t	υ	E (psi)	Vertical load (psi)
Plate 1	0.2	0.43	3560	0.009
Plate 2	0.2	0.43	3560	0.009
Plate 3	0.2	0.43	3560	0.129

Figure 8.26 Folded plate with gable.[27]

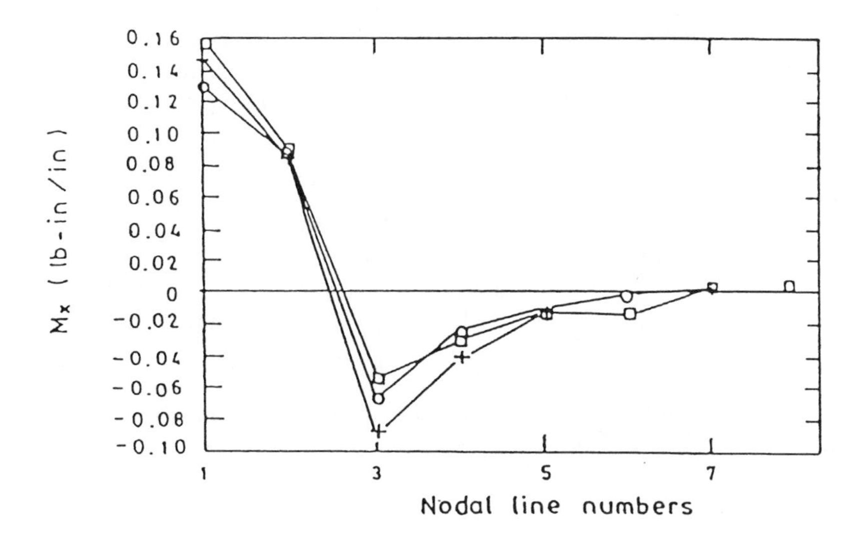

Figure 8.27 Moment at quarter span.[27]

8.2 BRIDGES

8.2.1 INTRODUCTION

In order to satisfy their engineering as well as aesthetical requirements, bridges can adopt rather complex structural forms. They may be slab bridges with or without beams or box girder bridges with or without transverse stiffeners. They may have one single span or multiple continuous spans. Their alignments may be straight lines, or simple or compound curves in plan. In accordance with the structural action, they may be classified as slab bridges, box girder bridges, truss bridges, arch bridges, cable-stayed bridges and suspension bridges. Nevertheless, it is not difficult to realize that bridges can be treated either as plates, folded plates or shells depending on their structural configurations.

A traditional method of analysis for right bridge decks is the orthotropic plate theory originated by Guyon(58) and Massonet,(59) and design charts have been prepared by various workers(60,61) using only the first term of the harmonic series. A set of design charts using fifteen terms of the series had been prepared by Cusens and Pama.(62) However, the method is less versatile because the deck thickness is assumed to be uniform throughout, and the use of design charts, apart from having some built-in approximations, will not allow a designer to go up to any accuracy that he might judge to be adequate.

The finite element method, finite difference method and grillage analysis are versatile but have been considered to be inefficient, due to the large number of unknowns involved in analyzing straight and curved bridge decks with simply supported ends.

The finite strip method is an ideal tool for analyzing bridges with the previously mentioned features. The analysis of right slab bridges by the finite strip method was first attempted by Cheung.(63) Later the method was extended to slab-beam bridges by Powell and Ogden(64) and by Cheung et al.(65). Further analysis of right bridges based on higher order strips was presented by Loo and Cusens.(13,66) In addition, a curved strip was applied by Cheung(67) to the solution of curved bridge decks. Prestressed forces were also included in the studies by Loo(68) and Abdullah et al.(69) End and intermediate diaphragms were the subjects of study carried out by Graves Smith.(70)

The spline finite strip was also used to analyze various types of bridges. Initially, Fan(18) analyzed right slab and box girder bridges. Carrying out suitable domain transformation, slab bridges of different geometric shapes were studied by Cheung et al.(71) They also reported applications of this method to box girder bridges.(72) Cheung studied bridges with haunched webs by both the classical(73) and the spline strips.(74)

Recently, Puckett(75) applied the compound strips, using both classical and spline strips, to analyze slab bridges.

The finite strip method is also a powerful tool for the dynamic analysis of various types of bridges. Cheung et al.[76] used a LO2 rectangular bending strip for the analysis of single and continuous span orthotropic slab bridges, while Babu and Reddy[77] studied the free vibration of skew bridges for a variety of skew angles and aspect ratios. The free vibration of right and curved box girder bridges was presented by Cheung and Cheung.[78] Li[23] conducted a detailed study on the applications of the spline strip in this area. The response of a bridge due to a moving vehicle, which is simulated as a moving mass or load, was evaluated by Srinivasan,[79] Smith[80] and Hutton.[81] Local buckling of a bridge deck was investigated by Delcourt.[43]

8.2.2 SLAB BRIDGES WITH OR WITHOUT STIFFENING GIRDERS

Depending on the thickness-span ratio, slab bridges can be treated either as Kirchhoff plates or shear-deformable (Mindlin) plates. Chapter 3 has extensively discussed this topic and readers may refer to it for details.

One has also to point out that there are certain special features commonly associated with these types of bridges but not with plates or folded plate structures, and therefore, some special considerations have to be made during the analyses. These special features include:

a) other than the end supports, the bridges may also be elastically supported by longitudinal beams and intermediate columns which need not be evenly spaced.

b) many bridges are stiffened in one or two directions by girders. Therefore, the actual structures have to be treated as girder-slab systems.

A number of novel techniques has been proposed to modify the finite strip method for accommodating these features, and detailed discussions on them will be made in the forthcoming sections.

8.2.2.1 *Analysis of column supported bridge decks*

Column or line supports with axial and rotational stiffnesses can be easily incorporated in the spline strip method by adopting the spring models as discussed in Section 3.4.4.

For bridges continuous over several line supports, the classical strip solution is rather straightforward if Eq. 2.11 of Section 2.2.1.1 is used for the series component of the shape function. The undetermined coefficients are determined by taking into account the bridge stiffness and span. It is also possible to treat each individual span of such bridges as simply supported, and continuity of slopes over the supports (at the nodal lines only) is subsequently restored. In this case the support moments are chosen as the redundant forces.[82]

If the bridge decks are supported by arbitrarily spaced columns, special treatments may be necessary for the classical strip. Two choices are available.

One approach is the force approach or flexibility approach. The released structure, which in the present case amounts to a bridge simply supported at its two ends only, is analyzed under the external loading and also under the redundant forces which are either unit point loads[66] or uniformly distributed rectangular patch loads[83] at the column locations, using the standard finite strip procedure and with as many right-hand side loading vectors as there are external load cases and redundant forces. This procedure produces a set of displacements Δ at the column locations for all the external loadings, and also another set of displacements, or flexibility coefficients f due to the unit reaction forces. In matrix notation,

$$[\Delta] = \begin{bmatrix} \Delta_{11} & \Delta_{12} & \cdots\cdots & \Delta_{1n} \\ \Delta_{21} & \Delta_{22} & \cdots\cdots & \Delta_{2n} \\ \cdots\cdots & \cdots\cdots & \cdots\cdots & \cdots\cdots \\ \Delta_{n1} & \Delta_{n2} & \cdots\cdots & \Delta_{nn} \end{bmatrix} ; \quad [f] = \begin{bmatrix} f_{11} & f_{12} & \cdots\cdots & f_{1n} \\ f_{21} & f_{22} & \cdots\cdots & f_{2n} \\ \cdots\cdots & \cdots\cdots & \cdots\cdots & \cdots\cdots \\ f_{n1} & f_{n2} & \cdots\cdots & f_{nn} \end{bmatrix} \tag{8.32a, b}$$

in which Δ_{ij} is the deflection at point i due to load vector j, and f_{ij} is the deflection at point i due to unit reaction force at j. It is possible also to include the effects due to elastic deformations of the columns and foundation settlements by modifying Eq. 8.2 to

$$[f] = \begin{bmatrix} (f_{11} + f'_{11}) & f_{12} & \cdots\cdots & f_{1n} \\ f_{21} & (f_{22} + f'_{22}) & \cdots\cdots & f_{2n} \\ \cdots\cdots & \cdots\cdots & \cdots\cdots & \cdots\cdots \\ f_{n1} & f_{n2} & \cdots\cdots & (f_{nn} + f'_{nn}) \end{bmatrix} \tag{8.33}$$

where f_{ii}' is the total deflection at the top of column i due to column shortening and foundation settlement. If the compatibility requirement of displacements at the column positions is now considered, it is obvious that we should have

$$[f]\,[R] + [\Delta] = [0] \tag{8.34}$$

from which the true column reactions [R] can be calculated.

The final displacements and internal forces can be evaluated by multiplying and combining with the results that were obtained in the first part of the analysis.

8.2.2.2 *Treatment of the stiffening girders*

A girder-slab system can, of course, be modelled by shell strips which include both the in-plane and out-of-plane stiffness. Since the formulation for such strips has been discussed in Section 8.1, it will not be repeated here. The only shortcoming of this approach is that a fairly large number of unknowns is required in the analysis. Another approach is to approximate a girder-slab system as an orthotropic slab with equivalent properties (Chu and

Krishnamoorthy).[84] It is simple but its application is limited by the fact that it is not easy, except for some special cases, to assign 'effective' stiffness parameters for the structural elements in the computation of equivalent orthotropic properties.

It is obvious that the compound strip method[26] (Section 8.1.9) can also be used to analyze such bridges. Each stiffener can be modelled by one or a number of one-dimensional members, of which stiffnesses are assumed to be concentrated along their centre lines. The top slab stiffness is formulated by the standard procedure. Following the principle outlined in Section 8.1.9, the stiffness of the stiffeners can be transformed in terms of the slab unknowns. The solution can then follow the usual procedures.

8.2.2.3 *Examples*

(i) A square constant thickness slab bridge[85,86] was analyzed for a central concentrated load, using 8LO2, 4HO2 strips, and 4x8 rectangular finite elements respectively for half the bridge. The distribution coefficients (Figure 8.28) computed for longitudinal and transverse moments are practically identical for the three cases.

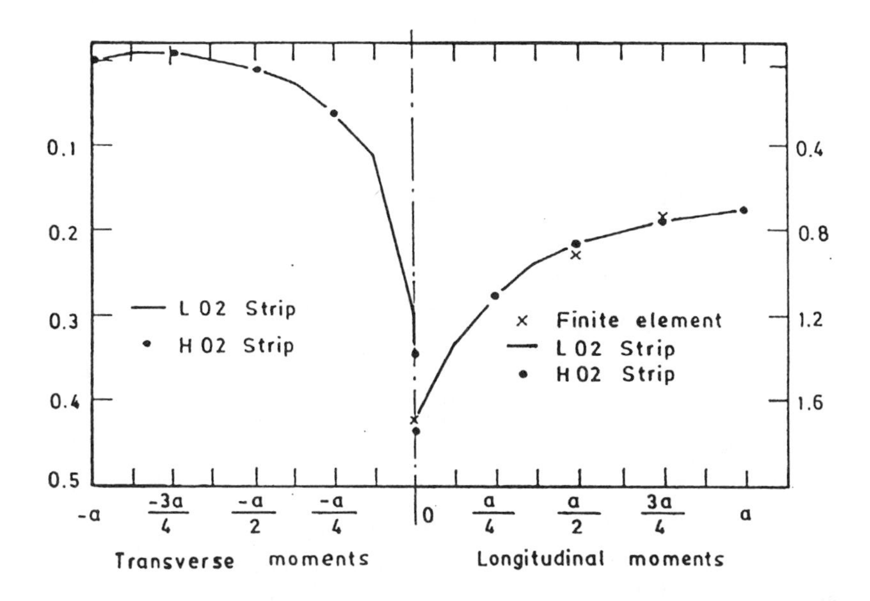

Figure 8.28 Distribution coefficients for transverse and longitudinal moments for a unit load at centre of a square bridge.[85,86]

(ii) A model bridge[65,85] with two beams and a variable thickness cross-section is shown in Figure 8.29. A central concentrated load of 556N was applied to the model and experimental results were compared with those of a LO2 finite strip analysis, in which half of the bridge was idealized by eight classical strips, and the flexural and torsional rigidities were assumed to be equal to that of a T-section with the flange equal to half the width of the bridge. Good agreement is observed for all comparison made in the longitudinal direction for plate moments as well as beam stresses (Figures 8.29b, 8.29c). Reasonable agreement is obtained for the curvature comparisons along the transverse centre line (Figure 8.29d), and indicates that the rotational stiffness of the beam might have been overestimated.

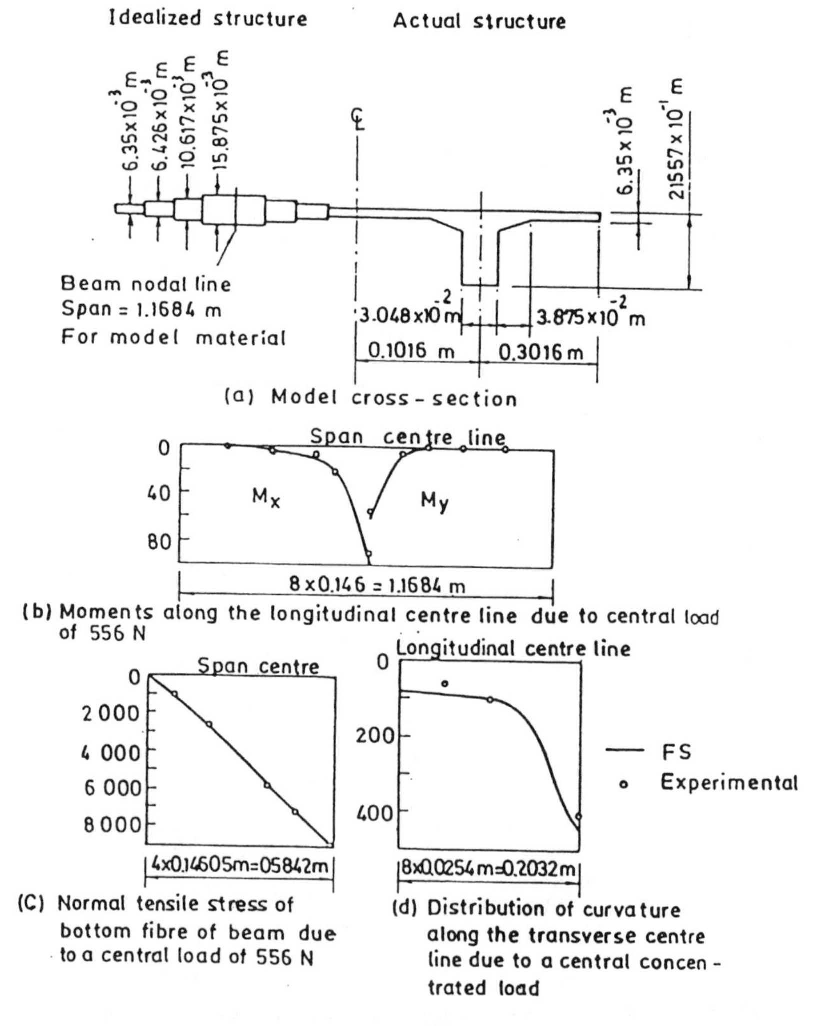

Figure 8.29 The model bridge with two beams.[65,85]

(iii) Another example[67,85] dealt with the analysis of uniform thickness isotropic curved bridges, and the theoretical and experimental results obtained by Coull and Das[87] were used as a check on the accuracy of the finite strip method. A model bridge slab made of asbestos-cement and with an included angle of 60° was loaded with central point loads at three different radial positions - the inner and outer edges, and at mid-radius respectively. A comparison of the bending moments along the central section is given in Figure 8.30 and, in general, there is good agreement between all sets of results. The finite strip method consistently gives higher moments under the load points when sufficient number of terms are used (eight terms in this case). However, if only three terms are used, the maximum values then are much closer to the theoretical results of Coull and Das,[87] who also took only three terms for their analysis, A point of interest is that the subsequent higher terms (e.g. after the third and fourth term) only affect results in the near vicinity of the point load.

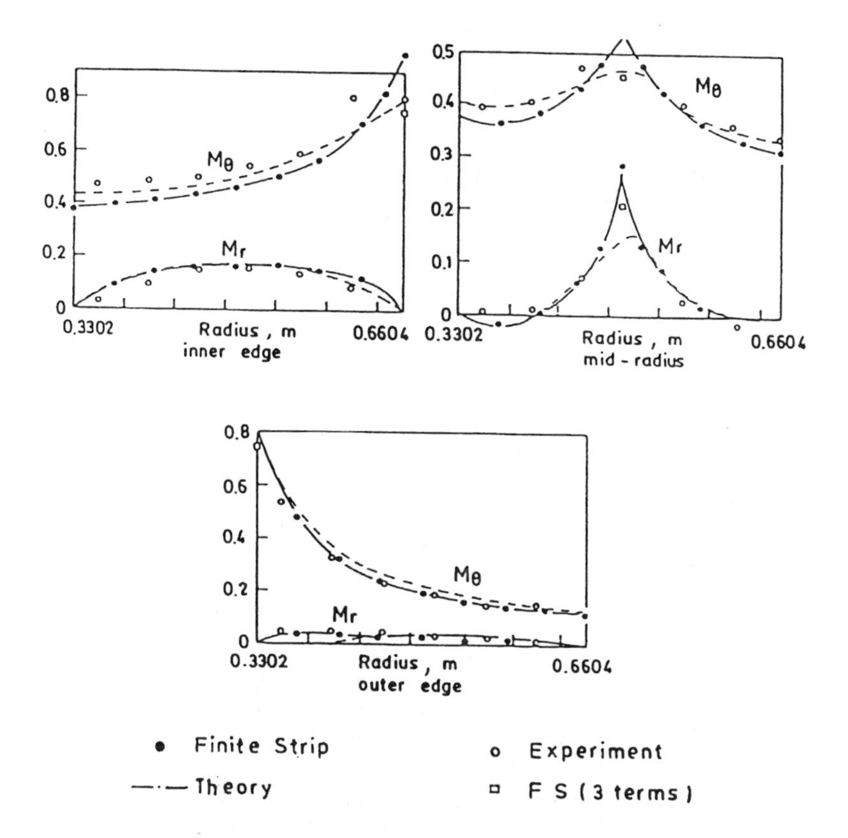

Figure 8.30 Distribution of tangential and radial bending moments at midspan due to unit load.[67,85]

It is also of considerable interest to note that the curved strip program can also be used to analyze right bridges by assigning a 'curved' bridge with a very small subtended angle and a very large radius. In a test example in which a square bridge of unit span and width was analyzed as a curved bridge with internal angle R_i = 199.5 , external radius R_e = 200.5, and subtended angle α = 0.005 radian, it was found that the maximum error in deflection is 0.7% while no difference can be detected for all comparisons made at mid-radius. No numerical instability has ever been experienced for other examples in which even higher radius values have been used.

(iv) Lobley Hill South Overbridge[88]

Lobley Hill South Overbridge (Figure 8.31) is a four-span continuous

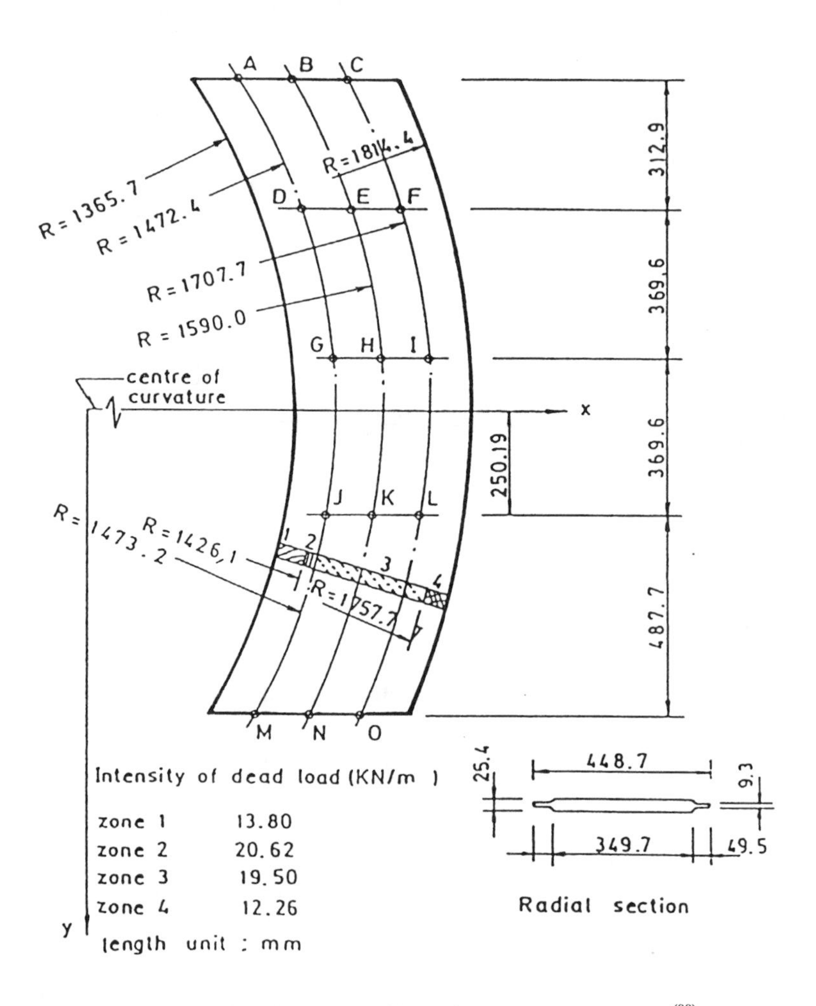

Figure 8.31 The detail of idealization of Lobley Hill South Overbridge model.[88]

bridge. The analysis[88] was applied only to its epoxy-sand model. The model bridge was 1:30 of full size. It had a width of 448.7mm and there were five rows of column support. The stiffness of the columns is given in Table 8.22.

Only the dead load case had been considered and the loading can be divided into four zones of different intensities (Figure 8.36) varying from 12.26kN/m^2 to 20.62kN/m^2 . The modulus of elasticity and Poisson's ratio were 1.62x10^7 kN/m^2 and 0.22 respectively.

The longitudinal bending moments obtained by the spline strips[23] are compared with those by Lim[88] as shown in Figure 8.32, and it can be seen that they are in good agreement. Note that Lim's computed results were based on a 9x26 mesh of elements with a total 702DOF while the spline finite strip mesh was 9x20 but with only 360DOF.

TABLE 8.22
Elastic stiffness of column supports of Lobley Hill South Overbridge (kN/mm).

Column	A	B	C	D	E
Elastic stiffness	33.80	33.62	33.45	33.62	33.27
Column	F	G	H	I	J
Elastic stiffness	32.75	32.33	31.87	31.52	31.00
Column	K	L	M	N	O
Elastic stiffness	30.65	30.47	33.97	33.80	33.80

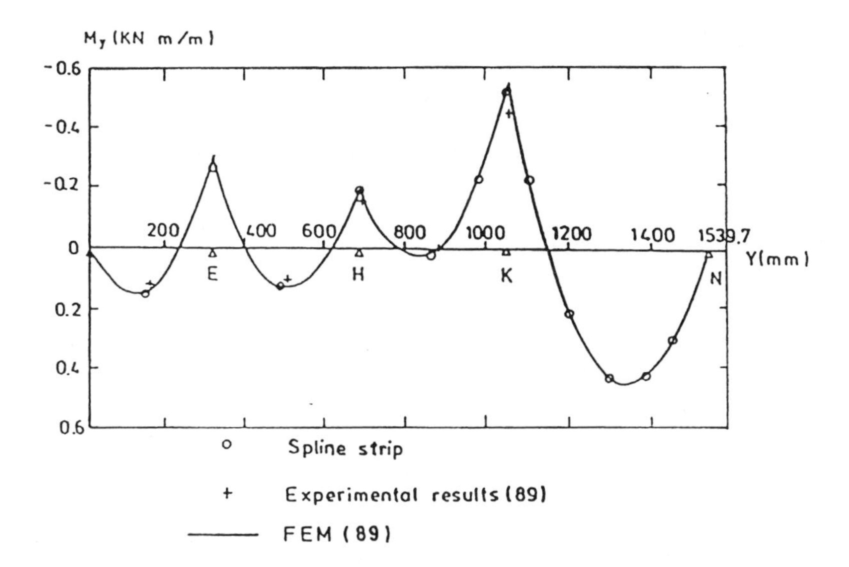

Figure 8.32 Moment of Lobley Hill South Overbridge.[23]

(v) 'S'-shaped continuous two-span bridge

A two-span 'S'-shaped continuous slab bridge as shown in Figure 8.33 was studied to demonstrate the versatility of the spline strip method.[71] Each curved side was described by using a cubic curve with four control points. It can be shown readily that the expressions of the two curves were defined by η $(-1 \leq \eta \leq 1)$ as

Curve 1:

$$x = 0.2250\eta^3 + 0.2250\eta^2 + 2.9750\eta + 2.9750$$
$$y = -0.6638\eta^3 - 0.5456\eta^2 + 0.0213\eta + 2.3281$$

Curve 2:

$$x = 3.0\eta + 3.0$$
$$y = 0.90\eta^3 - 0.90\eta^2 - 1.30\eta + 0.50$$

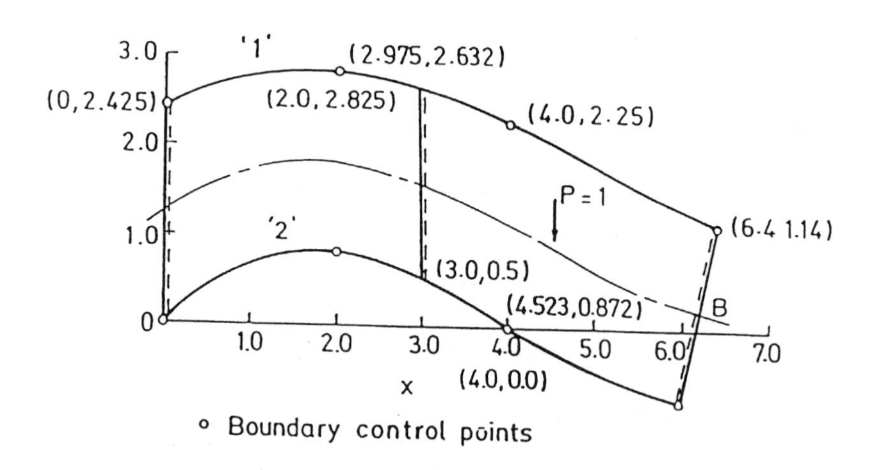

Figure 8.33 S-shaped slab bridge.[71]

The end and intermediate supports were assumed to be line supports. Analyses had been carried out for a concentrated load at point 'A' (Figure 8.33). The results are compared with those computed using the standard finite elements[89] as shown in Figure 8.34. It can be seen that the results of both the coarse mesh 8x4 (110 unknowns) and the fine mesh 24x6 (378 unknowns) compare favourably with the finite element ones which are based on 288 elements (525DOF).

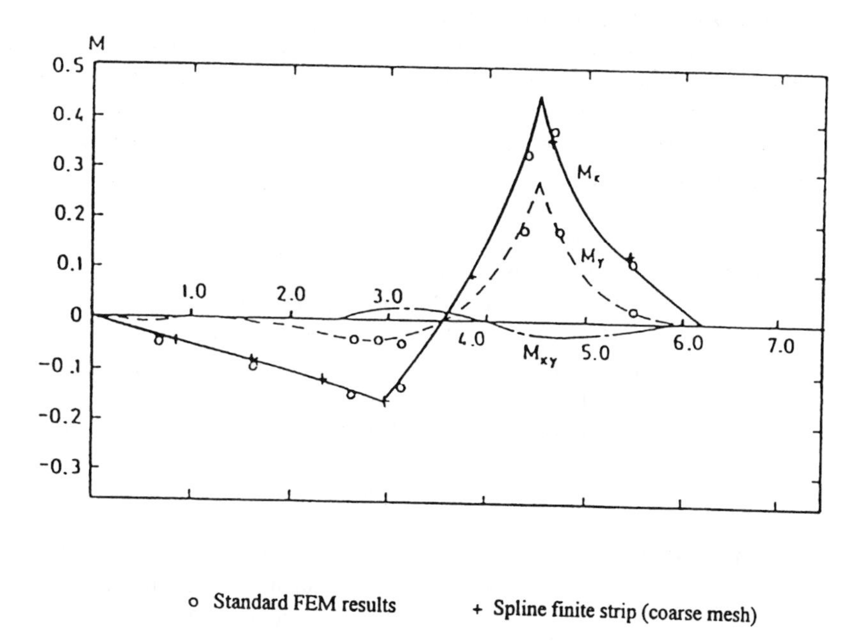

Figure 8.34 Bending moments of S-shaped continuous slab bridge.[71]

8.2.3 DEEP BOX GIRDERS WITH VERTICAL AND INCLINED WEBS

Box girder bridges can be treated as special forms of folded plate structures, but there are again certain special features commonly associated with these structures such that special attention and treatments are necessary in the analysis. These features will be discussed in conjunction with finite strip analysis.

(i) Due to the fact that the primary function of a box girder bridge is to carry traffic, the top flanges of a box girder bridge are acted upon by groups of point loads or patch loads representing the wheel loadings of trucks and other motor vehicles. Since the spline strip is based on the solution of a point load on a beam, it has a definite advantage over the classical strip in simulating such point loads. It is worthwhile here to say a few words to address the problems of slow convergence in the classical strip.

At the immediate vicinity of a point load, the stresses and moments obtained by the summation of a Fourier series tend to converge very slowly, and sometimes as many as fifty non-zero harmonics have to be taken before a 'converged' solution is arrived at. On the other hand, for stresses and moments at points a small distance away from the point load, hardly any changes will be observed after the first fifteen or so non-zero harmonics. Since the stresses and moments under a point load is mathematically infinite, it is probably not worth while to spend any significant amount of computer time to try to achieve 'convergence' at such a location.

In any case, the reinforcements in the bridge are designed to resist the total integrated force over some part of the cross-section, and not according to the exact stress distribution pattern.

As mentioned above, many terms of the series are required for bridge analysis due to the presence of concentrated loads; hence, boundary conditions other than simple supports at the two ends cannot be included because the terms of the series will then be coupled. Such coupling tends to create a stiffness matrix of very large bandwidth for concentrated load cases, and is therefore uneconomical.

(ii) Box girder bridges of longer spans very often are supported by intermediate discrete columns. Such columns are sometimes located at odd positions which are determined by site conditions or by the roadways passing underneath.

The methods of solution for column-supported slab bridges discussed in Section 8.2.2.1 can be applied directly to the present situation.

(iii) Internal diaphragms very often are present, especially for bridges of steel construction, in which stability of the various plate elements has emerged to be one of the critical factors in design. For concrete bridges, however, research work[(32,89)] has demonstrated that the boxes themselves possess sufficient torsional rigidity for adequate transverse load distribution, and therefore only end diaphragms or diaphragms over intermediate supports are regarded as absolutely essential. This is of some importance because the construction of concrete diaphragms can be a costly and time-consuming process, and their number should be limited to a minimum.

The analysis of box girder bridges with intermediate rigid diaphragms was carried out by Scordelis et al.,[(91)] using the elasticity method in conjunction with a flexibility approach. This procedure can be easily adapted for the finite strip method.

If the diaphragms are not perfectly rigid, a much more elaborate method by combining the finite element and finite strip may be used for the analyses.[(70)]

Diaphragms are treated as secondary structures and can be discretised into a number of finite elements, whereas the primary structure (the main structure) is modelled by finite strips (Figure 8.35). The stiffness matrices of the diaphragms are first to be formulated in terms of the element displacement parameters (Figure 8.35), and they can then be condensed in such a way that only terms related to nodes that are connected directly to the primary structure are retained. To permit the formulation of the stiffness matrix of the whole system, it will be more convenient to express the reduced stiffness matrices of the diaphragms in terms of the finite strip unknowns so that the two stiffness matrices of the primary and secondary structures can be added together. This can be achieved by carrying out appropriate transformation as in the standard approach. Note that the terms will couple up even for a single span simply-supported bridge.

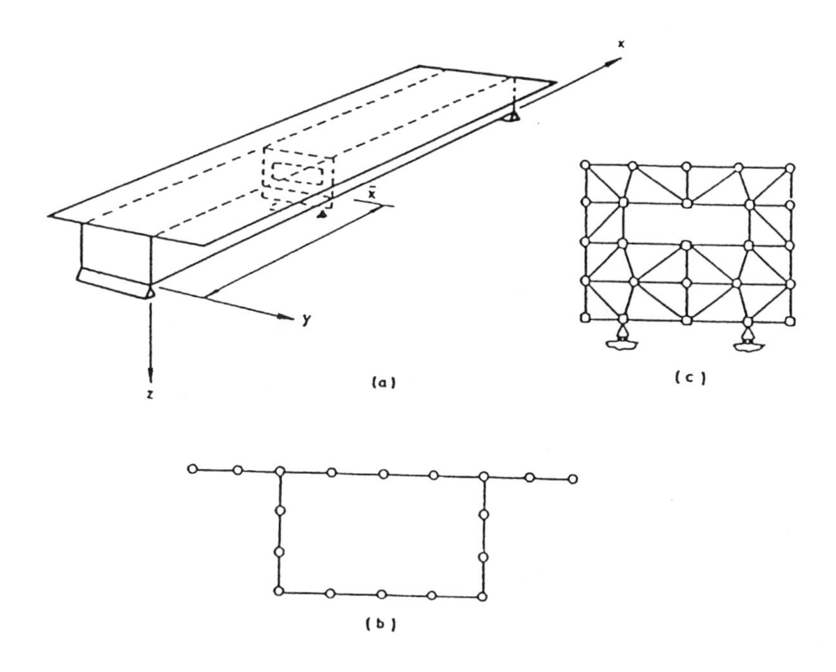

Figure 8.35 (a) Box girder with intermediate diaphragm. (b) Finite strip discretisation. (c) Finite element discretisation.[70]

8.2.3.1 *Examples*

(i) Right box girder bridge by higher order classical strip[13,85] (HO3)

The three-cell spine box girder under sixteen wheel loads, shown in Figure 8.36a, is used as an example to demonstrate the accuracy of the higher order strip HO3. The results are compared with those obtained from a lower order strip LO2 analysis. The strip simulation for the two analyses is given in Figure 8.36b. Twenty-five harmonics were used in both analyses. The transverse in-plane stress profiles at the section under the second axle are plotted in Figures 8.37. It can be seen that good agreement between the two cases is obtained. Since the transverse strain across the width of a lower order strip is constant, a stepped representation would result if the transverse in-plane stresses are plotted at the nodal lines.

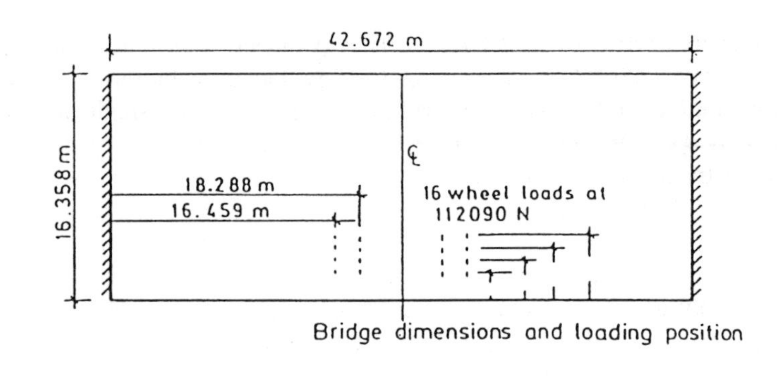

Figure 8.36 Simply supported three-cell spine box bridge: (a) Plan. (b) The finite strip mesh. (b) Loading and dimensions.[13,85]

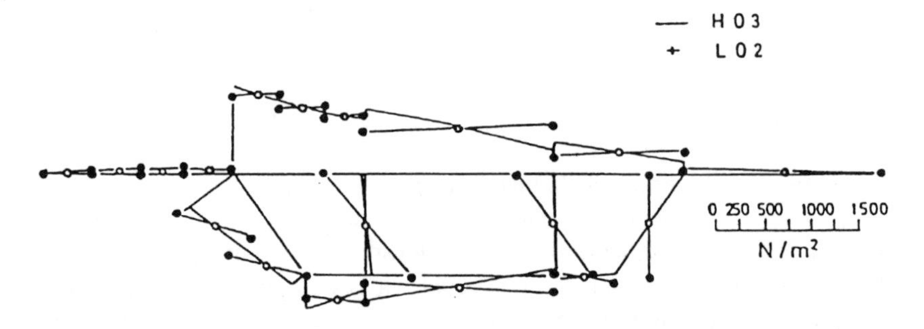

Figure 8.37 In-plane stress profiles for the three-cell spine box bridge under sixteen wheel loads (at section under the second axle).[13,85]

(ii) Non-circular curved bridge by spline strip[71]

This example demonstrates that the spline strip can be easily extended to the analysis of other forms of non-circularly curved bridges. The horizontal alignment of the concrete bridge shown in Figure 8.38 was defined by a cubic spline transition curve of length 10.3m (from point A to B) and a circular curve of length 27.9m (from B to C). The radius of the circular curve was 80m. Full details for the setting out of the curve are given in Figure 8.38. Four equally spaced points (A, D, E and C) were used as the control points for the geometry mapping. All structural members of the bridges were of thickness 300mm. The properties of the material were: Young's modulus = 0.26×10^8 kN/m^2 , Poisson's ratio = 0.2. The bridge was divided into 16 strips each consisting of 6 sections in the analysis. Two loading cases were studied: the uniformly distributed load and point loads at point F and G. The deflection obtained by the spline strip is in excellent agreement with the finite element one[91] (Figure 8.39).

AB : $y+2.667=(x+40)^3/24000$
BC : $(x-33.862)^2+(y+73.81)^2=6400$

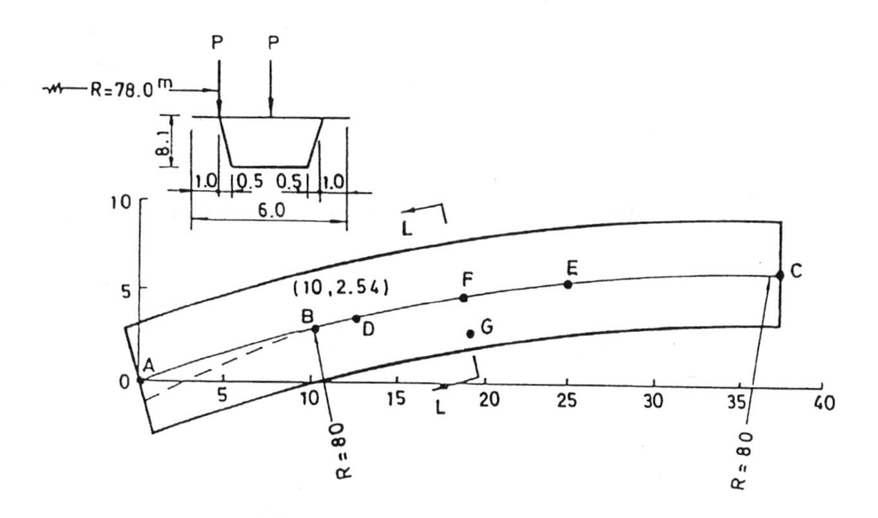

Figure 8.38 Compound curved bridge.[71]

(iii) Bridges with diaphragms[70]

The finite element-strip model for analysis of bridges with diaphragms was tested by an example of a concrete box girder bridge which was simply-supported at both ends. The bridge was 92m in length with a transverse diaphragm at 40m from one end. The diaphragm was 300mm in thickness and it was supported at the mid-point of its base (Figure 8.40). Point loads, each of 6000kN, were applied at 12, 28, 52, 72 and 80m. Other pertinent dimensions of the bridge are given in Figure 8.40. The longitudinal stresses (12 harmonic terms) are depicted in Figure 8.41. They are in good agreement with finite element results[92] which have been achieved by using 280 elements.

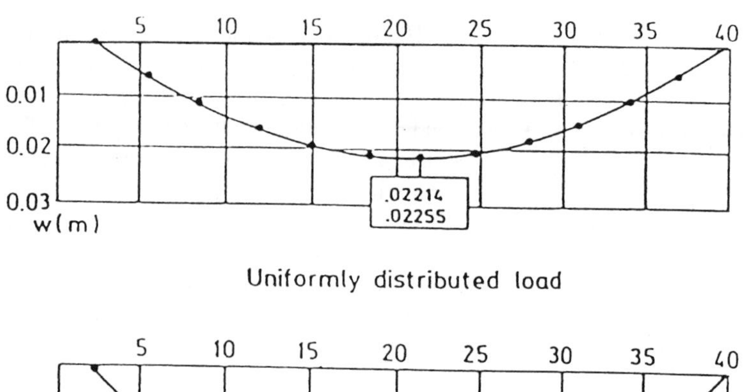

Figure 8.39 Top deck deflection along central line.[71]

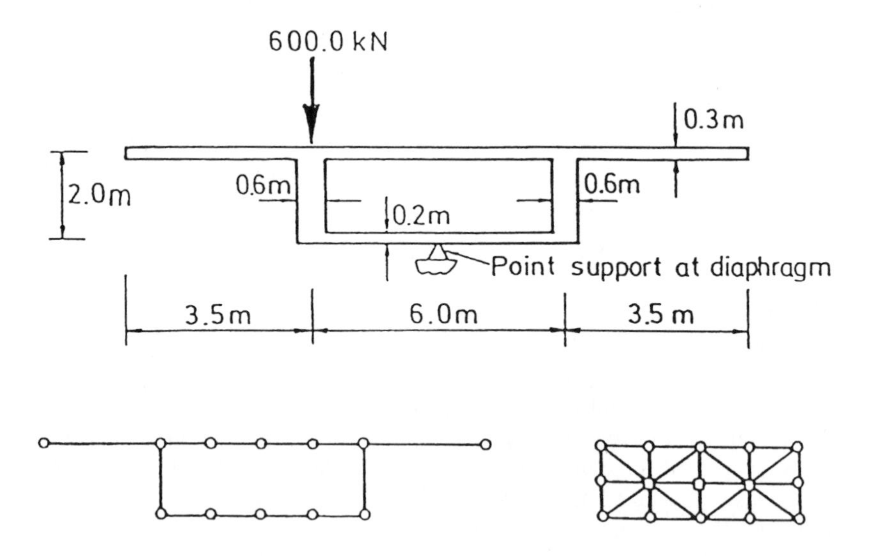

Figure 8.40 Cross-section and discretisation.[70]

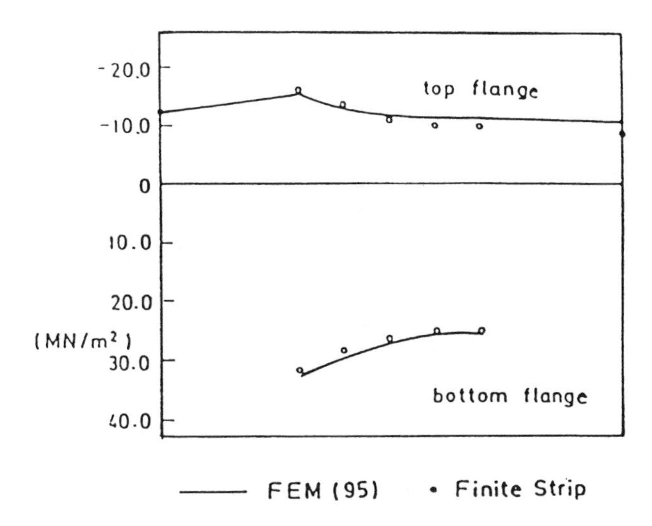

Figure 8.41 Longitudinal stress at 72m.[70]

(iv) Analysis of prestressed bridges[68,85] (LO2)

Many concrete bridges are prestressed, and the analysis of prestressed forces in bridge structure is very important. By following the treatment of curved cables proposed by Lin,[94] in which the prestressed force system is resolved into concentrated end loads and distributed upward pressure along the span, it is possible to compute the effects of prestressing with the standard finite strip procedure.

This simply supported deep beam with a curved frictionless post-tensioned cable is shown in Figure 8.42, and four equal strips were used to simulate the beam. The distribution of mid-span longitudinal stress is plotted in Figure 8.42. The agreement between the results of the finite strip analysis and those due to simple beam (prestressed) theory is excellent.

Another example of the analysis of a prestressed concrete bridge was presented by Abdullah and Abdul-Razzak.[69] The bridge had a four-cell cross-section with vertical exterior webs and short cantilever slabs (Figure 8.43). The span lengths were 150ft and 160ft. Diaphragms were located at each of the three supports. The properties of the materials assumed in the analysis were: an elastic modulus of 3600ksi and a Poisson's ratio of 0.17. Results of longitudinal stress are shown in Figure 8.44. Good agreement is found for the comparison of the finite strip results with those by MUPD14 (Scordelis et al).[95]

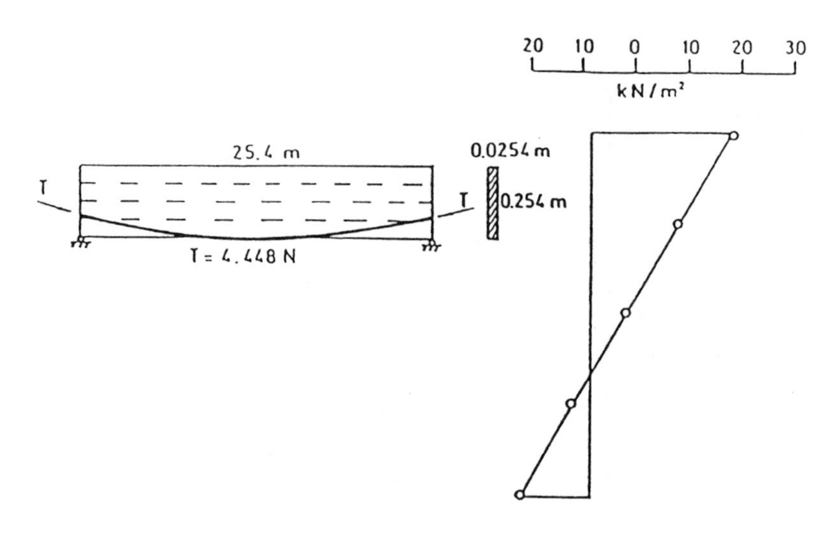

Figure 8.42 Midspan longitudinal stress profiles for beam with a curved post-tensioned cable.(68,85)

—— Simple beam theory; o o LO2 strips (4 strips and 19 harmonics)

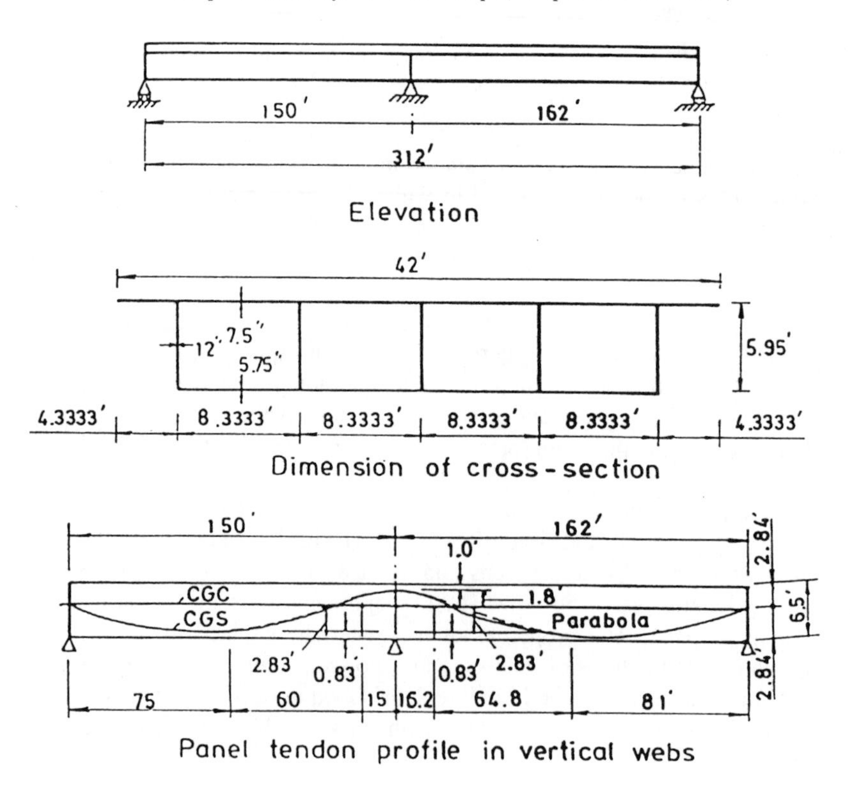

Figure 8.43 Four-celled prestressed bridge.(69)

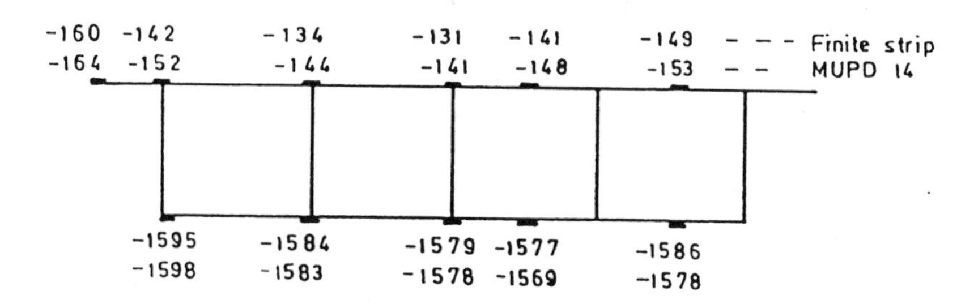

Figure 8.44 Longitudinal stress for the four-celled bridge.[71]

(v) Continuous haunched bridges

A three-span variable-depth single-cell box girder bridge shown in Figure 8.45 was analyzed by Au[96] using the isoparametric spline finite strip. The geometrical details of the bridge are given in the same figure as well. The horizontal alignment consists of a circular curve in the central span and two straight tangents at both end spans (Figure 8.45a). Two bearings were provided at each support. One of the bearings at Support 'B' was of fixed type while the others were of sliding types (Figure 8.45c). A total load of 1800kN equally distributed among the 16 wheels on four axles was applied, and the locations of the axles are depicted in Figure 8.45e.

A total of six spline strips (three for top flange, one for each web and one for bottom flange) were used. In the longitudinal direction, the strips were divided into eighteen sections, and, therefore, a total of 1800 degrees of freedom was employed in the analysis.

The distribution of the axial stress in the middle of the central span was compared to those obtained by finite elements using COSMOS,[97] and close agreement is observed (Figure 8.46).

(vi) Composite girder bridge with incomplete interaction

Arizum et al.[98] represented the shear connectors between the concrete decks and the steel girders of a composite curved bridge by springs. The longitudinal and transverse stiffnesses of the springs were determined from a push out test, and they were denoted by k_r and k_θ respectively. A curved

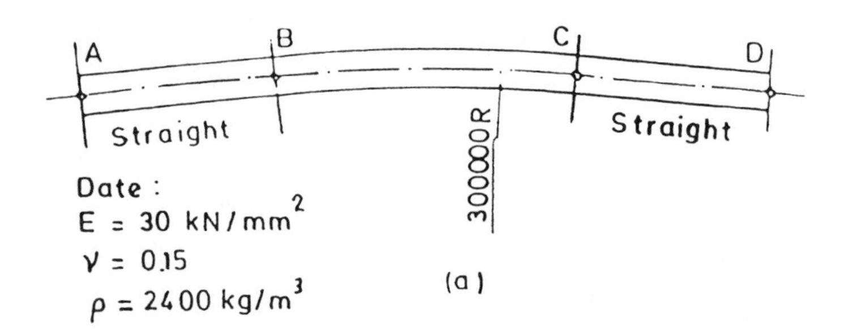

(a)

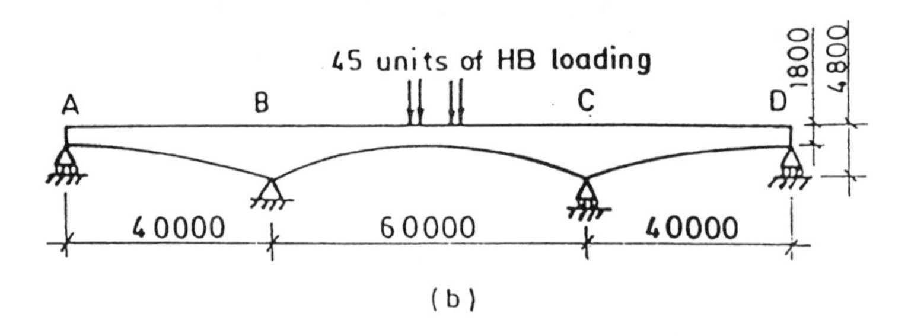

(b)

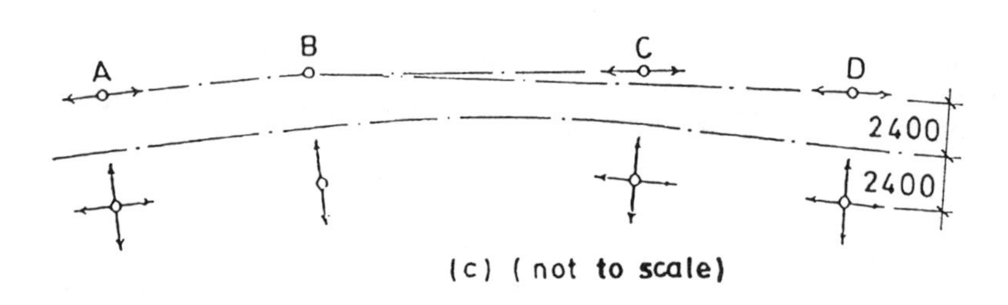

(c) (not to scale)

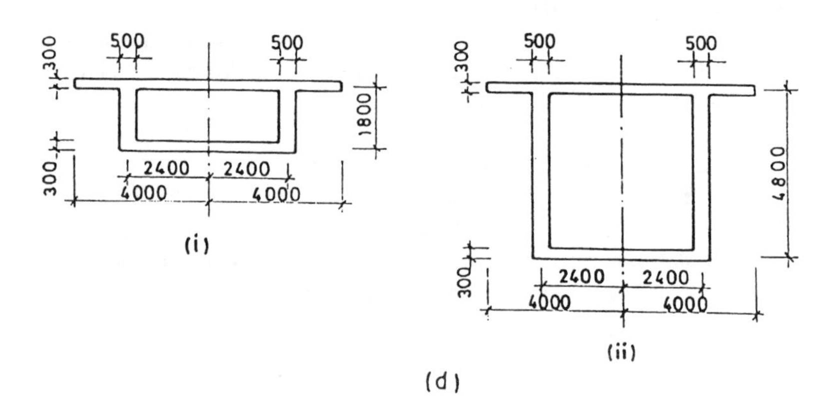

(d)

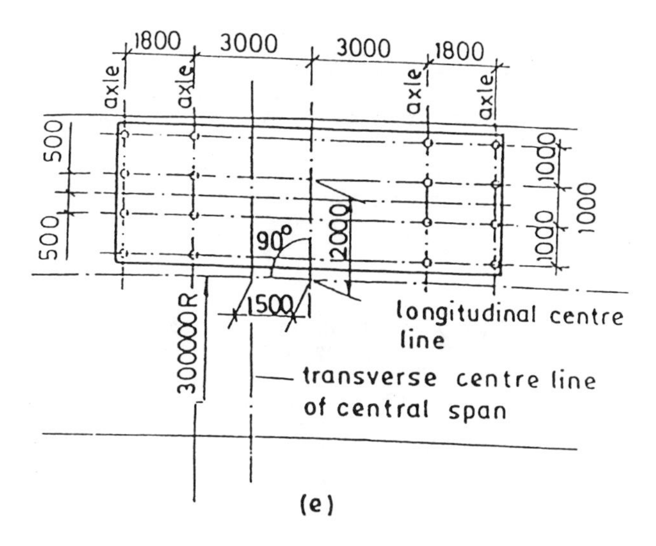

Figure 8.45 Three-span variable-depth box girder bridge: (a) Plan. (b) Developed sectional elevation along centre line. (c) Bearing layout. (d) Cross-sections - (i) cross-section at middle of central span and end supports. (ii) cross-section at interior supports. (e) Loading arrangement.(96)

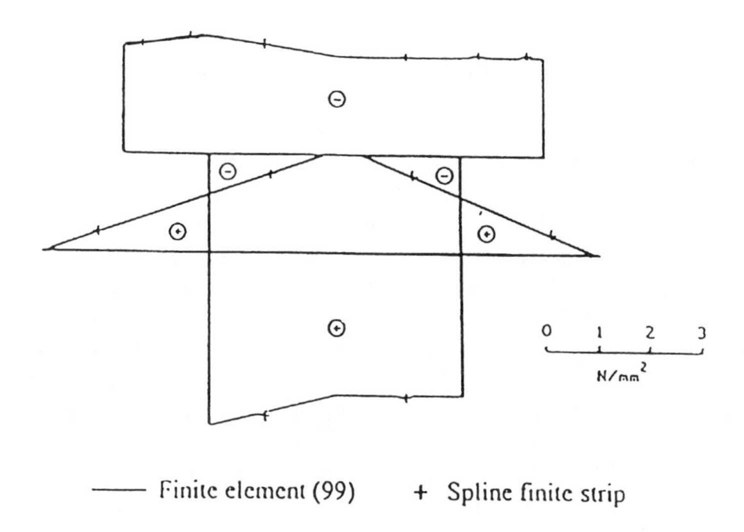

Figure 8.46 Three-span variable-depth box girder bridge: axial stress in middle of central span.(96)

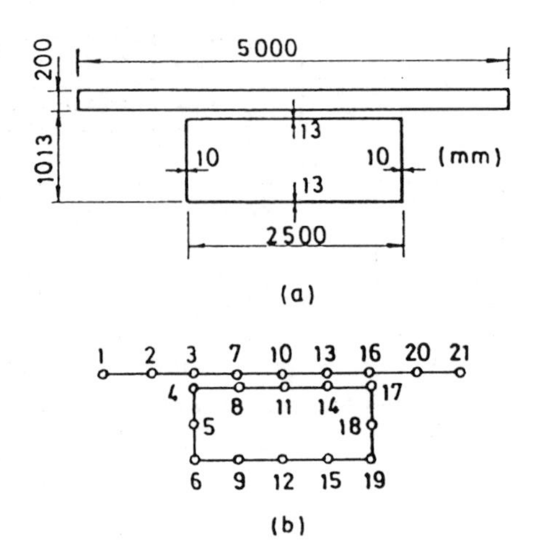

Figure 8.47 (a) Curved box girder bridge. (b) Numbering of nodes.[98]

composite girder with shear connectors arranged along the top flange of the box steel girder (Figure 8.47) was analyzed. The stiffnesses of the shear connectors were 30000kg/cm/cm and 20000kg/cm/cm respectively (Figure 8.48). The variations of the longitudinal and lateral slips are plotted in Figure 8.48.

8.2.4 DYNAMIC ANALYSIS: FREE VIBRATION AND EFFECTS DUE TO MOVING MASS

Following the same principle outlined in the previous section, one can form the mass matrices for various types of strips and the free vibration analysis can be carried out in the usual manner (Chapter 5). However, the dynamic behaviour of bridges under the action of moving vehicles is much more complicated. The displacements and stresses induced by a vehicle travelling across a bridge depend not only on the weight and position of the vehicle but also upon its velocity, mass distribution, stiffness, damping characteristics and surface of the bridge. Various schemes were developed for its solution. The classical finite strip solution offers good alternatives for dynamic analysis because of the small matrix used, since only a few terms of the series are required to produce accurately the few lower frequencies and the corresponding mode shapes. In particular, the matrix for a single span simply-supported bridge would be of a very small size due to the uncoupling between the terms of the series. On the other hand, the spline strip is more versatile in dealing with arbitrarily shaped bridges.

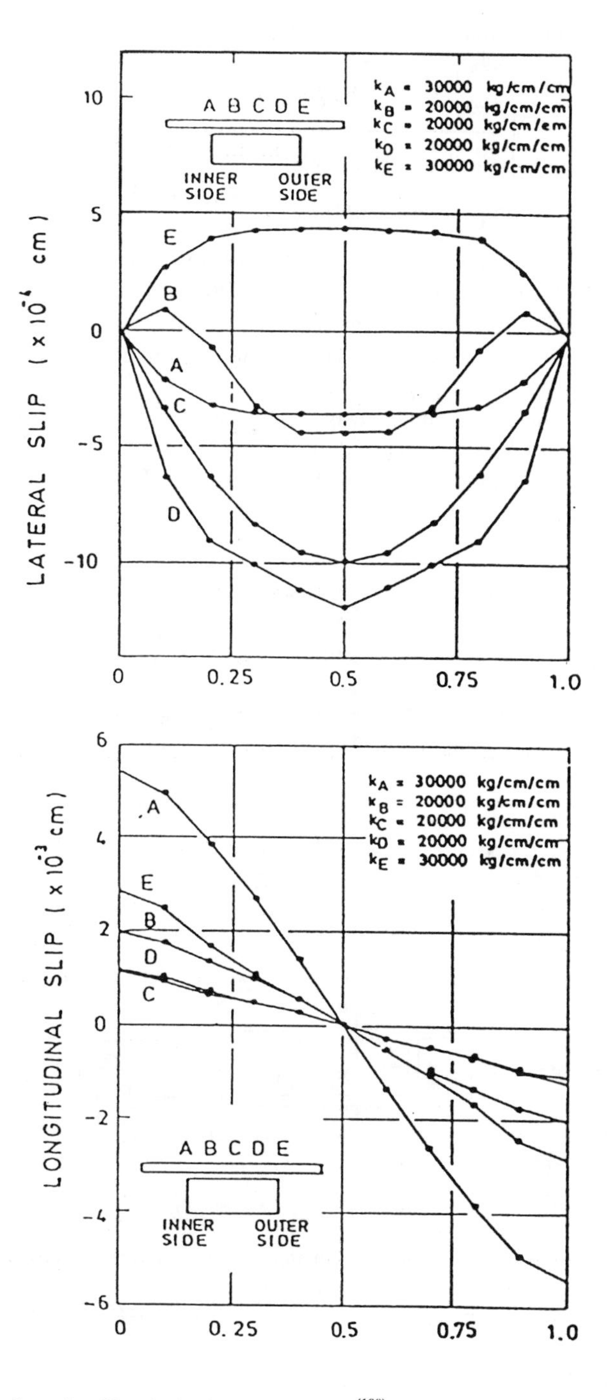

Figure 8.48 Lateral and longitudinal slip distribution.[100]

The equation of motion is described by:

$$[M]\left\{\frac{\partial^2\delta}{\partial\tau^2}\right\}+[\tilde{D}]\left\{\frac{\partial\delta}{\partial\tau}\right\}+[K]\{\delta\}=\{F(\tau)\} \tag{8.35}$$

$[\tilde{D}]$ is the damping matrix and is defined as:

$$[\tilde{D}]=\int [N]^T[\mu][N]\,dxdy \tag{8.36}$$

where [N] and [μ] are the shape function and viscosity matrices.

As the vehicles are moving across the bridge, the force vector will change with time. Having known the locations at any instant τ and tyre loads of the vehicles, one can obtain the force vector by using equations given in Chapter 3.

Eq. 8.35 can be uncoupled using modal analysis procedures and fairly accurate solutions can be obtained by taking only a couple of the lowest modes. The equations can be solved using the Runge-Kutta numerical integration procedure.

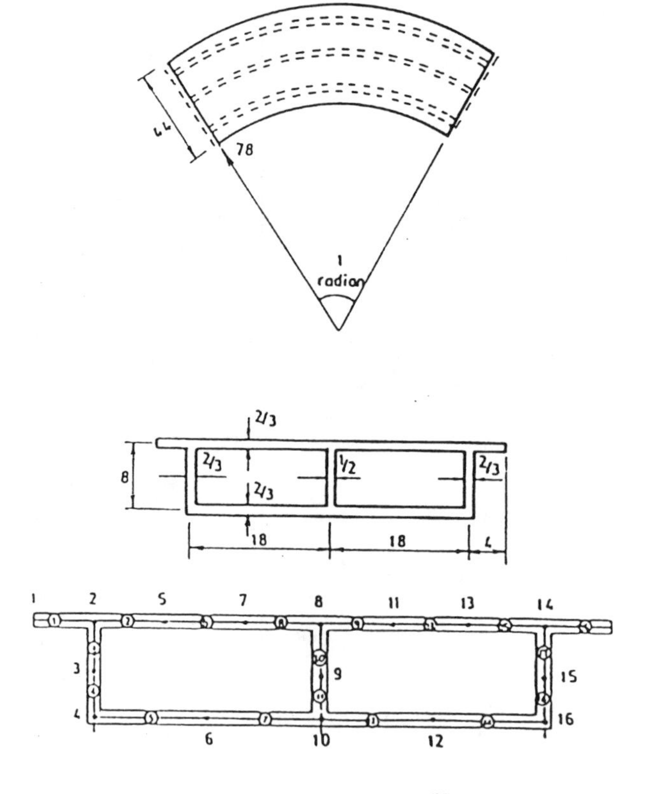

Figure 8.49 Curved box girder bridge (E = 1, υ = 0.16, ρ = 1).[85]

Vibration analysis[96] was also carried out for the continuous haunched bridge (Figure 8.45) employing the same mesh. The lowest four frequencies are tabulated in Table 8.23. The results are again in good agreement with those obtained by COSMOS[97] using 5160 degrees of freedom.

TABLE 8.23
Natural frequencies for the haunched box girder bridge.[96]

Mode Number	Natural frequency (Hz)	
	Spline strip	Finite element
1	2.235	2.221
2	4.032	3.988
3	4.902	4.817
4	5.562	5.382

The free vibration of a twin-cell curved box girder bridge (Figure 8.49) was studied. Solutions were obtained by both classical strips[78,85] and spline strips[99]. It is worth mentioning that the mid-point nodal lines of the three webs (Figure 8.49) can all be removed without prejudicing the accuracy of the solution, and that the use of only fifty-six equations per harmonic would be quite adequate for solving such a complicated structure.

Several frequencies and their corresponding mode shapes are shown in Figure 8.50.

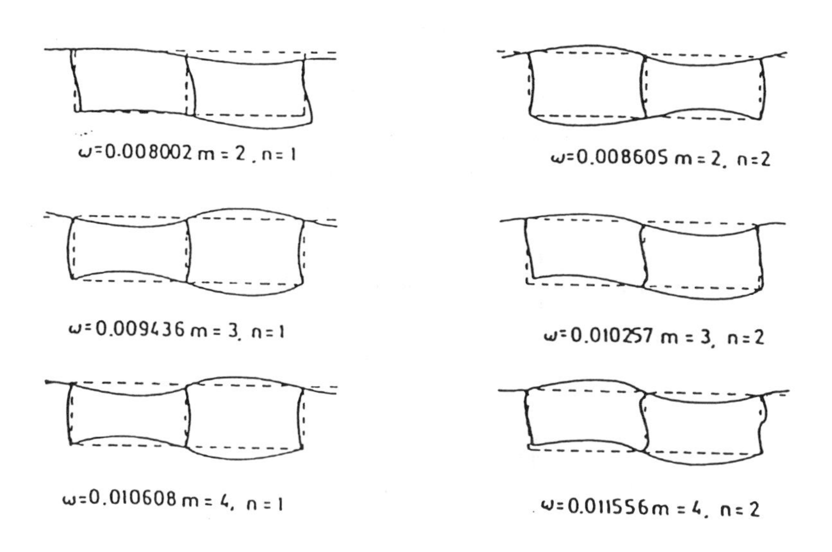

Figure 8.50 Modal shapes of a curved box girder bridge.[100]

Hutton and Cheung[81] adopted a more advanced model (Figure 8.51) in the analysis by representing the vehicle as a rigid body supported at two points by a suspension idealization that accounts for the effect of tyre stiffness and the friction of real suspension systems.

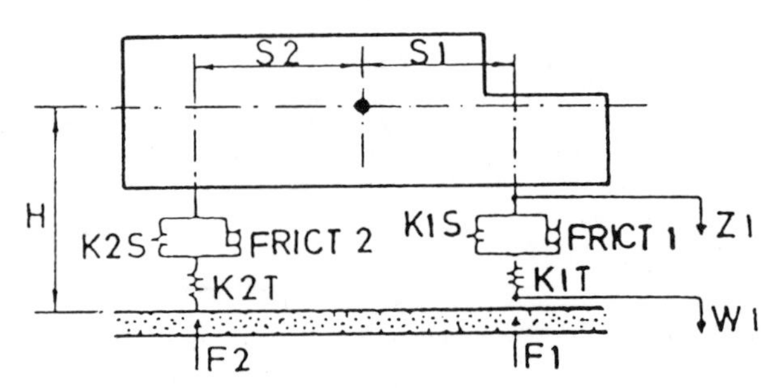

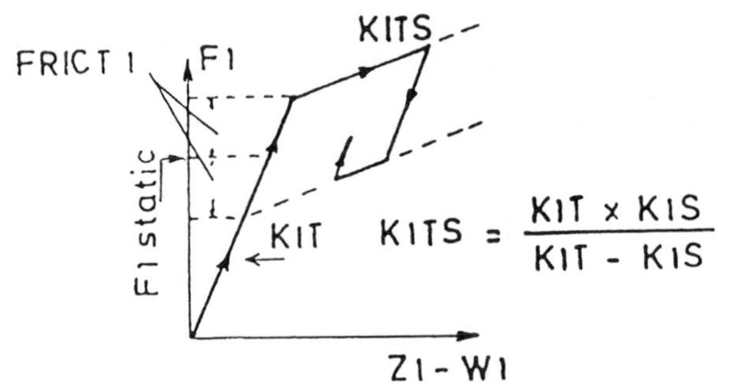

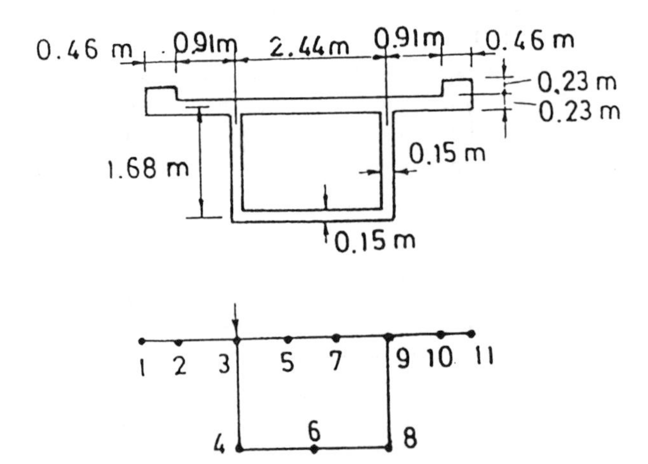

Figure 8.51 Vehicle characteristics and details of box girder bridge.[81]

The vehicle data (Figure 8.51) are as follows :

Total Mass = 27,250 kg
Pitching inertia = 52,170 kg-m^2
S1= S2 = 20m
H = 1829 m.

The paramters for the suspensions (Figure 8.51) are :

K1T = K2T = 668,720 - 22, 170,000 N/m
K1S = K2S = 3,939,000 N/m
FRICT 1= FRICT 2 = 20,028N

A box girder bridge with dimensions as given in Figure 8.51 was studied. The first six natural frequencies of the deck were 3.21, 11.53, 13.00, 18.73, 21.68 and 23.32Hz. The damping for each mode was constant and equivalent to 0.1% of the critical damping in the first mode. The Young's modulus, Poisson's ratio and density were 20.69GPa, 0.2 and 2570kg/m^3. The vehicle was assumed to be moving along nodal line '3' and the response was recorded at the mid-span of nodal line '1'.

The variation of the impact factor with the wavelength of the surface profile is shown in Figure 8.52. In the figure, ϕ_v is the ratio of the natural frequency of the vehicle on its tyres to the lowest natural frequency of the bridge deck.

8.2.5 LINEAR STABILITY

The linear stability analysis of bridge decks can be conveniently carried out by the finite strip method and the local buckling analysis of a stiffened bridge deck has been presented by Delcourt.[(100)]

However, it is also known that the collapse load of a box girder bridge can be smaller than the critical load obtained by a linear stability analysis, and this has been demonstrated by a number of incidents such as the collapse of the bridge on the Danube in Vienna in 1969. When the accident happened, according to Professor Salther's calculations[(101)] the panel was subjected to a total stress of

$$\sigma_{acc} = 219.7\text{MPa}$$

Thus it failed at a load level less than the critical load predicted by the finite strip method[(43)]

$$\sigma_{cri} = 251\text{MPa}$$

It appears that nonlinear analysis is necessary to provide a better prediction. The results due to a nonlinear finite strip analysis are discussed in Reference 100 and Chapter 11.

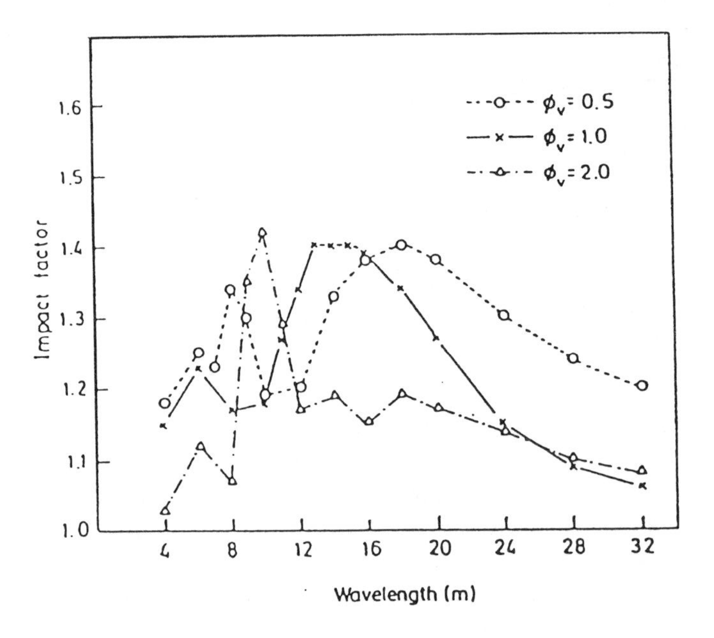

Figure 8.52 Variation of impact factor with surface profile.[81]

REFERENCES

1. Report of the Task Committee on Folded Plate Construction, Phase I report on folded plate construction, *J Struct Eng, ASCE*, 89, 365-405, 1963.
2. Goldberg, J. E. and Leve, H. L., Theory of prismatic folded plate structures, *IABSE Publications*, Zurich, Switzerland, 17, 59-86, 1957.
3. De Fries-Skene A. and Scordelis, A. C., Direct stiffness solution for folded plates, *J Struct Eng, ASCE*, 90, 15-47, 1964.
4. Scordelis, A. C., *Analysis of simply supported box girder bridges*, Structural Engineering and Structural Mechanics Report No SESM 66-17, Univ of California, Berkeley, 1966.
5. Chu, K. H. and Dudnik, E., Concrete box girder bridges analysed as folded plates, Concrete Bridge Design, *ACI Publications SP-23*, 221-246, 1969.
6. Goldberg, J. E., Castillo, M. and Keplar, R. K., An exact theory of horizontal curved girder structures, *Proc IASS Symposium on Folded Plates and Prismatic Structures*, Vienna, 1970.

7. Chu, K. H. and Pinjarker, S. G., Analysis of horizontally curved box girder bridges, *J Struct Eng, ASCE*, 97(10), 2481-2501, 1971.
8. Rehn, M. and Sved, G., Stiffness matrices of sector shaped thin shell elements, *Bulletin Int Association for Shell Structures*, No 45, 3-22, 1971.
9. Cheung, Y. K., Folded plate structures by finite strip method, *J Struct Eng, ASCE*, 95(12), 2963-2979, 1969.
10. Cheung, Y. K., Analysis of box girder bridges by finite strip method, Concrete Bridge Design, *ACI Publications SP-26*, 357-378, 1971.
11. Cheung, Y. K., Analysis of orthotropic prismatic folded plates, *Proc IASS Symposium of Folded Plates and Prismatic Structures, Vienna*, 1970.
12. William, K. J. and Scordelis, A. C., Analysis of eccentrically stiffened folded plates, *Proc IASS Symposium on Folded Plates and Prismatic Structures, Vienna*, 1970.
13. Loo, Y. C. and Cusens, A. R., Developments of the finite strip method in the analysis of cellular bridge decks, *Developments in Bridge Design and Construction* (ed. Rockey et al.), Crosby, Lockwood, 1971.
14. Meyer, C. and Scordelis, A. C., Analysis of curved folded plate structures, *J Struct Eng, ASCE*, 97(10), 2459-2480, 1971.
15. Cheung, M. S. and Cheung, Y. K., Analysis of curved box girder bridges by finite strip method, *Publications Int Association for Bridges and Structure Engineering*, Vol 31-I, 1-20, 1971.
16. Delcourt, C. and Cheung, Y. K., Finite strip analysis of continuous folded plates, *IABSE, Periodica*, 1-16, 1978.
17. Fan, S. C., Cheung, Y. K. and Wu, C. Q., Spline finite strip in structural analysis, *Proc of the Int Conf on Finite Element Methods*, Shanghai, 704-709, 1982.
18. Fan, S. C., *Spline finite strip in structural analysis*, PhD thesis, Department of Civil and Structural Engineering, University of Hong Kong, 1982.
19. Fan, S. C. and Cheung, Y. K., Analysis of shallow shells by spline finite strip method, *Engineering Structures*, 5(4), 255-263, 1983.
20. Lau, S. C. W., *Distortional buckling of thin-walled columns*, PhD thesis, School of Civil and Mining Engineering, University of Sydney, 1988.
21. Cheung, Y. K. and Zhu D. S., Large deflection analysis of arbitrary shaped thin plates, *Comp and Struct*, 26(5), 811-814, 1987.
22. Kwon, Y. B. and Hancock, G. J., Nonlinear elastic spline finite strip analysis for thin-walled sections, *Thin-walled Struct*, 12, 295-319, 1991.
23. Li, W. Y., *Spline finite strip analysis of arbitrarily shaped plates and shells*, PhD thesis, Department of Civil and Structural Engineering, University of Hong Kong, 1988.
24. Tham, L. G., Application of spline finite strip method in the analysis of space structure, *Thin-walled Struct*, 10(3), 235-246, 1990.

25. Kong, J. and Cheung, Y. K., Application of spline finite strip to the analysis of shear-deformable plates, *Comp and Struct*, 46(6), 985-988, 1993.
26. Puckett, J. A. and Gutkowski, R. M., Compound strip method for analysis of plate systems, *J of Struct Eng, ASCE*, 112, 121-138, 1986.
27. Maleki, S., Compound strip method for box girders and folded plates, *Comp and Struct*, 40(3), 527-538, 1991.
28. Chen, C. J., Gutkowski, R. M. and Puckett, J. A., Spline compound strip analysis of folded plate structures with intermediate supports, *Comp and Struct*, 39(3/4), 369-379, 1991.
29. Donnell, L. H., A discussion of thin shell theory, *Proc 5th Int Congress App Mech*, 1938.
30. Reissner, E., A new derivation of the equations of the deformation of elastic shells, *American J Math*, 63(1), 177-184, 1941.
31. Vlazov, V., General theory of shells and its application in engineering, *NASA TT F-99*, 1964.
32. Chapman, J. C., Dowling, P. J., Lim, P. T. K. and Billington, R. L., The structural behaviour of steel and concrete bridges, *Struct Eng*, 49(3), 111-120, 1971.
33. Megard, G., Planar and curved shell elements, *Finite element methods in stress analysis* (ed. I. Holand and K. Bell), Tapir-Trykk (Norway), 287-318, 1969.
34. Scordelis, A. C. and Lo, K. S., Computer analysis of cylindrical shells, *J of ACI*, 61(5), 539-561, 1964.
35. Lo, K. S. and Scordelis, A. C., Discussion of 'Analysis of continuous folded plate surface', *J of Struct Eng, ASCE*, 92, 281-286, 1969.
36. Beaufait, F. W. and Grey, G. A., Experimental analysis of continuous folded plates, *J of Struct Eng, ASCE*, 92(1), 11-19, 1966.
37. Beaufait, F. W., Analysis of continuous folded plate structures, *J of Struct Eng, ASCE*, 91(6), 117-140, 1965.
38. Kong, J., *Analysis of plate-type structures by finite strip, finite prism and finite layer method*, PhD thesis, Department of Civil and Structural Engineering, University of Hong Kong, 1994.
39. Wittrick, W. H. and Williams, F. W., Natural vibrations of thin, prismatic, flat-walled structures, *IUTAM Symposium on high speed computing of elastic structures*, Leige, 1970.
40. Maddox, N. R., Plumble, H. E. and King, W. W., Frequency analysis of a cylindrical curved panel with clamped and elastic boundaries, *J of Sound and Vib*, 12, 225-249, 1970.
41. Sewall, J. L., *Vibration analysis of cylindrically curved panels with simply supported or clamped edges and comparison with some experiments*, NASA TN D-3791.

42. Chen, R., *Vibration analysis of cylindrical panels and rectangular plates carrying a concentrated mass*, Douglas Aircraft Co Inc, California, Report, 1965.
43. Cheung, Y. K. and Delcourt, C., Buckling and vibration of thin, flat-walled structures continuous over several spans, *Proc of ICE, Part* 2, 93-103, 1977.
44. Biggs, J. M., *Introduction to structural dynamics*, McGraw-Hill, New York, 1964.
45. Li, W. Y., Tham, L. G., Cheung, Y. K. and Fan, S. C., Free vibration analysis of doubly curved shells by spline finite strip method, *J of Sound and Vib*, 140(1), 39-53, 1990.
46. Navaratna, D. R., Natural vibration of deep spherical shells, *J AIAA*, 4(11), 2056-2058, 1966.
47. Kalnins, A., Effect of bending on vibration of spherical shells, *J Acoust Soc Am*, 36, 74-81, 1964.
48. Rose Jr., E. W., Natural frequencies and mode shape for axisymmetric vibration deep spherical shells, *J Appl Mech*, 32, 553-561, 1965.
49. Hashish, M. G. and Abu-Sitta, S. H., Free vibration of hyperbolic cooling tower, *J of Eng Mech, ASCE*, 97(2), 252-269, 1971.
50. Yang, T. Y. and Kapania, R. K., Shell elements for cooling tower analysis, *J of Eng Mech, ASCE*, 109(5), 1270-1289, 1983.
51. Dawe, D. J., Finite strip buckling analysis of curved plate assemblies under biaxial loading, *Int J Solid Struct*, 13, 1141-1155, 1977.
52. Viswanathan, A. V. and Tamekuni, M., Elastic buckling analysis for composite stiffened panels and other structures subjected to biaxial inplane loads, *NASA CR-2216*, 1973.
53. Timoshenko, S., *Theory of elastic stability*, McGraw-Hill, 1936.
54. Stroud, W. J., Greene, W. H. and Anderson, M. S., Buckling loads for stiffened panels subjected to combined longitudinal compression and shearing loadings: results obtained with PASCO, EAL and STAGS computer program, *NASA Technical Memorandum, No 83194*.
55. Lee, P. K. K., Chung, H. W., Ho, D. and Tham, L. G., A folded plate space structure, *Proc of the 3rd Int Conf on Space Structures*, Guildford, 261-265, 1985.
56. Pultar, M., Billington, D. P. and Riera, J. D., Folded plates continuous over flexible support, *J of Struct Eng, ASCE*, 93(5), 253-277, 1967.
57. Mark, R. and Riera, J. D., Photoelastic analysis of folded plate structures, *J of Eng Mech, ASCE*, 93(4), 79-93, 1967.
58. Guyon, Y., Calcul des ponts larges a poutres multiples solidarisees par les entretoises, *Annuales des Ponts et Chausses*, 24(5), 1946.

59. Massonet, C., Method of calculation of bridge with several longitudinal beams taking into account their torsional resistance, *Publications, Int Assoc for Bridges and Structural Engineering*, 10, 1950.
60. Morice, P. B. and Little, G., *Analysis of right bridge decks subjected to abnormal loading*, Cement and Concrete Association, London, 1956.
61. Rowe, R. E., *Concrete bridge design*, C R Books Ltd, London, 1962.
62. Cusens, A. R. and Pama, R. P., *Design curves for the approximate determination of bending moments in orthotropic bridge decks*, Civil Engineering Department, University of Dundee, March, 1970.
63. Cheung, Y. K., Orthotropic right bridges by the finite strip method, presented in the *2nd Int Sym on Concrete Bridge Design, Chicago, March 1969.* Subsequently published in *Concrete Bridge Design, ACI Publications SP-26*, 182-205, 1971.
64. Powell, G. H. and Ogden, D. W., Analysis of orthotropic bridge decks, *J of Struct Eng, ASCE*, 95(5), 909-23, 1969.
65. Cheung, M. S., Cheung, Y. K. and Ghali, A., Analysis of slab and girder bridges by the finite strip method, *Building Sci*, 5, 95-104, 1970.
66. Loo, Y. C. and Cusens, A. R., A refined finite strip method for the analysis of orthotropic plates, *Proc of Inst of Civ Eng*, 40, 85-91, 1971.
67. Cheung, Y. K., The analysis of curvilinear orthotropic curved bridge decks, *Publications, Int Assoc of Bridges and Structural Engineering*, 29-II, 1969.
68. Loo, Y. C., *Developments and applications of the finite strip method in the analysis of right bridge decks*, PhD thesis, Department of Civil Engineering, University of Dundee, 1972.
69. Abdullah, M. A. and Abdul-Razzak, A. A., Finite strip analysis of prestressed box-girders, *Comp and Struct*, 36(5), 817-822, 1990.
70. Graves Smith, T. R., Gierlinski, J. T. and Walker, B., A combined finite strip/finite element method for analysing thin-walled structures, *Thin-walled Struct*, 3(2), 163-180, 1985.
71. Cheung, Y. K., Tham, L. G. and Li, W. Y., Application of spline finite strip method in the analysis of curved slab bridge, *Proc of Inst of Civ Eng, Part 2*, 81, 111-124, 1986.
72. Li, W. Y., Tham, L. G. and Cheung, Y. K., Curved box girder bridge, *J of Struct Eng, ASCE*, 114(6), 1324-1338, 1988.
73. Cheung, M. S. and Li, W., Analysis of haunched, continuous bridges by the finite strip method, *Comp and Struct*, 28(5), 621-626, 1988.
74. Cheung, M. S. and Li, W., Analysis of haunched, continuous bridges by spline finite strips, *Comp and Struct*, 36(2), 297-300, 1990.
75. Puckett, J. A., Application of the compound strip method for the analysis of slab-girder bridges, *Comp and Struct*, 22(6), 979-986, 1986.
76. Cheung, Y. K., Cheung, M. S. and Reddy, D. V., Frequency analysis of certain single and continuous span bridges, *Development in bridge design construction* (ed. Rockey et al.), Crosby Lockwood, 1972.

77. Babu, P. V. T. and Reddy, D. V., Frequency analysis of skew orthotropic plates by the finite strip method, *J of Sound and Vib*, 18, 465-474, 1971.
78. Cheung, Y. K. and Cheung, M. S., Free vibration of curved and straight beam-slab or box girder bridges, *Publications, Int Assoc for Bridges and Structural Engineering*, 32-II, 41-52, 1972.
79. Srinivasan, R. S. and Munaswany, K., Dynamic response analysis of stiffened slab bridge, *Comp and Struct*, 9, 559-566.
80. Smith, J. W., Finite strip analysis of the dynamic response of beam and slab highway bridges, *Earthquake Eng and Struct Dyn*, 1, 357-370, 1973.
81. Hutton, S. G. and Cheung, Y. K., Dynamic response of single span highway bridges, *Earthquake Eng and Struct Dyn*, 7, 543-553, 1979.
82. Ghali, A. and Tadros, G. S., *On finite strip analysis of continuous plates*, Research Report No CE-72-11, Department of Civil Engineering, University of Calgary, March, 1972.
83. Pama, R. P. and Cusens, A. R., A load distribution method for analysing statically indeterminate concrete bridge decks, *Concrete Bridge Design, ACI Publications SP-26*, 599-633, 1971.
84. Chu, K. H. and Krishnamoorthy, G., Use of orthotropic theory in bridge design, *J of Struct Eng, ASCE*, 35-77, 1962.
85. Cheung, Y. K., *Finite strip method in structural analysis*, Pergamon Press, 1976.
86. Cheung, M. S. and Cheung, Y. K., Static and dynamic behaviour of rectangular strips using higher order finite strips, *Building Sci*, 7(3), 151-158, 1972.
87. Coull, A. and Das, P. C., Analysis of curved bridge decks, *Proc of Inst of Civ Eng*, 37, 75-85, 1967.
88. Lim, P. T. K. and Moffatt, K. R., Finite element analysis of curved slab bridges with special reference to local stresses, *Developments in bridge design and construction*, 27-52, 1971.
89. Bazeley, G. P., Cheung, Y. K. , Irons, B. M. and Zienkiewicz, O. C., Triangular elements in plate bending conforming and non-conforming solution, *Proc Int Conf Matrix Method in Structural Mechanics*, Airforce Inst of Tech, Ohio, 547-576, 1965.
90. Sisodiya, R. G., Ghali, A., Cheung, Y. K., Diaphragm in single- and double-cell box girder bridges with varying angle of skew, *J of ACI*, 415-419, 1972.
91. Scordelis, A. C., Davis, R. E. and Lo, K. S., Load distribution in concrete box girder bridges, *Concrete Bridge Design, ACI Publications SP-23*, 117-136, 1969.
92. Chan, H. C. and Cheung, Y. K., On the formulation of higher order conforming flat shell elements, *Proc of the Int Conf on Finite Element Methods*, Shanghai, China, 121-125, 1982.
93. *QUEST Manual, HE CB/B/14, Highway Engineering Computing Board*, London, Dept of Transport, 1974.

94. Lin, T. Y., *Design of prestressed concrete structures*, 2nd Ed., John Wiley and Sons, New York, 1963.
95. Scordelis, A. C., Chan, E. C., Ketchum, M. A. and Van Der Walt, P., Computer programs for prestressed concrete box girder bridges, *Structural Engineering and Structural Mechanics, Department of Civil Engineering, Report No UCB/SESM-85/103*, University of California, Berkeley, 1985.
96. Au, F. T. K., *Spline finite strip method in the study of plates and shells with special reference to bridges*, PhD thesis, Department of Civil and Structural Engineering, University of Hong Kong, 1994.
97. Lashkari, M., *COSMOS/M User Guide*, 7th Ed., Structural Research and Analysis Corporation, 1992.
98. Arizum, Y., Oshiro, T. and Hamad, S., Finite strip analysis of curved composite girders with incomplete interaction, *Comp and Struct*, 15(1), 603-612, 1982.
99. Cheung, Y. K. and Li, W. Y., Free vibration analysis of arbitrarily curved box girder structures by spline finite strip method, *Proc of the Asian Pacific Conf on Comp Mechanics*, Hong Kong, Vol 2, 1139-1144, 1991.
100. Delcourt, C. R. C., *Linear and geometrically nonlinear analysis of flat-walled structures by the finite strip method*, PhD Thesis, Department of Civil Engineering, University of Adelaide, 1978.
101. Maquoi, R. and Massonet, C., Theorie non linearie de la resistance postcritique des grandes pontres in caisson raidies, *IABSE Publications, 31-II*, 91-140, 1971.

CHAPTER NINE

APPLICATIONS TO TALL BUILDINGS

9.1 GENERAL CONSIDERATION AND STRUCTURAL IDEALIZATION

The huge demand on space arising from human activities in a modern city can only be met by the construction of multi-story buildings which can multiply many times the usable area of the land. The role of the engineering profession is to design viable and safe structural systems for transferring horizontal as well as vertical loadings acting on the buildings to the ground.

Skeletal frames and walls have proved to be very effective in transmitting such loadings and they are usually coupled together to form different framing systems for the super-structures of the buildings (Figure 9.1). The super-structures are connected at their bases to substructures which consist of thick slabs or beams either resting directly on the subsoils (shallow foundations) or being supported by piles and piers (deep foundations). The whole system of a tall building is rather complicated and involves the interaction between the structures and the foundations and the soil, and therefore, it is not always necessary or practical to carry out a full analysis. As a matter of fact, certain simplifications are usually adopted in the analysis. One approach is to assume that the superstructure and substructure can be designed separately in most practical cases. The super-structure can be treated as a cantilever with the base either fixed firmly or restrained by springs representing the stiffnesses of the sub-structure and the forces at the base of the super-structure can be computed by any appropriate method. If these forces are then regarded as external loadings acting on the sub-structure, the sub-structure can then be designed to carry these loadings by following the standard procedure. As this approach is commonly accepted by design engineers, it is also adopted here. Since the present scope focuses on superstructure analysis, readers who are interested in substructure analysis should refer to textbooks on foundation design for details.

Most of the methods for the analysis of the superstructure are based on the assumption that the slab at each floor level can be treated as a rigid diaphragm which has a high in-plane rigidity but a negligible bending stiffness, and, therefore, the whole structure is only allowed to move rigidly in the horizontal plane.

It is further assumed that the structures will behave elastically in the working load range. This permits one to describe the stiffness of the structural members simply in terms of the Young's (elastic) modulus and Poisson's ratio.

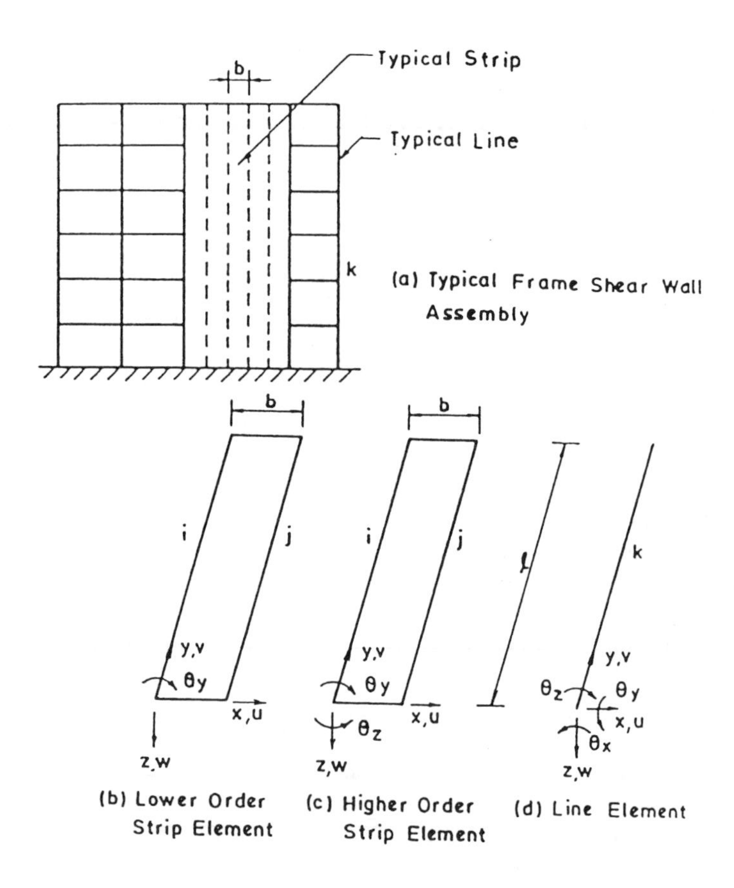

Figure 9.1 Typical frame shear wall assembly and the elements.

In the forties, Chitti[1] first derived closed form solutions for problems of plane coupled shear walls under lateral load. The work was followed up by various investigators[2-13] using the continuum method. On the other hand, McLeod[14,15] and Schwaighofer[16] idealized coupled frame shear wall structures by equivalent frames in which each shear wall was represented by a wide column with rigid arms. The third group[17,6,18,19,20] considered the structure to be composed of two parts, namely the shear wall and the frame. The walls and frames were considered to be acting as flexural and shear members respectively. Each part was made statically determinate by introducing the self equilibrating interactive forces, which were later solved by imposing the compatibility conditions between the two parts. Of course, the finite element method[21] is applicable to the analysis of tall buildings, but it is not a very attractive tool as the number of unknowns involved will be large and subsequently the computation cost will be high.

As the geometry of the structural members of buildings is fairly regular along the height, the finite strip method can provide an economical solution for the problem. Applications in various areas of analyses; including static, dynamic and stability, were developed by Cheung et al.[22-30] Excellent reviews on the development were prepared by Swaddiwudhipong,[31] Cheung[32,33] and Cheung and Chan.[34] Computer programs suitable for design office application were also available.[35,36,37] Guo[38] studied cores of non-uniform thickness by the finite strip model.

In the finite strip context,[31] the following idealizations in modelling of the structures are adopted:

(1) walls are divided into a number of strips. Each strip is characterized by a number of nodal displacement parameters representing the in-plane and out-of-plane stiffness. In the case of a lower order strip, there will be two nodal lines per strip and each nodal line has four degrees of freedom.

(2) skeletal frames can be treated by either one of the following methods:

(a) Columns can be represented by line elements which run the whole height of the building. The ends of a beam are either connected to walls or columns and the action of the beam is taken into account by transferring the standard beam stiffness to the adjacent strip or line elements through the compatibility conditions at its ends. Allowance for the effect of local deformation at the wall-beam junction which is significant in some cases can be made by modifying the stiffness of the beam.

(b) Each frame can be treated as an orthotropic continuum which has equivalent elastic properties, and therefore can be divided into finite strips as for a standard wall. The equivalent properties are computed by equating the relative displacements of a frame unit and those of the solid rectangular element (Figure 9.2). The end points of the beam and column members are the points of contraflexure. For the sake of convenience but also to preserve sufficient accuracy, the locations of the points of contraflexure are simply taken as the mid-points of the beams and columns. It can be shown readily that the equivalent elastic moduli[31] are

$$\overline{E}_x = \frac{b_1+b_2}{c_1+c_2}\frac{E/t}{(b_1/A_{b1}+b_2/A_{b2})} \tag{9.1}$$

$$\overline{E}_y = \frac{c_1+c_2}{b_1+b_2}\frac{E/t}{(c_1/A_{c1}+c_2/A_{c2})} \tag{9.2}$$

$$\overline{G} = \frac{3E/t}{\frac{c_1+c_2}{b_1+b_2}\left[\frac{b_1^3}{I_{b1}}+\frac{b_2^3}{I_{b2}}\right]+\frac{b_1+b_2}{c_1+c_2}\left[\frac{c_1^3}{I_{c1}}+\frac{c_2^3}{I_{c2}}\right]} \tag{9.3}$$

and the equivalent density is

$$\overline{\rho} = \frac{(b_1A_{b1}+b_2A_{b2}+c_1A_{c1}+c_2A_{c2})\rho/t}{(b_1+b_2)(c_1+c_2)} \tag{9.4}$$

with $\overline{\upsilon}_x = \overline{\upsilon}_y = 0$ as generally used in the idealization. In the equation E and ρ refer to the actual Young's modulus and density of the members.

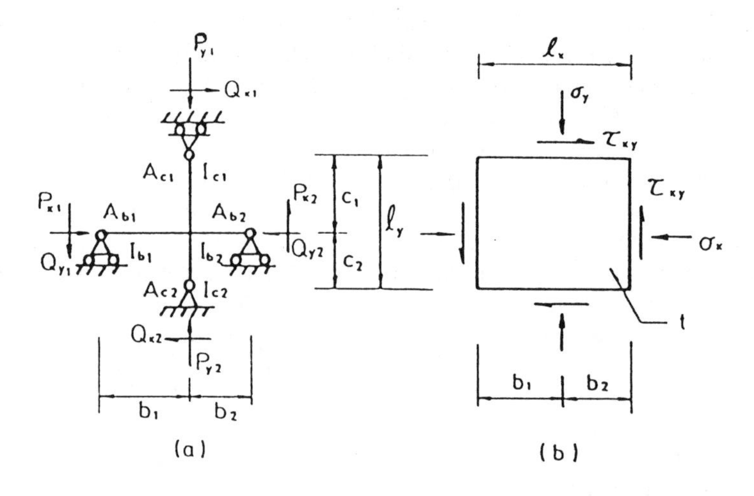

Figure 9.2 Equivalent wall modelling: (a) Actual joint. (b) Equivalent wall segment.

For the case of a panel of spandrel beams, such as those used in connecting two shear walls, the equivalent continuum properties are[31]

$$\overline{E}_x = \frac{EA_b}{t(c_1+c_2)} \quad ; \overline{E}_y = 0 \quad ; \overline{G} = \frac{12EI_b}{t(c_1+c_2)(b_1+b_2)^2} \tag{9.5}$$

Analysis based on the above model can only give the stresses of the equivalent wall and this is of little use for frame design. Fortunately, one can easily calculate the member's forces from these stresses. Equating the stress of the equivalent wall with the actual forces on the frame's members, one can show that the forces in each member are simply the sum of the stresses,[37] that is

$$P_{y1} = t\int_{l_x} \sigma_y \, dx \quad ; \quad P_{y2} = t\int_{l_x} \sigma_y \, dx \tag{9.6}$$

$$P_{x1} = t\int_{l_y} \sigma_x \, dx \quad ; \quad P_{x2} = t\int_{l_y} \sigma_x \, dx \tag{9.7}$$

$$Q_{x1} = t\int_{l_x} \tau_{xy} \, dx \quad ; \quad Q_{x2} = t\int_{l_x} \tau_{xy} \, dx \tag{9.8}$$

$$Q_{y1} = t\int_{l_y} \tau_{xy} \, dx \quad ; \quad Q_{y2} = t\int_{l_y} \tau_{xy} \, dx \tag{9.9}$$

Similarly, the average moments are

$$M_{c1} = c_1\, t\int_{l_x} \tau_{xy} \, dx \quad ; \quad M_{c2} = c_2\, t\int_{l_x} \tau_{xy} \, dx \tag{9.10}$$

$$M_{b1} = b_1\, t\int_{l_y} \tau_{xy} \, dx \quad ; \quad M_{b2} = b_2\, t\int_{l_y} \tau_{xy} \, dx \tag{9.11}$$

9.2 STRIP ELEMENTS

Solid walls and equivalent walls representing the frame action can be treated as flat shells which have their horizontal movements constrained by the floor slabs. Two approaches can be adopted for modelling such behaviour:

(a) shell with modified displacement functions
(b) standard flat shell strip

9.2.1 SHELLS WITH MODIFIED DISPLACEMENT FUNCTIONS[31]

It would be more convenient if one can take account of the rigid floor effect while formulating the displacement functions. The constrained displacements can be expressed as follows:

$$u_{im} = u_{jm} \tag{9.12a}$$

$$w_{im} = w_{jm} - b\theta_{yim} \tag{9.12b}$$

$$\theta_{yim} = \theta_{yjm} \tag{9.12c}$$

in which b is the width of the strip and other variables are the nodal displacement parameters.

(a) Lower order strip (SLR)[31]

In view of Eqs. 9.12a to 9.12c, the displacement functions for a typical lower order, rectangular strip (SLR) as shown in Figure 9.1 can be expressed as

$$u = \sum_{m=1}^{r} u_{im} U_m \tag{9.13}$$

$$v = \sum_{m=1}^{r} [(1-\frac{x}{b}) v_{im} + \frac{x}{b} v_{jm}] V_m \tag{9.14}$$

$$w = \sum_{m=1}^{r} [w_{im} + x\theta_{yim}] W_m \tag{9.15}$$

in which U_m , V_m and W_m are the longitudinal functions satisfying *a priori* at least the essential (geometric) boundary conditions at the two ends. Thus for the lower order (SLR) strip, the displacement parameters have been reduced from eight to five.

(b) Higher order strip (SHR)[31]

To ensure full compatibility at the base of the wall when it is supported on a beam or portal frame, it is necessary to introduce two extra parameters corresponding to the rotation about the z-axis (Figure 9.1). The shape functions are:

$$u = \sum_{m=1}^{r} u_{im} U_m \tag{9.16}$$

$$v = \sum_{m=1}^{r} [(1-3\bar{x}^2+2\bar{x}^3)v_{im}+x(1-2\bar{x}+\bar{x}^2)\,\theta_{zim}+(3\bar{x}^2-2\bar{x}^3)v_{jm}+x(\bar{x}^2-\bar{x})\theta_{zjm}]V_m \tag{9.17}$$

$$w = \sum_{m=1}^{r} [w_{im} + x\theta_{yim}]\,W_m \tag{9.18}$$

where $\bar{x} = x/b$.

(c) Longitudinal displacement function[31]

Assuming the super-structure to be fixed at the base, the mode shape resulting from the free vibration of a cantilever beam is an admissible choice for the longitudinal function (Eq. 2.8 of Section 2.2.1.1)

$$Y_m^f = \sin(\mu_m y) - \sinh(\mu_m y) - \alpha_m [\cos(\mu_m y) - \cosh(\mu_m y)] \tag{9.19}$$

where $\alpha_m = \dfrac{\sin(\mu_m)+\sinh(\mu_m)}{\cos(\mu_m)+\cosh(\mu_m)}$ and $\mu_m = 1.875, 4.694, \ldots \dfrac{(2m-1)\pi}{2}$

The superscript ' f ' denotes that the function is for a fixed end beam.

U_m^f and W_m^f are chosen as

$$U_1^f = W_1^f = y \tag{9.20}$$

and

$$U_m^f = W_m^f = Y_{m-1}^f \qquad \text{for } m=2,3,\ldots \tag{9.21}$$

Function V_m^f generally takes up the form

$$V_1^f = y \tag{9.22a}$$

$$V_m^f = \frac{dY_{m-1}^f}{dy} \qquad \text{for } m=2,3,\ldots \tag{9.22b}$$

The first term (Eq. 9.22a) is used to account for the end shear, since all the other terms will produce zero end shear.

It was also shown[31] that for the stability problem, the buckling shape for a cantilever beam yielded better results with less computational effort and is thus preferable. The buckling shape is (Eq. 2.21 of Section 2.2.1.2):

$$U_m^{fb} = W_m^{fb} = 1 - \cos\left(\frac{(2m-1)\pi y}{2}\right) \tag{9.23}$$

The superscript 'fb' denotes that the function is based on the buckling mode of a beam with a fixed end. The function V_m^{fb} may take either the form

$$V_m^{fb} = \frac{dU_m^{fb}}{dy} \tag{9.24}$$

or

$$V_m^{fb} = \sin\left(\frac{(2m-1)\pi y}{2}\right) \tag{9.25}$$

In the case where rotation at the base is allowed,[31] the functions U_m and W_m are now taken as

$$U_m^r = W_m^r = U_m^f \tag{9.26}$$

The superscript 'r' denotes that the function is for a beam with a hinged end. The function $V_m^r = dU_m^f/dy$ is usually applicable for such a case.

If both rotation and displacement are allowed at the base, a new set of functions for U_m and W_m may be created by adding a rigid body translation term to the functions

$$U_m^{rd} = W_1^{rd} = 1 \tag{9.27}$$

$$U_m^{rd} = W_1^{rd} = U_{m-1}^f \qquad (m=2,3, ...) \tag{9.28}$$

The superscript 'rd' denotes that the function is for a beam with free ends. Function V_m, in this case, is best expressed as

$$V_m^{rd} = \frac{dU_m^{rd}}{dy} \tag{9.29}$$

(d) Stiffness matrices of strip element

Based on the well-known set of formulae for the finite strip method (Section 9.2), one can derive the stiffness of the strip element. It can be shown readily that the in-plane and out-of-plane stiffnesses are uncoupled,

$$[K]_{mn}^e = \begin{bmatrix} [K]_{mn}^p & [0] \\ [0] & [K]_{mn}^b \end{bmatrix} \tag{9.30}$$

in which $[K]_{mn}^p$ and $[K]_{mn}^b$ are the in-plane and out-of-plane stiffness matrices respectively. In case of a lower order strip, we have[31]

$$[K]_{mn}^p = \frac{1}{6b}\begin{bmatrix} 6b^2I_1 & -6bI_1 & 6bI_1 \\ -6bI_1 & 2b^2I_2 + 6I_1 & b^2I_2 - 6I_1 \\ 6bI_1 & b^2I_2 - 6I_1 & 2b^2I_2 + 6I_1 \end{bmatrix} \tag{9.31}$$

where $I_1 = \sum G_r t_r [U_m' U_n']_r$

$$I_2 = \sum \frac{E_{yr}}{(1-\upsilon_x\upsilon_y)} t_r [V_m' V_n']_r$$

and

$$[K]_{mn}^b = \frac{b}{6}\begin{bmatrix} 6I_4 & 3bI_4 \\ 3bI_4 & 2b^2I_4 + 24I_3 \end{bmatrix} \tag{9.32}$$

where $I_3 = \sum D_{xyr} [W_m' W_n']_r$; and $I_4 = \sum D_{yr} [W_m'' W_n'']_r$.

Similarly, the stiffness matrices of a higher order strip can be derived accordingly.

Similar to other space structure analysis, the strip stiffness matrix has to be transformed from local coordinates to global coordinates (Figure 9.3).

However, the rigid section assumption will mean identical lateral displacements and rotation about the vertical axis for all strip elements. The transformation matrix for a lower order strip is of the form[32]

$$\begin{Bmatrix} u_{im} \\ v_{im} \\ v_{jm} \\ w_{im} \\ \theta_{yim} \end{Bmatrix} = \begin{bmatrix} \cos\beta & \sin\beta & x_i \sin\beta - z_i \cos\beta & 0 & 0 \\ 0 & 0 & 0 & 1 & 0 \\ 0 & 0 & 0 & 0 & 1 \\ -\sin\beta & \cos\beta & x_i \cos\beta + z_i \sin\beta & 0 & 0 \\ 0 & 0 & 0 & 0 & 0 \end{bmatrix} \begin{Bmatrix} u'_{om} \\ w'_{om} \\ \theta'_{yom} \\ v'_{im} \\ v'_{jm} \end{Bmatrix} \qquad (9.33)$$

where O is a chosen reference point and the primed and unprimed systems refer to global and local coordinate systems respectively.

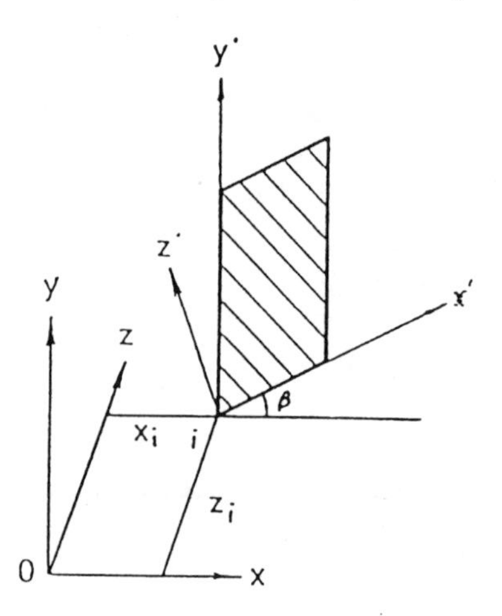

Figure 9. 3 Rigid section transformation.

9.2.2 STANDARD FLAT SHELL STRIP[32]

The behaviour of the wall can be modelled by the lower order flat shell strip described in Section 9.2. Of course, the stiffness and load matrices have to be modified to take into account the rigid floor effect on the displacement patterns. Thus if the horizontal deflections and rotation of a given reference point on the cross-section with coordinates x_s ,z_s for a given mode m are denoted by u_{sm} , w_{sm} , θ_{szm}, then the deflections and the rotation of other points with coordinates x_p , z_p may be written as

$$u_{pm} = u_{sm} - (z_p - z_s)\,\theta_{szm} \;;\; w_{pm} = w_{sm} - (x_p - x_s)\,\theta_{szm} \;;\; \theta_{pzm} = \theta_{szm} \qquad (9.34a, b, c)$$

The above relations can be obtained from Eq. 9.33 by setting $\beta = 90°$.

The vertical deflection and the other two rotations are not related. In matrix form,

$$\begin{Bmatrix} u_{pm} \\ v_{pm} \\ w_{pm} \\ \theta_{pxm} \\ \theta_{pym} \\ \theta_{pzm} \end{Bmatrix} = \begin{bmatrix} 1 & 0 & 0 & 0 & 0 & -(z_p - z_s) \\ 0 & 1 & 0 & 0 & 0 & 0 \\ 0 & 0 & 1 & 0 & 0 & -(x_p - x_s) \\ 0 & 0 & 0 & 1 & 0 & 0 \\ 0 & 0 & 0 & 0 & 1 & 0 \\ 0 & 0 & 0 & 0 & 0 & 1 \end{bmatrix} \begin{Bmatrix} u_{sm} \\ v_{sm} \\ w_{sm} \\ \theta_{sxm} \\ \theta_{sym} \\ \theta_{szm} \end{Bmatrix} \tag{9.35}$$

9.3 LINE ELEMENTS

9.3.1 LOWER ORDER LINE ELEMENT (LLE)[31]

A line element spanning the whole height of the building is used to represent the long column and a typical one with its displacement parameters is shown in Figure 9.4.

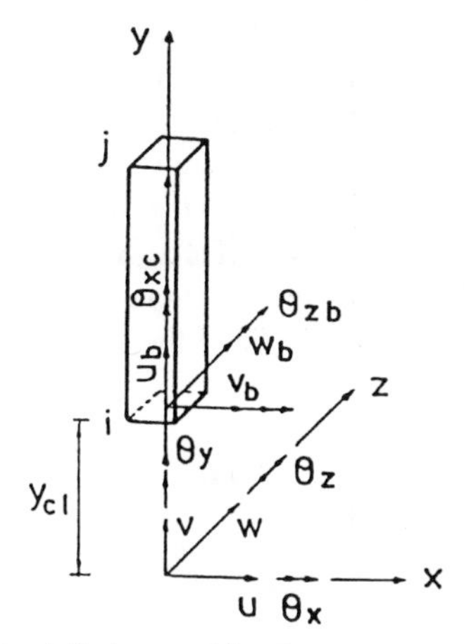

Figure 9.4 Line element displacement function.

The displacement field may be expressed as

$$u = \sum_{m=1}^{r} u_{km} U_m \tag{9.36a}$$

$$v = \sum_{m=1}^{r} v_{km} V_m \tag{9.36b}$$

$$w = \sum_{m=1}^{r} w_{km} W_m \tag{9.36c}$$

$$\theta_x = \sum_{m=1}^{r} \theta_{xkm} W_m \tag{9.36d}$$

$$\theta_y = \sum_{m=1}^{r} w_{km} W_m' \tag{9.36e}$$

$$\theta_z = \sum_{m=1}^{r} u_{km} U_m' \tag{9.36f}$$

From Figure 9.4, it can be seen that the deformation of a long column in a frame can be split into two parts:
(a) an overall deformation (given by dotted line in Figure 9.4) which can be represented by a few terms of a series given in Eqs. 9.36a to 9.36f.
(b) a local deformation mode which gives rise to contraflexure points in all beam and column members and which can never be accommodated by a few terms of the series.

This problem can be overcome if we assume that the long column is in fact made up of many short columns, but with their end displacements constrained by the 'long' column displacement functions. Since the bending displacement field for each short column (using a standard frame element) is a cubic polynomial, any local deformation mode can be represented accurately. Thus in place of the conventional finite element formulation, the stiffness of a 'long' column is obtained through the transformation of the stiffness matrices of all individual frame members which make up the long column.

For the adopted coordinate systems of the standard frame and line elements as depicted in Figure 9.4, the relationship of the displacement parameters of the two systems when the node of the frame is at height y_{c1} from the base can be easily verified to be

$$u_{c1} = \sum_{m=1}^{r} v_{km} V_m(y_{c1}) \tag{9.37a}$$

$$v_{c1} = \sum_{m=1}^{r} u_{km} U_m(y_{c1}) \tag{9.37b}$$

$$w_{c1} = \sum_{m=1}^{r} w_{km} W_m(y_{c1}) \tag{9.37c}$$

$$\theta_{xc1} = \sum_{m=1}^{r} w_{km} W_m'(y_{c1}) \tag{9.37d}$$

$$\theta_{yc1} = \sum_{m=1}^{r} \theta_{xkm} W_m (y_{c1}) \tag{9.37e}$$

$$\theta_{zc1} = \sum_{m=1}^{r} u_{km} U_m'(y_{c1}) \tag{9.37f}$$

where the subscript 'c' refers to those of the frame element parameters.

Similar expressions at the other end of the column can also be obtained in the same fashion. They can be inclusively expressed as

$$q_c = \sum_{m=1}^{r} T_{cm}\, q_{km} \tag{9.38}$$

in which

$$q_c = [u_{c1}\ \ v_{c1}\ \ w_{c1}\ \ \theta_{xc1}\ \ \theta_{yc1}\ \ \theta_{zc1}\ u_{c2}\ \ v_{c2}\ \ w_{c2}\ \ \theta_{xc2}\ \ \theta_{yc2}\ \ \theta_{zc2}]^T$$

and

$$q_{km} = [u_{km}\ \ v_{km}\ \ w_{km}\ \ \theta_{xkm}]^T$$

are the nodal displacement vectors for the column and line elements respectively.

Based on the principle of virtual work, it can be shown that the stiffness matrices in terms of the 'long' column parameters are:

$$k_{cmn} = T_{cm}{}^T\, k_c\, T_{cn} \tag{9.39}$$

where k_c is the stiffness in terms of the parameters of the two ends of the column.

9.3.2 HIGHER ORDER LINE ELEMENT (HLE)[31]

For a higher order line element (Figure 9.4), two extra nodal displacement parameters, θ_{xkm} and θ_{zkm}, are introduced in addition to the existing parameters for the lower order one. The number of unknowns and the size of the problem will obviously be greater than the previous case. The displacement field is then

$$u_{c1} = \sum_{m=1}^{r} v_{km}\, V_m(y_{c1}) \tag{9.40a}$$

$$v_{c1} = \sum_{m=1}^{r} u_{km}\, U_m(y_{c1}) \tag{9.40b}$$

$$w_{c1} = \sum_{m=1}^{r} w_{km}\, W_m(y_{c1}) \tag{9.40c}$$

$$\theta_{xc1} = \sum_{m=1}^{r} \theta_{ykm}\, W_m(y_{c1}) \tag{9.40d}$$

$$\theta_{yc1} = \sum_{m=1}^{r} \theta_{xkm}\, W_m'(y_{c1}) \tag{9.40e}$$

$$\theta_{zc1} = \sum_{m=1}^{r} \theta_{zkm}\, U_m'(y_{c1}) \tag{9.40f}$$

The relationship between the general column nodal parameters and the line element parameters can be established as for the lower order line element. Therefore, the necessary transformation can be carried out appropriately.

9.4 DISCRETE BEAM MODEL[31]

For a typical beam element with the nodal displacement parameters shown in Figure 9.5, the compatibility conditions at the end of the beam (at the height y_b from the base) which is connected to the nodal line i require that

$$u_b = \sum_{m=1}^{r} u_{im} U_m(y_b) \tag{9.41a}$$

$$v_b = \sum_{m=1}^{r} v_{im} V_m(y_b) \tag{9.41b}$$

$$w_b = \sum_{m=1}^{r} w_{im} W_m(y_b) \tag{9.41c}$$

$$\theta_{yb} = \sum_{m=1}^{r} \theta_{yim} W_m(y_b) \tag{9.41d}$$

The compatibility requirements of θ_{xb} and θ_{zb} depend on the type of element the beam is connected to and will be dealt with separately.

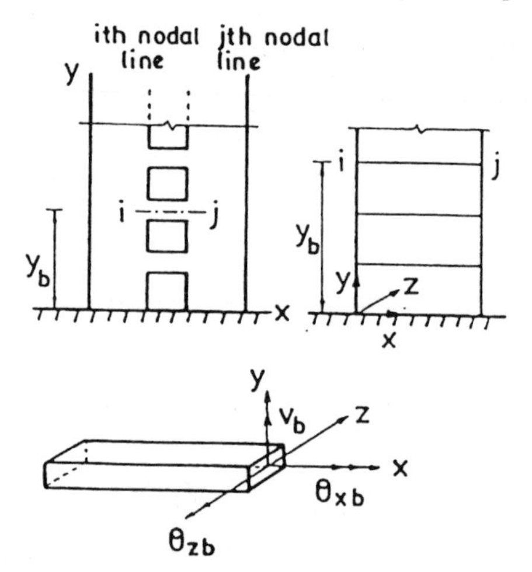

Figure 9.5 Discrete beam element.

(a) connecting to lower order strip (SLR)

$$\theta_{xb} = - \sum_{m=1}^{r} w_{im} W_m'(y_b) \tag{9.42a}$$

$$\theta_{zb} = - \sum_{m=1}^{r} u_{im} U_m'(y_b) \tag{9.42b}$$

(b) connecting to higher order strip (SHR)

$$\theta_{xb} = - \sum_{m=1}^{r} w_{im} W_m'(y) \tag{9.43a}$$

$$\theta_{zb} = \frac{1}{2} \sum_{m=1}^{r} [\theta_{zim} V_m (y_b) - u_{im} U_m'(y_b)] \tag{9.43b}$$

(c) connecting to a line element

$$\theta_{xb} = - \sum_{m=1}^{r} \theta_{xim} W_m(y_b) \tag{9.44a}$$

$$\theta_{zb} = - \sum_{m=1}^{r} \theta_{zim} U_m(y_b) \tag{9.44b}$$

It is obvious that the above expressions are similar to the ones presented in the previous section, that is

$$q_b = \sum_{m=1}^{r} T_{bm} q_{im} \tag{9.45}$$

in which q_b and q_{im} are the nodal displacement vectors for the beam and line elements respectively. Consequently, we have

$$k_{bmn} = T_{bm}^T k_b T_{bn} \tag{9.46}$$

where k_b is the stiffness in terms of the parameters of the two ends of the beam.

9.5 EXAMPLES

9.5.1 STATIC EXAMPLES

A study on the static behaviour of a complex ten-story frame-shear wall structure shown in Figure 9.6 by finite strip was conducted by Swaddiwudhipong.[31] A horizontal concentrated load of 100kN and a uniformly distributed load of intensity 1kN/m were acting at the top of joint 'j' and over the whole area of Wall No. 1 respectively. The beams and columns were modelled by the discrete beam model and line elements respectively. The floor slabs of the building were assumed to have only in-plane rigidity. The member properties, which vary with the height of the structure, are tabulated in Table 9.1. The deflections and rotation at the centre of the buildings are shown in Figure 9.7.

TABLE 9.1
Geometric properties of various parts of structures shown in Figure 9.6.

	Wall thickness mm	Cross-sectional dimension	
		Beam mmxmm	Column mmxmm
Upper 5 stories	200	200 x 500	200 x 200
Lower 5 stories	250	250 x 500	250 x 250

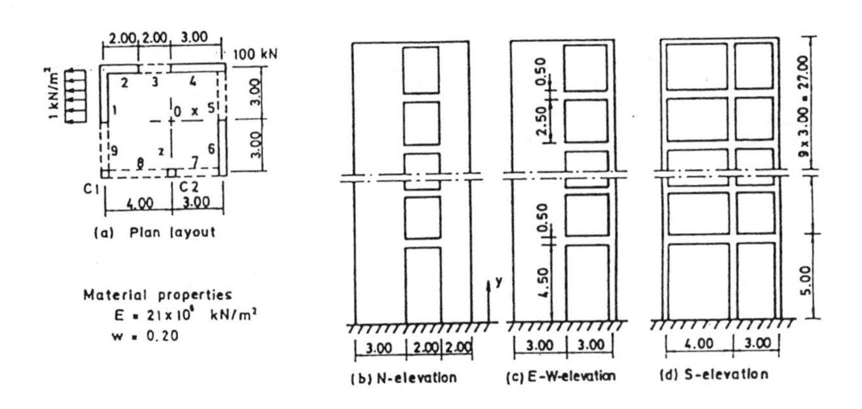

Figure 9.6 General coupled frame-shear wall multi-story structure.[31]

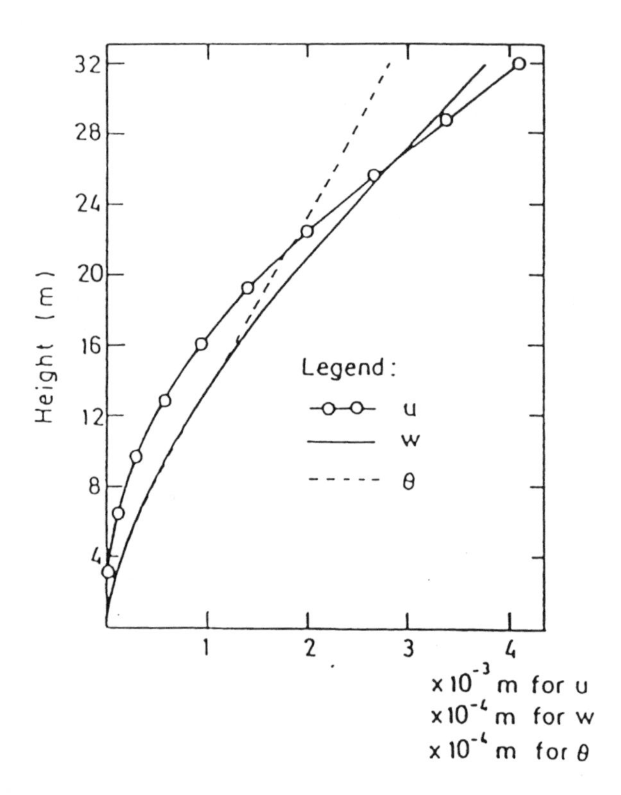

Figure 9.7 Deflections along x, z and rotation about y-axis at centre of the building.[31]

A model of a solid shear wall structure of 12.7mm in thickness supported on a portal frame (Figure 9.8) was analyzed.[31] The structure was subjected to a concentrated force of 89N at its top. Tests on the model were conducted by Chung[39] and he reported the values of the Young's modulus and Poisson's ratio to be $0.317 \times 10^4 N/mm^2$ ($0.46 \times 10^6 lbf/in^2$) and 0.377 respectively. The results based on the higher order strip (SHR) are shown in Figure 9.9.

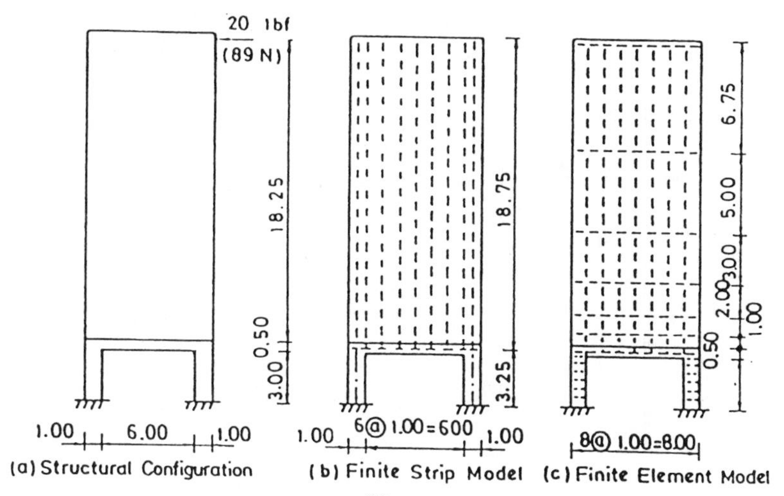

Figure 9.8 Solid shear wall on portal frame.[31]

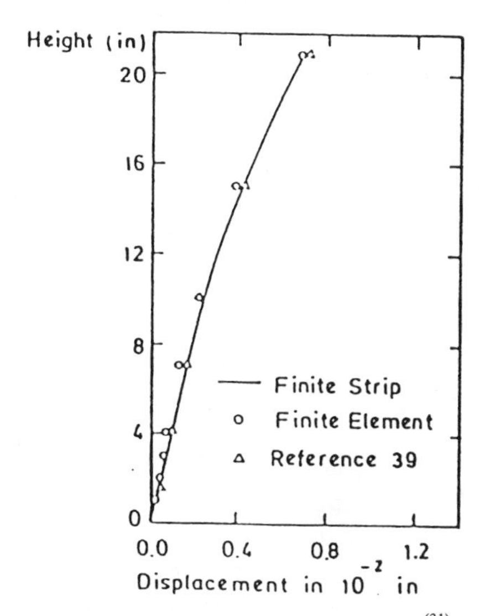

Figure 9.9 Deflection profile for solid shear wall on portal frame.[31]

The wind load analysis of an office tower, which forms part of a commercial complex, was carried out by Swaddiwudhipong.[31] The structural system, which was 32 stories high with two stories below the ground level, was made up of a central transportation core wall system interacting with the peripheral frame system. The typical layout of the structural system with all the dimensions is shown in Figure 9.10. The sizes of the structural members are given in detail in Reference 31 and will not be repeated here.

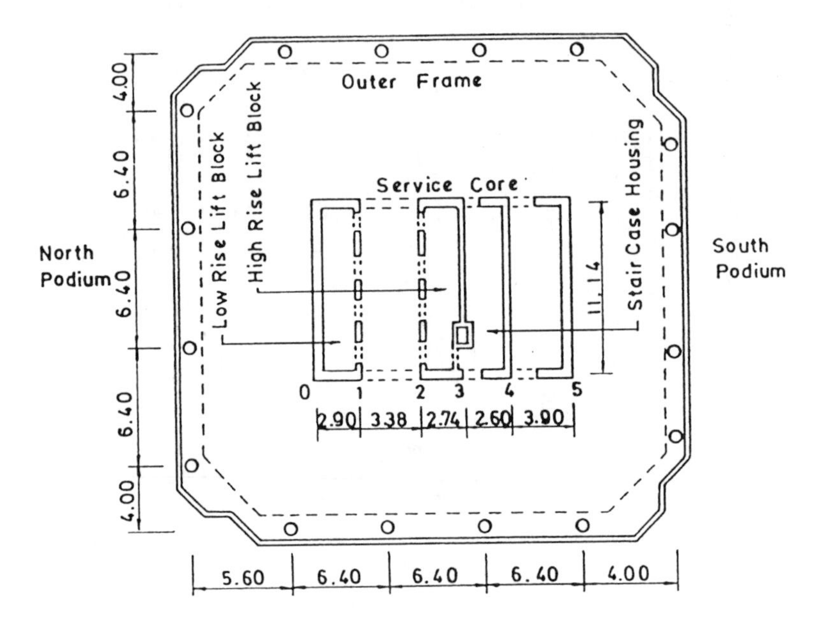

Figure 9.10 Framing plan of an office tower.(31)

In the overall analysis, the core wall system was subdivided into small strip elements and the connecting beams between the walls were treated as the equivalent continuum media. They were represented by the lower order strip elements of rigid section type. The columns in the peripheral frame system were, on the other hand, idealized as the lower order line element coupled together by the connecting beams which were treated by the discrete beam model. The layout of the discretised structure with the element and node numbering systems was depicted in Figure 9.11.

A simplified diagram of the statically equivalent wind pressure was used in the present study and this is shown in Figure 9.12. In the analysis, the wind pressure was assumed to be acting against and perpendicular to the north and east sides of the structure separately, one at a time. The Young's modulus of elasticity and the Poisson's ratio of concrete were assumed to be $0.31 \times 10^8 kN/m^2$ and 0.2 respectively. The displacements at each floor level can be obtained in the analysis, and displacements for some chosen floors are tabulated in Table 9.2. As the building is unsymmetrical along the East-West axis, the torsional rotations are significant.

After the finite strip analysis of the overall system, the outer frame was re-analyzed for a more accurate local variation of the stress-resultants. A

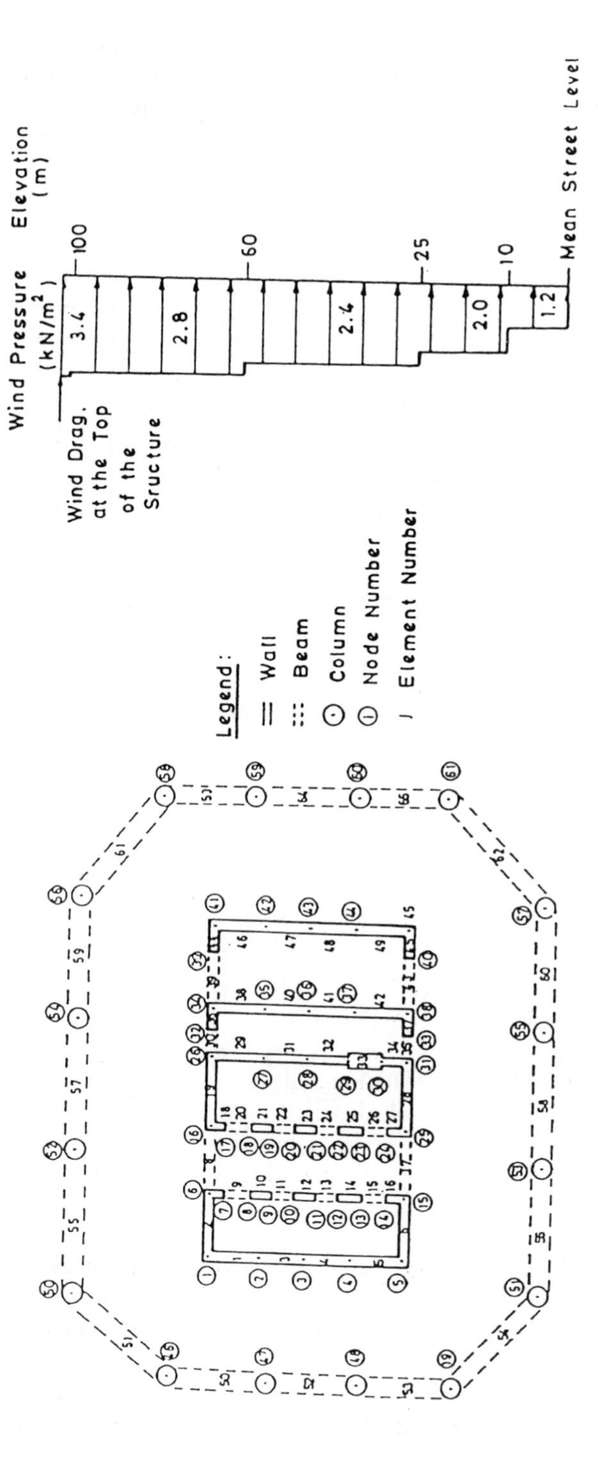

Figure 9.11 Element and node numbering system for finite strip analysis.[31]

Figure 9.12 Wind pressure along the height of the tower.[31]

simplification was made in which the outer frame on each side was treated individually as a plane frame subjected to the known displacement at each floor level obtained from the first analysis. The distribution of the longitudinal stress acting at the shear wall due to wind load from the east direction is shown in Figure 9.13.

TABLE 9.2
Displacements of the tower.

Floor No	North wind load	East wind load	
	u (mm)	w (mm)	$\theta \times 10^5$
Top	68.234	69.572	14.045
Roof	65.764	67.180	15.592
29	63.364	64.711	17.022
25	53.317	54.703	19.059
20	38.961	40.827	10.088
15	25.638	27.393	3.562
10	15.175	15.797	3.376
5	6.271	6.723	1.808
1	1.773	2.058	0.680
G	0.514	0.620	0.245

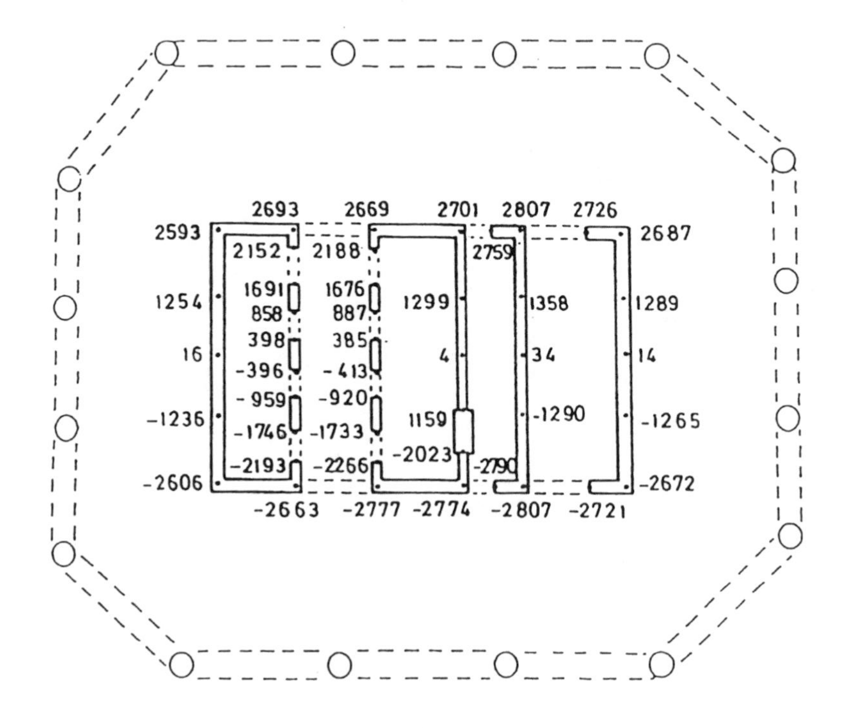

Figure 9.13 Longitudinal stress distribution in (kN/m^2) at 4th floor (wind load from east direction).[(31)]

9.5.2 VIBRATION AND STABILITY EXAMPLES

The model (Figure 9.14) previously studied by Heidebrecht and Raina[40] was used as a benchmark example.[22] In this case the shear centre and the centroid do not coincide; hence, coupled flexural and torsional modes of vibration are to be expected. Heidebrecht and Raina[40] considered the model in terms of Vlasov thin-walled beam theory[41] by assuming that the cross-section was rigid. Table 9.3 compares the finite strip results[22] with those obtained by Heidebrecht and Raina.[40] As will be noted, the finite strip frequencies are somewhat lower than the results obtained on the basis of thin-walled beam theory. Such a trend would be expected as the present procedure includes the effect of shear stresses which are likely to be of significance in the model under consideration.

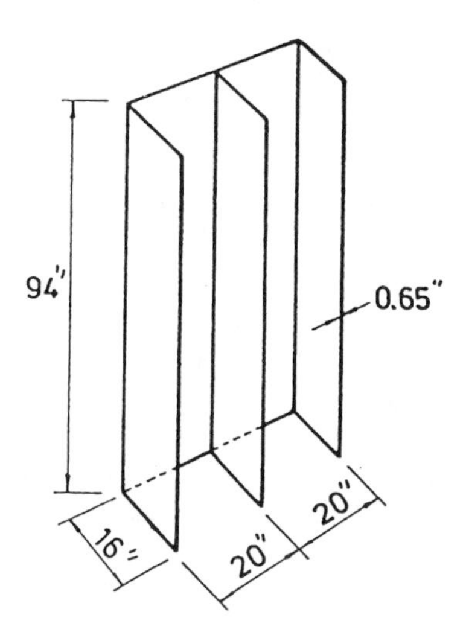

Figure 9.14 Non-planar shear wall model.[31]

TABLE 9.3
Natural frequencies of non-planar shear wall model.

Predominant Mode	Torsion	Bending		Torsion
		Weak	Strong	
Heidebrecht & Raina[40]	30.2		125.0	186.5
Number of terms				
3	30.29	37.49	99.55	165.4
10	30.24	37.40	98.23	163.6

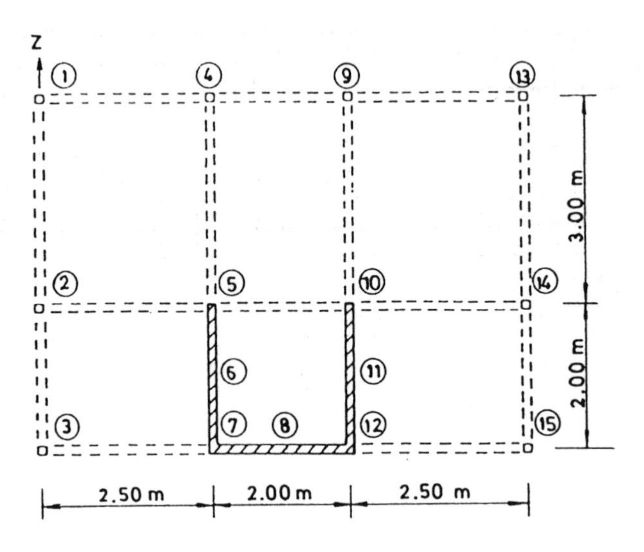

Figure 9.15 Structural plan of a three-dimensional coupled frame shear wall structure.[31]

The vibration behaviour of an eleven-story three-dimensional coupled frame shear wall structure (Figure 9.15) was investigated by Swaddiwudhipong[31] using the SLR. The story height, except for the first story which was 5.00m, was 3.50m.

The cross-sectional area of the columns, the beams and the thickness of the walls were 300mmx300mm; 200mmx500mm and 150mm respectively. The Young's modulus, Poisson's ratio and mass density were 0.21×10^{11}N/m^2 , 0.2 and 2400N-sec^2/m^4 respectively. The spring constants corresponding to various degrees of foundation rigidity are given in Table 9.4.

The angular frequencies for the different cases are tabulated in Table 9.5. The mode shapes corresponding to Case E4 using five terms in the displacement series are shown in Figures 9.16.

TABLE 9.4
Foundation spring stiffness at bases of a general frame shear wall structure (Subscript c stands for column).[31]

CASE	Spring constant in 10^6N/m					
	k_{cv}	$k_{c\theta x}$; $k_{c\theta z}$	k_5 ; k_{10}	k_6 ; k_{11}	k_7 ; k_{12}	k_8
E1	472.50	1.772	464.91	1711.80	123.87	123.87
E2	472.50	1.772	46.49	171.18	12.39	12.39
E3	47.25	0.177	464.91	1711.80	123.87	123.87
E4	47.25	0.177	46.49	171.18	12.39	12.39

TABLE 9.5
The effect of foundation flexibility on natural frequencies of a general frame shear wall structure.(31)

CASE	Number of Terms	Angular frequencies in radian/second				
		1	2	3	4	5
Fixed	3	4.757	5.183	8.914	17.212	21.098
	4	4.738	5.169	8.873	15.623	20.707
	5	4.690	5.153	8.840	15.574	20.694
E1	5	4.136	4.242	8.171	13.920	18.400
E2	5	3.500	3.962	7.141	13.506	16.917
E3	5	3.504	3.506	7.625	13.827	18.222
E4	3	2.773	3.272	6.807	14.325	17.156
	4	2.754	3.218	6.700	13.876	16.866
	5	2.737	3.175	6.600	13.390	16.721

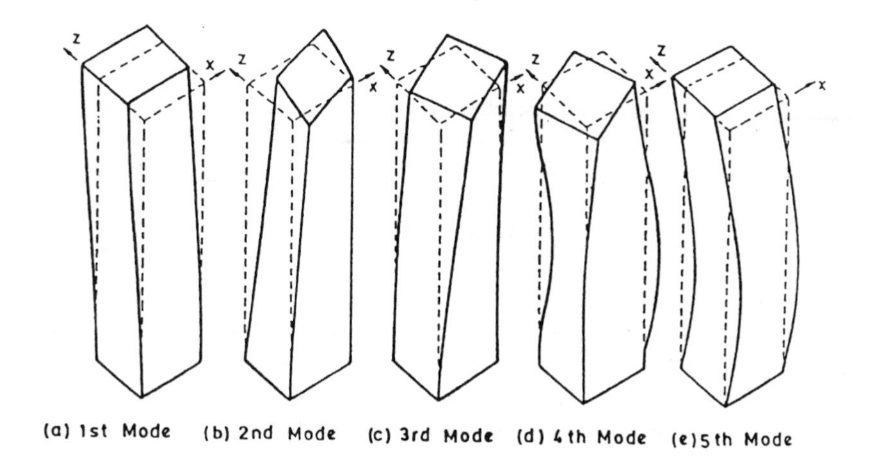

Figure 9.16 Mode shape of a general frame shear wall structure on flexible bases of Type E4.(31)

The stability of an unequal width coupled shear wall (Figure 9.17) due to uniformly distributed axial stresses in the walls was studied by Swaddiwudhipong.(31) Both the finite strip (SLR and SHR strips) and finite element analyses were carried out and the results are compared in Table 9.6. There is almost no discrepancy among the results.

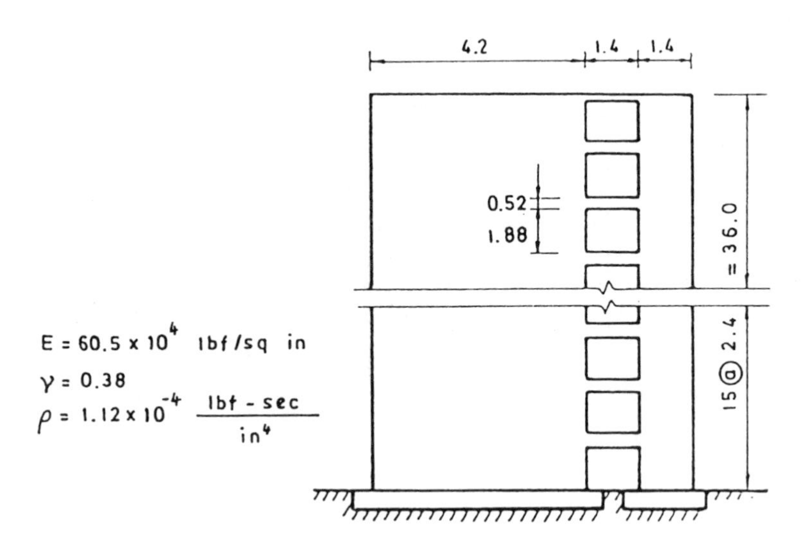

Figure 9.17 Coupled plane shear wall structure with unequal width.(31)

TABLE 9.6
Buckling load for unequal width coupled shear wall (Fixed base).(31)

Type of strips	Spandrel beam model	No of terms		
		1	2	3
SLR SHR	Discrete	4597.7 4542.0	4596.7 4540.8	4995.9 4540.0
Finite element		4587.4		

REFERENCES

1. Chitti, L., On the cantilever composed of a number of parallel beams interconnected by cross bars, *Philosophical Magazine*, Vol 38, No 285, 685-699, 1947.
2. Beck, H., Contribution to the analysis of coupled shear walls, *J of ACI*, Vol 59, No 8, 1055-1070, 1962.
3. Rosman, R., Contribution to the static analysis of horizontally loaded transverse walls in tall buildings - I, *Der Bauingenieur*, Vol 35, No 4, 133-136, 1960.
4. Rosman, R., Contribution to the static analysis of horizontally loaded transverse walls in tall buildings - III, *Der Bauingenieur*, Vol 37, No 8, 303-308, 1962.
5. Rosman, R., Approximate analysis of shear walls subject to lateral loads, *J of ACI*, Vol 61, No 6, 717-733, 1964.
6. Rosman, R., Lateral loaded systems consisting of walls and frames, *Proc Symposium on tall buildings with particular reference to shear wall structures,* University of Southampton, April 1966, 273-289, 1967.

7. Rosman, R., Static of non-symmetric shear wall structures, *Proc of Inst of Civil Eng, Supplement XII*, Paper 73935, 211-214, 1971.
8. Rosman, R., Stability and dynamics of shear-wall frame structures, *Building Sci*, Vol 9, No 1, 55-63, 1974.
9. Coull, A. and Chantaksinopas, B., Design curves for coupled shear wall on flexible bases, *Proc of Inst of Civil Eng, Part 2*, Vol 57, 595-618, 1974.
10. Mukherjee, P. R. and Coull, A., Free vibrations of open-section shear walls, *Earthquake Eng and Struct Dyn*, Vol 5, No 1, 81-101, 1977.
11. Tso, W. K. and Biswas, J. K., General analysis of nonplanar coupled shear walls, *J of Struct Eng, ASCE*, 99(3), 365-380, 1973.
12. Tso, W. K. and Chan, P. C. K., Flexible foundation effect on coupled shear walls, *J of ACI*, Vol 69, 11, 678-683, 1972.
13. Danay, A., Gluck, J. and Gellert, J. M., A generalised continuum method for dynamic analysis of asymmetric tall buildings, *Earthquake Eng and Struct Dyn*, Vol 4, No 2, 1975.
14. McLeod, I. A., Lateral stiffness of shear walls with openings, *Proc Symposium on tall buildings with particular reference to shear wall structures,* University of Southampton, April 1966, 223-252, 1967.
15. McLeod, I. A., Structural analysis of wall systems, *Struct Eng*, Vol 55, No 11, 487-495, 1977.
16. Schwaighofer, J. and Microys, H. F., Analysis of shear walls using standard computer programs, *J of ACI*, Vol 66, No 12, 1005-1007, 1969.
17. Cardan, B., Concrete shear walls combined with rigid frames in multi-story buildings subject to lateral loads, *J of ACI*, Vol 58, Bo 3, 299-316, 1961.
18. Khan, F. R. and Sbarounis, J. A., Interaction of shear walls and frames, *J of Struct Eng, ASCE*, 90(3), 285-335, 1964.
19. Heidebrecht, A. C. and Stafford Smith, B., Approximate analysis of tall wall-frame structures, *J of Struct Eng, ASCE*, 99(2), 199-221, 1973.
20. Ruthenberg, A., Tso, W. K. and Heidebrecht, A. C., Dynamic properties of asymmetric wall-frame structures, *Earthquake Eng and Struct Dyn*, Vol 5, No 1, 41-51, 1977.
21. Zienkiewicz, O. C., *The finite element method*, Fourth Edition, McGraw-Hill, London, 1989.
22. Cheung, Y. K., Hutton, S. G. and Kasemset, C., Frequency analysis of coupled shear wall assemblies, *Earthquake Eng and Struct Dyn*, Vol 5, 191-201, 1977.
23. Cheung, Y. K., Kasemset, C. and Swaddiwudhipong, S., Vibration and stability of tall frame structure, *Int Conf on Computer Applications in Developing Countries, Bangkok*, Vol 2, 977-989, 1977.
24. Cheung, Y. K. and Kasemset, C., Approximate frequency analysis of shear wall frame structures, *Earthquake Eng and Struct Dyn*, Vol 6, 221-229, 1978.

25. Cheung, Y. K., Tall buildings - analysis and design, *Proc of 3rd CI Conf, Assoc of Singapore*, 1978.
26. Cheung, Y. K. and Swaddiwudhipong, S., Analysis of frame shear wall structures using finite strip elements, *Proc of Inst of Civil Eng, Part 2*, Vol 65, 517-535, 1978.
27. Cheung, Y. K. and Swaddiwudhipong, S., Free vibration of frame shear wall structures using finite strip elements, *Earthquake Eng and Struct Dyn*, Vol 7, 355-367, 1979.
28. Cheung, Y. K. and Swaddiwudhipong, S., Finite strip analysis of tall buildings, *Eng J of Singapore*, Vol 5, No 1, 9-26, 1979.
29. Cheung, Y. K. and Swaddiwudhipong, S., Analysis of tall frame structure using finite strip method, *Int J of Struct*, Vol 1, No 4, 115-128, 1981.
30. Cheung, Y. K. and Swaddiwudhipong, S., A general method for tall building analysis, *Asian Regional Conf on Tall Buildings and Urban Habitat, Kuala Lumpur*, Technical Session 3, August, 1-18, 1982.
31. Swaddiwudhipong, S., *A unified approximate analysis of tall buildings using generalised finite strip method*, PhD Thesis, Department of Civil Engineering, University of Hong Kong, 1979.
32. Cheung, Y. K., Tall buildings 2, *Handbook of Structural Concrete*. ed. F. K. Kong, R. H. Evans, E. Cohen and F. Roll, Pitman, Chapter 38, 1983.
33. Cheung, Y. K., Computer analysis of tall buildings, *Proc of 3rd Int Conf on Tall Buildings, Hong Kong and Guangzhou*, 8-15, 1984.
34. Cheung, Y. K. and Chan, H. C., Changing scene of the Hong Kong skyline and advancement in tall building analysis, *Proc of Int Conf on Tall Buildings - Reach for the Sky, Kuala Lumpur, Malaysia*, 28-30 July, 33-40, 1992.
35. Cheung, Y. K., Fu, Zizhi and Tham, L. G., TBFSM - a program for the analysis of tall buildings by the finite strip method, *J of HKIE*, Vol 11, No 5, 35-43, 1983.
36. Cheung, Y. K., Luo, Songfa and Fu, Zizhi, A program system for the analysis of tall building by finite strip method, *Proc of 3rd Int Conf on Tall Buildings, Hong Kong and Guangzhou*, 624-630, 1984.
37. Cheung, Y. K., Man, K. F. and Tham, L. G., Tall building analysis by Mini-TBFSM, *Proc of 3rd Int Conf on Tall Buildings, Hong Kong and Guangzhou, December*, 216-219, 1984.
38. Guo, D. J., Cheung, Y. K. and Tham, L. G., Post-processor for finite strip method in the analysis of non-uniform thickness core, *Proc of 4th Int Conf on Tall Buildings, Hong Kong and Shanghai*, Vol 1, 299-303, 1988.
39. Chung, W. L., *Model analysis of solid shear wall on portal frame under lateral load*, BSc Dissertation, Department of Civil Engineering, University of Hong Kong, Hong Kong, 1977.
40. Heidebrecht, A. C. and Raina, R. K., Frequency analysis of thin-walled shear walls, *J of Eng Mech, ASCE*, 90, 239-251, 1971.
41. Vlazov, V. Z., *A general theory of shells*, Moscow, 1949.

CHAPTER TEN

APPLICATIONS TO LAYERED SYSTEMS IN STRUCTURAL AND GEOTECHNICAL ENGINEERING

10.1 SANDWICH AND LAMINATED PLATES

10.1.1 SANDWICH PLATES

The analysis of conventional sandwich plates with two stiff layers separated by a weak core has been the topic of extensive investigation, and several reference books[1,2] have been written on the subject. On the other hand, multi-layer plates with n stiff layers and n-1 alternating weak cores have not received much attention.

Lungren and Salama[3] developed a rectangular hybrid element and studied the stability problem of elastic plates, using the multi-layer sandwich plate theory given by Liaw and Little[4] in which the stiff layers were treated as membranes capable of resisting in-plane forces only, and a common shear angle was adopted for all the core layers. However, it had been pointed out by Kao and Ross[5] that the theory was not valid for multi-layer sandwich beams in which the shear strengths of the individual core layers are different. The above statement was confirmed by Khatua and Cheung[6,7] for plate problems, using rectangular and triangular finite elements based on displacement models, and for axisymmetric shell problems.[18]

The idea of a common shear angle is done away with by prescribing arbitrary in-plane displacements in all stiff layers. While this allows for a more accurate analysis and covers a wider range of problems, the number of DOF for a finite element becomes fairly large because the number of in-plane DOF is now proportional to the number of stiff layers present. Thus the finite strip analysis can be used advantageously because of the significant reduction in the total number of unknowns. The applications of the finite strip model in the analysis of such sandwich plates were first discussed by Chan and Cheung.[8] They treated the stiff layers as thin membranes and the cores as a material with only shear stiffness, and presented a number of examples on the static as well as vibration analyses of multi-layered flat plates. Ibrahim and Monforton[9] further reported the results of their study on the analysis of cylindrical folded plate structures by this approach. Chan and Foo[10] carried out the stability analysis for such multi-layered sandwich plates.

Nevertheless, such a model is only suitable for sandwich plates which comprise core layers of uniform thickness with negligible normal stiffness. To overcome these limitations, a general method derived by coupling the finite strip and prism (finite strip-prism model) was developed by Cheung et al.[11,12,13,14,15] Based on this model, examples of static, vibration and buckling problems were presented to demonstrate its versatility. For analyses of simply supported rectangular plates, they[16] also demonstrated that the computational efforts could be significantly reduced if the finite layer model was used. Both models will be presented in detail in the following sections.

10.1.1.1 *Finite strip model*

(I) General theory

(a) Assumptions

(1) The stiff layers have relatively high modulus of elasticity and they can withstand both bending and direct forces. However, it has been found that the bending moments in the stiff layers are usually quite small unless fairly thick stiff layers are used, and as a result it is normally safe enough to treat the stiff layers as membranes.

(2) The cores are considered to have zero normal stress stiffness and provide resistance to transverse shear only.

(3) There is no normal strain in the thickness direction, with the result that all stiff layers have the same vertical deflection, although each stiff layer is allowed to have independent in-plane displacements so that each core layer may take up a different shear angle (Figure 10.1).

(4) The materials can be isotropic, orthotropic, or in general anisotropic.

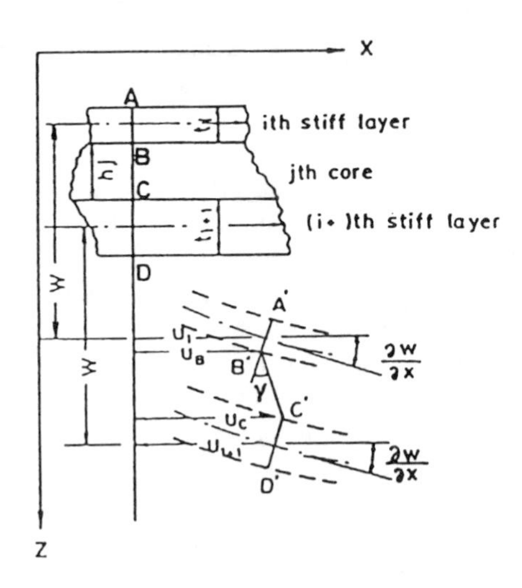

Figure 10.1 Deformation of stiff layers and cores.

(b) Stress-strain relationships (Figure 10.2)

(1) Stiff layer

The bending and stretching of the i-th stiff layer has the same stress-strain relationships as for an ordinary thin plate, and is simply[8]

$$M_{xi} = -[(d^b)_{xi} \frac{\partial^2 w}{\partial x^2} + (d^b)_{li} \frac{\partial^2 w}{\partial y^2}] \tag{10.1a}$$

$$M_{yi} = -[(d^b)_{li} \frac{\partial^2 w}{\partial x^2} + (d^b)_{yi} \frac{\partial^2 w}{\partial y^2}] \tag{10.1b}$$

$$M_{xyi} = -2(d^b)_{xyi} \frac{\partial^2 w}{\partial x \partial y} \tag{10.1c}$$

where $(d^b)_{xi} = \dfrac{E_{xi} t_i^3}{12(1-\upsilon_{xi}\upsilon_{yi})}$; $(d^b)_{yi} = \dfrac{E_{yi} t_i^3}{12(1-\upsilon_{xi}\upsilon_{yi})}$; $(d^b)_{li} = \upsilon_{yi}\,(d^b)_{xi} = \upsilon_{xi}$ $(d^b)_{yi}$; $(d^b)_{xyi} = \dfrac{E_{xyi} t_i^3}{12}$ and the suffix i denotes the i-th layer.

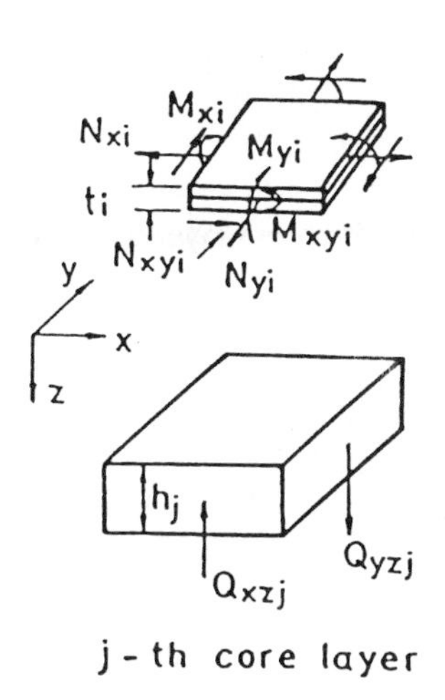

Figure 10.2 Stresses in typical plate layer.

(2) Shear deformation in the j-th core layer[8]

Only shear deformation needs to be considered for the case of a weak core layer since all the other stress (strain) components are assumed to be zero. The relationships between shear force, shear strain, and displacements are somewhat more complicated and are worthy of some lengthy discussions.

From Figure 10.1 it is seen that the shear strain in the xz-plane of the j-th core is represented by the angle γ, which is in turn made up of two parts; one part is due to bending of the plate and is equal to the slope $\partial w/\partial x$, while the other part is due to shearing of the j-th core layer and is equal to $(u_c-u_b)/h_j$, that is

$$\gamma_{xzj} = \frac{u_c - u_b}{h_j} + \frac{\partial w}{\partial x} \tag{10.2}$$

It is now necessary to relate u_c and u_b to the nodal displacement parameters u_i and u_{i+1} . Again, from Figure 10.1,

$$u_c = u_{i+1} + \frac{t_{i+1}}{2}\frac{\partial w}{\partial x} \tag{10.3}$$

$$u_b = u_i - \frac{t_i}{2}\frac{\partial w}{\partial x} \tag{10.4}$$

Therefore, substituting Eqs. 10.3 and 10.4 into Eq. 10.2,

$$\gamma_{xzj} = \frac{1}{h_j}[(u_{i+1} + \frac{t_{i+1}}{2}\frac{\partial w}{\partial x}) - (u_i - \frac{t_i}{2}\frac{\partial w}{\partial x})] + \frac{\partial w}{\partial x} \tag{10.5}$$

Simplifying Eq. 10.5,

$$\gamma_{xzj} = \frac{C_j}{h_j}[\frac{u_{i+1} - u_i}{C_j} + \frac{\partial w}{\partial x}] \tag{10.6}$$

where $C_j = h_j + \frac{t_{i+1}+t_i}{2}$.

The shear force Q of the j-th core layer can now be expressed in terms of the shear strain as

$$Q_{xzj} = (G_{xzj}\, h_j)\, \gamma_{xzj} = (d^s)_{xzj}\, \gamma_{xzj} \tag{10.7}$$

Similar expression can be derived for the shearing in the yz plane, and we have

$$\gamma_{yzj} = \frac{C_j}{h_j}[\frac{v_{i+1} - v_i}{C_j} + \frac{\partial w}{\partial y}] \tag{10.8}$$

and

$$Q_{yzj} = (G_{yzj}\, h_j)\, \gamma_{yzj} = (d^s)_{yzj}\, \gamma_{yzj} \tag{10.9}$$

(II) Formulation of the characteristic matrices

A lower order multi-layer sandwich strip will be developed herein. There are two in-plane displacements u_i and v_i (linear polynomial of x) and two out-of-plane displacement parameters w_i and θ_i (cubic polynomial of x) for each individual stiff layer at nodal line i. Due to assumption (2) w_i and θ_i will be the same for all stiff layers. Thus the number of unknown parameters for a nodal line of a strip with n stiff layers will be (2+2n) times the number of terms used in the series part (Y_m) of the displacement function. Note that in static analysis the end condition at a support should either be simply supported or clamped, while for vibration analysis all six types of end conditions listed in Section 2.2.2.1 can be included.

The deformation of a multi-layer sandwich strip can be viewed as the combined result of three separate processes, which include the bending of the stiff layers, the stretching of the stiff layers, and the shearing of the cores which couples the different stiff layers together. The bending of the stiff layers is directly related to the sum of the individual flexural rigidities since the bending displacement parameters are the same for all the layers.

Schematically the formulation of the stiffness matrix is as shown in Figure 10.3. The submatrix $[K_{lmk}{}^{b}]$ is the total bending stiffness of all the stiff layers and is related to the nodal displacement parameters w_1 and θ_1 or w_2 and θ_2, and $[K_{lmk}{}^{p}]$ is the in-plane stiffness of each individual stiff layer and is related to the displacement parameters u_{1k}, v_{1k} or u_{2k}, v_{2k} (here the suffix k refers to the k-th stiff layer and not in the usual context to the k-th terms of the series), in which l and m (1 and 2) are the nodal line numbers. These submatrices can be obtained by the established procedures for finite strips. The effect of the shearing of the cores is to couple the different stiff layers together or, in other words, to fill up some of the empty space of the schematic diagram in Figure 10.3.

$$
\begin{array}{ccccc|ccccc}
w_1;\theta_1 & u_{11};v_{11} & u_{12};v_{12} & & u_{1n};v_{1n} & w_2;\theta_2 & u_{21};v_{21} & u_{22};v_{22} & & u_{2n};v_{2n} \\
[K^{b}_{11}] & x & x & x & x & [K^{b}_{12}] & x & x & x & x \\
x & [K^{p}_{111}] & x & & & x & [K^{p}_{121}] & x & & \\
x & x & [K^{p}_{112}] & x & & x & x & [K^{p}_{122}] & x & \\
x & & & . & x & x & & & . & x \\
x & & & x & [K^{p}_{11n}] & x & & & x & [K^{p}_{12n}] \\
\hline
[K^{b}_{21}] & x & x & x & x & [K^{b}_{22}] & x & x & x & x \\
x & [K^{p}_{211}] & x & & & x & [K^{p}_{221}] & x & & \\
x & x & [K^{p}_{212}] & x & & x & x & [K^{p}_{222}] & x & \\
x & & & . & x & x & & & . & x \\
x & & & x & [K^{p}_{21n}] & x & & & x & [K^{p}_{22n}]
\end{array}
$$

Figure 10.3 Schematic diagram of stiffness matrix.

Listing the nodal displacement parameters as

$$\{\delta\}=\{w_1\ \theta_1\ u_{11}\ v_{11}\ u_{12}\ v_{12} \ldots u_{1n}\ v_{1n}\ w_2\ \theta_2\ u_{21}\ v_{21}\ u_{22}\ v_{22} \ldots u_{2n}\ v_{2n}\}^{T} \tag{10.10}$$

and the shear strains for the n-1 core layer as

$$\{\varepsilon\} = \begin{Bmatrix} \gamma_{xz1} \\ \gamma_{yz1} \\ \gamma_{xz2} \\ \gamma_{yz2} \\ \cdot \\ \cdot \\ \cdot \\ \gamma_{xzn-1} \\ \gamma_{yzn-1} \end{Bmatrix} \tag{10.11}$$

the strain matrix $[B^s]$ is easily established through the use of Eqs. 10.6 and 10.8. Thus

$$\{\varepsilon\} = \begin{Bmatrix} \frac{C_1}{h_1}(\frac{u_2-u_1}{C_1}+\frac{\partial w}{\partial x}) \\ \frac{C_1}{h_1}(\frac{v_2-v_1}{C_1}+\frac{\partial w}{\partial y}) \\ \frac{C_2}{h_2}(\frac{u_3-u_2}{C_1}+\frac{\partial w}{\partial x}) \\ \frac{C_2}{h_2}(\frac{v_3-v_2}{C_1}+\frac{\partial w}{\partial y}) \\ \cdot \\ \cdot \\ \cdot \\ \frac{C_{n-1}}{h_{n-1}}(\frac{u_n-u_{n-1}}{C_1}+\frac{\partial w}{\partial x}) \\ \frac{C_{n-1}}{h_{n-1}}(\frac{v_n-v_{n-1}}{C_1}+\frac{\partial w}{\partial y}) \end{Bmatrix} \tag{10.12}$$

$$= [B^s]\{\delta\}$$

The stiffness matrix due to the shearing of the core can now be readily obtained through the familiar relationship of

$$[K^s] = \int [B^s]^T [D^s] [B^s] \, dxdy \tag{10.13}$$

where $[D^s]$ is a diagonal matrix of the shear constants (d^s) of the cores.

The mass matrix is obtained in the usual manner through the use of Eq. 5.12, and the arrangement of the submatrices is also depicted in Figure 10.3 in which $[K_{1m}{}^b]$ and $[K_{lmk}{}^P]$ should be replaced by the bending mass matrix $[M_{lm}{}^b]$ and in-plane mass matrix $[M_{lmk}{}^P]$ respectively. For the $[M_{lm}{}^b]$ the overall mass distribution of the plate, obtained by summing up the mass of every layer including the cores, is used. For $[M_{lmk}{}^P]$ the mass of half of the thickness of the adjacent core layers and the mass of the k-th stiff layer are included. There is no coupling effect between the in-plane and out-of-plane mass matrices, and therefore all the crosses in Figure 10.3 are equal to zero.

Similarly, the geometric stiffness matrix can be obtained through the use of Eq. 5.33.

(III) Numerical examples

(a) Bending problems

Three-layer isotropic and orthotropic sandwich plates with various boundary conditions were analyzed. Only half of the plates with five strips were used wherever symmetry existed. The results for simply supported plates are listed in Table 10.1 along with available published results, and good agreement is observed for all cases. In Table 10.2, central deflections and moments and maximum edge moments are given for several boundary conditions, and serve to demonstrate the versatility of the method.

TABLE 10.1

Three-layer simply supported isotropic and orthotropic sandwich plates under uniform load.[17]

Terms	Central deflection (m)		Central bending moment M_x (kN-m/m)		Central bending moment M_y (kN-m/m)	
	Isotropic	Orthotropic	Isotropic	Orthotropic	Isotropic	Orthotropic
1	0.0007642	0.0012689	4.962	7.677	5.229	3.265
3	-0.0000248	-0.0000595	-0.157	-0.280	-0.457	-0.420
5	0.0000047	0.0000117	0.031	0.040	0.012	0.100
7	-0.0000016	-0.0000041	-0.011	-0.012	-0.037	-0.037
Σ	0.0007435	0.001217	4.825	7.425	4.838	2.908
Series solution[4]	0.0007395	-	4.79	-	4.79	-
Finite element[7]	0.0007361	0.001213	4.7789	7.4433	4.7789	-

Note : $A = 10m$; $B = 10m$; $t_1 = t_2 = 0.028m$; $h_1 = 0.75m$; $q = 1kN/m^2$

Isotropic case :

$E_{x1} = E_{x2} = E_{y1} = E_{y2} = 10^7$ kN/m^2; $G_{xz1} = G_{zy1} = 3x10^4$kN/m^2;

$\upsilon_{x1} = \upsilon_{x2} = \upsilon_{y1} = \upsilon_{y2} = 0.3$

Orthotropic case :

$E_{x1} = E_{x2} = 10^7$ kN/m^2; $E_{y1} = E_{y2} = 4x10^6$kN/m^2; $E_{xy1} = E_{xy2} = 1.875x10^6$kN/m^2

$\upsilon_{x1} = \upsilon_{x2} = 0.3$; $\upsilon_{y1} = \upsilon_{y2} = 0.12$; $G_{xz1} = 3x10^4$kN/m^2; $G_{yz1} = 0.2x10^4$kN/m^2

(b) Vibration problems

Natural frequencies for rectangular five-layer sandwich plates were obtained by this method and compared with published results (Table 10.3).[17] The advantage of the finite strip method over the conventional finite element method is more obvious for such eigenvalue problems since a much smaller matrix is needed for the finite strip method formulation.

(c) Stability problems

The applications of finite strip model in stability analysis was reported by Chan and Foo.[10] Critical stresses for sandwich plates with various number of layers and aspect ratios were computed. The plates also had different support conditions. The results compared favourably with other published results.

TABLE 10.2
Three-layer isotropic sandwich plate under uniform load.[17]

Boundary conditions	Central deflection (m)	Central moment (kN-m/m)		Edge moment (kN-m/m)
		M_x	M_y	M_y
s-c-s-s	0.0006036	3.828	3.709	-6.409
c-c-s-s	0.0005449	3.437	3.185	-5.105
c-c-c-s	0.0004647	2.66	2.74	-4.99
c-c-c-c	0.0004244	2.36	2.368	-4.361
c-s-c-s	0.000529	3.13	3.13	-5.745

(Note: Dimensions and elastic properties are the same as Table 10.1)

TABLE 10.3
Natural frequencies (cps) of a five-layer simply supported sandwich plate.[17]

Modal Number	m'	1	2	1	3	2	3	4	1	2
	n'	1	1	2	1	2	2	1	3	3
Reference 4		19	38	60	69	78	109	115	128	145
Finite strip		20	38	60	68	78	107	109	127	144

Note : $n=3$; $A=1.83m$; $B=1.22m$; $t_1=t_2=t_3=0.00028m$; $h=0.00318m$
$E_{x1}=E_{x2}=E_{x3}=E_{y1}=E_{y2}=E_{y3}=6.89x10^7 kN/m^2$
$G_{xz1}=G_{xz2}=134.45x10^3 kN/m^2$; $G_{yz1}=G_{yz2}=51.71x10^3 kN/m^2$
$\upsilon_{x1}=\upsilon_{y1}=\upsilon_{x2}=\upsilon_{y2}=\upsilon_{x3}=\upsilon_{y3}=0.33$
$\rho_{s1}=\rho_{s2}=\rho_{s3}=2.767kN/s^2/m^4$; $\rho_{c1}=\rho_{c2}=0.123kN/s^2/m^4$

(d) Shear angles of cores

An isotropic five-layer sandwich plate with shear constants of the two cores having a ratio of ten was analyzed. The dimensions and properties of the plate are the same as the ones shown in Table 10.1 except that $G_{xz1} = G_{yz1} = 30000kN/m^2$ and $G_{xz2} = G_{yz2} = 3000kN/m^2$. The in-plane displacements of the stiff layers in the x-direction, u, at the quarter point of the plate are plotted and shown in Figure 10.4. It can be seen that the shear angles of the two cores differ significantly. This shows that the increase in displacement parameters (u_i and v_i for each individual stiff layer) to allow for this effect is justified, and that the assumption of common shear angle for all cores can be quite erroneous for certain cases.

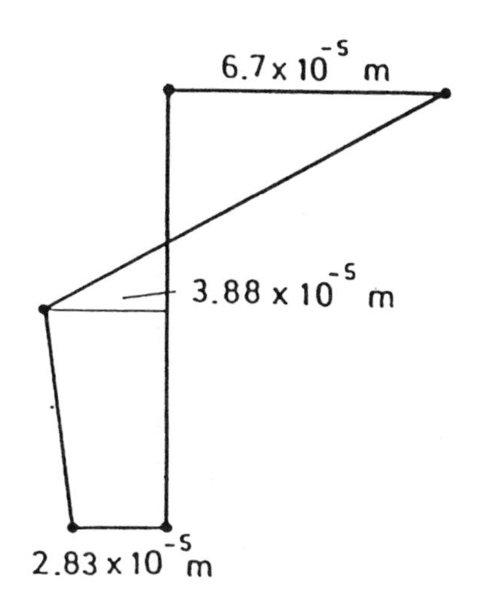

Figure 10.4 Displacement u of stiff layer.[17]

10.1.1.2 *Finite strip-prism model*

In the previous section, we have outlined a finite strip method for approximate analyses of sandwich plates in which the thickness of each core is uniform throughout its width. However, there are many sandwich plates which have cores being formed into shapes such that their thicknesses are not uniform, and the analysis of such panels can no longer be carried out in a straightforward manner by the finite strip method. Nevertheless, such prefabricated sandwich panels can be analyzed by coupling the finite strip and prism methods. The coupled scheme provides a general solution for such plates by assuming that the facings are flat shells (see Section 10.2) with both in-plane and bending stiffness whereas the cores are three-dimensional solids with normal as well as transverse stiffnesses. This approximation permits the facings and cores to be modelled respectively by finite strips and prisms (Figure 10.5). The derivation of the characteristic matrices of the strips and prisms has already been outlined in the previous chapters (Chapters 3 to 6), and the equations necessary for the computation of these matrices are listed in Table 10.4 for easy reference.

Having the stiffness matrices of the strips and prisms formed in terms of the nodal parameters, the combined stiffness matrix for the sandwich system can be obtained by enforcing the displacement continuity conditions along the interface between the core and facing. In the case of a simply supported panel, the equilibrium equation is

$$ {}^{f}K_{mn}^{\ b}\, {}^{f}\delta_{m}^{\ b} + {}^{f}K_{mn}^{\ p}\, {}^{f}\delta_{m}^{\ p} + {}^{c}K_{mn}\, {}^{c}\delta_{m} = F_{m} \tag{10.14} $$

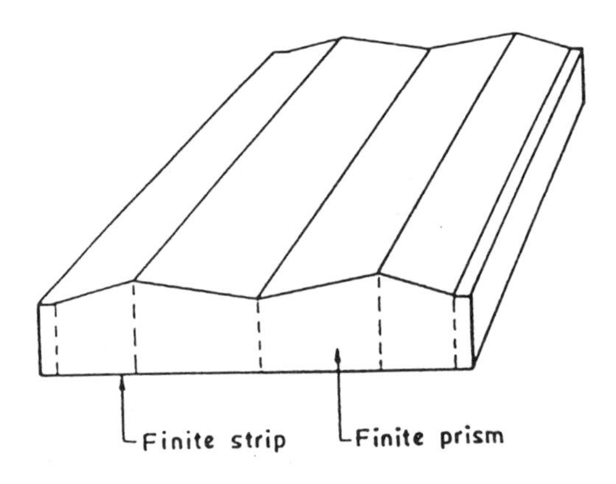

Figure 10.5 Finite-strip-prism model for sandwich panels.

TABLE 10.4
Equations for computation of the characteristic matrices.
a) facing

	Equation/Table	Section
Bending stiffness ${}^{f}K_{mn}^{\ b}$	Eq. 3.14	3.3.1
Inplane stiffness ${}^{f}K_{mn}^{\ p}$	Eq. 4.13	4.3.1
Bending force ${}^{f}F_{m}^{\ b}$	Eq. 3.15	3.3.1
Inplane force ${}^{f}F_{m}^{\ p}$	Eq. 4.15	4.3.1
Bending mass ${}^{f}M_{mn}^{\ b}$	Eq. 5.14	5.2.1.1
Inplane mass ${}^{f}M_{mn}^{\ p}$	Table 5.8	5.2.2
Geometric ${}^{f}K_{mn}^{\ G}$	Eq. 5.33	5.5

b) Core

	Equation	Section
Stiffness ${}^{c}K_{mn}$	Eq. 6.11	6.2.2
Mass ${}^{c}M_{mn}$	Eq. 5.12	5.2

(the superscripts f, c, b and p represent facing, core, bending and inplane respectively)

The matrix equation for stability analysis is

$$ {}^{f}K_{mn}{}^{b}\,{}^{f}\delta_m{}^{b} + {}^{f}K_{mn}{}^{p}\,{}^{f}\delta_m{}^{p} + {}^{c}K_{mn}\,{}^{c}\delta_m + \lambda\,{}^{f}K_{mn}{}^{G}\,\delta_m = 0 \tag{10.15} $$

Furthermore, the free vibration analysis can be carried out by using the equation

$$ {}^{f}K_{mn}{}^{b}\,{}^{f}\delta_m{}^{b} + {}^{f}K_{mn}{}^{p}\,{}^{f}\delta_m{}^{p} + {}^{c}K_{mn}\,{}^{c}\delta_m - \omega^2\,{}^{f}M_{mn}{}^{b}\,{}^{f}\delta_m{}^{b} - \omega^2\,{}^{f}M_{mn}{}^{p}\,{}^{f}\delta_m{}^{p} - \omega^2\,{}^{c}M_{mn}\,{}^{c}\delta_m = 0 \tag{10.16} $$

A comprehensive study, including static, vibration as well as buckling analyses, on the behaviour of a commercial type of formed face sandwich panel was carried out. The dimensions of the panels and the mesh divisions adopted are depicted in Figure 10.6, whereas the material constants are tabulated in Table 10.5. Comparison of the results with other available published results are made in Table 10.6.

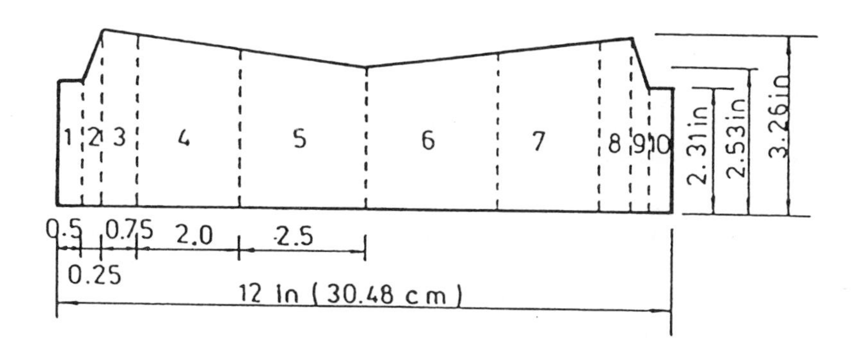

Figure 10.6 Mesh details of formed sandwich panel. (Span = 45in ; loading = 14400psi.)

TABLE 10.5
Material constants of the panel.

Properties of core	Properties of facings
$E_c = 679$ psi $\upsilon_c = 0.177$	$E_f = 29.5\times10^6$ psi $\upsilon_f = 0.267$ $t_f = 0.02$in

TABLE 10.6
Results of coupled model.

Mid-point deflection[12]	Buckling load[12]	Free vibration[12]
Coupled model: 0.176in	Coupled model: 12.1 kip	Coupled model: Mode 1 : 62.9 Mode 2 : 253.7
Harstock[19]: Eq. 57 : 0.179in Eq. 34 : 0.220in	Split rigidity[1]: 9.6 kip	

10.1.1.3 *Finite layer model*

The finite layer model can also be applied to the analysis of sandwich plates which are comprised of layers of uniform thickness. The characteristic matrices of each layer can be formed by substituting the material constants into the appropriate equations. For example, the geometric stiffness matrix for a finite layer model can be obtained by direct substitution of the shape functions into Eq. 5.39. In the case of linear approximation across the depth, the geometric matrix is

$$[K_{mnpq}]^G = \int [G_{mn}]^T [S] [G_{pq}] \, dvol \tag{10.17}$$

where

$$[S] = \begin{bmatrix} \sigma_x^o & \tau_{xy}^o \\ \tau_{xy}^o & \sigma_y^o \end{bmatrix}$$

$$[G_{ij}] = \begin{bmatrix} (1-\frac{z}{c})\cos\frac{i\pi x}{L_x}\sin\frac{j\pi y}{L_y} & (\frac{z}{c})\cos\frac{i\pi x}{L_x}\sin\frac{j\pi y}{L_y} \\ (1-\frac{z}{c})\sin\frac{i\pi x}{L_x}\cos\frac{j\pi y}{L_y} & (\frac{z}{c})\sin\frac{i\pi x}{L_x}\cos\frac{j\pi y}{L_y} \end{bmatrix}$$

Note that for pure axial compression the geometric stiffness matrix is decoupled and the problem is simply reduced to the determination of critical load for each mode shape. As an example, the buckling loads for a square sandwich plate (20inx20in) under different loadings are determined by the finite layer model[16]. The plate consists of three layers with a thick core (0.5in) and two thin facings (0.02in). The Young's modulus of the facing is taken to be $0.3\text{x}10^8$lb/in , whereas the effect of the core stiffness on the plate behaviour was studied by varying the core modulus. The buckling loads due to axial compression and shear loadings are tabulated in Table 10.7.

TABLE 10.7
Buckling loadings of sandwich plates.[16]
a) axial compression

E_f/E_c	Mode (1,1)			Mode (1,3)		
	Cheung[16]	Reference 20	Reference 1	Cheung[16]	Reference 20	Reference 1
100	3942	3817	3808	16370	16060	15930
1000	2684	2682	2645	5700	5773	5613
10000	674	677	653	788	780	751

b) shear loading

E_f/E_c	Cheung[16]	Reference 20	Reference 2
100	909		765
1000	6410	6140	5729
10000	17440	17210	17090

Chong[21] took into account improper curing by modelling the core by three layers: one weaker core sandwiched between two stronger layers (Figure 10.7a). The results of the analysis of a 3mx1m rectangular panel are plotted in Figure 10.7b. The Poisson's ratio and the mean Young's modulus of the core are 0.25 and 5175kN/m . The total thickness of the panel is 7.7cm.

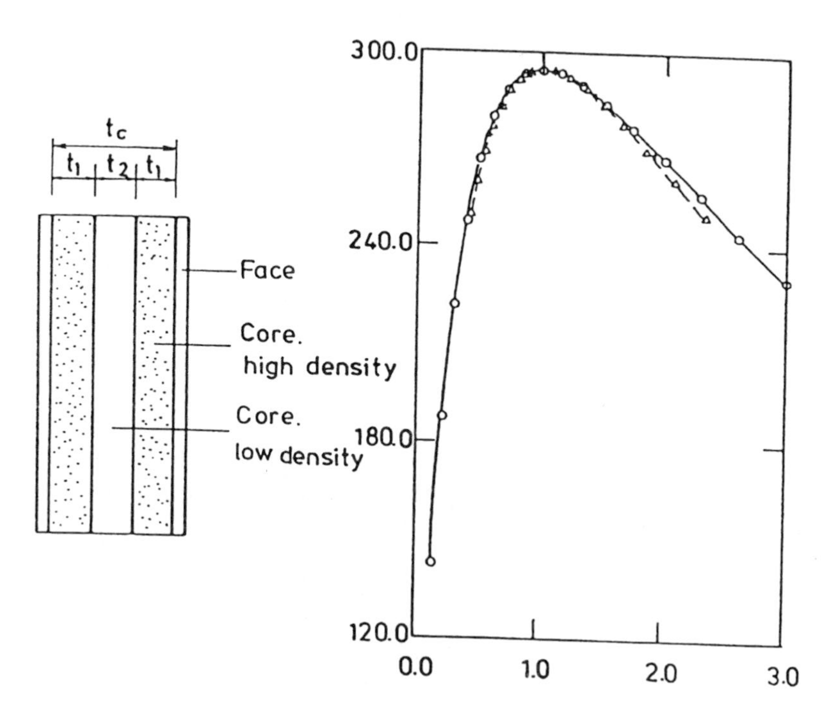

Figure 10.7 (a) Sandwich core with variable thickness. (b) Buckling load of short sandwich panels with different core thickness ratio and variable thickness.[21]

o $t_2/t_1 = 1.0$; △ $t_2/t_1 = 2.0$; + $t_2/t_1 = 3.0$

10.1.2 LAMINATED PLATES

Laminated plates are plates with another form of composite arrangement consisting of thin laminas which are firmly bonded together. The lamina usually consists of stiff fibres embedded in a matrix material. The fibre materials include glass, boron, silicon carbide, carbon graphite, beryllium and steel, etc. The common matrix materials are phenolics, polyesters, aluminum and various epoxies. Therefore, the laminas can be assumed to possess a plane of elastic symmetry parallel to the xy plane, and the axes of material symmetry are also parallel and normal to the fibre direction. Changing the orientation of the laminas results in corresponding changes in strength and stiffness, and, therefore, new materials that can precisely satisfy the requirements can be produced.

Using equivalent moduli for the material constants, the analyses of such plates by the finite strip method were first reported by Hinton[22]. Further studies were reported by Chai.[23] Craig and Dawe[24] predicted the vibration frequencies of rectangular laminated plates. Examples of plates with various support conditions were presented. The dynamic stability of rectangular layered plates due to periodic in-plane loads was studied by Srinivasan et al.[25] On the other hand, Kong[26] proposed a local-global model based on finite prisms for vibration analysis.

10.1.2.1 *Equivalent stiffness model*

On the assumption that each layer is in a state of plane stress, the stress-strain relationships at a general point for the *l*-th layer become:

$$\begin{Bmatrix} \sigma_x \\ \sigma_y \\ \tau_{xy} \end{Bmatrix} = \begin{bmatrix} Q_{11} & Q_{12} & Q_{16} \\ Q_{21} & Q_{22} & Q_{26} \\ Q_{61} & Q_{62} & Q_{66} \end{bmatrix} \begin{Bmatrix} \varepsilon_x \\ \varepsilon_y \\ \gamma_{xy} \end{Bmatrix} \; ; \quad \begin{Bmatrix} \tau_{xz} \\ \tau_{yz} \end{Bmatrix} = \begin{bmatrix} C_{44} & C_{45} \\ C_{54} & C_{55} \end{bmatrix} \begin{Bmatrix} \gamma_{xz} \\ \gamma_{yz} \end{Bmatrix} \qquad (10.18a, b)$$

where C_{ij} (i,j = 4,5) are the ply shear stiffness and

$$Q_{ij} = C_{ij} - \frac{C_{i3} - C_{3j}}{C_{33}} \qquad (i,j = 1,2,6)$$

The constitutive equations for the laminated plate can be obtained by carrying out appropriate integration through the thickness:

$$\begin{Bmatrix} N_x \\ N_y \\ N_{xy} \\ M_x \\ M_y \\ M_{xy} \end{Bmatrix} = \begin{bmatrix} A_{11} & & & & & \\ A_{12} & A_{22} & & & & \\ A_{16} & A_{26} & A_{66} & & & \\ B_{11} & B_{12} & B_{16} & D_{11} & & \\ B_{12} & B_{22} & B_{26} & D_{12} & D_{22} & \\ B_{16} & B_{26} & B_{66} & D_{16} & D_{26} & D_{66} \end{bmatrix} \begin{Bmatrix} \frac{\partial u}{\partial x} \\ \frac{\partial v}{\partial y} \\ \frac{\partial u}{\partial y} + \frac{\partial v}{\partial x} \\ -\frac{\partial^2 w}{\partial x^2} \\ -\frac{\partial^2 w}{\partial y^2} \\ -2\frac{\partial^2 w}{\partial x \partial y} \end{Bmatrix} \qquad (10.19)$$

In the equation, N_x , N_y and N_{xy} are the membrane direct and shearing stress resultants per unit length and M_x , M_y and M_{xy} are the bending and twisting moments per unit length. The stiffness coefficients are defined as

$$A_{ij} = \int_{-h/2}^{h/2} Q_{ij}\, dz \; ; \quad B_{ij} = \int_{-h/2}^{h/2} zQ_{ij}\, dz \; ; \quad D_{ij} = \int_{-h/2}^{h/2} z^2 Q_{ij}\, dz$$

The shear constitutive relations for the plate are also given in terms of shear forces as

$$\begin{Bmatrix} Q_x \\ Q_y \end{Bmatrix} = [S] \begin{Bmatrix} \gamma_{xz} \\ \gamma_{yz} \end{Bmatrix} \tag{10.20}$$

where $[S] = \begin{bmatrix} K_1^2 A_{44} & K_1 K_2 A_{45} \\ K_1 K_2 A_{45} & K_2^2 A_{55} \end{bmatrix}$ and K_1 and K_2 are shear correction factors to allow for the fact that the transverse shear stresses are not constant in each ply of the composite laminate.

10.1.2.2 *Examples*

Anisotropic square plates were studied by Craig et al.[24] The plates were fully clamped and the material constants are as follows:

$$E_L/E_T = 10 \quad ; \quad G_{LT}/E_T = G_{TT}/E_T = 0.25 \quad ; \quad \upsilon_{LT} = 0.3.$$

The subscripts 'L' and 'T' denote the longitudinal and transverse directions respectively. The shear correction factor was taken as 5/6. Two plate thicknesses were investigated and the results were tabulated in Table 10.8.

Two symmetric cross-ply square laminated plates with three and nine layers of composite materials were subjected to uniaxial constant stresses.[26] The lamina properties and the geometry of the plates were assumed to be

$$E_{11} = 40\, E_{22} = 40\, E_{33} \quad ; \quad G_{12} = G_{13} = 0.6E_{22} \quad ; \quad G_{23} = 0.5E_{22}$$

$$\upsilon_{12} = \upsilon_{13} = \upsilon_{23} = 0.25 \quad ; \quad \text{span-to-thickness ratio} = 10$$

The plates were simply supported on all edges. The buckling stresses are listed in Table 10.9. The results of the elasticity solution is obtained by a finite difference scheme developed by Noor.[28]

TABLE 10.8

Frequencies of anisotropic plates.[24]

No of strips	No of series terms	$p\frac{A^2}{t}\{\frac{12(1-\upsilon_{LT}\upsilon_{TL})}{E_L}\}^{1/2}$					
		t/A = 0.1			t/A = 0.01		
		Mode 1	Mode 2	Mode 3	Mode 1	Mode 2	Mode 3
2	1	15.74	24.72	36.73	23.34	38.98	63.85
	2	14.70	22.90	28.89	21.59	34.11	55.24
	3	14.50	22.50	27.24	21.39	33.49	52.40
	4	14.39	22.40	26.98	21.34	33.46	51.70
	5	14.32	22.29	26.74	21.31	33.38	51.59
3	1	15.74	24.71	36.46	23.32	38.95	62.15
	2	14.70	22.89	28.89	21.52	33.81	53.41
	3	14.50	22.48	27.24	21.31	33.12	50.76
	4	14.39	22.39	26.97	21.26	33.09	50.57
	5	14.31	22.27	26.73	21.23	33.01	50.47
Reference 27		21.35	33.18	50.72	21.35	33.18	50.72

TABLE 10.9
Critical buckling coefficients for simply supported square laminated plates ($\overline{N} = \sigma_x A^2/E_2h^2$ with A/h = 10).[26]

Number of layers	Spline finite strip	Elasticity solution[28]
3	23.3430	22.8807
9	25.7917	25.3436

Chai et al.[23] had carried out analysis to optimise the ply angle of antisymmetrical composite laminated plates subjected to in-plane compression. The results of their analysis are depicted in Figures 10.8. They also published results for other support conditions and loading cases.

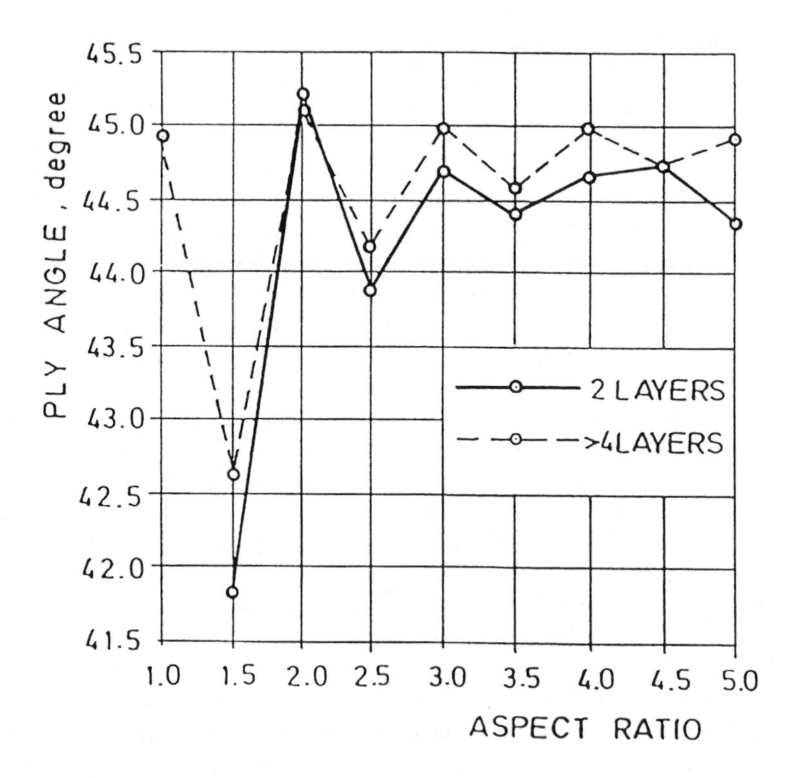

Figure 10.8 Optimum ply angle versus aspect ratio for SSSS antisymmetrically laminated angle-ply plate.[23]

10.2 GEOTECHNICAL PROBLEMS

10.2.1 INTRODUCTION

The behaviour of soils under loadings is another area that one may find interesting applications of the finite strip method. It is understood that the behaviour of soil is very complicated and advanced constitutive models will be required for its exact description. Nevertheless, the high computational cost that is normally incurred in the solution process has greatly limited their applications. Therefore, certain approximations may have to be made to simplify the analyses. One of the approaches is to model the soils as elastic materials of which the deformation properties can be described by elastic moduli and Poisson's ratios. The soil is further assumed to be composed of a number of layers with deformation properties which are uniform in both horizontal directions. These properties can however vary along the depth. Such a model, although an approximate one, is able to provide reasonably good predictions of the soil behaviour. It is obvious that this model can allow, with only slight modifications, the application of the principles discussed in the previous chapters for the analysis.

The finite prism model was first employed to analyze a layered pavement system.[29,30] The tyre print was assumed to be of rectangular shape and the pressure had a parabolic distribution. The subsoil was modelled as a number of eight-noded prisms of length much greater than the length of the tyre print. Since the ends of the prisms were assumed to be simply supported, the solution was reduced to a two-dimensional one. Cheung and Fan[31] as well as Fan[32] then attempted solutions for soils under surface patch loadings by a number of rectangular finite layers such that each layer can be assigned an elastic modulus and Poisson's ratio. They assumed that the layers were simply supported along their edges and further reduced the solution into a one-dimensional one. However, loadings are not applied in many cases directly to the soil but through footings having considerable stiffness which cannot be ignored in the analyses. It is, therefore, necessary that a solution which is able to reflect the contribution of the footing as well must be devised. Tham and Cheung[33] studied the behaviour of strip footings resting on layered soil by the finite strip-layer model. Parikh and Pathak[34] developed a finite element-finite layer model for the analysis of plates on elastic foundations. Based on the flexibility approach,[35] a solution was offered by Tham,[36] Cheung[37] and Man.[38] The flexibility matrix of the soil layer was first established and it was then inverted to obtain the stiffness matrix. This stiffness matrix was then added to the stiffness matrix of the footing to form the total stiffness matrix of the whole system. The solution was then obtained by following the usual procedure. Examples including rectangular and circular footings were reported. In case of circular footings, cylindrical layers were used.

Analysis of piles embedded in layered soils has also been tackled in a similar manner by Cheung and his collaborators.[39,40,41,42,43] They employed the infinite layer model to prepare design charts for the estimation of the displacements as well as stresses (moments) of single piles embedded in layered soils. The soil layers were modelled by layers with infinite extent in the horizontal directions and the piles by beam elements. Interaction between two similar piles was also studied. The interaction coefficients were computed by an iterative scheme. The method was further extended to include the contribution of the pile cap. An iterative procedure was also developed by Guo[44] to take into consideration the slip along the interface between pile and soil.

Consolidation of soils, which is the process of dissipation of pore water present within the soil matrix, can also be analyzed by the finite strip model. An advanced theory which couples the 'effective' stress-strain and flow relations was widely accepted for the description of the whole process. It is usually referred to as the Biot's consolidation theory.[45] In the theory, the stresses and strains are related according to Hooke's Law based on the effective material constants whereas the flow is assumed to obey Darcy's Law.[46] The governing equations thus derived include equilibrium relations of the effective stresses and the flow continuity relations. Based on this theory, various numerical schemes had been proposed to solve such problems and the finite strip solutions were presented by Booker[47,48] (transformation approach) and Cheung (separation of variable approach).[49,50]

Other areas of applications were also reported. They include the analyses of the dynamical behaviour of soil[51,52,53,54,55] as well as field problems.[56,57]

10.2.2 SOILS UNDER SURFACE LOADINGS

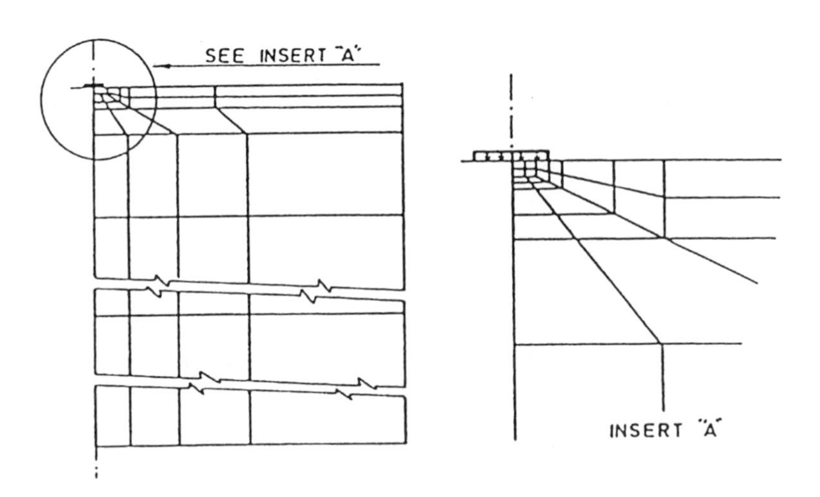

Figure 10.9 Finite prism mesh for isotropic half-space problem.[29]

The finite prism model was used by Cheung et al.[30] in the analysis of a three layer road pavement subject to two passing single axles with dual tyres, and using the mesh shown in Figure 10.9. The stiffness matrix of a typical prism is given by Eq. 6.11. Because of symmetry, calculations were made for only one half of the pavement.

Each wheel load was considered to act as a parabolic distribution over a rectangular tyre print with a length equal to twice the width. The thickness of each layer was arbitrarily taken as equal to the tyre width. Since it was intended to examine the effects of a single edge load a fine mesh was used directly under the outer wheel load while a coarse mesh was used elsewhere (Figure 10.10).

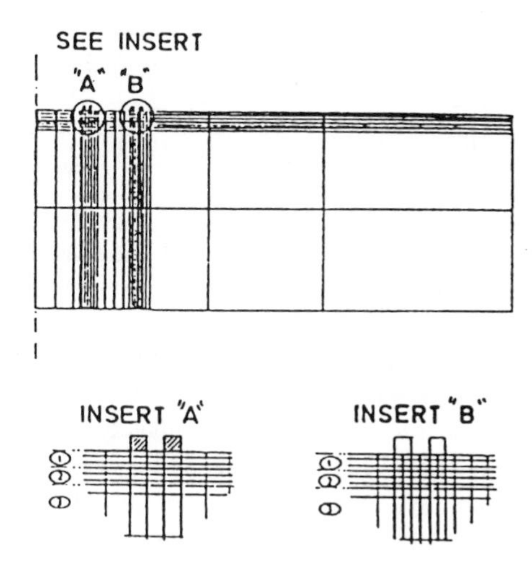

Figure 10.10 Finite prism mesh for road pavement.[30]

The following parameters were used in the study:

(1) Three sets of moduli ratios in the upper layers with the values 2:1:0.5, 5:1:0.2 and 10:1:0.1.

(2) Two sets of loadings with the wheels very far away from the edge in one case and the outer wheel right by the edge in the other.

In Figure 10.11 contours of vertical stress were plotted for two sets of modular ratios 2:1:0.5 and 10:1:0.1 under both sets of loadings. Thirty-one non-zero terms have been used in the analysis.

A layered soil can also be modelled by the finite layer method.[31,32] The subsoil is imagined to be divided into a number of layers which are assumed to be simply supported along the vertical boundaries. The boundaries are placed sufficiently far away from the loading area such that their effect on the stress distribution in the region near the load would be minimal. It is obvious that the stiffness matrix and load vector for each layer can be derived by the method as outlined in Section 6.3.1.

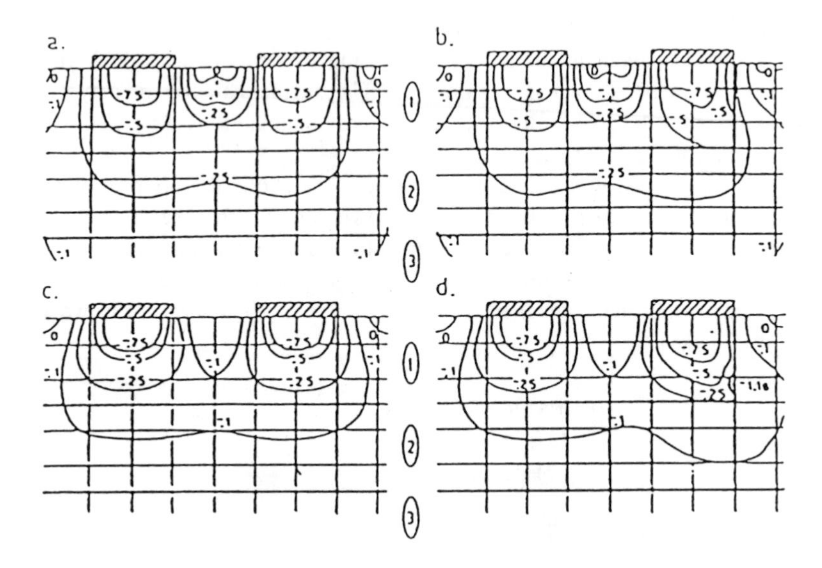

Figure 10.11 Vertical stress under wheel loads. (a) 2:1:0.5 wheels far from edge. (b) 2:1:0.5 outer wheel on edge. (c) 10:1:0.1 wheels far from edge. (d) 10:1:0.1 outer wheel on edge.[30]

A typical two-layered system was studied by Cheung and Fan.[31,32] The modular ratio for the layers was 5:1 and the Poisson's ratio was taken as 0.35 for both layers. The circular loading was approximated as an equivalent square patch load. The meshes employed in the analysis are depicted in Figure 10.12. Comparison of the finite layer model results with other published results[29,58] on vertical stresses depicted in Figure 10.13 demonstrates the accuracy of the method.

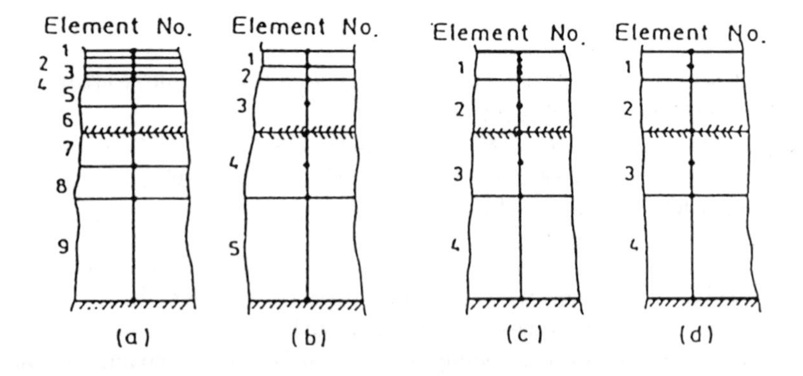

Figure 10.12 Meshes for the finite layer analysis of two-layered soil.

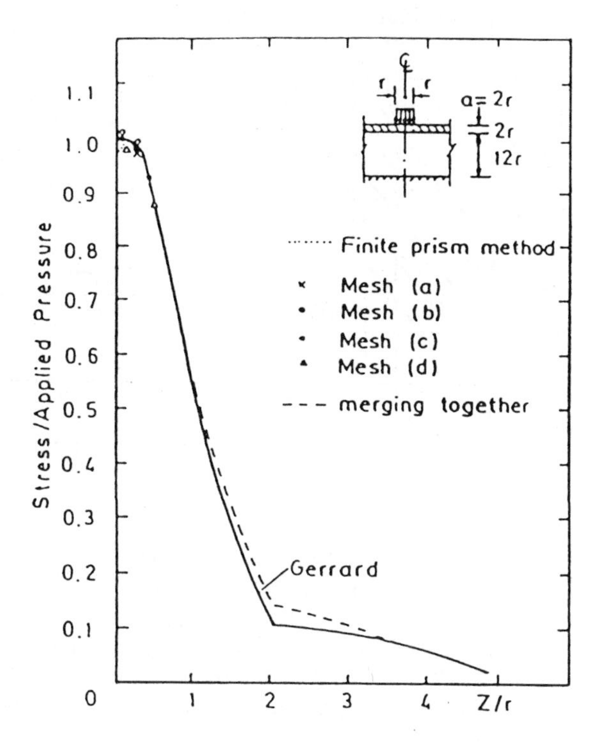

Figure 10.13 Comparison of vertical stresses for different mesh divisions.[31,32]

Combining with the approach developed by Cheung and Zienkiewicz[35] in the mid-sixties for the analysis of a plate on homogeneous soil, the finite layer model can be employed to analyze footings resting on layered soils. The footing is sub-divided into a regular rectangular mesh with elements of size 'axb'. Instead of using the Boussinesq equations,[60] the flexibility coefficients of the layered soil are obtained by applying a patch load of unit intensity over an area 'axb' at the surface of the subsoils (Figure 10.14). The loading is supposed to be applied on each node in turn to generate the flexibility matrix, but we can easily realise that it is unnecessary as the coefficients possess translation property(Man[38]), that is

$$f_{ij}^{mn} = f_{oo}^{i-m,j-n}$$

where f_{ij}^{mn} is the flexibility coefficient between nodes with nodal numbering (i,j) and (m,n) (Figure 10.14). Therefore, a single analysis by applying the load at the centre of the footing will be sufficient to generate all the coefficients.

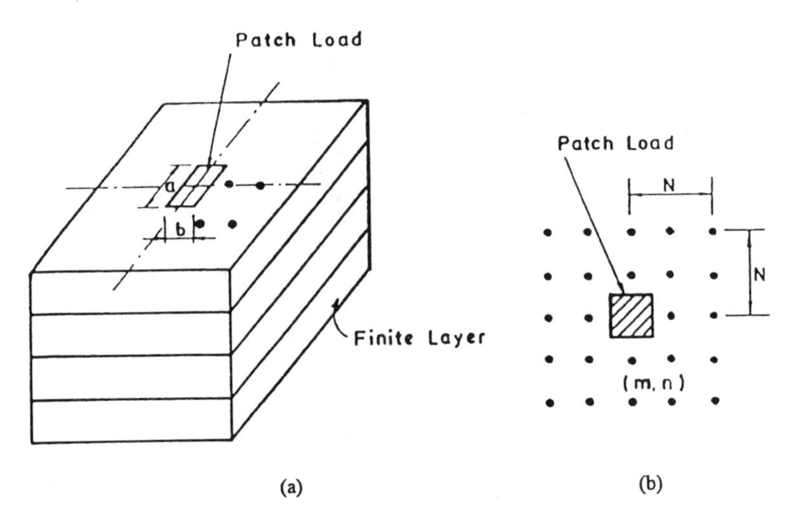

Figure 10.14 (a) The patch load on layered soil. (b) Pattern for coupling of the soil and footing.

The flexibility matrix can thus be formed and inverted to obtain the stiffness matrix. The total stiffness matrix of the whole system can be obtained by combining the plate and soil stiffness matrices, and the solution can then be sought in the usual way.

It is obvious that the best way to present the solutions is to express the settlement in terms of dimensionless influence factors. In the case of a homogeneous soil, the settlement of the footing (s_h) is defined as[36]:

$$s_h = \frac{P}{BE} I_o I_\upsilon I_\gamma$$

where E is the elastic modulus of the soil layer,
B is the width of the footing,
P is the total applied load,
I_o is the settlement influence factor of rigid footing for soil of zero Poisson's ratio (Figure 10.15a),
I_υ is the correction factor of Poisson's ratio (Figure 10.15b),
I_γ is the correction factor for the footing rigidity (Figure 10.15c).

The rigidity of the footing is given as :

$$\gamma = \log\left[\frac{1000E_p(1-\upsilon_s^2)t^3}{\pi E_s(1-\upsilon_p^2)B^3}\right]$$

where t is the thickness of the footing,
E_p and E_s are the elastic moduli of the footing and soil respectively,
υ_p and υ_s are the Poisson's ratios of the footing and soil respectively.

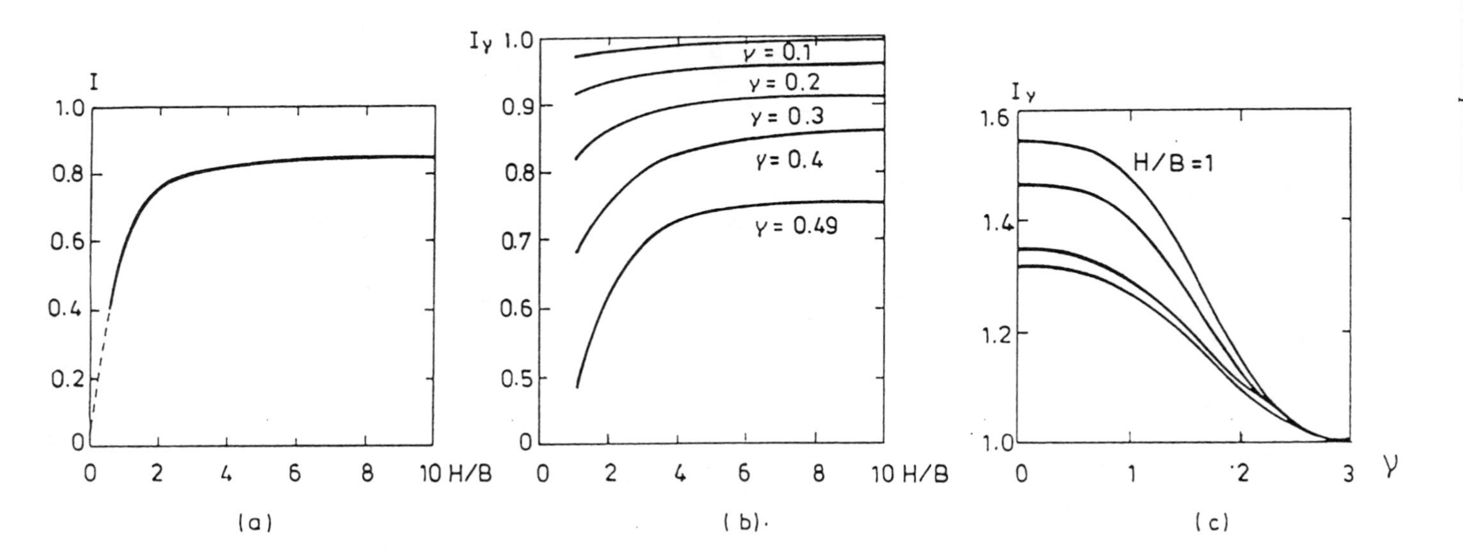

Figure 10.15 (a) Settlement influence factor for rigid square footing under uniform load. (b) Correction factor for Poisson's ratio. (c) Correction factor for footing rigidity.[36]

Circular footings can also be analyzed by a similar method.[37,38] In this case, it is more appropriate to use the infinite layer. The footing was divided into a number of concentric rings, and the stiffness matrix for the footing was established accordingly. The solution procedure was similar to that of the rectangular footing, except that the flexibility coefficients for the subsoil were obtained by applying in turn a unit pressure on each ring. Similar to the rectangular footing, the settlement is defined as:[37,38]

$$s_h = \frac{P}{DE} I_o I_\upsilon I_\gamma$$

where E is the elastic modulus of the soil layer,
D is the diameter of the footing,
P is the total applied load,
I_o, I_υ and I_γ are the settlement influence factors and correction factors.

Figure 10.16 shows the influence factors for computing the settlement of a circular footing.

10.2.3 ANALYSIS OF PILES BY INFINITE LAYER MODEL

Guo[44] conducted a study on the application of the infinite layer model (Section 2.2.1.4) in pile analysis. Solid bar elements with displacement shape functions taking the same form as those for the infinite layer are used to model the pile (Figure 10.17).

In order to satisfy the rigid section assumption, the displacement functions are chosen as follows:

$$\begin{Bmatrix} u_p \\ v_p \\ w_p \end{Bmatrix} = \sum_{m=0}^{M} \begin{bmatrix} N_{pm1} & N_{pm2} & N_{pm3} \end{bmatrix} \begin{Bmatrix} \delta_{pm1} \\ \delta_{pm2} \\ \delta_{pm3} \end{Bmatrix}$$

in which

$$N_{pmi} = \begin{bmatrix} Z_i^u H_p^u \cos m\theta & 0 & 0 \\ 0 & Z_i^v H_p^v \sin m\theta & 0 \\ 0 & 0 & Z_i^w H_p^w \cos m\theta \end{bmatrix}$$

$$\delta_{pmi}{}^T = \begin{bmatrix} \delta_{pmi}^u & \delta_{pmi}^v & \delta_{pmi}^w \end{bmatrix}$$

$$H_p^u(r) = H_p^v(r) = \begin{cases} r/r_o & m = 0 \\ 1 & m \neq 0 \end{cases} \; ; \; H_p^w(r) = \begin{cases} r/r_o & m = 0 \\ 1 & m \neq 0 \end{cases}$$

Note that only one term is used in the radial shape function.

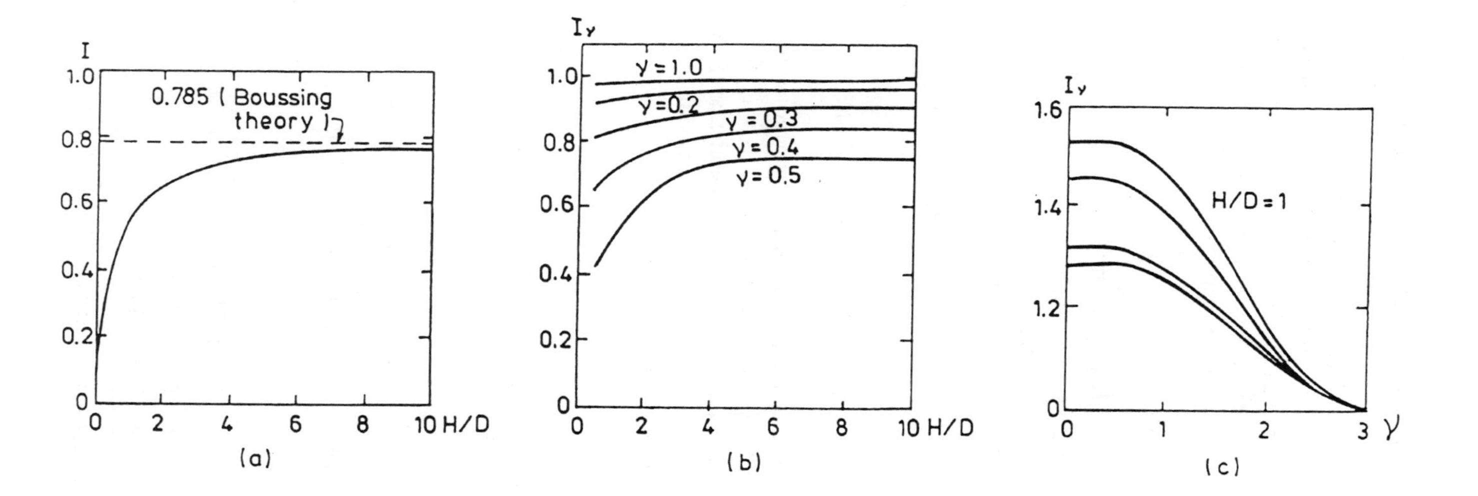

Figure 10.16 (a) Settlement influence factor for rigid circular footing under uniform load. (b) Correction factor for Poisson's ratio. (c) Correction factor for footing rigidity. ($\gamma = \log \left[\frac{1000E_p(1-\upsilon_s^2)t^3}{\pi E_s(1-\upsilon_p^2)D^3} \right]$ where t is the thickness of the footing)[37,38]

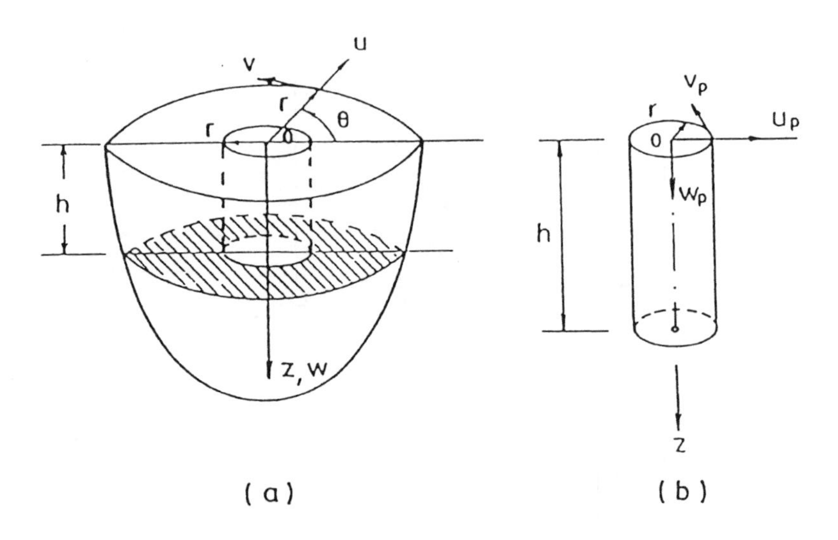

Figure 10.17 (a) Infinite layer element in cylindrical coordinate system. (b) Bar element.

As in the case of surface loading on layered soil (Section 10.2.2), the settlement of the pile under an axial load is expressed in terms of an influence factor (I_A):

$$\delta_A = \frac{P}{LE_s} I_A$$

where δ_A = settlement of the pile head
P = applied axial load
L = pile length
E_s = soil elastic modulus
I_A = influence factor

The influence factor (Figure 10.18) depends on the slenderness ratio, pile stiffness factor, the stiffness of the underlying layer as well as the thickness of the deposit. Note that the pile stiffness factor is defined as:

$$K_A = \frac{E_p A_p}{E_s \pi d^2 / 4}$$

where K_A = axial stiffness factor of the pile,
E_p = elastic modulus of the pile, and
A_p = area of the pile section.

Having the pile stiffness factor determined, the influence factor can be obtained from Figure 10.18 accordingly.

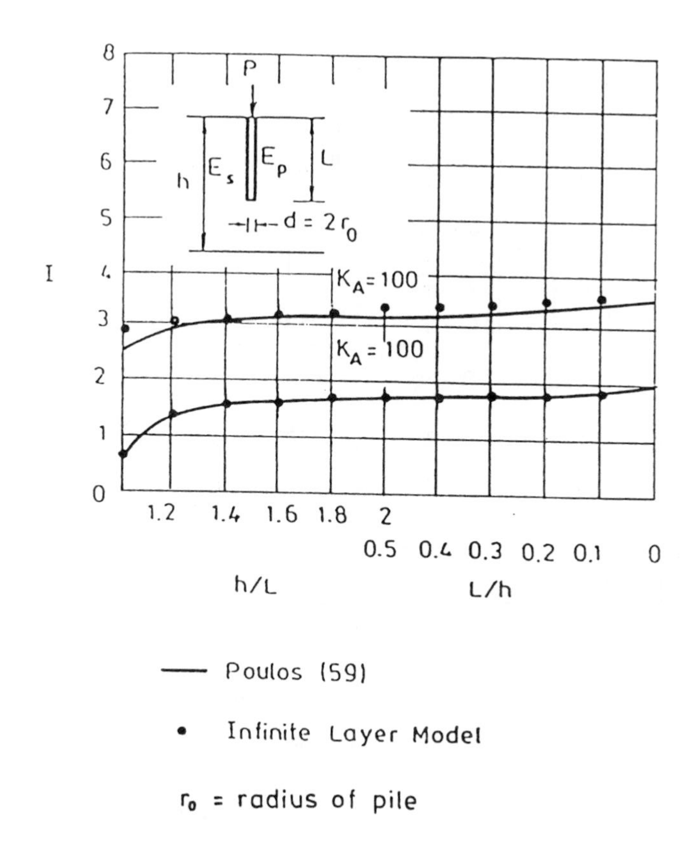

Figure 10.18 Influence factor of pile embedded in soil of thickness h.[44]

An iterative procedure was also suggested by Guo[44] to compute the interaction factor between two similar piles, and the results of some typical cases are depicted in Figure 10.19. The factor is considered to be the most important parameter for the analysis of pile groups.

It can be shown readily that similar expressions can also be written for horizontal as well as torsional loadings. Similar design curves were presented by Guo.[44]

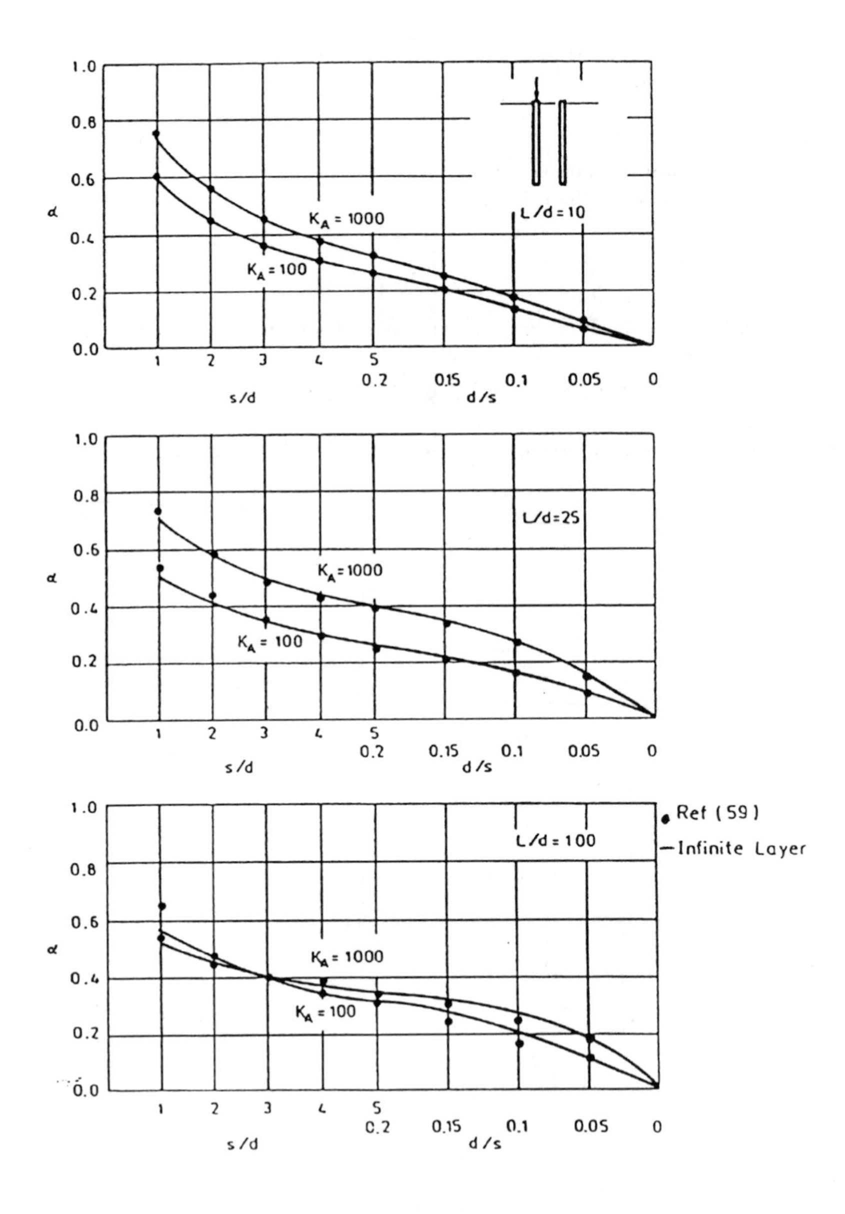

Figure 10.19 Axial interaction factor for piles embedded in homogeneous soil.[44]

A detailed study was conducted to investigate the effect of the pile cap rigidity on the behaviour of pile groups[44] with the pile caps being modelled by finite elements. The cap rigidity is defined as

$$K_R = \frac{E_c t^3}{12(1-\upsilon_c^2)s^2} \frac{\Delta_{single}}{P_{ave}}$$

where E_c = modulus of the pile cap
υ_c = Poisson's ratio of the pile cap
t = thickness of the pile cap
s = pile spacing
Δ_{single} = settlement of a single pile
P_{ave} = average load per pile.

The piles were embedded in a two-layered soil and arranged in a 3x3 pattern at a spacing with spacing/diameter ratio of 3 (Figure 10.20). The pile stiffness with reference to the stiffness of the top soil layer was 1000. A uniformly distributed load was applied on the surface of the pile cap. The loads carried by the piles and their settlements are plotted in Figures 10.21 and 10.22.

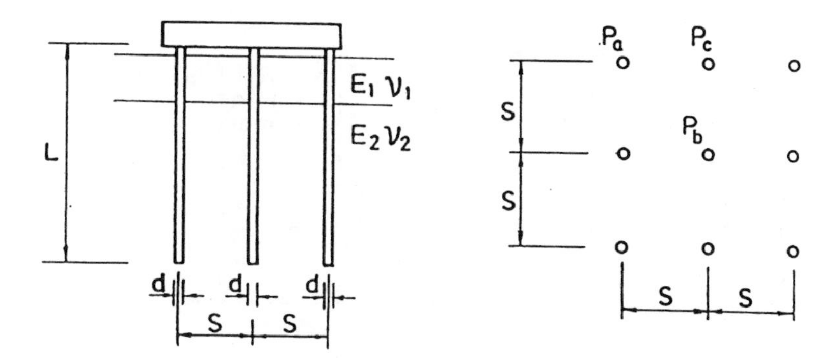

Figure 10.20 3x3 pile group embedded in a two-layered soil.
($h_1/L = 0.4$; $E_2/E_1=5$; $\upsilon_s = 0.35$; $L/d = 20$; $s/d = 3$)

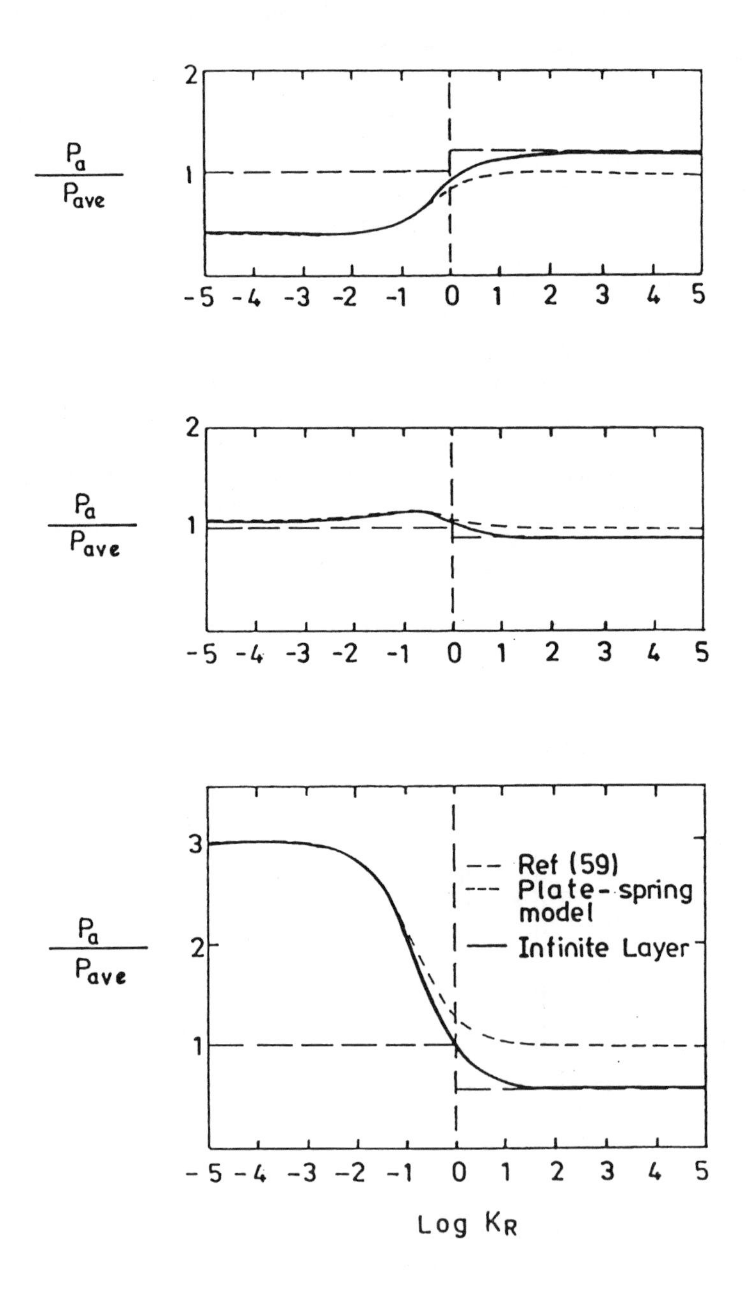

Figure 10.21 Pile load distribution for a 3x3 pile group embedded in a two-layered soil.[44]

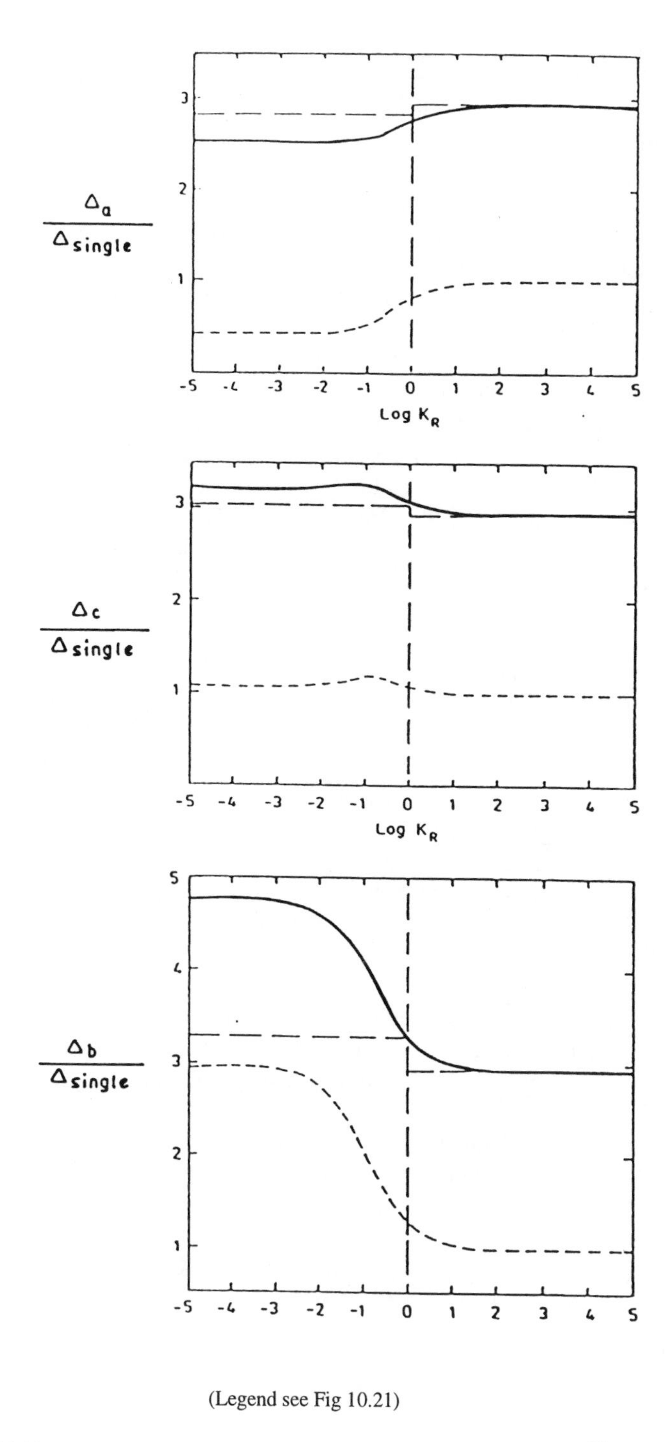

(Legend see Fig 10.21)

Figure 10.22 Settlement for a 3x3 pile group embedded in a two-layered soil.[44]

REFERENCES

1. Allen, H. G., *Analysis and design of structural sandwich panels*, Pergamon Press, London, 1969.
2. Plantema, F. J., *Sandwich construction*, John Wiley, New York, 1966.
3. Lungren, H. R. and Salama, A. E., Buckling of multilayer plates by finite elements, *J of Eng Mech, ASCE*, 97(2), 477-494, 1971.
4. Liaw, B. D. and Little, R. W., Theory of bending multilayer sandwich plates, *J AIAA*, 5(2), 301-304, 1967.
5. Kao, J. S. and Ross, R. J., Bending of multilayer sandwich beams, *J AIAA*, 6(8), 961, 1968.
6. Khatua, T. P. and Cheung, Y. K., Bending and vibration of multilayer sandwich beams and plates, *Int J for Num Meth in Eng*, 6, 11-24, 1973.
7. Khatua, T. P. and Cheung, Y. K., Triangular element for multilayer sandwich plates, *J of Eng Mech, ASCE*, 98(5), 1225-1238, 1972.
8. Chan, H. C. and Cheung, Y. K., Static and dynamic analysis of multilayered sandwich plates, *Int J Mech Sci*, 14, 399-406, 1972.
9. Ibrahim, I. M. and Monforton, G. R., Finite strip laminated sandwich roof analysis, *J of Struct Eng, ASCE*, 105(5), 905-919, 1979.
10. Chan, H. C. and Foo, O., Buckling of multi-layer sandwich plates by the finite strip method, *Int J Mech Sci*, 19, 447-456, 1977.
11. Tham, L. G., Chong, K. P. and Cheung, Y. K., Flexural bending and axial compression of architectural sandwich panels by combined finite-prism-strip method, *J Reinforced Plastic and Composites*, 1, 16-28, 1982.
12. Cheung, Y. K., Chong, K. P. and Tham, L. G., Vibration, bending and compression of formed sandwich panels, *Proc Int Conf on Finite Element*, Shanghai, China, 126-131, 1982.
13. Tham, L. G., Chong, K. P. and Cheung, Y. K., Thermal and buckling behaviour of sandwich panels by finite-prism, finite-strip and finite-layer method, *Proc Int Conf on Finite Element*, Shanghai, China, 797-802, 1982.
14. Chong, K. P., Cheung, Y. K. and Tham, L. G., Free vibration of formed sandwich panel, *J of Sound and Vib*, 81(4), 575-582, 1982.
15. Chong, K. P., Tham, L. G. and Cheung, Y. K., Thermal behaviour of formed sandwich plate by finite-prism-strip method, *Comp and Struct*, 15(3), 321-324, 1982.
16. Cheung, Y. K., Tham, L. G. and Chong, K. P., Buckling of sandwich plate by finite layer method, *Comp and Struct*, 15(2), 131-134, 1982.
17. Cheung, Y. K., *Finite strip method in structural analysis*, Pergamon Press, 1976.
18. Kasemset, C., Cheung, Y. K., Khatua, T. P., Y. K., Curved multi-layered element for axisymmetric shells, *J of Eng Mech, ASCE*, 103(1), 139-151, 1972.

19. Harstock, J. A. and Chong, K. P., Analysis of sandwich panels with formed faces, *J of Struct Eng*, ASCE, 1976.
20. Academia Sinica, Solid Mechanics Group, Beijing, China, *Bending, stability and vibration of sandwich plates*, 1977 (in Chinese).
21. Chong, K. P., Lee, B. and Lavdas, P. A., Analysis of thin-walled structures by finite strip and finite layer methods, *Thin-walled Struct*, 2(1), 1984.
22. Hinton, E., Flexural of composite laminates using the thick finite strip method, *Comp and Struct*, 7, 217-220,
23. Chai, G. B., Ooi, K. T. and Khong, P. W., Buckling strength optimisation of laminated composite plates, *Comp and Struct*, 46(1), 77-82, 1993.
24. Craig, T. J. and Dawe, D. J., Flexural vibration of symmetrically laminated composite rectangular plates including transverse shear effects, *Int J Solids Struct*, 22(2), 155-169, 1986.
25. Srinivasan, R. S. and Chellapandi, P., Dynamic stability of rectangular laminated composite plates, *Comp and Struct*, 24(2), 233-238, 1986.
26. Kong, J., *Analysis of plate type structures by finite strip, finite prism and finite layer methods*, PhD thesis, University of Hong Kong, Department of Civil and Structural Engineering, 1994.
27. Ashton, J. E. and Whitney, J. M., *Theory of laminated plates*, Technomic, Stanford, 1970.
28. Noor, A. K., Stability of multi-layered composite plates, *Fibre Sci Tech*, 8, 81-89, 1975.
29. Cheung, Y. K., Numerical analysis of pavements, *Proc of Symp on Recent Development in the Analysis of Soil Behaviour and their Application to Geotechnical Structures*, Univ of New South Wales, July, 1976.
30. Cheung, Y. K., Yeo, M. F. and Cumming, D. A., Three dimensional analysis of flexible pavements with special reference to edge loads, *Proc of the First Int Conf of the Road Eng Assoc of Asia and Australia, Bangkok, 1976*, 468-483, 1976.
31. Cheung, Y. K. and Fan, S. C., Analysis of pavements and layered foundations by finite layer method, *Proc of the Third Int Conf on Num Methods in Geomechanics, Aachen, Germany, 2-6 April, 1979*, 1129-1135, 1979.
32. Fan, S. C., *Displacements and stresses analyses in layered elastic systems: a finite layer approach*, BSc (Eng) final year project, Department of Civil Engineering, University of Hong Kong, 1978.
33. Tham, L. G. and Cheung, Y. K., Infinite rigid pavement on layered foundation, *Proc of the Eighth Canadian Congress of Applied Mech, Moncton, 7-12 June, 1981*, 875-876, 1981.

34. Parikh, S. K. and Pathak, S. P., A coupled finite element finite layer formulation for analysis of foundation problems, *Proc of the Fifth Int Conf on Num Methods in Geomechanics, Nagoya, Japan, 1-5 April, 1985*, 221-228, 1985.
35. Cheung, Y. K. and Zienkiewicz, O. C., Plates and tanks on elastic foundation: an application of finite element method, *Int J of Solids and Struct*, Vol 1, 451-461, 1965.
36. Tham, L. G., Man, K. F. and Cheung, Y. K., Analysis of footing resting on non-homogeneous soil by double spline method, *Comp and Geot*, 5, 249-268, 1988.
37. Cheung, Y. K., Tham, L. G. and Man, K. F., Analysis of circular footing resting on layered soil, *Proc of the 2nd East Asia-Pacific Conf on Struct Eng and Construction*, Jan 11-13, 1989, Chiang Mai, Thailand, 715-720, 1989.
38. Man, K. F., *Elastic solution for rectangular and circular plates on non-homogeneous soil foundation*, MPhil Thesis, Department of Civil and Structural Engineering, University of Hong Kong, 1988.
39. Cheung, Y. K., Tham, L. G. and Guo, D. J., Applications of finite strip and layer methods in mirco-computers, *Proc of the Fifth Int Conf on Num Methods in Geomechanics, Nagoya, Japan, 1-5 April, 1985*, 1755-1762, 1985.
40. Cheung, Y. K., Guo, D. J., Tham, L. G. and Cao, Z. Y., Infinite ring element, *Int J of Num Meth for Eng*, 23, 385-396, 1986.
41. Guo, D. J., Tham, L. G. and Cheung, Y. K., Infinite layer for the analysis of a single pile, *Comp and Geot*, 3, 229-249, 1987.
42. Tham, L. G., Cheung, Y. K. and Guo, D. J., Analysis of pile by infinite layer method, *Proc of 4th Int Conf on Tall Buildings, Hong Kong and Shanghai*, 372-376, 1988.
43. Cheung, Y. K., Tham, L. G. and Guo, D. J., Analysis of pile group by infinite layer method, *Geotechnique*, 38(3), 415, 431, 1988.
44. Guo, D. J., *Infinite layer method and its application to the analysis of pile systems*, PhD thesis, Department of Civil and Structural Engineering, University of Hong Kong, 1988.
45. Biot, M. A., General theory of three-dimensional consolidation, *J of Applied Physics*, 12, 155-164, 1941.
46. Darcy, H., *Les fontaines publiques de la ville de Dijon*, Dalmont, Paris, 1856.
47. Booker, J. R. and Small, J. C., Finite layer analysis of consolidation, Part 1, *Int J for Num and Analy Meth in Geom, 6, 151-172, 1982.*
48. Booker, J. R. and Small, J. C., Finite layer analysis of consolidation, Part 2, *Int J for Num and Analy Meth in Geom, 6, 173-194, 1982.*

49. Tham, L. G. and Cheung, Y. K., Consolidation of layered soil by least square finite layered method, *Proc of the Eighth Canadian Congress of Applied Mech, Moncton, 7-12 June, 1981*, 843-844, 1981.
50. Cheung, Y. K. and Tham, L. G., Numerical solutions for Biot's consolidation of layered soil, *J of Eng Mech, ASCE*, 109(3), 669-679, 1983.
51. Cao, Z. and Cheung, Y. K., Dynamic analysis of prismatic structures embedded in infinite soil medium, *Proc of the Fifth Int Conf on Num Methods in Geomechanics, Nagoya, Japan, 1-5 April, 1985*, 1441-1448, 1985.
52. Cheung, Y. K., Cao, Z. and Wu, S. Y., Dynamic analysis of prismatic structures surrounded by an infinite fluid medium, *Earthquake Eng and Struct Dyn*, 13, 351-360, 1985.
53. Cao, Z., Cheung, Y. K. and Tong, Z., A semi-analytical method for structural-external fluid dynamic interaction problems, *Acta Mechanica Sinica*, 17(5), 389-399, 1985 (in Chinese).
54. Cao, Z. and Tong, Z., The semi-analytical element method for dynamic interaction problems in buried structures, *J of Civil Engineering*, 19(3), 11-26, 1986 (in Chinese).
55. Cheung, Y. K. and Zhu, J. X., Dynamic interaction analysis of a circular cylindrical shell of finite length in a half-space, *Earthquake Eng and Struct Dyn*, 21, 799-809, 1992.
56. Chakrabarti, S., Heat conduction in plates by finite strip method, *J of Eng Mech, ASCE*, 106(2), 233-244, 1980.
57. Tham, L. G., *Numerical solutions for time-dependent problems*, PhD Thesis, Department of Civil Engineering, University of Hong Kong, 1981.
58. Gerrard, C. M., Tables of stresses, strains and displacements in two layer elastic systems under various traffic loads, *Special Report No 3*, Australian Road Research Board, 1969.
59. Poulos, H. G. and Davis, E. H., *Pile foundation analysis and design*, John Wiley & Sons, 1980.
60. Boussinesq, J., *Application des potentials a l'Etude de l'Equilibre at du mouvement des solids elastique*, Gauthier-Villars, Paris.

CHAPTER ELEVEN

NON-LINEAR ANALYSIS

11.1 INTRODUCTION

The problems discussed in the previous chapters are described by linear governing equations. In the context of solid mechanics, it implies that the second order terms of the strain-displacement can be ignored and (or) the material properties are independent of the stresses. In field problems, the permeabilities are independent of the potential.

However, in reality most problems are to a more or less extent of a non-linear nature. For example, a non-linear material response can result from elasto-plastic material behaviour such that the property matrix is no longer constant but will be given in terms of the strains (displacements). As a result, the stiffness matrices of the non-linear problem will depend on the displacements/strains. It is obvious that the solution of such a problem cannot be obtained by simply inverting the stiffness matrix. On the other hand, iterative means and special algorithms have to be developed for solving such problems. The applications of the finite strip method to non-linear problems were first attempted in the late eighties and since then the developments have been quite remarkable. For such non-linear problems, the finite strip method has a definite advantage over the finite element method because of its smaller number of degrees of freedom involved and consequently the much smaller amount of computational efforts.

11.2 SOLUTION FOR NON-LINEAR SYSTEM

The equilibrium equation for a system exhibiting non-linear behaviour can be written as:

$$K(\delta)\,\delta + f = 0 \qquad (11.1)$$

In most cases, the non-linear term is the stiffness matrix of the system. However, there are cases that the forces acting on the system also depend on the displacements, leading to an additional degree of non-linearity.

While the solution of a linear system can be accomplished without difficulty in a direct manner, this is not possible for non-linear systems. A variety of solution schemes have been developed to solve such problems. As in-depth reviews of such schemes can be found in many articles or textbooks, it will only be possible to give a brief discussion here on some commonly adopted methods including the direct iteration method, Newton-Raphson method and arc-length iteration methods.

11.2.1. DIRECT ITERATION

Direct iteration starts with the form:

$$K(\delta)\,\delta + f = 0 \tag{11.2}$$

If it is assumed that $\delta = \delta^o$ in the calculation of $K(\delta)$, we can have a better approximation for δ by obtaining

$$\delta^1 = -\,K(\delta^o)^{-1}\,f \tag{11.3}$$

By repeating the process, we have

$$\delta^n = -K(\delta^{n-1})^{-1}\,f \tag{11.4}$$

and the process will be terminated when the error

$$\varepsilon = \sqrt{(\delta^n - \delta^{n-1})^2} \tag{11.5}$$

is sufficiently small. The scheme is illustrated in Figure 11.1. It is obvious that the stiffness matrix has to be inverted and updated in each iteration step, and considerable computing time will be required to obtain the solution.

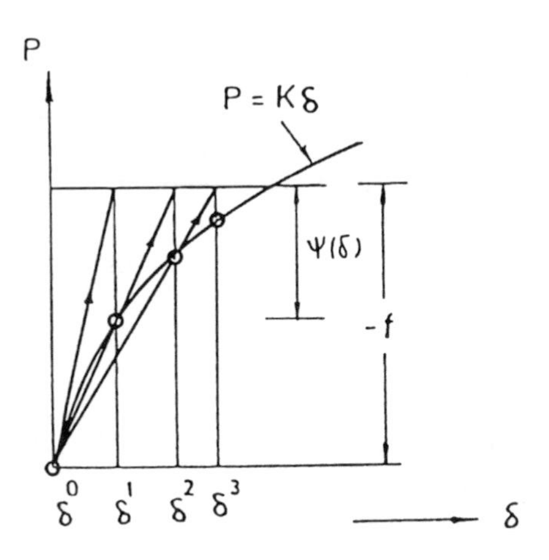

Figure 11.1 Direct iteration method.

11.2.2 NEWTON-RAPHSON METHOD

Referring to Figure 11.1, one can define the residual force $\psi(\delta)$, such that

$$\psi(\delta) = P(\delta) + f = K(\delta)\,\delta + f \tag{11.6}$$

An approximate solution δ^n can be improved to a better solution δ^{n+1} by using the Taylor expansion,

$$\psi(\delta^{n+1}) = \psi(\delta^n) + \left(\frac{d\Psi}{d\delta}\right)_n \Delta\delta^n = 0 \tag{11.7}$$

with

$$\delta^{n+1} = \delta^n + \Delta\delta^n \tag{11.8}$$

In the above equation, it can be shown readily that

$$(\frac{d\Psi}{d\delta})_n = K_T(\delta^n) \tag{11.9}$$

represents the tangential stiffness at displacement δ^n. Rewriting Eq. 11.7, we have

$$\begin{aligned}\Delta\delta^n &= -\,(K_T(\delta^n))^{-1}\psi(\delta^n)\\ &= -\,(K_T(\delta^n))^{-1}(P^n + f)\end{aligned} \tag{11.10}$$

The process is illustrated in Figure 11.2 and it is repeated until the residual force is sufficiently small. Again inversion and formation of the stiffness matrix is required for each iteration step.

The convergency of this algorithm is faster than the direct iteration method which has no guarantee on the convergence and stability of the solution, particularly if δ^o has not been chosen appropriately.

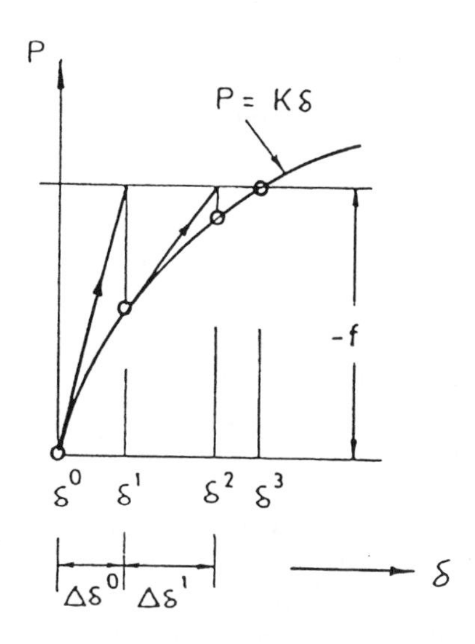

Figure 11.2 Newton-Raphson method.

To save computing time required for the inversion of the stiffness matrix, one can carry out the iteration by modifying the Newton-Raphson method. The iteration can be carried out by taking

$$K_T(\delta^n) = K_T(\delta^o) \tag{11.11}$$

Eq. 11.10 becomes

$$\Delta\delta^n = -\,(K_T(\delta^o))^{-1}(P^n + f) \tag{11.12}$$

Figure 11.3 illustrates the scheme. As the inversion of the stiffness matrix is not required for each step, the computing time can be reduced. On the other hand, the convergence is much slower. In general, this process still has an overall economy.

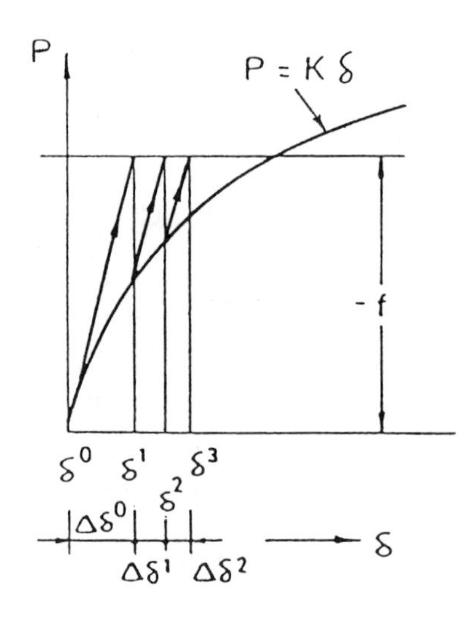

Figure 11.3 Modified Newton-Raphson method.

11.2.3 ARC-LENGTH ITERATION METHOD

The above-mentioned methods work fairly well for most cases, however, they may encounter difficulties in obtaining solutions for snap-back and snap-through problems. Therefore, various arc-length iteration strategies were proposed to overcome the difficulties, for example, the schemes proposed by Riks(1) and Crisfield(2) respectively. The new schemes also require less number of iterations to achieve convergent results. These schemes are illustrated in Figures 11.4 and 11.5.

Riks(1) proposed that the iteration is constrained to follow a plane perpendicular to the tangent at the beginning of each step (Figure 11.4) and that the $\Delta\lambda_i$ (Figure 11.4) has to be computed for each new iteration.

To improve the rate of convergence, Crisfield(2) devised a new iterative process by following a spherical path (Figure 11.5). The schemes can be implemented by taking the following steps (Zhu(3)):

1. It can be shown readily that Eq. 11.6 can be rewritten into two equations:

$$K\,\delta_p = P \tag{11.13}$$

$$K\,\Delta\delta_{\psi i} = \psi(\delta^{i-1}) \tag{11.14}$$

Solving these equations, we can obtain the displacement δ_p due to P and the displacement increment $\Delta\delta_{\psi i}$ due to $\psi(\delta^{i-1})$. Except for the

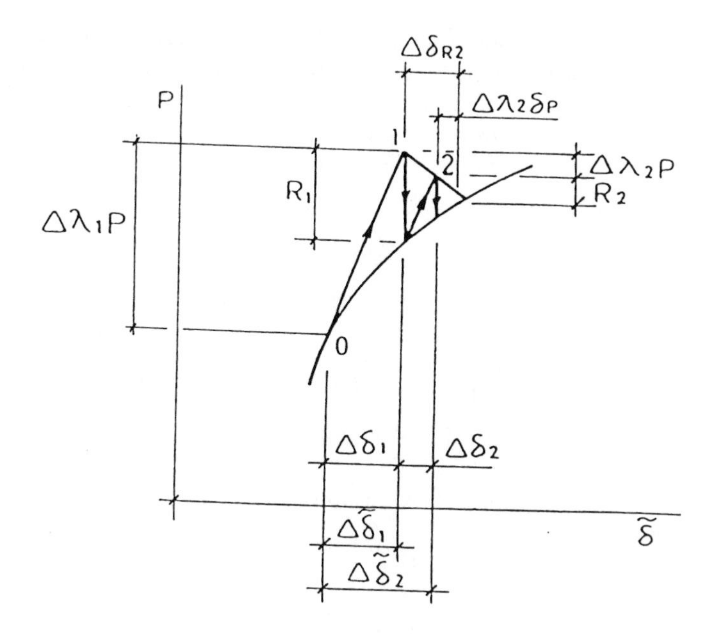

Figure 11.4 Arc length iteration method (Riks' approach).

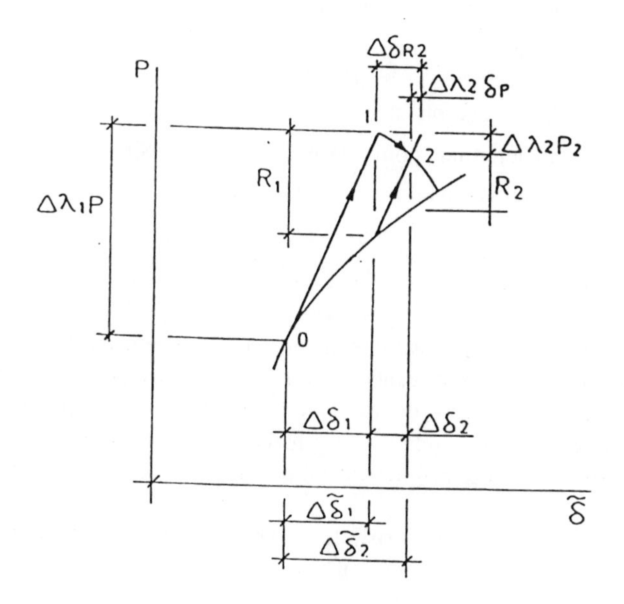

Figure 11.5 Arc length iteration method (Crisfield's approach).

displacement dependent load, Eq. 11.13 needs to be solved only once in every increment step and Eq. 11.14 needs to be solved in every iteration. Certainly, Eqs. 11.13 and 11.14 can be solved simultaneously in one elimination-back substitution procedure.

2. After every iteration, the displacement increment obtained would be (Figure 11.6)

$$\Delta \tilde{\delta}_i = \Delta\delta_i + \Delta \tilde{\delta}_{i-1} = \Delta\lambda_i \, \delta_p + \Delta\delta_{\psi i} + \Delta \tilde{\delta}_{i-1} \qquad (11.15)$$

in which $\Delta\lambda_i$ has not yet been determined but would be evaluated by applying the following constraint

$$(\Delta \tilde{\delta}_i^{\,T})(\Delta \tilde{\delta}_i) = (\Delta l)^2 \qquad (11.16)$$

where Δl is the specified arc-length (Figure 11.5).
Substituting Eq. 11.15 into Eq. 11.16, one can obtain a second order equation with one variable $\Delta\lambda_i$ as follows

$$c_1(\Delta\lambda_i)^2 + c_2\Delta\lambda_i + c_3 = 0 \qquad (11.17)$$

where
$$c_1 = \delta_p^{\,T}\delta_p$$
$$c_2 = 2(\Delta\delta_{Ri} + \Delta \tilde{\delta}_{i-1})^T\delta_p$$
$$c_3 = (\Delta\delta_{Ri} + \Delta \tilde{\delta}_{i-1})^T(\Delta\delta_{Ri} + \Delta \tilde{\delta}_{i-1}) - (\Delta l)^2$$

Two values of $\Delta\lambda_i$ are possible. The correct choice of $\Delta\lambda_i$ is such that the scalar product of the incremental displacement vector $\Delta \tilde{\delta}_i$ and $\Delta \tilde{\delta}_{i-1}$ is a positive quantity. When both choices of $\Delta\lambda_i$ yield a positive result, the correct root is the one nearest to the linear solution $\Delta\lambda_i = -c_3/c_2$.

3. For the first iteration step in every increment step $\Delta\lambda_i$ needs to be specified to determine point "1" in Figure 11.6. Since $\Delta \tilde{\delta}_o = 0$ and $\Delta\delta_{Ri} = 0$, the constraint equation (Eq. 11.17) becomes

$$(\Delta\lambda_i)^2 \, \delta_p^{\,T} \, \delta_p = (\Delta l)^2$$
$$\Delta\lambda_i = \pm\Delta l / \sqrt{\delta_p^T\delta_p} \qquad (11.18)$$

From that $\Delta\lambda_i$ can be evaluated and the sign is chosen by calculating the determinant of the tangent stiffness matrix or by considering the sign of the incremental work done. This choice plays a role in passing a limit point or bifurcation point, and it makes $\Delta\lambda_i$, in fact, a control variable in every increment step.

In view of the fact that Crisfield's scheme still requires very small steps to negotiate a curve in the vicinity of the limit points, Zhu[3] proposed predictor schemes for estimating new points based on previous computed values in the iterative process. He recommended that a three-point (quadratic) extrapolation will be sufficient for most cases but a four-point (cubic) one will be necessary if the behaviour is highly non-linear. He also pointed out that further increasing the order of extrapolation may lead to instability. In Figure 11.6, a two-dimensional representation of Zhu's strategy is shown.

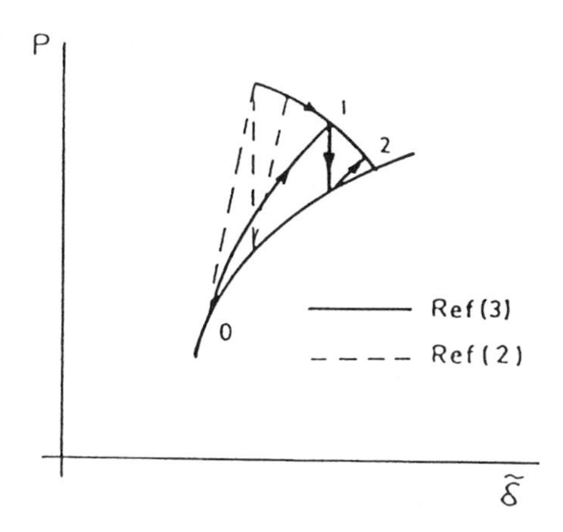

Figure 11.6 Arc length iteration method (Zhu's approach).

11.3 MATERIAL NON-LINEARITY

Mofflin[4] first incorporated plasticity in the finite strip method. Based on the plane strain assumption in bending of reinforced concrete structures, the finite strip method was introduced into the material non-linear analysis of reinforced concrete beams and slabs.[5,6,7] The layered concept for material non-linear finite element analysis was adopted earlier and the finite strip was similarly divided into imaginary concrete layers in thickness and a number of segments in the longitudinal direction for numerical integration requirement. Reinforcement was further replaced by a smeared steel layer with the original reinforcement area having stiffness only in the direction of the reinforcement. In every iteration of the non-linear analysis, the strain state at any point of each layer can be found in terms of the strip displacements. Its stress state is then calculated in accordance with an appropriate constitutive model of concrete. The yielding of the concrete was defined by the Von Mises' criterion, and the plastic strain increments were evaluated by the Prandlt-Reuss flow rule for isometric hardening material. The progressive failure for composite laminated plates was later studied by Cheung[8] using the classical strips. The failure was described by Lee's strength criterion.[9] The effects of the fiber orientation and number of plies on the capacity of the plates were studied.

The buckling of steel columns at increasing temperature was studied by Olawale and Plank.[10] The behaviour of the steel was defined by a Ramberg-Osgood stress-strain relationship. The results obtained showed that the method can provide a sound basis for predicting the collapse behaviour of steel columns in fire. Further study on the topic was carried out by Burgess.[11]

11.3.1 CONSTITUTIVE RELATIONSHIPS AND FORMATION OF THE CHARACTERISTIC MATRICES

In this section, the constitutive relationships for strain hardening materials are discussed. Under uniaxial load, the stress-strain curve for these materials is shown in Figure 11.7. Unlike the ideal plastic materials in which the yield stress is constant, here the yield stress depends on the hardening parameters κ. Mathematically, the yield surface can be defined in terms of the stresses and hardening parameter as:

$$F(\sigma,\kappa) = 0 \tag{11.19}$$

where σ are the stresses. As an example, for a material obeying the Von Mises' criterion and showing no hardening, the yield surface is

$$F(\sigma,\kappa) = \left[\frac{1}{2}(\sigma_1-\sigma_2)^2+\frac{1}{2}(\sigma_2-\sigma_3)^2+\frac{1}{2}(\sigma_3-\sigma_1)^2+3\sigma_{12}^2+3\sigma_{23}^2+3\sigma_{31}^2\right]^{1/2}-\sigma_y$$

where σ_i and σ_{ij} are the normal and shear stress components in a general three-dimensional stress state. σ_y is the yield stress.

According to the associated flow rule, the plastic strain can be defined as:

$$d\varepsilon_p = \lambda \frac{\partial F}{\partial \sigma} \tag{11.20}$$

in which λ is a proportionality constant.

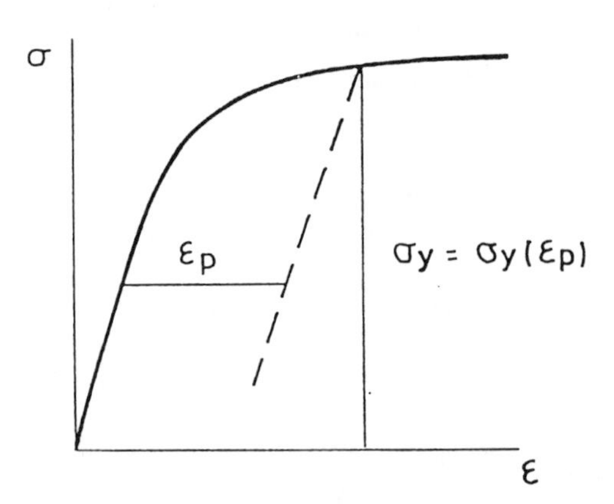

Figure 11.7 Stress-strain for a strain hardening material.

The strain increment can be written in terms of two components: the elastic and the plastic parts,

$$d\varepsilon = d\varepsilon_e + d\varepsilon_p \tag{11.21}$$

The elastic strain increments are related to stress increments by

$$d\varepsilon_e = D^{-1}\, d\sigma \tag{11.22}$$

Combining Eqs. 11.21 and 11.22, one can show that

$$d\varepsilon = D^{-1}\, d\sigma + \lambda\, \frac{\partial F}{\partial \sigma} \tag{11.23}$$

In the state of yielding, the stress states satisfy the yield criterion given by Eq. 11.19. Taking the total differential, we have

$$dF = \frac{\partial F}{\partial \sigma_1} d\sigma_1 + \frac{\partial F}{\partial \sigma_2} d\sigma_2 + ...+ \frac{\partial F}{\partial \kappa} d\kappa \tag{11.24}$$

In matrix form, the right hand side can be written as

$$\left\{\frac{\partial F}{\partial \sigma}\right\}^T d\sigma - A\,\lambda \tag{11.25}$$

where $A = - \frac{\partial F}{\partial \kappa} d\kappa\ \frac{1}{\lambda}$.

Eqs. 11.23 and 11.25 can be written in a single symmetric matrix form as

$$\begin{Bmatrix} d\varepsilon \\ 0 \end{Bmatrix} = \begin{bmatrix} D^{-1} & \frac{\partial F}{\partial \sigma} \\ (\frac{\partial F}{\partial \sigma})^T & -A \end{bmatrix} \begin{Bmatrix} d\sigma \\ \lambda \end{Bmatrix} \tag{11.26}$$

Eliminating λ, the stress-strain relationships become

$$d\sigma = D_{ep}\, d\varepsilon \tag{11.27}$$

$$D_{ep} = D - D \left\{\frac{\partial F}{\partial \sigma}\right\} \left\{\frac{\partial F}{\partial \sigma}\right\}^T D[A + \left\{\frac{\partial F}{\partial \sigma}\right\}^T D \left\{\frac{\partial F}{\partial \sigma}\right\}]^{-1} \tag{11.28}$$

Modifying Eqs. 3.14 and 4.13 by replacing the property matrix with D_{ep}, one can obtain the bending and in-plane stiffness matrices. Of course, the stiffness matrices are no longer stress independent.

In order to evaluate these matrices, it is necessary to divide each strip (layer) into a number of substrips (sublayers) along the thickness and carry out integration along the length of the substrips (sublayers) at a number of integration points. The stresses at each integration point are first assumed to initiate the iteration process for the solution, and the process will only cease when the results satisfy some chosen convergence criteria.

The load vectors are identical to those derived in Chapters 3 and 4, and no special treatment is required in most cases.

11.3.2 EXAMPLES

In order to illustrate the method, Cheung[6] carried out the analysis for a simply supported concrete plate under a uniformly distributed load. The width and thickness were 72in and 2in respectively. The plate was reinforced in both directions by 3/16in bars at spacing of 3in and 2.5in. The properties of the concrete and steel are described in detail in Reference 7.

In the analysis, the cross-section was divided into ten concrete layers and two orthogonal steel layers. Figure 11.8 shows the load-deflection curve of the plate.

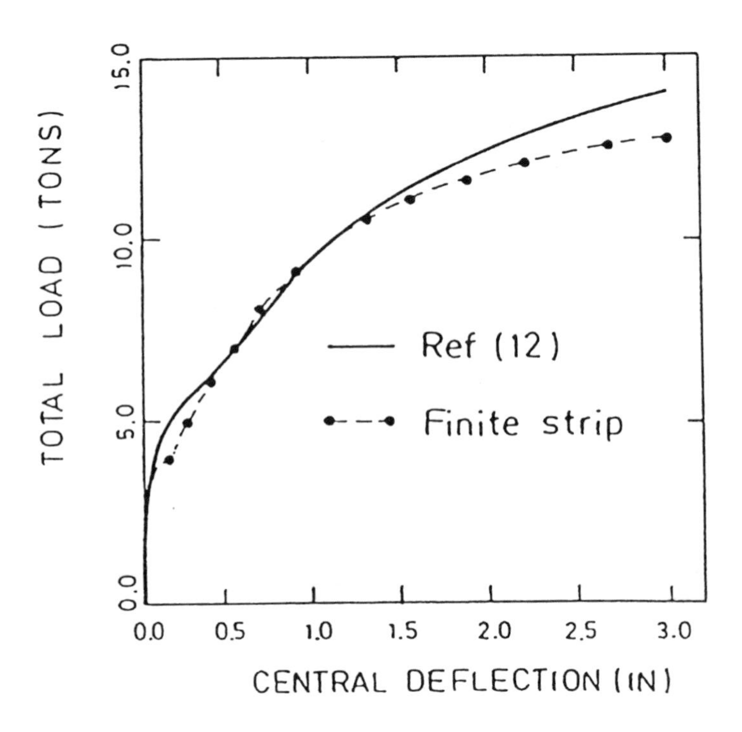

Figure 11.8 Load deflection curve of simply supported concrete plate.[6]

Olawale et al.[10] carried out the analysis on the buckling of 203x203 universal steel channels under increasing temperature. The stress-strain behaviour of the material was described by the Ramberg-Osgood model. Figure 11.9 shows the variation of the critical stress for channels with various aspect ratios with temperature.

Progressive failure for a square laminated plate (a x a) was studied by Cheung.[8] The laminated plate consisted of four equally thick layers which were orientated at 0°/90°/0°/90°. Lee's criteria[9] were adopted to define the failure of the materials. The strength properties were given in Reference 9. The loading is defined as:

$$q = P \sin\frac{\pi x}{a} \sin\frac{\pi y}{a}$$

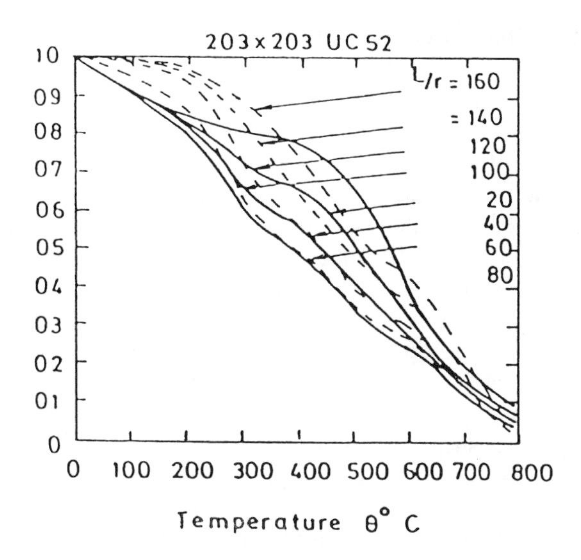

Figure 11.9 The reduction in strength with increasing temperature for columns with different slenderness ratios.[10]

Due to symmetry, only half of the plate had to be modelled. The plate was divided into four strips and four symmetrical terms were taken for the longitudinal function. The maximum load-capacity was shown in Figure 11.10 as a function of the length-to-thickness ratio(s). The figure also shows the results of the finite element method based on the higher order shear theory (HSDT) and second order theory (FSDT).[13]

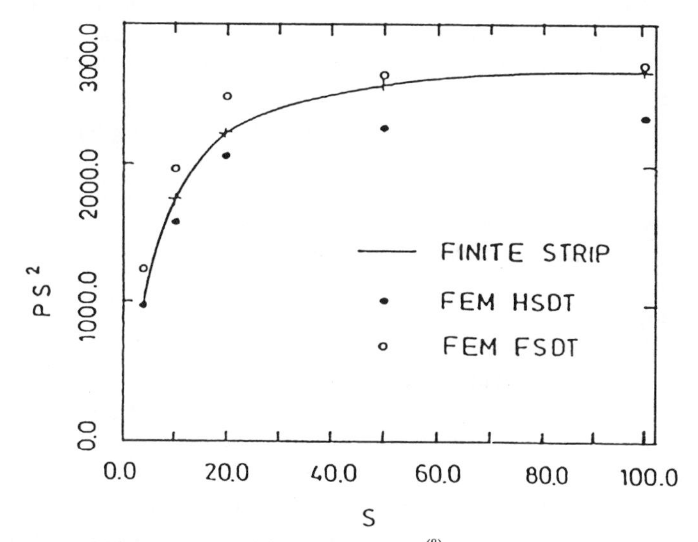

Figure 11.10 Ultimate strength of the laminated plate.[8]

11.4 GEOMETRIC NON-LINEARITY

11.4.1 INTRODUCTION

The first report on finite strip analysis of geometrically non-linear structures was made by Delcourt in her thesis.[14] During the last decade, many more papers were published and they have demonstrated that the finite strip method has greater efficiency than the finite element method, particularly in case of plate bending problems.

Plank[15] reported on the analyses of inelastic stability of stiffened panels carried out by the finite strip method. The elastic moduli were assumed to be non-linear and defined by an equation suggested by Ylinen.[16] Further work on inelastic stability was reported by Bradford and his co-worker,[17,18] using the classical finite strips augmented with bubble functions. Lau and Hancock[19] applied the spline finite strip method to the inelastic buckling analysis which could take into account the non-linear nature of the stress-strain curves and strain-hardening of the material and residual stress distribution.

The finite strip method for a plate undergoing large deflection has been formulated on the basis of the non-linear relationship of strain and displacement which took into account the interaction between out-of-plane and in-plane deformation.[20,21,22,23] Selection of the non-linear terms in the strain-displacement relation can be somewhat different depending on the primary components of displacements chosen for a strip to simulate a plate or a web member in a box girder.

Gierlinski and Graves-Smith[22] discussed a particular difficulty with the finite strip method when analyzing structures whose non-linear load deflection curves contain maxima. The commonly used method to solve this problem is to replace load incrementation with displacement incrementation. However, such a strategy did not appear to work with classical finite strips, because the continuous series representation of displacements does not allow localised displacements to be imposed on structures. They proposed to overcome this difficulty by using an improved iteration strategy of selecting variable load increments on both the loading and the unloading curves. The ability of the method for predicting all types of elastic non-linear behaviour, namely, stiffening and softening behaviour, was demonstrated through examples of square plates, square web panels and shallow cylindrical roofs subjected to various types of loads.

The large deflection elastic-plastic analyses of plate structures and cylindrical shells under uniform loads were carried out by Abayakoon, Olson and Anderson[24] and by Kumar et al.[25] The work has been later extended by Khalil, Olson and Anderson[26] to investigate the large deflection, elastic-plastic dynamic response of air-blast loaded stiffened plates. The loads have been modelled as uniformly distributed time dependent pressures. The method is ideally suited to this type of load because the distribution can be represented fairly well by a single mode. The results of the analysis involving only a few modes and a few finite strips were encouraging in comparison to the experiments.

Azizian and Dawe have applied the geometrically non-linear finite strip method to the large deflection problems of Mindlin plates(27) and laminated plates.(28) Further work was carried out by Dawe et al.(29,30,31) Sekulovic and Milasinovic(32) presented a finite strip analysis of plates and folded plates taking into account the geometrical non-linearities and the effects of creep.

Cheung et al.(33) applied the finite strip to model the non-linear behaviour of cable-stayed bridges.

Kwon(34) and Kwon and Hancock(35) further developed the spline finite method to handle local, distortional and overall buckling modes in post-buckling range, and the interaction between the various modes. The method also allowed for geometric imperfections, arbitrary loading and complicated support conditions. Advanced theories based on the convective curvilinear coordinate system were used to carry out non-linear analysis for plates and shells by Zhu.(36,37)

11.4.2 INELASTIC STABILITY PROBLEMS

It has been shown in a previous discussion that the analysis of elastic stability problems can be reduced to an eigenvalue solution for the following matrix equation

$$\left[K - \lambda K_g\right] = 0 \tag{11.29}$$

where λ is the load factor. K and K_g are the stiffness and geometric stiffness matrices respectively.

Having chosen the displacement functions for the interpolation of the displacements, the matrices can be formed by the procedure as discussed in Chapter 5.

In the plastic range, the stiffness matrix is no longer constant but depends on the stress level. Though standard procedures can be adopted to form the two matrices, the search of the eigenvalues is less straightforward and special treatments will be required to obtain the critical loads.

Pifko and Isakson(38) presented a trial and error process to determine λ. In this method, a trial stress level is assumed. The K and K_g are computed at the trial stress level and λ can be determined by the usual eigenvalue algorithm . If λ thus determined is equal to one, the trial stress level is the critical one. If it is greater or less than one, the trial stress level will be increased or decreased and the process has to be repeated until the solution is obtained.

Gupta(39) proposed a method which was based on the Sturm sequence property employing a bisection strategy. A trial stress level and λ are assumed for Eq. 11.29. The usual procedure of reduction to an upper triangular form is performed. The number of negative diagonal elements in the triangulated matrix is counted. If there is no negative term, the stress level is greater than the critical one and it should be increased in the next iteration step. On the other hand, the stress level will be decreased if there are negative terms. The process is repeated until convergence on the trial stress level is achieved.

11.4.3 FORMULATION OF LARGE DEFORMATION PROBLEMS IN CARTESIAN COORDINATES

If a structure undergoes large deformation, the second order terms of the strains cannot be ignored. Consequently, the membrane and bending actions are coupled. The strains in the shell with initial imperfection at mid-surface are given by (Kwon[34])

$$\{\varepsilon_M\} = \begin{Bmatrix} \varepsilon_x \\ \varepsilon_y \\ \gamma_{xy} \end{Bmatrix} = \begin{Bmatrix} \frac{\partial u}{\partial x} - \frac{\partial \tilde{u}}{\partial x} \\ \frac{\partial v}{\partial y} - \frac{\partial \tilde{v}}{\partial y} \\ \frac{\partial u}{\partial y} - \frac{\partial \tilde{u}}{\partial y} + \frac{\partial v}{\partial x} - \frac{\partial \tilde{v}}{\partial x} \end{Bmatrix} +$$

$$\begin{Bmatrix} \frac{1}{2}(\frac{\partial u}{\partial x})^2 + \frac{1}{2}(\frac{\partial v}{\partial x})^2 + \frac{1}{2}(\frac{\partial w}{\partial x})^2 - \frac{1}{2}(\frac{\partial \tilde{u}}{\partial x})^2 - \frac{1}{2}(\frac{\partial \tilde{v}}{\partial x})^2 - \frac{1}{2}(\frac{\partial \tilde{w}}{\partial x})^2 \\ \frac{1}{2}(\frac{\partial u}{\partial y})^2 + \frac{1}{2}(\frac{\partial v}{\partial y})^2 + \frac{1}{2}(\frac{\partial w}{\partial y})^2 - \frac{1}{2}(\frac{\partial \tilde{u}}{\partial y})^2 - \frac{1}{2}(\frac{\partial \tilde{v}}{\partial y})^2 - \frac{1}{2}(\frac{\partial \tilde{w}}{\partial y})^2 \\ \frac{\partial u}{\partial x}\frac{\partial u}{\partial y} + \frac{\partial v}{\partial x}\frac{\partial v}{\partial y} + \frac{\partial w}{\partial x}\frac{\partial w}{\partial y} - \frac{\partial \tilde{u}}{\partial x}\frac{\partial \tilde{u}}{\partial y} - \frac{\partial \tilde{v}}{\partial x}\frac{\partial \tilde{v}}{\partial y} - \frac{\partial \tilde{w}}{\partial x}\frac{\partial \tilde{w}}{\partial y} \end{Bmatrix} \tag{11.30}$$

and the curvature-displacement relations are taken as

$$\{\varepsilon_F\} = \begin{Bmatrix} \rho_x \\ \rho_y \\ \rho_{xy} \end{Bmatrix} = \begin{Bmatrix} -\frac{\partial^2 (w-\tilde{w})}{\partial x^2} \\ -\frac{\partial^2 (w-\tilde{w})}{\partial y^2} \\ -2\frac{\partial^2 (w-\tilde{w})}{\partial x \partial y} \end{Bmatrix} \tag{11.31}$$

where $\tilde{u}$, $\tilde{v}$ and $\tilde{w}$ are the initial geometric imperfections.

Eqs. 11.30 and 11.31 can be expressed in a condensed form

$$\{\varepsilon\} = [[B_L] + [B_n(\delta)]]\,\{\delta\} \tag{11.32}$$

where $[B_L]$ denotes the linear strain matrix and $[B_n(\delta)]$ is a non-linear strain matrix which is a function of the displacements. Taking variation for Eq. 11.32 with respect to δ to get the incremental strain-displacement relation

$$d\{\varepsilon\} = [[B_L] + [B_n(\delta)]]\, d\{\delta\} + \frac{d[B_n(\delta)]}{d\{\delta\}}\{\delta\} d\{\delta\}$$

$$=[B_L + B_N(\delta)]]\, d\{\delta\} = [\,\overline{B}\,] d\{\delta\} \tag{11.33}$$

where $B_N(\delta) = B_n(\delta) + \frac{d[B_n(\delta)]}{d\{\delta\}}\{\delta\}$

Using the virtual work principle, the stiffness matrix can be derived:

$$K = K_L + K_N \tag{11.34}$$

where $K_L = \int [B_L]^T[D][B_L]dvol$

$$K_N = \int [B_L]^T[D][B_N(\delta)]dvol + \int [B_N(\delta)]^T[D][B_L]dvol$$

In addition, the geometric stiffness matrix can be written as:

$$K_\sigma d\{\delta\} = \int d[B_N(\delta)]^T\{\sigma\}dvol \tag{11.35}$$

The total stiffness matrix is

$$K = K_L + K_N + K_\sigma \tag{11.36}$$

Material non-linearity can also be easily taken into account by adopting a property matrix which depends on the stresses as discussed in the previous section.

A simplified version of the displacement-strain relationships, which only include the second order term of w, was adopted by Azizian et al.[27] in their analyses for plates without initial imperfections. Eqs. 11.30 and 11.31 are modified respectively as:

$$\{\varepsilon_M\} = \begin{Bmatrix} \varepsilon_x \\ \varepsilon_y \\ \gamma_{xy} \end{Bmatrix} = \begin{Bmatrix} \dfrac{\partial u}{\partial x} \\ \dfrac{\partial v}{\partial y} \\ \dfrac{\partial u}{\partial y} + \dfrac{\partial v}{\partial x} \end{Bmatrix} + \begin{Bmatrix} \dfrac{1}{2}(\dfrac{\partial w}{\partial x})^2 \\ \dfrac{1}{2}(\dfrac{\partial w}{\partial y})^2 \\ \dfrac{\partial w}{\partial x}\dfrac{\partial w}{\partial y} \end{Bmatrix} \tag{11.37}$$

$$\{\varepsilon_F\} = \begin{Bmatrix} \rho_x \\ \rho_y \\ \rho_{xy} \end{Bmatrix} = \begin{Bmatrix} -\dfrac{\partial^2 w}{\partial x^2} \\ -\dfrac{\partial^2 w}{\partial y^2} \\ -2\dfrac{\partial^2 w}{\partial x \partial y} \end{Bmatrix} \tag{11.38}$$

In the case of moderately thick plates, bending strains (Eq. 11.38) are replaced by Eq. 3.55 (Azizian et al.)[28]

The explicit forms for the strain vectors can be easily written down for each type of strip. However, the shape functions of the classical strips may have to be modified to account for the boundary conditions. For example, if there is edge shortening, the shape functions should be modified to[29]

$$u = \varepsilon(\frac{A}{2} - x) + \sum_{i=1}^{m} f_i^u(y)\sin(\frac{i\pi x}{L})$$

$$v = \begin{cases} \alpha\varepsilon y + \sum_{i=o}^{m} f_i^v(y)\cos(\frac{i\pi x}{L}) & \text{free to expand laterally} \\ \alpha\varepsilon y + \sum_{i=o}^{m} f_i^v(y)\sin(\frac{i\pi x}{L}) & \text{restrained} \end{cases}$$

$$w = \sum_{i=1}^{m} f_i^{w}(y) \sin\left(\frac{i\pi x}{L}\right)$$

where ε is the amount of edge shortening.

Formulation for a cylindrical shell was carried out by Kumar et al.[25] In the cylindrical coordinate system, the strain-displacement relations can be defined as follows:

$$\varepsilon_x = \frac{\partial u}{\partial x} - z\frac{\partial^2 w}{\partial x^2} + \frac{1}{2}\left(\frac{\partial w}{\partial x}\right)^2$$

$$\varepsilon_y = \frac{\partial v}{\partial y} + \frac{w}{R} - z\left(\frac{\partial^2 w}{\partial y^2} - \frac{1}{R}\frac{\partial v}{\partial y}\right) + \frac{1}{2}\left(\frac{\partial w}{\partial y}\right)^2$$

$$\gamma_{xy} = \frac{\partial u}{\partial y} + \frac{\partial v}{\partial x} - z\left(2\frac{\partial^2 w}{\partial x \partial y} - \frac{1}{R}\frac{\partial v}{\partial x}\right) + \frac{1}{2}\frac{\partial w}{\partial x}\frac{\partial w}{\partial y}$$

in which R is the radius of the shell. u and x, v and y, and w and z are the displacements and coordinates along the longitudinal, circumferential and radial directions respectively.

11.4.4 FORMULATION OF LARGE DEFORMATION PROBLEMS IN CONVECTIVE CURVILINEAR COORDINATES[3]

The formulation of the problems can also be carried out in a convective curvilinear coordinate system as shown in Figure 11.11. In this coordinate

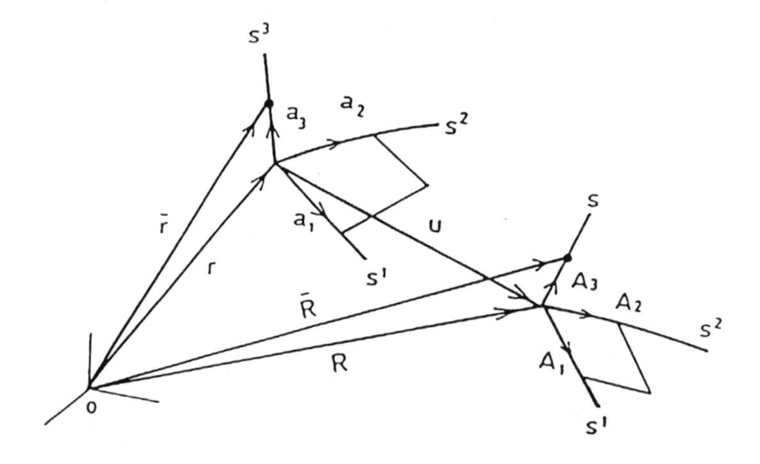

Figure 11.11 Convective curvilinear coordinate system.

system, the location of a particle of the shell is specified by the curvilinear coordinates s^1, s^2, s^3. $s^3 = 0$ are curvilinear coordinate lines embedded in the middle surface of the shell and s^3 is normal to the middle surface in the undeformed state. In the case of circular cylindrical shells, one may define the meridian and latitude lines at the undeformed state as s^1 and s^2 respectively. s^3 is normal to this s^1-s^2 plane. The corresponding displacements are u, v and w. Zhu[3] has shown that the strain-displacement relations are:

$$\varepsilon_{11} = \frac{\partial u}{\partial s_1} + \frac{1}{2}\left[\left(\frac{\partial u}{\partial s_1}\right)^2 + \left(\frac{\partial v}{\partial s_1}\right)^2 + \left(\frac{\partial w}{\partial s_1}\right)^2\right] - s^3 \frac{\partial^2 w}{\partial s_1^2} \tag{11.39a}$$

$$\varepsilon_{22} = \frac{\partial v}{\partial s_2} + Kw + \frac{1}{2}\left[\left(\frac{\partial u}{\partial s_2}\right)^2 + \left(\frac{\partial v}{\partial s_2} + Kw\right)^2 + \left(\frac{\partial w}{\partial s_2} - Kv\right)^2\right] - s^3\left(\frac{\partial^2 w}{\partial s_2^2} - K\frac{\partial v}{\partial s_2}\right) \tag{11.39b}$$

$$\varepsilon_{12}+\varepsilon_{21} = \frac{\partial u}{\partial s_2} + \frac{\partial v}{\partial s_1} + \frac{\partial u}{\partial s_1}\frac{\partial u}{\partial s_2} + \frac{\partial v}{\partial s_1}\left(\frac{\partial v}{\partial s_2} + Kw\right) + \frac{\partial w}{\partial s_1}\left(\frac{\partial w}{\partial s_2} - Kv\right) - 2s^3\left(\frac{\partial^2 w}{\partial s_1 \partial s_2} - K\frac{\partial v}{\partial s_1}\right) \tag{11.39c}$$

If the initial imperfections are denoted by $\tilde{w}$, the relations will be modified to:

$$\varepsilon_{11} = \frac{\partial u}{\partial s_1} + \frac{1}{2}\left[\left(\frac{\partial u}{\partial s_1}\right)^2 + \left(\frac{\partial v}{\partial s_1}\right)^2 + \frac{\partial w}{\partial s_1}\left(\frac{\partial w}{\partial s_1} + 2\frac{\partial \tilde{w}}{\partial s_1}\right)\right] - s^3 \frac{\partial^2 w}{\partial s_1^2} \tag{11.40a}$$

$$\varepsilon_{22} = \frac{\partial v}{\partial s_2} + Kw + \frac{1}{2}\left[\left(\frac{\partial u}{\partial s_2}\right)^2 + \left(\frac{\partial v}{\partial s_2} + Kw + 2K\tilde{w}\right)\left(\frac{\partial v}{\partial s_2} + Kw\right) + \left(\frac{\partial w}{\partial s_2} + 2\frac{\partial \tilde{w}}{\partial s_2} - Kv\right)\left(\frac{\partial w}{\partial s_2} - Kv\right)\right] - s^3\left(\frac{\partial^2 w}{\partial s_2^2} - K\frac{\partial v}{\partial s_2}\right) \tag{11.40b}$$

$$\varepsilon_{12}+\varepsilon_{21} = \frac{\partial u}{\partial s_2} + \frac{\partial v}{\partial s_1} + \frac{\partial u}{\partial s_1}\frac{\partial u}{\partial s_2} + \frac{\partial v}{\partial s_1}\left(\frac{\partial v}{\partial s_2} + Kw + K\tilde{w}\right) + \frac{1}{2}\left(\frac{\partial w}{\partial s_1} + 2\frac{\partial \tilde{w}}{\partial s_1}\right)\left(\frac{\partial w}{\partial s_2} - Kv\right) + \frac{1}{2}\left(\frac{\partial w}{\partial s_2} + 2\frac{\partial \tilde{w}}{\partial s_2} - Kv\right) - 2s^3\left(\frac{\partial^2 w}{\partial s_1 \partial s_2} - K\frac{\partial v}{\partial s_1}\right) \tag{11.40c}$$

When one gives the displacements a set of increments from $\Omega^{(N)}$ equilibrium state to the $\Omega^{(N+1)}$ state, the displacements at the $\Omega^{(N+1)}$ state can be expressed in terms of the displacements at $\Omega^{(N)}$ state

$$u = u_o + \Delta u \tag{11.41a}$$

$$v = v_o + \Delta v \tag{11.41b}$$

$$w = w_o + \Delta w \tag{11.41c}$$

The strains can be given as

$$\varepsilon = \varepsilon_o + \varepsilon_1 + \varepsilon_2 \tag{11.42}$$

where ε_o, ε_1 and ε_2 represent the 0th, 1st and 2nd order terms of the displacement increment. The explicit forms for these strains of the spline finite strip are given as follows:

$$\varepsilon_o = \left\{ \begin{array}{c} \frac{\partial u_o}{\partial s_1} + \frac{1}{2}(\frac{\partial u_o}{\partial s_1})^2 + \frac{1}{2}(\frac{\partial v_o}{\partial s_1})^2 + \frac{1}{2}(\frac{\partial w_o}{\partial s_1} + 2\frac{\partial \tilde{w}}{\partial s_1})^2 \frac{\partial w_o}{\partial s_1} \\ \hdashline \frac{\partial v_o}{\partial s_2} + Kw_o + \frac{1}{2}(\frac{\partial u_o}{\partial s_2})^2 + \frac{1}{2}(\frac{\partial v_o}{\partial s_2} + Kw_o)(\frac{\partial v_o}{\partial s_2} + Kw_o + 2K\tilde{w}) + \\ \frac{1}{2}(\frac{\partial w_o}{\partial s_2} + 2\frac{\partial \tilde{w}}{\partial s_2} - Kv_o)(\frac{\partial w_o}{\partial s_2} - Kv_o) \\ \hdashline \frac{\partial u_o}{\partial s_2} + \frac{\partial v_o}{\partial s_1} + \frac{\partial u_o}{\partial s_1}\frac{\partial u_o}{\partial s_2} + \frac{\partial v_o}{\partial s_1}(\frac{\partial v_o}{\partial s_2} + Kw_o + 2K\tilde{w}) + \\ \frac{1}{2}(\frac{\partial w_o}{\partial s_1} + 2\frac{\partial \tilde{w}}{\partial s_1})(\frac{\partial w_o}{\partial s_2} - Kv_o) + \frac{1}{2}(\frac{\partial v_o}{\partial s_2} + 2\frac{\partial \tilde{w}}{\partial s_2} - Kv_o)\frac{\partial w_o}{\partial s_1} \\ \hdashline -\frac{\partial^2 w_o}{\partial s_1{}^2} \\ \hdashline -(\frac{\partial^2 w_o}{\partial s_2{}^2} - K\frac{\partial v_o}{\partial s_2}) \\ \hdashline -2(\frac{\partial^2 w_o}{\partial s_1 \partial s_2} - K\frac{\partial v_o}{\partial s_1}) \end{array} \right\} \tag{11.43}$$

$$\varepsilon_1 = \left\{ \begin{array}{c} \frac{\partial \Delta u}{\partial s_1} + \frac{\partial u_o}{\partial s_1}\frac{\partial \Delta u}{\partial s_1} + \frac{\partial v_o}{\partial s_1}\frac{\partial \Delta v}{\partial s_1} + (\frac{\partial w_o}{\partial s_1} + \frac{\partial \tilde{w}}{\partial s_1})\frac{\partial \Delta w}{\partial s_1} \\ \hdashline \frac{\partial \Delta v}{\partial s_2} + K\frac{\partial \Delta w}{\partial s_2} + \frac{\partial u_o}{\partial s_2}\frac{\partial \Delta u}{\partial s_2} + (\frac{\partial v_o}{\partial s_2} + Kw_o + K\tilde{w})(\frac{\partial \Delta v}{\partial s_2} + K\Delta w) + \\ (\frac{\partial w_o}{\partial s_2} + \frac{\partial \tilde{w}}{\partial s_2} - Kv_o)(\frac{\partial \Delta w}{\partial s_2} - K\Delta v) \\ \hdashline \frac{\partial \Delta u}{\partial s_2} + \frac{\partial v_o}{\partial s_1} + \frac{\partial \Delta u_o}{\partial s_1}\frac{\partial \Delta u}{\partial s_2} + \frac{\partial v_o}{\partial s_1}(\frac{\partial \Delta v_o}{\partial s_2} + K\Delta w) + \frac{\partial \Delta v}{\partial s_1}(\frac{\partial v_o}{\partial s_2} + Kw_o + K\tilde{w}) + \\ (\frac{\partial w_o}{\partial s_1} + \frac{\partial \tilde{w}}{\partial s_1})(\frac{\partial \Delta w}{\partial s_2} - K\Delta v) + (\frac{\partial w_o}{\partial s_2} + \frac{\partial \tilde{w}}{\partial s_2} - Kv_o)\frac{\partial \Delta w}{\partial s_1} \\ \hdashline -\frac{\partial^2 \Delta w}{\partial s_1{}^2} \\ \hdashline -(\frac{\partial^2 \Delta w}{\partial s_2{}^2} - K\frac{\partial \Delta v}{\partial s_2}) \\ \hdashline -2(\frac{\partial^2 \Delta w}{\partial s_1 \partial s_2} - K\frac{\partial \Delta v}{\partial s_1}) \end{array} \right\} \tag{11.44}$$

$$\varepsilon_2 = \left\{ \begin{array}{c} \frac{1}{2}(\frac{\partial \Delta u}{\partial s_1})^2 + \frac{1}{2}(\frac{\partial \Delta v}{\partial s_1})^2 + \frac{1}{2}(\frac{\partial \Delta w}{\partial s_1})^2 \\ \hdashline \frac{1}{2}(\frac{\partial \Delta u}{\partial s_2})^2 + \frac{1}{2}(\frac{\partial \Delta v}{\partial s_2} + K\Delta w)^2 + \frac{1}{2}(\frac{\partial \Delta w}{\partial s_2} - K\Delta v)^2 \\ \hdashline \frac{\partial \Delta u}{\partial s_1}\frac{\partial \Delta u}{\partial s_2} + \frac{\partial \Delta v}{\partial s_1}(\frac{\partial \Delta v}{\partial s_2} + K\Delta w) + \frac{\partial \Delta w}{\partial s_1}(\frac{\partial \Delta w}{\partial s_2} - K\Delta v) \\ \hdashline 0 \\ \hdashline 0 \\ \hdashline 0 \end{array} \right\} \tag{11.45}$$

Substituting the displacement shape functions into these equations, we can show readily that:

$$\varepsilon_o = (B + \tfrac{1}{2} A_o L)\, \delta \tag{11.46}$$

$$\varepsilon_1 = (B + A_1 L)\, \Delta\delta \tag{11.47}$$

$$\varepsilon_2 = \tfrac{1}{2} A_2 L\, \Delta\delta \tag{11.48}$$

In the above equations,

B =

$$\left[\begin{array}{cccccc} N_3'\phi_1 & N_4'\phi_2 & & & & \\ & & N_3\phi_3' & N_4\phi_4' & KN_3\phi_5 & KN_4\phi_6 \\ N_3\phi_1' & N_4\phi_2' & N_3'\phi_3 & N_4'\phi_4 & & \\ & & & & -N_3''\phi_5 & -N_4''\phi_6 \\ & & KN_3\phi_3' & N_4\phi_4' & -N_3\phi_5'' & -N_4\phi_6'' \\ & & 2KN_3'\phi_3 & 2KN_4'\phi_4 & -2N_3'\phi_5' & -2N_4'\phi_6' \end{array}\right. \tag{11.49}$$

$$\left.\begin{array}{cccccc} N_5'\phi_7 & N_6'\phi_8 & & & & \\ & & N_5\phi_9' & N_6\phi_{10}' & KN_5\phi_{11} & KN_6\phi_{12} \\ N_5\phi_7' & N_6\phi_8' & N_5'\phi_9 & N_6'\phi_9 & & \\ & & & & -N_5''\phi_{11} & -N_6''\phi_{12} \\ & & KN_5\phi_9' & KN_6\phi_{10}' & -N_5\phi_{11}'' & -N_6\phi_{12}'' \\ & & 2KN_5'\phi_9 & 2KN_6'\phi_{10} & -2N_5'\phi_{11}' & -2N_6'\phi_{12}' \end{array}\right]$$

L =

$$\left[\begin{array}{cccccc} N_3'\phi_1 & N_4'\phi_2 & & & & \\ & & N_3'\phi_3 & N_4'\phi_4 & & \\ & & & & N_3'\phi_5 & N_4'\phi_6 \\ N_3'\phi_1 & N_4'\phi_2 & & & & \\ & & N_3\phi_3' & N_4\phi_4' & KN_3\phi_5 & KN_4\phi_6 \\ & & -KN_3\phi_3 & -KN_4\phi_4 & N_3\phi_5' & N_4\phi_6' \end{array}\right. \tag{11.50}$$

$$\left.\begin{array}{cccccc} N_5'\phi_7 & N_6'\phi_8 & & & & \\ & & N_5'\phi_9 & N_6'\phi_{10} & & \\ & & & & N_5'\phi_{11} & N_6'\phi_{12} \\ N_5'\phi_7 & N_5'\phi_7 & & & & \\ & & N_5\phi_9' & N_6\phi_{10}' & KN_5\phi_{11} & KN_6\phi_{12} \\ & & -KN_5\phi_9 & -KN_6\phi_{10} & N_5\phi_{11}' & N_6\phi_{12}' \end{array}\right]$$

$$A_o = \begin{bmatrix} (L_1\delta_o)^T & 0 \\ 0 & (L_2\delta_o)^T \\ (L_2\delta_o)^T & (L_1\delta_o)^T \\ 0 & 0 \\ 0 & 0 \\ 0 & 0 \end{bmatrix} ; \; A_1 = \begin{bmatrix} (L_1\delta_1)^T & 0 \\ 0 & (L_2\delta_1)^T \\ (L_2\delta_1)^T & (L_1\delta_1)^T \\ 0 & 0 \\ 0 & 0 \\ 0 & 0 \end{bmatrix} ; \; A_2 = \begin{bmatrix} (L_1\Delta\delta)^T & 0 \\ 0 & (L_2\Delta\delta)^T \\ (L_2\Delta\delta)^T & (L_1\Delta\delta)^T \\ 0 & 0 \\ 0 & 0 \\ 0 & 0 \end{bmatrix}$$

$\delta = [u^i \ (\frac{\partial u}{\partial s_1})^i \ v^i \ (\frac{\partial v}{\partial s_1})^i \ w^i \ (\frac{\partial w}{\partial s_1})^i \ u^{i+1} \ (\frac{\partial u}{\partial s_1})^{i+1} \ v^{i+1} \ (\frac{\partial v}{\partial s_1})^{i+1} \ w^{i+1}$ $(\frac{\partial w}{\partial s_1})^{i+1}]^T$. $\Delta\delta$ is the increment of δ.

$$\delta_o = [u^i \ (\frac{\partial u}{\partial s_1})^i \ v^i \ (\frac{\partial v}{\partial s_1})^i \ (w^i+2\,\tilde{w}^{\,i}) \ (\frac{\partial w}{\partial s_1})^i \ u^{i+1} \ (\frac{\partial u}{\partial s_1})^{i+1} \ v^{i+1} \ (\frac{\partial v}{\partial s_1})^{i+1} \ (w^{i+1}+2\,\tilde{w}^{\,i+1}) \ (\frac{\partial w}{\partial s_1})^{i+1}]^T$$

$$\delta_1 = [u^i \ (\frac{\partial u}{\partial s_1})^i \ v^i \ (\frac{\partial v}{\partial s_1})^i \ (w^i+\tilde{w}^{\,i}) \ (\frac{\partial w}{\partial s_1})^i \ u^{i+1} \ (\frac{\partial u}{\partial s_1})^{i+1} \ v^{i+1} \ (\frac{\partial v}{\partial s_1})^{i+1} \ (w^{i+1}+\tilde{w}^{\,i+1}) \ (\frac{\partial w}{\partial s_1})^{i+1}]^T$$

in which $\tilde{w}^{\,i}$ and $\tilde{w}^{\,i+1}$ are the deformed value normal to the middle surface of the initial imperfections at the respective nodal lines and which are normal to middle surface. The stiffness matrix and load vector can thus be formed accordingly. The characteristic matrices for other types of finite strips can also be obtained by substituting the displacement functions into Eqs. 3.29 or 3.30.

11.4.5 EXAMPLES

Inelastic stability analysis of simply supported square plates was carried out using finite elements by Pifko and Isakson.[38] This problem was also studied by Lau and Hancock[40] using the spline finite strip method and by Bradford[18] using the special forms of the finite strip method augmented with bubble functions. The results are tabulated in Table 11.1. In the analyses, the stress-strain curve was based on the Ramberg-Osgood representation. The material properties adopted were: $E = 10^7$ psi; $\upsilon = 0.5$; $\sigma_{0.7} = 10^5$psi; $\sigma_{0.85} = 0.906\sigma_{0.7}$, where $\sigma_{0.7}$ and $\sigma_{0.85}$ are the stress corresponding to $E_s = 0.7E$ and $E_s = 0.85E$, respectively.

TABLE 11.1
Inelastic critical stresses for simply supported square plate under uniform compression (critical stresses are in psi, the dimensions of the plates are 20inx20in)

Thickness (in)	Reference 43	Reference 41	Reference 19		Exact
			1 strip	2 strips	
0.77867	65002	65002	65004	65000	65000
1.36678	105002	105000	105001	105000	105000
1.76752	115000	115000	115000	115000	115000
2.39053	125002	125000	125000	125000	125000

The analysis of a cylindrical shell-roof problem was carried out by Kumar et al.[25] The geometry of the roof is shown in Figure 11.12. The shell material was assumed to be elastic-perfectly plastic with the following parameters: Young's modulus = 21000 x 10^3 kN/m^2 ; Poisson's ratio = 0.0 ; Yield stress = 4.1 x 10^3 kN/m^2 ; Maximum load = 3.0 kN/m^2.

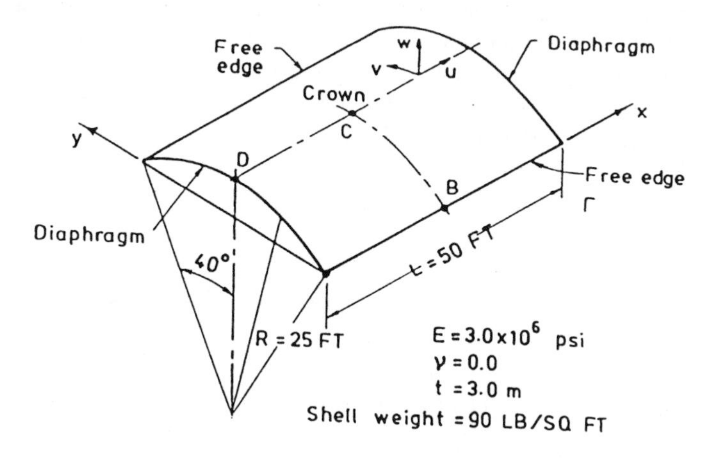

Figure 11.12 Cylindrical shell roof.(25)

In the investigation, twelve strips were used to model one symmetrical half of the shell. The results were compared to those obtained by Figueriras and Owen.(41)

The vertical displacement at the centre of the free edge was plotted against the load in Figure 11.13. It was observed that yielding took place at a load of 1.2kN/m^2; however, the response still remains quite linear up to 2kN/m^2.

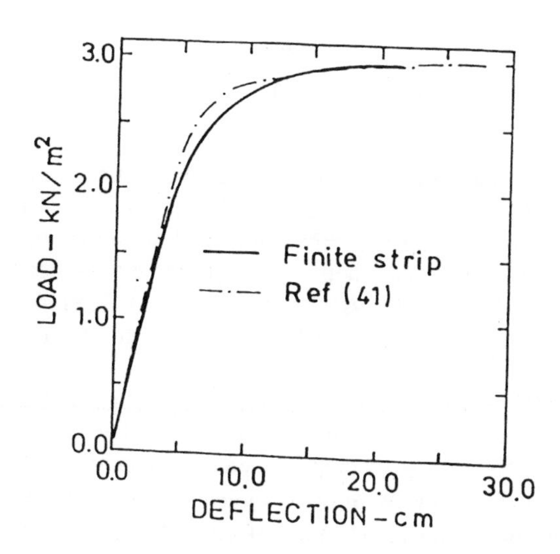

Figure 11.13 Vertical displacement of the free edge of the cylindrical shell roof.(25)

Zhu and Cheung[36] carried out analyses for a series of circular and elliptical plates using spline finite strip models. The mesh division was taken as 4x4 for a quarter of the plate. Geometric transformation was carried out using the domain transformation method as described in Chapter 7. Typical results for elliptical plates are plotted in Figure 11.14.

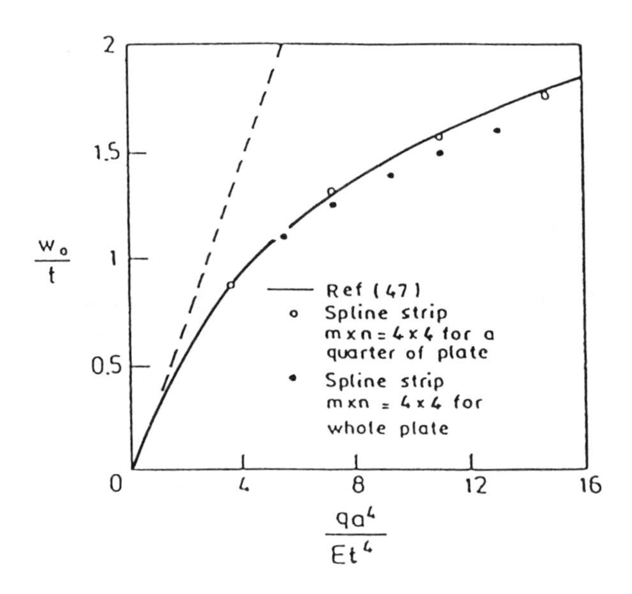

Figure 11.14 Load deflection curve for a clamped elliptical plate with clamped edges under uniform load ($\upsilon = 0.3$).[36]

Olson[42] also studied the response of a cylindrical shell under the action of blast loading by the finite strip method. An analysis was carried out for a cylindrical shell of thickness 0.125in with a mesh as shown in Figure 11.15. The radius and length of the shells were 3in and 6in respectively. Other parameters adopted in the analysis were reported in Reference 42. Figure 11.16 shows the variation of the radial displacements for the nodal lines 1 and 3 with time.

Kwon and Hancock[35] carried out the analysis for a simply supported square plate with a small initial imperfection. The plate was compressed by uniform strain. Loaded edges were assumed to be compressed by frictionless platens and unloaded edges were free to move. Uniform compressive strain was applied using an initial strain ε_o with fixed ends for the longitudinal in-plane displacement.

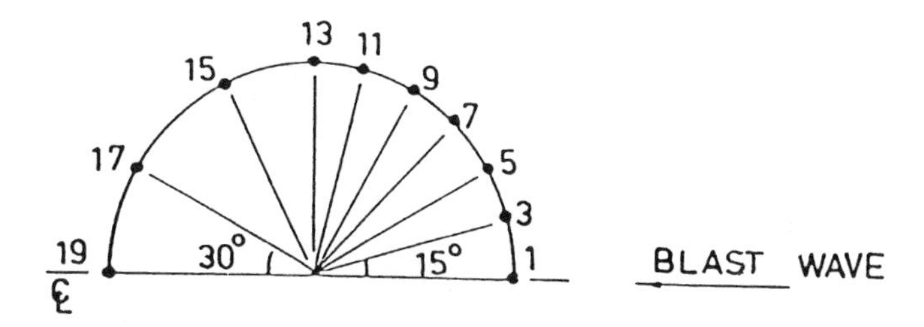

Figure 11.15 Finite strip grid for blast loaded cylindrical shell.[42]

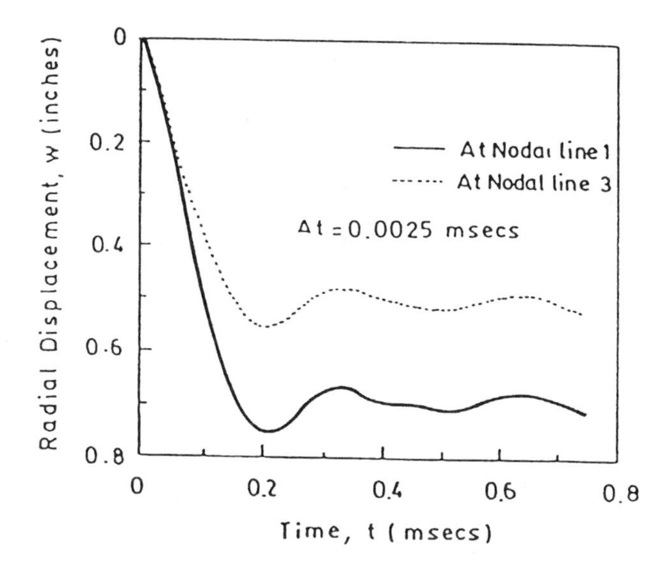

Figure 11.16 Transient response of blast loaded cylindrical shell.[42]

Graphs of average stress versus deflection were shown in Figure 11.17 for increasing numbers of strips and sections. The results showed good agreement with Yamaki's solution.[43] Results produced by uniform compressive stress rather than strain are shown in the same figure.

A square laminated plate simply supported at its edges was analyzed by Lam et al.[30] for out-of-plane deflection. In-plane lateral movement was allowed at the loaded edges and the unloaded edges were free to move. The plate consisted of eight layers in a 0/90° assembly. The properties of the material forming the plate were: $E_L/E_T = 40$; $G_{LT}/E_T = 0.5$; $G_{TT}/E_T = 0.6$; $\upsilon_{LT} = 0.25$ (The subscript 'L' and 'T' denote the longitudinal and transverse directions respectively). Figure 11.18 shows the variation of the longitudinal

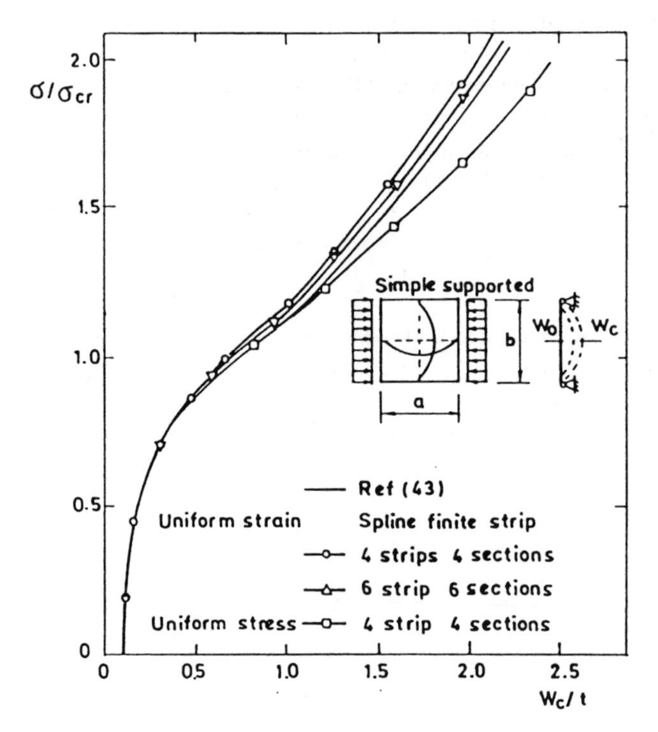

Figure 11.17 Square plate subjected to axial compression.[35]

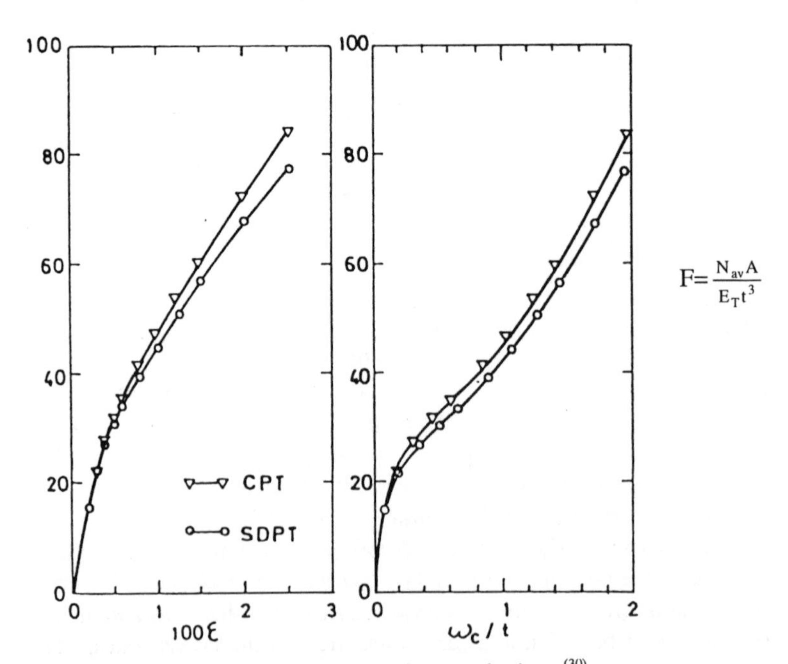

Figure 11.18 Response of eight-layer cross-ply square laminate.[30]

force with the applied end-shortening strain and with the central deflection. In the figure, 'CPT' and 'SDPT' represent the results obtained by formulation based on classical plate theory and Mindlin plate theory respectively.

A perfect circular cylindrical shell under axial compression was studied by Zhu.[37] The geometric parameters of the shell are as follows:
Radius (R) = 100mm; thickness (h) = 0.247mm; length (L) = 71.9mm.

The Young's modulus and Poisson's ratio are 5.56 GPa and 0.3 respectively. The boundary conditions at the two ends of the shell were

$$v = w = \frac{dw}{dx} = 0 \quad ; \quad u = \text{constant}$$

where u, v and w are the meridional, circumferential and normal displacements. The computational results were shown in Figure 11.19 and compared with those of References 44 and 45. It can be seen that the results computed by the spline finite strip method are nearest to the experimental results.

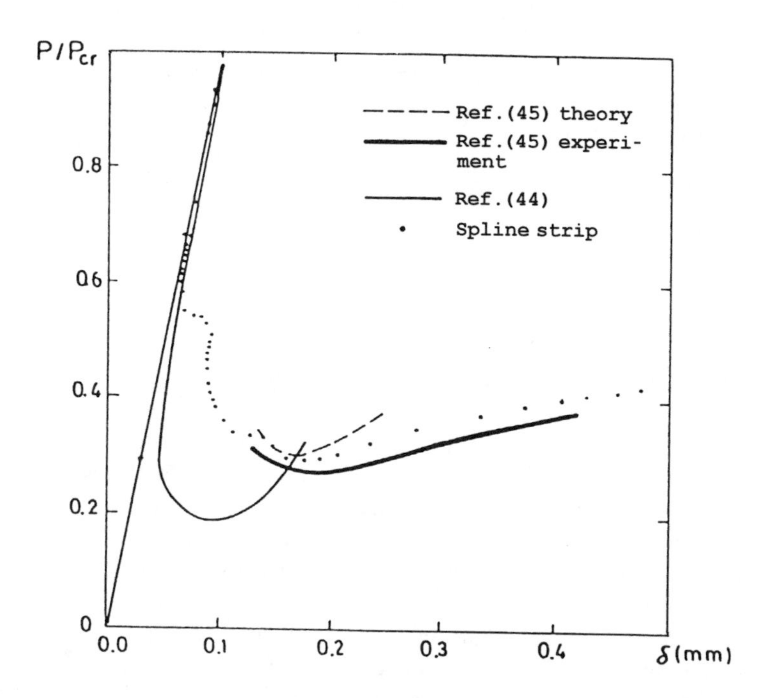

Figure 11.19 Load versus edge-short curve for a perfect circular cylindrical shell under axial compression.[37]

In the case of a shell under the action of pressure load, the loading may become dependent on the displacements if the deformations are large. Zhu[3] had carried out extensive study on this class of problem by applying the spline finite strip method. In one of the examples, a cylindrical shell was assumed to be simply supported on its straight edges and free on the curved ones. The surface area of the shell was 320 x 60 mm^2; thickness (t) = 1 mm; Young's modulus = 5.5GPa and Poisson's ratio = 0.3. The shell was subject to a parabolically distributed pressure with its maximum equal to 0.01N/mm^2. The load-deformation curves were plotted in Figure 11.20.

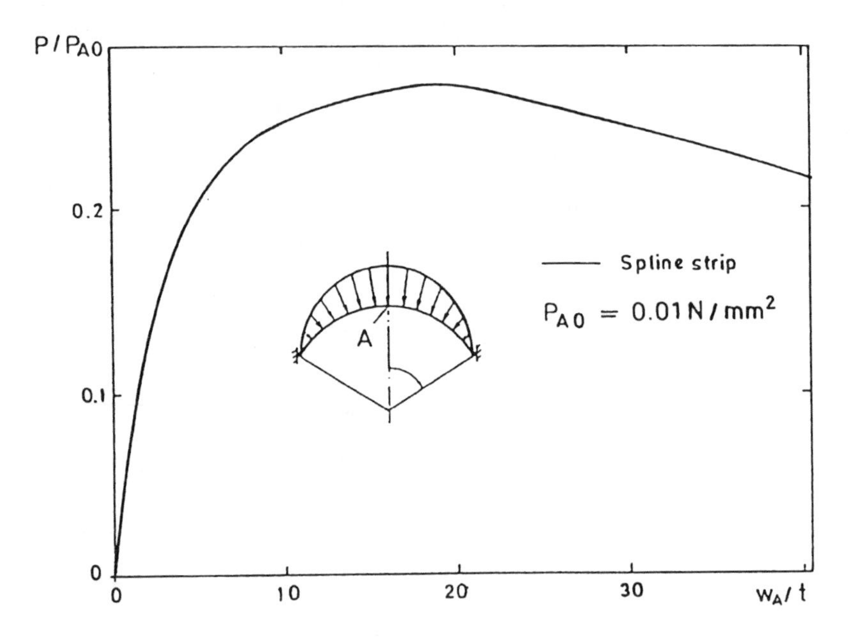

Figure 11.20 Load-central deflection curve of cylindrical shell.[3]

The effect of the imperfection of the shell was studied as well[3]. The cylinder was made of Mylar foil with Young's modulus = 5GPa and Poisson's ratio = 0.3. The dimensions of the shell are given in Figure 11.21.

The computation was performed for a characteristic wave number of 8. Making use of the periodicity and symmetry, only a cylindrical panel of length L/2 and subtended angle π/n was modelled. The initial imperfection $\tilde{w}$ imposed was similar to the buckling configuration and was close to a sinusoidal surface. The panel was idealized by five spline sections in the circumferential direction and five higher order shell strips in the other direction. The results of the analysis are given in Figure 11.21 in the form of equilibrium path curves for different $z = \tilde{w}_{max}/t$, where $\tilde{w}_{max}$ is the maximum amplitude of $\tilde{w}$ and t is the thickness of the shell.

Zhu[3] also demonstrated that his iteration strategy is superior to other iteration methods by comparing the number of iteration steps and iterations to achieve the converged results in the 0.1254-0.1160 load range. The comparison is tabulated in Table 11.2

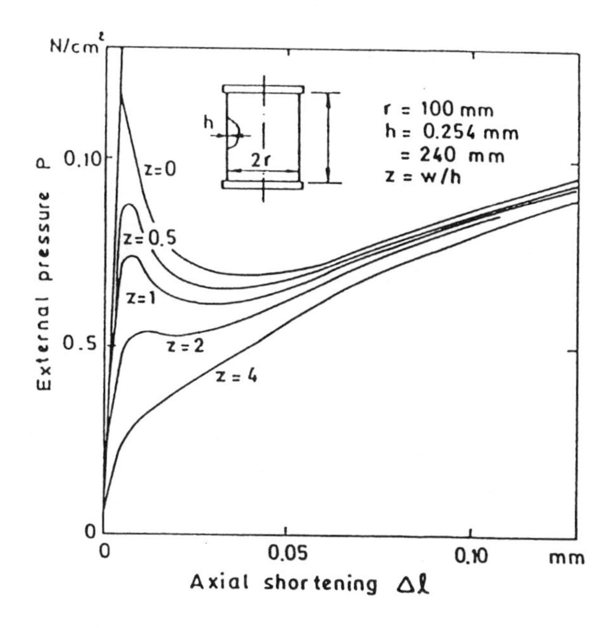

Figure 11.21 Postbuckling curves of an imperfect shell.[3]

TABLE 11.2
A comparison of convergence between Zhu's and Crisfield's approaches.[3]

Method	Number of iteration steps	Total number of iterations
Zhu's approach	4	10
Crisfield's approach	18	65

11.5 NON-LINEAR VIBRATION

The understanding of geometrically non-linear, forced vibration of plates with time-varying loads is important for many engineering structures. The non-linearity is mainly contributed by the membrane action of the plate when it undergoes large out-of-plane deformations.

The basic equations for the non-linear vibration of plates were established by Herrmann.[46] These are the dynamic analogues of the Von Karman equations modified by the transverse inertia term. Based on these equations, numerous studies have been conducted on the non-linear vibration of circular, rectangular and other shaped plates including anisotropic and laminated plates under various boundary conditions. References include Chia,[47] Nayfeh[48] and Sathyamoorthy.[49] Kong [50] carried out analysis on the response of plates under various forms of loads by spline strips. The Ritz vector was adopted to improve the efficiency in the solution process. Lau and Cheung[51,52] developed an incremental harmonic balance method for the analysis of highly non-linear structural vibration problems. This method is capable of treating highly non-linear problems with complicated responses due to the fact that the in-plane displacement and inertia are taken into account and an incremental/iterative procedure is adopted. The non-linear free vibration, forced vibration and internal resonance responses of the plates have been computed by this method. It was also applied to the non-linear vibration analysis of laminated plates and the dynamic instability of plates. Also using spline finite strips, Zhu[(3)] solved the non-linear vibration problems by the incremental harmonic balance method. The mathematical formulation of the problems is rather complicated but fortunately it is clearly explained by Lau,[(53)] Iu[(54)] and Zhu.[(3)] Therefore, readers may refer to the theses for details.

11.5.1 EXAMPLES

A square plate under a uniformly distributed load was analyzed by Kong.[(50)] Half of the plate was divided into four strips and eight sections. Time marching was carried out by using the Newmark scheme with $\Delta\tau = 0.001$sec. Initially, the plate was at rest: displacement, velocity and acceleration were zero. A load varying sinusoidally with time was applied, such that

$$q(x,y,\tau) = 2.441\text{x}10^{-2}\sin\left(\frac{\pi\tau}{0.1257}\right) q_o \text{ kg/cm}^2$$

q_o is taken as $4.882\text{x}10^{-4}$. The results were plotted in Figure 11.22.

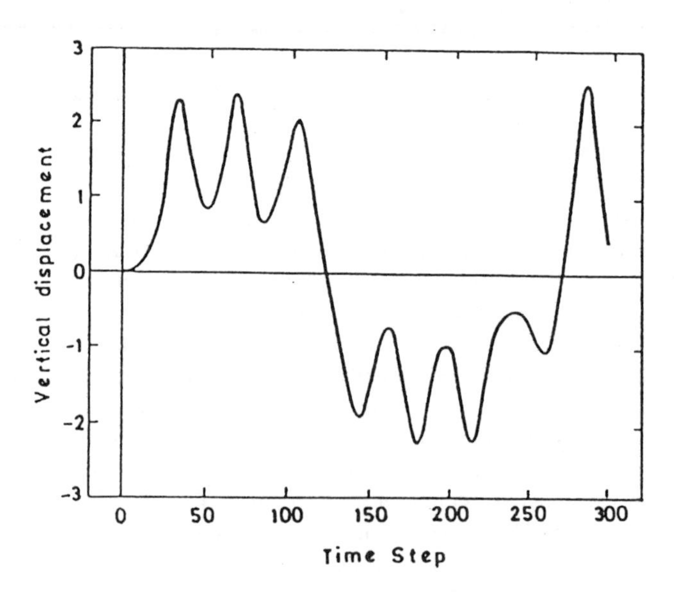

Figure 11.22 Vertical displacement response of square plate subjected to a sinusoidal load of maximum amplitude $50q_o$.[(50)]

The highly non-linear free vibration of a square plate with initial stress was studied.[3] The plate parameters were assumed to be

$$\frac{1}{\alpha^2}\sqrt{\frac{D}{\rho t}} = 1 \quad ; \quad \frac{t}{\alpha} = 0.01 \quad ; \quad \upsilon = 0.3$$

where α is the edge length of the square plate and D is the bending stiffness. The boundary conditions were

$$w = u = v = 0 \qquad \text{at } x = 0, \alpha \text{ or } y = 0, \alpha$$

The plate was subject to an initial force with an intensity equal to

$$S_i^o = \alpha_i \frac{\pi^2 D}{\alpha^2}$$

where α_i is a dimensionless constant. This initial internal force was distributed uniformly along the y-axis direction or both the x-axis and y-axis directions.

The solution process started from an initial solution near the corresponding known linear solution with a sufficiently small amplitude. For the non-linear fundamental frequency, the initial solution could be taken as the fundamental linear frequency for the problem with the initial stress and the generalized coordinate corresponding to the $\cos\omega t$ term, and the first reduced basis is a small value with all the others being zero. It is to be noted that starting from any specified frequency, the computation will automatically converge to some nearby linear frequency. This characteristic is helpful in the analysis of a system in which the linear frequency is not accurately known in advance.

The backbone curves with various initial internal forces in the x- and y directions are shown in Figure 11.23. A_{1c} represents the first harmonic dimensionless amplitudes at the centre point 'c' respectively. It can be seen that an initial compressive force produces a softening effect on the non-linear resonance frequency but an initial tensile force will, on the contrary, produce a hardening effect. The curves obtained by Crawford and Atluri[55] using perturbation techniques are shown in the same figure. They themselves have pointed out that the upper part of the curves obtained are not consistent with experimental ones. The figure also shows that the results for the case without initial stress are identical to those obtained by Yamaki et al.[56] One can note that the results for $\alpha_x = \alpha_y = \pm 1$ are identical to those for $\alpha_x = \pm 2$ while $\alpha_y = 0$.

The forced vibrations with damping and different initial stresses for a square plate were computed. The parameters and boundary conditions of the plate were the same as those of the previous example. The plate was excited by a uniformly distributed force with an intensity equal to

$$q = 1.5\,\frac{Dh}{\alpha^4}\cos\omega t$$

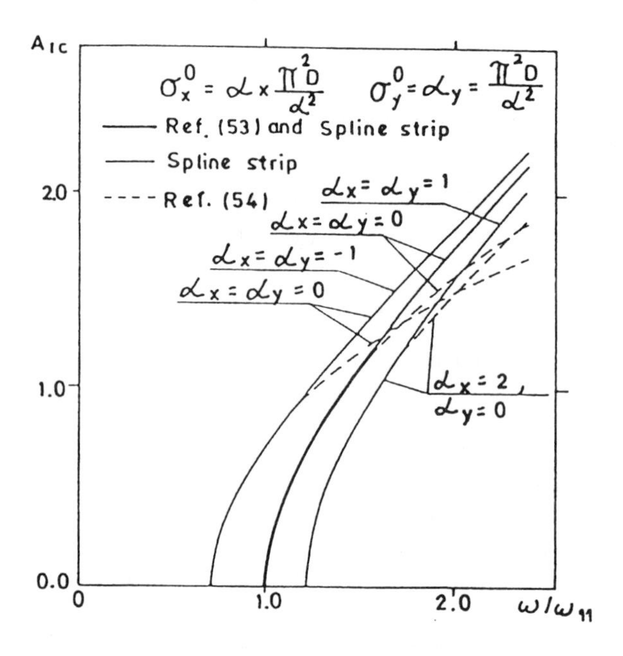

Figure 11.23 Backbone curve of square plate with initial stress.[3]

Due to the presence of damping forces which are related to the velocities of the vibrating system, a cosine harmonic term in the Fourier expansion of the displacements will produce a sine harmonic term. For consistency, it was necessary to include the sine harmonic terms in the expansion. Hence, the expression for the out-of-plane motion was written as

$$\frac{w(x,y,\tau)}{t} = \sum_{i=0}^{3} A_{ci} \cos i\omega\tau + \sum_{i=0}^{3} A_{si} \sin i\omega\tau = A_o + \sum_{i=0}^{3} A_i \cos(i\omega\tau + \phi_i)$$

The two branches of the undamped responses were connected here through a gradual change of the phase difference of each harmonic amplitude.

The computation was carried out for a constant damping ratio $\xi_I = 0.05$. The amplitude response curves with the different initial stresses were illustrated in Figure 11.24. It can be seen from the figure that the initial tensile stress reduces the amplitude of the vibration with a translation of the backbone curve to the direction of increasing frequency but the initial compressive stress produces an opposite effect. This trend is in agreement with the conclusion that initial tensile stress increases the stiffness of plates but initial compressive stress reduces it.

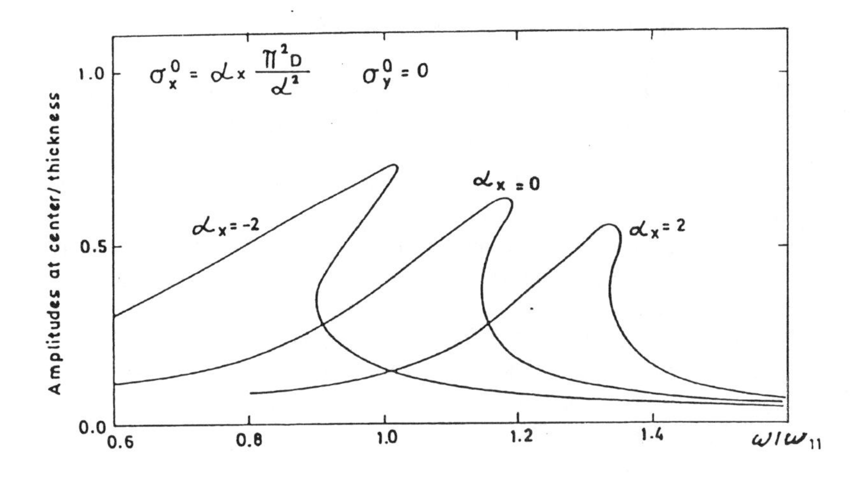

Figure 11.24 Damped vibration of a square plate with initial stress.[(3)]

REFERENCES

1. Riks, E., An incremental approach to the solution of snapping and buckling problems, *Int J of Solids and Struct*, 15, 529-551, 1979.
2. Crisfield, M. A., A fast incremental-iterative solution procedure that handles snap-through, *Comp and Struct*, 13, 55-72, 1981.
3. Zhu, D. S., *Nonlinear static and dynamic analysis of plates and shells by spline finite strip method*, PhD thesis, Department of Civil and Structural Eng, University of Hong Kong, 1988.
4. Mofflin, D. S., *Plate buckling in steel and aluminum*, PhD thesis, University of Cambridge, England, 1983.
5. Cheung, M. S., Ng, S. F. and Zhong, B., Finite strip analysis of beams and plates with material nonlinearity', *Comp and Struct*, 33, 1031-1035, 1989.
6. Cheung, M. S. and Li, W., Finite strip method for material nonlinear analysis of reinforced concrete slabs, *Comp and Struct*, 35, 603-607, 1990.
7. Ng, S. F., Cheung, M. S. and Zhong, B., Finite strip method for analysis of structures with material nonlinearity, *J of Struct Eng, ASCE*, 117, 489-550, 1991.

8. Cheung, M. S., Akhras, G. and Li, W., Progressive failure of composite plates by finite strip method, *Comp Meth in App Mech and Eng*, 124, 49-61, 1995.
9. Lee, J. D., Three-dimensional finite element analysis of damage accumulation in composite laminate, *Comp and Struct*, 15, 335-350, 1982.
10. Olawale, A. O. and Plank, R. J., The collapse analysis of steel columns in fire using a finite strip method, *Int J for Num Meth in Eng*, 26, 2755-2764, 1988.
11. Bugress, I. W., Olawale, A. O. and Plank, R. J., Failure of steel columns in fire, *J of Fire Safety*, 18, 183-201, 1992.
12. Taylor, S., Maher, D. R. H. and Hayes, B., Effect of the arrangement of reinforcement on the behaviour of reinforced concrete slabs, *Mag Con Research*, 18, 85-94, 1966.
13. Tolson, S. and Zabaras, N., Finite element analysis of progressive failure in laminated composite plates, *Comp and Struct*, 38(3), 361-376, 1991.
14. Delcourt, C. R., *Linear and geometrically nonlinear analysis of flat-walled structures by the finite strip method*, PhD thesis, University of Adelaide, Australia, 1978.
15. Plank, R. J., *The initial buckling of thin walled structures under combined loading*, PhD thesis, University of Birmingham, 1973.
16. Ylinen, A., Lateral buckling of an I-beam in pure bending beyond the limit of proportionality, *Proc of 2nd Conf on Dimensioning and Strength Calculations, Hungarian Academy of Sciences, Budapest*, 157-167, 1965.
17. Azhari, M. and Bradford, M. A., Local buckling by complex finite strip method using bubble finite strips, *J of Eng Mech, ASCE*, 120, 43-57, 1994.
18. Bradford, M. A. and Azhari, M., Inelastic local buckling of plates and plate assemblies using bubble functions, *Eng Struct*, 17, 95-103, 1995.
19. Lau, S. C. W. and Hancock, G. J., Inelastic buckling analyses of beams, columns and plates using the spline finite strip method, *Thin-walled Struct*, 7, 213-238, 1989.
20. Cheung, M. S. and Li, W., A modified finite strip method for geometrically nonlinear analysis of plates, *Comp and Struct*, 33, 1031-1035, 1989.
21. Dawe, D. J. and Azizian, Z. G., The performance of Mindlin plate finite strips in geometrically nonlinear analysis, *Comp and Struct*, 23, 1-14, 1986.
22. Gierlinski, J. T. and Graves-Smith, T. R., The geometric nonlinear analysis of thin-walled structures by finite strips, *Thin-walled Struct*, 2, 27-50, 1984.

23. Langyel, P. and Cusens, A. R., Finite strip method for the geometrically nonlinear analysis of plate structures, *Int J for Num Method in Eng*, 19, 331-340, 1983.
24. Abayakoon, S. B. S., Olson, M. D. and Anderson, D. L., Large deflection elastic-plastic analysis of plate structures by the finite strip method, *Int J for Num Meth in Eng*, 28, 331-358, 1989.
25. Kumar, P., Olson, M. D. and Anderson, D. L., Large deflection elastic-plastic analysis of cylindrical shells using the finite strip method, *Int J for Num Meth in Eng*, 31, 837-857, 1991.
26. Khalil, M. R., Olson, M. D. and Anderson, D. L., Nonlinear dynamic analysis of stiffened plates, *Comp and Struct*, 29, 929-941, 1988.
27. Azizian, Z. G. and Dawe, D. J., Geometrically nonlinear analysis of rectangular Mindlin plates using the finite strip method, *Comp and Struct*, 21, 423-436, 1985.
28. Azizian, Z. G. and Dawe, D. J., Analysis of the large deflection behaviour of laminated composite plates using the finite strip method, *Composite Struct*, 3, 677-691, 1985.
29. Dawe, D. J., Lam, S. S. E. and Azizian, Z. G., Nonlinear finite strip analysis of rectangular laminates under end shortening, using classical plate theory, *Int J for Num Meth in Eng*, 35, 1087-1110, 1992.
30. Lam, S. S. E., Dawe, D. J. and Azizian, Z. G., Nonlinear analysis of rectangular laminates under end shortening, using shear deformation plate theory, *Int J for Num Meth in Eng*, 36, 1045-1064, 1993.
31. Dawe, D. J., Lam, S. S. E. and Azizian, Z. G., Finite strip post-buckling analysis of composite prismatic plate structures, *Comp and Struct*, 48, 1011-1023, 1993.
32. Sekulovic, M. and Milasinovic, D., Nonlinear analysis of plate and folded structures by the finite strip method, *Eng Computations*, 4, 41-47, 1987.
33. Cheung, M. S., Li, W. and Jaeger, L. G., Nonlinear analysis of cable-stayed bridges by the finite strip method, *Comp and Struct*, 29, 687-692, 1988.
34. Kwon, Y. B., *Post-buckling behaviour of thin-walled channel sections*, PhD thesis, School of Civil and Mining Engineering, University of Sydney, 1992.
35. Kwon, Y. B. and Hancock, G. J., A nonlinear elastic spline finite strip analysis for thin-walled section, *Thin-walled Struct*, 12, 295-319, 1991.
36. Zhu, D. S. and Cheung, Y. K., Large deflection analysis of arbitrary shaped thin plates, *Comp and Struct*, 26, 811-814, 1987.
37. Zhu, D. S. and Cheung, Y. K., Postbuckling analysis of shells by spline finite strip method, *Comp and Struct*, 31, 357-364, 1989.
38. Pifko, A. and Isakson, G., A finite element method for the plastic buckling analysis of plates, *J of AIAA*, 7, 1950-1957, 1969.
39. Gupta, K. K., On a numerical solution of the plastic buckling problem of structures, *Int J for Num Meth in Eng*, 12, 941-947, 1978.

40. Lau, S. C. W. and Hancock, G. J., Inelastic buckling analysis of beams, columns and plates using the spline finite strip method, *Research Report 553, School of Civil and Mining Engineering, University of Sydney, Australia*, 1987.
41. Figueriras, J. A. and Owen, D. R. J., Analysis of elasto-plastic and geometrically non-linear anisotropic plates and shells, *Finite element software for plates and shells* (ed. Hinton and Owen), Pineridge Press, Swansea, UK, 1984.
42. Olson, M. D., Efficient modelling of blast loaded stiffened plate and cylindrical shell structures, *Comp and Struct*, 40(5), 1139-1149, 1991.
43. Yamaki, N., Postbuckling behaviour of rectangular plates with small initial curvature loaded in edge compression, *J App Mech, ASME*, 407-414, 1959.
44. Matsui, T. and Matsuoka, O., A new finite element scheme for instability analysis of thin shell, *Int J for Num Meth in Eng*, 10, 145-170, 1976.
45. Yamaki, N., *Elastic stability of circular cylindrical shells*, Elsevier, Amsterdam, 1984.
46. Herrmann, G., *Influence of large amplitudes on flexural motions of elastic plates*, NACA TN, 3578, 1956.
47. Chia, C. Y., *Nonlinear analysis of plates*, McGraw-Hill, NY, 1980.
48. Nayfeh, A. H. and Mook, D. T., *Nonlinear oscillations*, Wiley-Interscience, 1979.
49. Sathyamoorthy, M., Nonlinear vibration of plates - a review, *Shock Dig*, 15(6), 3-16, 1983.
50. Kong, J., *Analysis of plate-type structures by finite strip, finite prism and finite layer method*, PhD thesis, Department of Civil and Struct Eng, University of Hong Kong, 1994.
51. Cheung, Y. K., Lau, S. L. and Wu, S. Y., Incremental time-space finite strip method for nonlinear structural vibrations, *Earthquake Eng and Struct Dyn*, 10, 239-253, 1982.
52. Lau, S. L., Cheung, Y. K. and Wu, S. Y., Nonlinear vibration of thin elastic plates, Part 1: Generalised incremental Hamilton's principle and element formulation. *J App Mech, ASME*, 51, 837-844, 1984.
53. Lau, S. L., *Incremental harmonic balance method for nonlinear structural vibration*, PhD thesis, Department of Civil and Structural Engineering, University of Hong Kong, 1982.
54. Iu, V. P., *Nonlinear vibration analysis of multilayer sandwich structures by incremental finite elements*, PhD thesis, Department of Civil and Structural Engineering, University of Hong Kong, 1985.
55. Crawford, J. and Atluri, S., Nonlinear vibrations of a flat plate with initial stresses, *J of Sound and Vib*, 43(1), 117-129, 1975.
56. Yamaki, N. and Chiba, M., Nonlinear vibrations of a clamped rectangular plate with initial deflection and initial edge displacement- Part 1: Theory. *Thin-walled Struct*, 1, 3-29, 1983.

CHAPTER TWELVE

TRANSFORMATION APPROACH: FOURIER AND HANKEL TRANSFORMS

12.1 INTRODUCTION

Another way to reduce the order of differential equations so that they can become more tractable is by applying appropriate transformations. Green's functions have been used extensively in the boundary element method to reduce the dimension of the differential equations by one or two. As noted in Chapter 10, the finite layer or prism models can be used to analyze three-dimensional layered soil problems by interpolating in one or two directions respectively. An alternative approach using suitable transformations to achieve the same results was suggested by Gerrard[(1)]. If the problem is defined in the Cartesian coordinate system, single or double Fourier transforms are applied to the variables, governing equations, and boundary conditions as well as the continuity conditions between the interface of the layer. For axisymmetrical problems, however, Hankel transforms are more suitable as the problems are defined in terms of polar coordinate variables. For time-dependent problems, a Laplace transform has to be carried out in the time domain.

Solutions will be first sought in the transformed space in terms of the transformed variables. To obtain the field quantities, inverse transform has to be carried out analytically or numerically.

The effect of surface loads on layered soils was reported by Gerrard and his associates[(1,2,3,4,5,6)] using Laplace and Hankel transforms. Booker[(7)] then formulated the problem of consolidation for a single layer of homogeneous isotropic clay by applying a Fourier transform followed by a Laplace transform. The solution was extended to layered soil problems by Small and Booker.[(8,9)] Rowe et al. presented a solution for a soil with a crust[(10,11)] and soil with an elastic modulus varying exponentially with depth.[(12)] Further attempts were made to solve problems with materials exhibiting viscoelastic behaviour.[(13,14)]

A detailed analysis of consolidation problems was reported soon after the publication of the original papers on elasticity by Booker and Small.[(15,16)] In their paper, results for layered soils under strip, circular and rectangular loadings were presented. Harnpattanapanich et al.[(17,18)] presented solutions for consolidation of Gibson soils. Secondary consolidation was also considered.[(19)] The advantages of the direct numerical scheme for the inversion of the Laplace transform were discussed in References 19 and 20. The consolidation of clay under embankment loading was the subject of a study conducted by Small and Zhang.[21] The settlement due to the drawdown of a point sink was studied by Booker et al.[22,23]

This was later extended to a surface raft on consolidating soils,[24,25] while further developments involving pile foundations were carried out by Small and his collaborators.[26,27,28,29] They included single pile and pile group as well as pile group interacting with the pile cap. This method also found applications in the analyses of under-reamed anchored plates and thermal effect due to a decaying source.

Reviews on the developments of the method were made in a number of papers[30-36] and readers would find them informative and useful.

Another area of development of the method is its applications to pollutant transport problems. The earliest paper on this subject was reported by Rowe and Booker[37] on a one-dimensional transport problem. It was accompanied by a series of other papers.[38-48]

12.2 PLANE STRAIN ANALYSIS[8]

Let the plane strain problem under static load be defined in the x-z plane. It is obvious that there is no deformation in the y-direction. As was discussed in Chapter 4, the displacement field can be defined in terms of the displacements in the x and z directions. In addition, the stress (strain) field variables consist of two direct stresses (strains) and one shear stress (strain). Applying Fourier transforms to these variables, we can define our problem in the transformed variables. The Fourier transforms to be applied are:

$$\mathbf{U} = [U, W]^T = \frac{1}{2\pi}\int_{-\infty}^{+\infty} [iu\,,\, w]^T e^{-i\alpha x}\,dx \tag{12.1}$$

$$\mathbf{S} = [S_{xx}\,,\, S_{zz}\,,\, T_{xz}]^T = \frac{1}{2\pi}\int_{-\infty}^{+\infty} [\sigma_{xx}\,,\, \sigma_{zz}\,,\, i\tau_{xz}]^T e^{-i\alpha x}\,dx \tag{12.2}$$

$$\mathbf{E} = [E_{xx}\,, E_{zz}\,, G_{xz}]^T = \frac{1}{2\pi}\int_{-\infty}^{+\infty} [\varepsilon_{xx}\,,\, \varepsilon_{zz}\,,\, i\gamma_{xz}]^T e^{-i\alpha x}\,dx \tag{12.3}$$

where u and w are the x and z displacements respectively, α is the argument of the transform,
σ_{xx} , σ_{zz} , τ_{xz} are the stress variables, and
ε_{xx} , ε_{zz} , γ_{xz} are the strain variables.

The state equations for an anisotropic elastic body which exhibits transverse isotropy can be re-written as:

a) The stress-strain relationship:

$$\begin{Bmatrix} E_{XX} \\ E_{ZZ} \\ E_{XZ} \end{Bmatrix} = \begin{bmatrix} A & -B & 0 \\ -B & C & 0 \\ 0 & 0 & F \end{bmatrix} \begin{Bmatrix} S_{XX} \\ S_{ZZ} \\ S_{XZ} \end{Bmatrix} \tag{12.4}$$

$$\mathbf{E} = R\,\mathbf{S}$$

where $A = \dfrac{(1-\upsilon_h^2)}{E_h}$; $B = \dfrac{\upsilon_{hv}(1-\upsilon_v)}{E_h}$; $C = \dfrac{1}{E_v} - \dfrac{\upsilon_{hv}^2}{E_h}$; $F = \dfrac{1}{G_{hv}}$

(the subscripts 'h' and 'v' signify the horizontal and vertical directions respectively).

b) The equations of equilibrium:

$$\begin{bmatrix} \alpha & 0 & -\frac{\partial}{\partial z} \\ 0 & -\frac{\partial}{\partial z} & -\alpha \end{bmatrix} \begin{Bmatrix} S_{xx} \\ S_{zz} \end{Bmatrix} = 0 \tag{12.5}$$

$$\mathbf{M}(\alpha,z)\,\mathbf{S} = 0$$

where $\mathbf{M}(\alpha,z) = \begin{bmatrix} \alpha & 0 & -\frac{\partial}{\partial z} \\ 0 & -\frac{\partial}{\partial z} & -\alpha \end{bmatrix}$

c) The strain-displacement relationship:

$$\begin{Bmatrix} E_{xx} \\ E_{zz} \\ E_{xz} \end{Bmatrix} = \begin{bmatrix} \alpha & 0 \\ 0 & -\frac{\partial}{\partial z} \\ -\frac{\partial}{\partial z} & -\alpha \end{bmatrix} \begin{Bmatrix} u \\ w \end{Bmatrix} \tag{12.6}$$

$$\mathbf{E} = -\mathbf{N}(\alpha,z)\,U$$

Note that $\mathbf{N}(\alpha,z) = M^T(\alpha,-z)$.

The solution for the problem can be obtained by introducing a stress function (Φ). This function is related to the stresses as follows:

$$S_{xx} = \frac{\partial^2\Phi}{\partial z^2} \quad ; \quad S_{zz} = -\alpha^2\,\Phi \quad ; \quad T_{xz} = \alpha\frac{\partial\Phi}{\partial z}$$

It is noted that the equilibrium equations are satisfied automatically, and the stress-strain relationship becomes:

$$\alpha U = A\,\frac{\partial^2\Phi}{\partial z^2} + B\,\alpha^2\,\Phi \tag{12.7a}$$

$$\frac{\partial W}{\partial z} = -B\,\frac{\partial^2\Phi}{\partial z^2} - C\,\alpha^2\,\Phi \tag{12.7b}$$

$$\frac{\partial U}{\partial z} - \alpha\,W = F\,\alpha^2\,\frac{\partial^2\Phi}{\partial z^2} \tag{12.7c}$$

Eliminating U and W from the equations, we have

$$A\,\frac{\partial^4\Phi}{\partial z^4} + \alpha^2\,(2B\text{-}F)\,\frac{\partial^2\Phi}{\partial z^2} + \alpha^4\,C\,\Phi = 0 \tag{12.8}$$

It can be shown readily that the solution for the above equation is:

$$\Phi = L_a\cosh(pz) + M_a\cosh(qz) + L_b\sinh(pz) + M_b\sinh(qz) \tag{12.9}$$

where $\frac{p}{\alpha} = \frac{-(2B-F)+\sqrt{(2B-F)-4AC}}{2C}$ and $\frac{q}{\alpha} = \frac{-(2B-F)-\sqrt{(2B-F)-4AC}}{2C}$

Having established the solution for the stress function, the flexibility relationship for a typical layer (i-th layer) can be written as

$$\begin{Bmatrix} W_k \\ U_k \\ -W_l \\ -U_l \end{Bmatrix}^i = \begin{bmatrix} A\frac{\Omega_a+\Omega_b}{2\alpha} & A\frac{\chi_a+\chi_b}{2\alpha}-\frac{B}{\alpha} & -A\frac{\Omega_a-\Omega_b}{2\alpha} & A\frac{\chi_b-\chi_a}{2\alpha} \\ & A\frac{\psi_a+\psi_b}{2\alpha} & -A\frac{\chi_b-\chi_a}{2\alpha} & A\frac{\psi_b-\psi_a}{2\alpha} \\ & & A\frac{\Omega_a+\Omega_b}{2\alpha} & -A\frac{\chi_a+\chi_b}{2\alpha}+\frac{B}{\alpha} \\ & & & A\frac{\psi_a+\psi_b}{2\alpha} \end{bmatrix}^i \begin{Bmatrix} N_k \\ T_k \\ N_l \\ T_l \end{Bmatrix}^i$$

$$\tilde{\delta}^i = F^i\,\tilde{p}^i \tag{12.10}$$

where $\tilde{\delta} = \{W_k, U_k, -W_l, -U_l\}^T$ and $\tilde{p} = \{N_k, T_k, N_l, T_l\}^T$. N and T are the normal and shear stress coefficient. The subscripts k and l define the faces of the layer. In addition,

$$\Omega_a = \frac{pq(p^2-q^2)S_pS_q}{D_a} \; ; \; \Omega_b = \frac{pq(p^2-q^2)C_pC_q}{D_b} \; ; \; \chi_a = \frac{pq}{\alpha^2}\frac{D_b}{D_a} \; ;$$

$$\chi_b = \frac{pq}{\alpha^2}\frac{D_a}{D_b} \; ; \; \psi_a = \frac{\alpha^2(p^2-q^2)C_pC_q}{D_a} \; ; \; \psi_b = \frac{\alpha^2(p^2-q^2)S_pS_q}{D_b}$$

$D_a = \alpha^3\,(pS_p\,C_q - qS_q\,C_p)$; $D_b = \alpha^3\,(pC_p\,S_q - qC_q\,S_p)$

$C_p = \cosh(ph)$; $C_q = \cosh(qh)$; $S_p = \sinh(ph)$; $S_q = \sinh(qh)$.

In the case of isotropic materials, the roots (p,q) are identical and the above expressions are modified as:

$$\Omega_a = \frac{2p^3}{\alpha^3}\frac{(C_{2p}-1)}{(S_{2p}+2ph)} \; ; \quad \Omega_b = \frac{2p^3}{\alpha^3}\frac{(C_{2p}+1)}{(S_{2p}-2ph)}$$

$$\chi_a = \frac{p^2}{\alpha^2}\frac{(S_{2p}-2ph)}{(S_{2p}+2ph)} \; ; \quad \chi_b = \frac{p^2}{\alpha^2}\frac{(S_{2p}+2ph)}{(S_{2p}-2ph)}$$

$$\psi_a = \frac{2p}{\alpha}\frac{(C_{2p}+1)}{(S_{2p}+2ph)} \; ; \quad \psi_b = \frac{2p}{\alpha}\frac{(C_{2p}-1)}{(S_{2p}-2ph)}$$

The flexibility relation for the whole layered system can be formed by incorporating the continuity conditions for displacements and stresses along the layer interfaces. Mathematically, the conditions for two adjacent layers are:

$$W^a = W^b \; ; \; U^a = U^b \tag{12.11}$$
$$N^a = N^b \; ; \; T^a = T^b \tag{12.12}$$

The superscripts a, b denote values just above and below the layer interfaces respectively.

The flexibility equation thus formed may be written as

$$\tilde{\delta} = F\,\tilde{P} \tag{12.13}$$

where $\tilde{\delta} = \sum_{i=1}^{n} \tilde{\delta}^i$; $\tilde{P} = \sum_{i=1}^{n} \tilde{p}^i$; $F = \sum_{i=1}^{n} F^i$

The above equation can be solved by subjecting it to the boundary conditions at the surface and base:

Base (z = H)	Rough - Rigid Smooth - Rigid	$U = 0$; $W = 0$ $T_{xz} = 0$; $W = 0$
Surface (z = 0)	Specified surface traction	$T_{xz} = F_x$ $S_{zz} = F_z$

where $(F_x , F_z) = \frac{1}{2\pi}\int_{-\infty}^{+\infty} [if_x , f_z]^T e^{-i\alpha x} dx$, and f_x , f_z are the applied surface tractions.

It is recommended that the displacement coefficients are to be determined by:

$$\begin{Bmatrix} W_{av} \\ U_{av} \end{Bmatrix} = \frac{1}{\alpha}\begin{bmatrix} A\Omega_a + B & A\chi_a - B \\ A\chi_a - B & A\psi_a \end{bmatrix}\begin{Bmatrix} N_{av} \\ T_{av} \end{Bmatrix} \tag{12.14}$$

where $W_{av} = \dfrac{W_k - W_l}{2}$; $U_{av} = \dfrac{U_k + U_l}{2}$; $N_{av} = \dfrac{N_k + N_l}{2}$; $T_{av} = \dfrac{T_k - T_l}{2}$.

A three layered system (Figure. 12.1) was considered by Booker[8]. The loading was applied in the region $-a \le x \le a$, and the depths of the layers were respectively 2a, 2a and 10a. The material properties of each of the layers were tabulated in Table 12.1. Results of the analysis are plotted in Figure 12.1 in terms of dimensionless parameters.

TABLE 12.1
Material properties of three layered soil system.

	Case 1			Case 2		
	Layer 'A'	Layer 'B'	Layer 'C'	Layer 'A'	Layer 'B'	Layer 'C'
E/E_c	25	5	1	5	25	1
υ	0.3	0.4	0.5	0.4	0.3	0.5

(E_c is the modulus of Layer 'C')

12.3 CIRCULAR LOADING[9]

Hankel transforms can be applied to the variables for problems of layered soils under circular loadings:

$$W = \int_0^{\infty} r\,w\,J_o(\alpha r)\,dr \tag{12.15}$$

$$U_r , U_\theta = \int_0^{\infty} r\,(u , u_\theta)\,J_1(\alpha r)\,dr \tag{12.16}$$

$$\varepsilon = \int_0^{\infty} \alpha[\mathbf{N}_1 (\alpha,r,z)\,J_o(\alpha r) + \mathbf{N}_2 (\alpha,r,z)\,J_1(\alpha r)]\mathbf{U}\,d\alpha \tag{12.17}$$

where $U = [U_r , U_\theta, W]$; $\varepsilon = [\varepsilon_{rr} , \varepsilon_{\theta\theta} , \varepsilon_{zz} , \gamma_{rz} , \gamma_{\theta z} , \gamma_{r\theta}]$

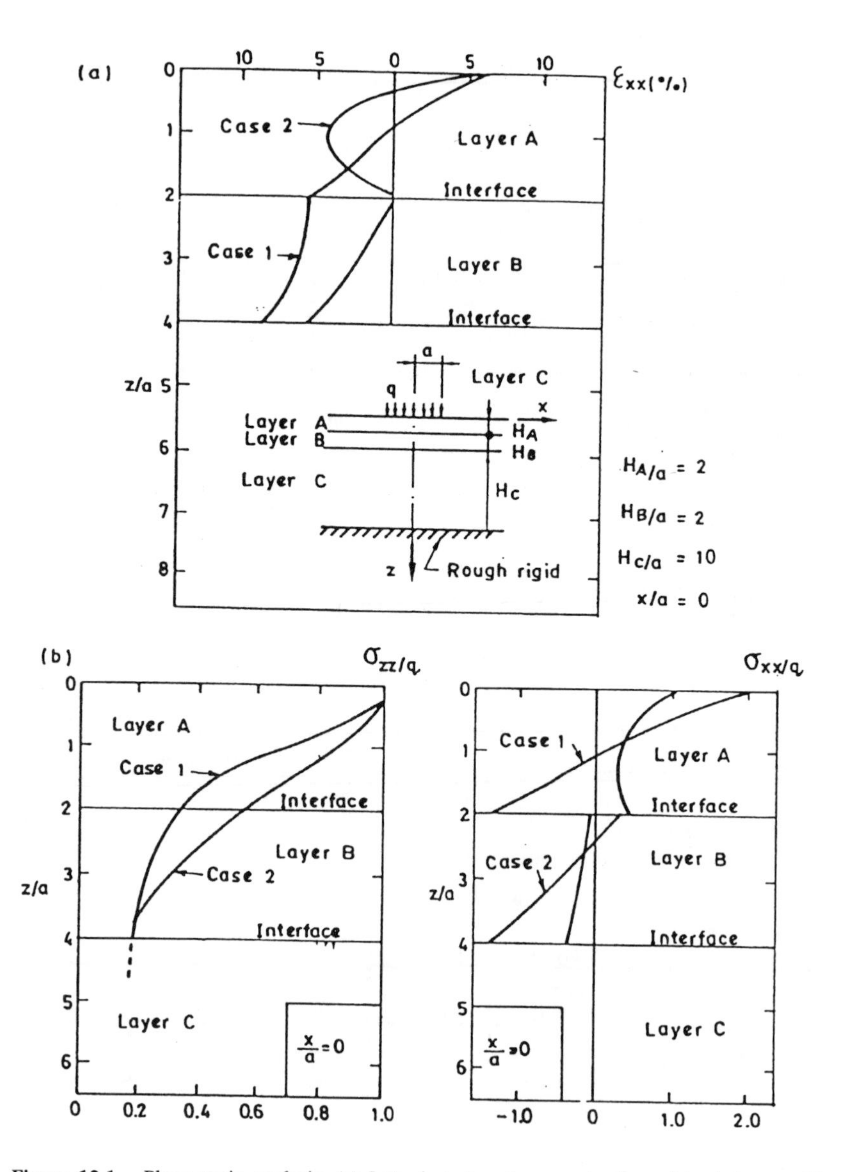

Figure 12.1 Plane strain analysis: (a) Lateral strains along centre line of loaded area. (b) Stresses along centre line of loaded area.[9]

$$\mathbf{N}_1(\alpha,r,z) = \begin{bmatrix} \alpha & 0 & 0 & 0 & 0 & 0 \\ 0 & 0 & 0 & 0 & 0 & \alpha \\ 0 & 0 & \frac{\partial}{\partial z} & 0 & 0 & 0 \end{bmatrix}^T ; \mathbf{N}_2(\alpha,r,z) = \begin{bmatrix} -\frac{1}{r} & \frac{1}{r} & 0 & \frac{\partial}{\partial z} & 0 & 0 \\ 0 & 0 & 0 & 0 & \frac{\partial}{\partial z} & -\frac{2}{r} \\ 0 & 0 & 0 & -\alpha & 0 & 0 \end{bmatrix}^T$$

The equations of equilibrium are given by:

$$-\alpha[\alpha\, a\, U_r + c\,\frac{\partial W}{\partial z}] - \frac{\partial}{\partial z}[\alpha W - \frac{\partial U_r}{\partial z}](\frac{f}{2}) = 0 \qquad (12.18a)$$

$$-\alpha[\alpha\, W - \frac{\partial U_r}{\partial z}] - \frac{\partial}{\partial z}[\alpha\, c\, U_r + d\,\frac{\partial W}{\partial z}] = 0 \qquad (12.18b)$$

$$-\alpha^2(\frac{a-b}{2})U_\theta + \frac{f}{2}\,\frac{\partial^2 U_\theta}{\partial z^2} = 0 \qquad (12.18c)$$

For an anisotropic material with $E_r = E_\theta = E_h$; $E_z = E_v$; $G_{rz} = G_{r\theta} = G_{hv}$,

$$a = \frac{(1-\upsilon_{hv}{}^2)E_h}{(1+\upsilon_h)(1-\upsilon_h-2\upsilon_{hv}^2)}\,;\; b = \frac{(\upsilon_h+\upsilon_{hv}{}^2)E_h}{(1+\upsilon_h)(1-\upsilon_h-2\upsilon_{hv}^2)}\,;$$

$$c = \frac{\upsilon_{hv}E_h}{(1-\upsilon_h-2\upsilon_{hv}^2)}\,;\; d = \frac{(1-\upsilon_h)E_v}{(1-\upsilon_h-2\upsilon_{hv}^2)}\,;\; f = 2G_{hv}.$$

It is noted that U_θ becomes uncoupled and the solution for Eq. 12.18c can be sought independently.

Solution for Eq. 12.18a and Eq. 12.18b:

Introducing the following variables:

$$M = \alpha\, a\, U_r + c\,\frac{\partial W}{\partial z} \qquad (12.19a)$$

$$T = [-\alpha W + \frac{\partial U_r}{\partial z}](\frac{f}{2}) \qquad (12.19b)$$

$$N = \alpha\, c\, U_r + d\,\frac{\partial W}{\partial z} \qquad (12.19c)$$

one can show that the variables are related to a stress function Φ as follows:

$$M = \frac{\partial^2 \Phi}{\partial z^2}\,;\quad T = \alpha\,\frac{\partial \Phi}{\partial z}\quad;\quad N = -\alpha^2\,\Phi$$

In this way, Eqs. 12.19a, 12.19b and 12.19c can be modified to:

$$\alpha U_r = A\,\frac{\partial^2 \Phi}{\partial z^2} + B\,\alpha^2\,\Phi \qquad (12.20a)$$

$$\frac{\partial W}{\partial z} = -B\,\frac{\partial^2 \Phi}{\partial z^2} - C\,\alpha^2\,\Phi \qquad (12.20b)$$

$$\frac{\partial U_r}{\partial z} - \alpha\, W = \alpha\, F\,\frac{\partial \Phi}{\partial z} \qquad (12.20c)$$

However, in this case, $A = \frac{d}{(ad-c^2)}$; $B = \frac{c}{(ad-c^2)}$; $C = \frac{a}{(ad-c^2)}$; $F = \frac{f}{2}$.

The first two equations have the same form as those of Eqs. 12.7a to 12.7b, while the third equation is similar to Eq. 12.7c with U_r replacing U. A flexibility relation can be established in the same manner as in the previous case and the solution can be obtained by imposing the boundary and continuity conditions as tabulated in Table 12.2.

TABLE 12.2
Boundary and continuity conditions for circular loadings.
(a) Boundary conditions

(1) U_r and U_z		
Base (z = H)	Rough - Rigid Smooth - Rigid	$U_r = 0$; $W = 0$ $S_{rz} = 0$; $W = 0$
Surface (z = 0)	Specified surface traction	$S_{zz} = F_{zz} = N$ $S_{rz} = F_{rz}$
(2) U_θ		
Base (z = H)	Rough Smooth	$U_\theta = 0$ $S_{\theta z} = 0$
Surface (z = 0)	Specified surface traction	$S_{\theta z} = F_{\theta z}$

(b) Continuity conditions

$(U_r , U_z)^a = (U_r , U_z)^b$ $(S_{rz} , S_{zz})^a = (S_{rz} , S_{zz})^b$	$U_\theta^a = U_\theta^b$ $S_{\theta z}^a = S_{\theta z}^b$

where $S_{rz} , S_{\theta z} , F_{rz} , F_{\theta z} = \int_0^\infty rJ_1(\alpha r) [\sigma_{rz} , \sigma_{\theta z} , f_{rz} , f_{\theta z}]dr$

$S_{zz} , F_{zz} = \int_0^\infty rJ_o(\alpha r) [\sigma_{zz} , f_{zz}]dr$

$f_{rz} , f_{\theta z} , f_{zz}$ are the applied surface tractions

Solution for Eq. 12.18c:

The solution for Eq. 12.18c is:

$$U_\theta = C_1 \sinh(\kappa z) + C_2 \cosh(\kappa z) \tag{12.21}$$

where $\kappa = \alpha \sqrt{\frac{a-b}{f}}$, C_1 and C_2 are undetermined constants.

Recalling the stress-strain relationship, we have

$$S_{z\theta} = (\frac{f}{2})\,\kappa\,[\,C_1 \cosh(\kappa z) + C_2 \sinh(\kappa z)\,] \tag{12.22}$$

Similarly, a flexibility relation can be established from Eqs. 12.21 and 12.22:

$$\begin{Bmatrix} U_{\theta k} \\ -U_{\theta l} \end{Bmatrix} = \frac{F}{\kappa} \begin{bmatrix} \cot anh(2\kappa h) & -\operatorname{cos}ech(2\kappa h) \\ -\operatorname{cos}ech(2\kappa h) & \cot anh(2\kappa h) \end{bmatrix} \begin{Bmatrix} S_{z\theta k} \\ S_{z\theta l} \end{Bmatrix} \tag{12.23}$$

The solution for the above equation can be obtained by imposing the boundary and continuity conditions as given in Table 12.2.

A two-layered soil system (Figure 12.2) was used to demonstrate the method.[9] The soil had anisotropic properties with constants as given in Table 12.3. The second layer was infinite in depth. The results are depicted in Figure 12.2 and it is noted that they are in good agreement with those obtained by Gerrard.[6]

TABLE 12.3
Material properties of the two-layered system.[9]

	E_h/E_v	f/E_v	υ_h	υ_{hv}	υ_{vh}	H_A/a	$(E_v)_A/(E_v)_B$
Case 1							
Layer A	1.0	0.8	0.25	0.25	0.25	1.5	5.0
Layer B	2.0	0.9	0.25	0.35	0.175		
Case 2							
Layer A	3.0	1.0	0.1	0.9	0.3	1.5	5.0
Layer B	2.0	0.9	0.25	0.35	0.175		

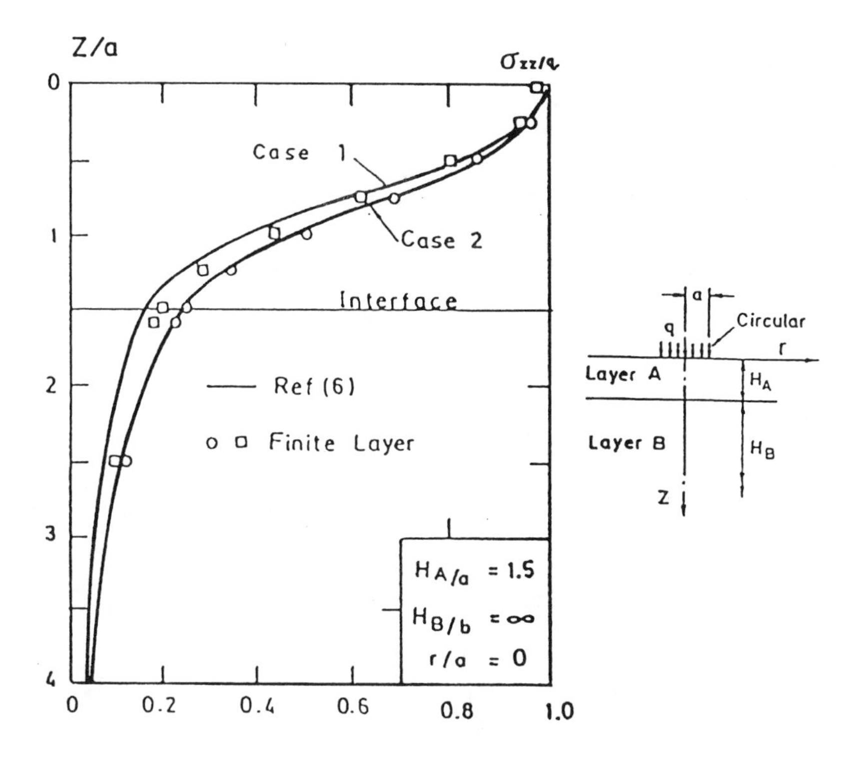

Figure 12.2 Two-layered axisymmetrical system: Stresses on axis beneath uniform circular loading.[9]

12.4 THREE-DIMENSIONAL PROBLEM[9]

Problems with rectangular loadings acting on the surface of layered soils can be treated by applying the following double Fourier transforms to the field quantities:

$$(U, V, W) = \frac{1}{4\pi^2} \int_{-\infty}^{+\infty} \int_{-\infty}^{+\infty} (iu, iv, w)\, e^{-i(\alpha x+\beta y)}\, dxdy \qquad (12.24)$$

$$(S_{xx}, S_{yy}, S_{zz}, T_{xy}, T_{yz}, T_{zx}) = \frac{1}{4\pi^2}\int_{-\infty}^{+\infty}\int_{-\infty}^{+\infty}(\sigma_{xx}, \sigma_{yy}, \sigma_{zz}, i\tau_{xy}, i\tau_{yz}, i\tau_{zx})\, e^{-i(\alpha x+\beta y)}\, dxdy \quad (12.25)$$

To obtain the solution, it is necessary to define a new set of coordinate systems (ξ, η) having an angle ε with the original x-y axes such that

$$\xi = x\cos\varepsilon + y\sin\varepsilon \quad (12.26a)$$

$$\eta = -x\sin\varepsilon + y\cos\varepsilon \quad (12.26b)$$

and redefine α and β in terms of parameter, $\hat{\rho}$, as

$$\alpha = \hat{\rho}\cos\varepsilon \;;\; \beta = \hat{\rho}\sin\varepsilon \quad (12.26c)$$

Correspondingly, there will be a new set of displacements, strains and stresses given as:

a) displacements: $\tilde{u}$, $\tilde{v}$, $\tilde{w}$

b) strains: $\tilde{\varepsilon}_{\xi\xi}, \tilde{\varepsilon}_{\eta\eta}, \tilde{\varepsilon}_{zz}, \tilde{\gamma}_{\xi\eta}, \tilde{\gamma}_{\eta z}, \tilde{\gamma}_{z\xi}$

c) stresses: $\tilde{\sigma}_{\xi\xi}, \tilde{\sigma}_{\eta\eta}, \tilde{\sigma}_{zz}, \tilde{\tau}_{\xi\eta}, \tilde{\tau}_{\eta z}, \tilde{\tau}_{z\xi}$

where $(\tilde{u}, \tilde{v}, \tilde{w}) = (-iU(\alpha,\beta,z), -iV(\alpha,\beta,z), W(\alpha,\beta,z))e^{-i(\alpha x+\beta y)}$, and the stress as well as strain fields are also defined accordingly.

In the new coordinate system, the stress-strain relationships are:

$$\tilde{\varepsilon}_{\xi\xi} = A\,\tilde{\sigma}_{\xi\xi} - B\,\tilde{\sigma}_{zz} \quad (12.27a)$$

$$\tilde{\varepsilon}_{zz} = -B\,\tilde{\sigma}_{\xi\xi} + C\,\tilde{\sigma}_{zz} \quad (12.27b)$$

$$\tilde{\gamma}_{\xi z} = F\,\tilde{\tau}_{\xi z} \quad (12.27c)$$

$$\tilde{\varepsilon}_{\eta\eta} = 0 \quad (12.27d)$$

A, B, C and F are as defined in Eq. 12.4. The equations of equilibrium can now be written as:

$$\frac{\partial\tilde{\sigma}_{\xi\xi}}{\partial\xi} + \frac{\partial\tilde{\tau}_{\xi z}}{\partial z} = 0 \quad (12.28a)$$

$$\frac{\partial\tilde{\sigma}_{zz}}{\partial z} + \frac{\partial\tilde{\tau}_{\xi z}}{\partial\xi} = 0 \quad (12.28b)$$

$$\frac{\partial\tilde{\tau}_{\eta\xi}}{\partial\xi} + \frac{\partial\tilde{\tau}_{\eta z}}{\partial z} = 0 \quad (12.28c)$$

Finally, the stress-strain, displacement-strain and equilibrium equations are all given in terms of the transformed variables (U_ξ, U_η, U_z are the transformed displacements in ξ, η and z directions.) (Table 12.4).

The first two equations of the equilibrium equations and the stress-strain relationship are equivalent to those of the plane strain case, and they can be solved in the same manner.

The uncoupled term may be transformed to give:

$$-\hat{\rho}\, S_{\xi\eta} + \frac{\partial S_{\eta z}}{\partial z} = 0 \quad (12.29)$$

TABLE 12.4
Stress-strain, displacement-strain and equilibrium equations for the three-dimensional problem in terms of the transformed variables.[12]

(a) $\tilde{E} = \mathbf{R}\,\tilde{S}$ where $\tilde{E} = [E_{\xi\xi}\ \ E_{zz}\ \ G_{\xi z}]^T$ $\tilde{S} = [S_{\xi\xi}\ \ S_{zz}\ \ T_{\xi z}]^T$ $R = \begin{bmatrix} A & -B & 0 \\ -B & C & 0 \\ 0 & 0 & F \end{bmatrix}$ $S_{\xi\xi} = \cos^2\varepsilon\, S_{xx} + 2\cos\varepsilon \sin\varepsilon\, S_{xy} + \sin^2\varepsilon\, S_{yy}$; $E_{\xi\xi} = \cos^2\varepsilon\, E_{xx} + 2\cos\varepsilon \sin\varepsilon\, E_{xy} + \sin^2\varepsilon\, E_{yy}$ $T_{\xi z} = \cos\varepsilon\, S_{xy} + \sin\varepsilon\, S_{yz}$; $G_{\xi z} = 0.5(\cos\varepsilon\, G_{xy} + \sin\varepsilon\, G_{yz})$
(b) $\mathbf{M}(\rho,z)\,\tilde{S} = 0$ where $\mathbf{M}(\rho,z) = \begin{bmatrix} \hat{\rho} & 0 & -\frac{\partial}{\partial \zeta} \\ 0 & -\frac{\partial}{\partial \zeta} & -\hat{\rho} \end{bmatrix}$
(c) $\tilde{E} = -\mathbf{N}(\hat{\rho},z)\,\tilde{U}$ where $\mathbf{N} = \mathbf{M}^T$; $\tilde{U} = [U_\xi\ \ U_z]^T$
(d) $T = S_{\xi z} = \hat{\rho}\,\frac{\partial \Phi}{\partial z}$ $N = S_{zz} = -\hat{\rho}^{\,2}\Phi$ $M = S_{\xi\xi} = \frac{\partial^2 \Phi}{\partial z^2}$
(e) $\hat{\rho}\ U_\xi = A\,\frac{\partial^2 \Phi}{\partial z^2} + B\,\alpha^2\,\Phi$ $\frac{\partial U_z}{\partial z} = -B\,\frac{\partial^2 \Phi}{\partial z^2} - C\,\alpha^2\,\Phi$ $\frac{\partial U_\xi}{\partial z} - \hat{\rho}\ U_z = F\,\hat{\rho}\,\frac{\partial \Phi}{\partial z}$
(f) $\tilde{\delta}^i = F^i\,\tilde{p}^i$ $\tilde{\delta}^i = [U_{zp}\ \ U_{\xi p}\ \ -U_{zm}\ \ -U_{\xi m}]^T$; $\tilde{p}^i = [N_p\ \ T_p\ \ N_m\ \ T_m]^T$

while the stress-strain relationship relating to $\tilde{u}$ may be used to obtain the following two equations:

$$\frac{\partial U_\eta}{\partial z} = \frac{f}{2}\,S_{\eta z} \qquad (12.30a)$$

$$\hat{\rho}\ U_\eta = \frac{2}{a-b}\,S_{\xi\eta} \qquad (12.30b)$$

It therefore follows that

$$\frac{\partial^2 S_{\eta z}}{\partial z^2} = \hat{\rho}^{\,2}\,\frac{a-b}{f}\,S_{\eta z} \qquad (12.31)$$

The general solution may then be written:

$$S_{\eta z} = C_1 \cosh(\kappa z) + C_2 \sinh(\kappa z) \tag{12.32}$$

where C_1 and C_2 are constants and

$$\kappa = \hat{\rho}\sqrt{\frac{a-b}{f}} \tag{12.33}$$

Hence

$$U_\eta = \frac{2}{f\kappa}[C_1 \cosh(\kappa z) + C_2 \sinh(\kappa z)] \tag{12.34}$$

which leads to the flexibility relation

$$\begin{Bmatrix} U_{\eta k} \\ -U_{\eta l} \end{Bmatrix} = \frac{F}{\kappa}\begin{bmatrix} \cot\text{anh}(2\kappa h) & -\text{cosech}(2\kappa h) \\ -\text{cosech}(2\kappa h) & \cot\text{anh}(2\kappa h) \end{bmatrix}\begin{Bmatrix} S_{\eta zk} \\ S_{\eta zl} \end{Bmatrix} \tag{12.35}$$

As usual, the problem has to be solved by satisfying the following conditions:

a) Boundary conditions:

(a) Boundary conditions (U_ξ , U_z)		
Base (z=H)	Rough -rigid Smooth - rigid	$U_\xi = 0$, $U_z = 0$ $S_{\xi z} = 0$, $U_z = 0$
Surface (z = 0)	Specified surface tractions	$S_{zz} = N = F_{zz}$ $S_{\xi z} = F_{\xi z}$
(b) Boundary conditions (U_η)		
Base (z=H)	Rough Smooth	$U_\eta = 0$ $S_{\eta z} = 0$
Surface (z = 0)	Specified surface tractions	$S_{\eta z} = F_{\eta z}$

where $F_{\xi z} = \frac{1}{4\pi^2}\int_0^\infty \int_0^{2\pi} i(f_{xz}\cos\varepsilon + f_{yz}\sin\varepsilon)\, e^{i\hat{\rho}\xi}\hat{\rho}\,d\hat{\rho}\; d\varepsilon$

$F_{\eta z} = \frac{1}{4\pi^2}\int_0^\infty \int_0^{2\pi} i(-f_{xz}\sin\varepsilon + f_{yz}\cos\varepsilon)\, e^{i\hat{\rho}\xi}\hat{\rho}\,d\hat{\rho}\, d\varepsilon$

$F_{zz} = \frac{1}{4\pi^2}\int_0^\infty \int_0^{2\pi} f_{zz}\, e^{i\hat{\rho}\xi}\hat{\rho}\,d\hat{\rho}\, d\varepsilon$; and f_{xz} , f_{yz} , f_{zz} are the applied surface tractions.

b) Continuity conditions:

$$U_z^a = U_z^b \quad ; \quad U_\xi^a = U_\xi^b$$
$$(S_{zz}\, , S_{\xi z})^a = (S_{zz}\, , S_{\xi z})^b$$
$$U_\eta^a = U_\eta^b \quad ; \quad S_{\eta z}^a = S_{\eta z}^b$$

Solutions for a two-layered system (Figure 12.3) were obtained by Booker[9]. The soils were assumed to be anisotropic with material constants given in Table 12.5. The vertical and horizontal stresses along the centre axis of the loading are depicted in Figures 12.3.

12.5 VISCOELASTIC MATERIAL (CREEP)[14]

The stress-strain relationship for a material exhibiting viscoelastic behaviour can be expressed in terms of Laplace transformed field quantities:

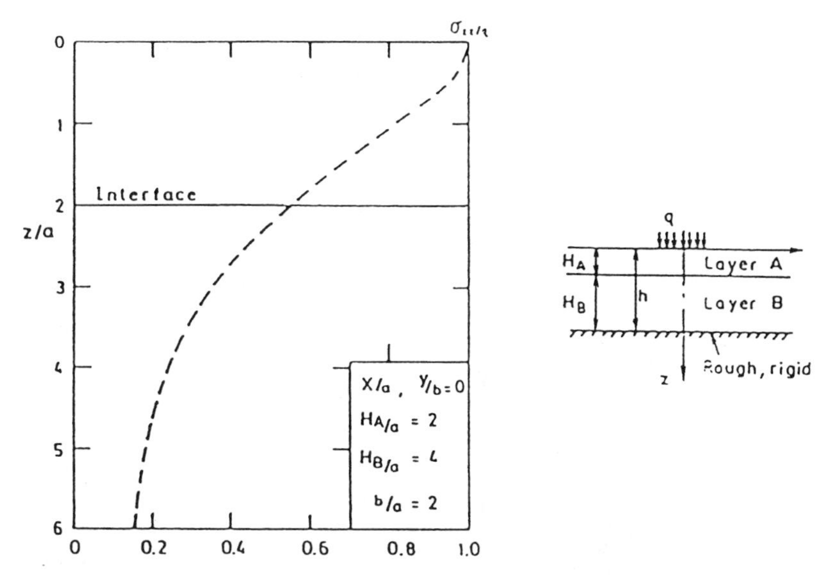

Figure 12.3 Two-layered system under rectangular load: Stresses beneath centre point of the loading.[14]

TABLE 12.5
Material properties of the two-layered system.[14]

	E_h/E_v	f/E_v	υ_h	υ_{hv}	υ_{vh}	H_B/a	H_A/a	$(E_v)_A/(E_v)_B$
Layer 'A'	1.5	0.9	0.25	0.3	0.2			
Layer 'B'	3.0	1.0	0.1	0.9	0.3	4	2.0	0.25

$$\bar{\varepsilon}_{XX} = \bar{A}\ \bar{\sigma}_{XX} - \bar{B}\ \bar{\sigma}_{ZZ}\ ;\ \bar{\varepsilon}_{ZZ} = -\bar{B}\ \bar{\sigma}_{XX} + \bar{A}\ \bar{\sigma}_{ZZ}\ ;\ \bar{\varepsilon}_{XZ} = \bar{F}\ \bar{\sigma}_{XZ}$$

where $\bar{A} = \dfrac{\bar{\kappa}+4\bar{G}/3}{4\bar{G}(\bar{\kappa}-\bar{G}/3)}$; $\bar{B} = -\dfrac{\bar{\kappa}-2\bar{G}/3}{4\bar{G}(\bar{\kappa}-\bar{G}/3)}$; $\bar{F} = \dfrac{1}{\bar{G}}$.

In the equations, the bar denotes a Laplace transform, that is

$$\bar{\sigma}_{XX} = \int_0^\infty \sigma_{xx}\, e^{-st}\, dt$$ (where s is the argument of the transform)

$$\bar{\kappa} = \frac{1}{sJ_V}$$ (J_v = volumetric creep function)

$$\bar{G} = \frac{1}{sJ_D}$$ (J_D = deviatoric creep function)

It can be shown readily that other state equations can also be written in terms of the Laplace transformed variables and they are similar to those described in the previous sections. Hence, the problem can be solved by the same procedures as in the plane strain case.

The behaviour of a two-layered viscoelastic soil system under strip loading was studied[14] (Figure 12.4). The upper layer exhibited viscoelastic behaviour whereas the bottom layer was elastic. Figure 12.5 shows the variation of the stresses with depth.

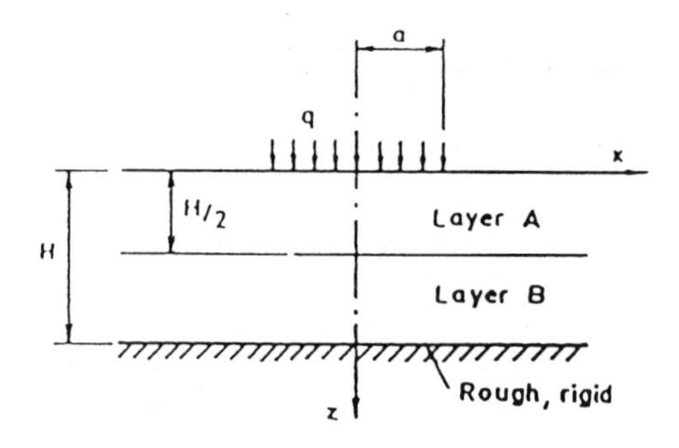

Figure 12.4 Two-layered soils.(14)

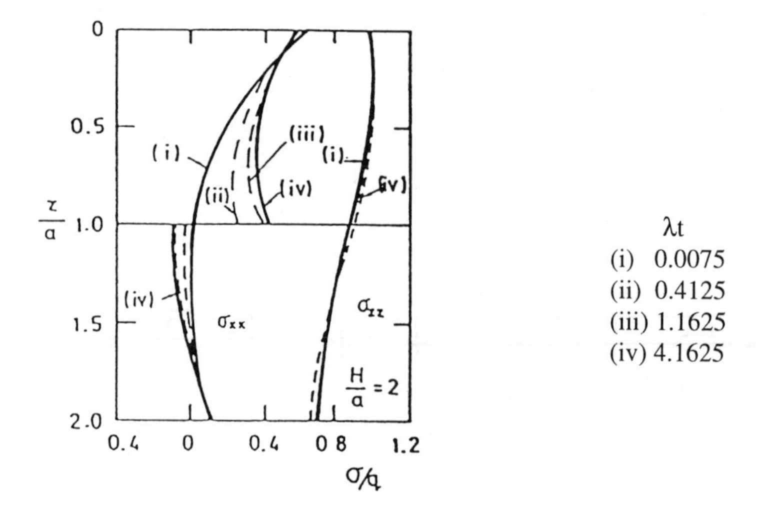

Figure 12.5 Stresses on centre line beneath strip loading.(14)

12.6 CONSOLIDATION ANALYSIS(15,16)

Explicit solutions for consolidation problems, even in the transformed form, are not easy (and sometimes impossible) to obtain. For these problems, it is then more convenient to employ polynomials for the approximation of the transformed variables. Functionals of the transformed variables based on a variational approach are established and the solution process can then follow the standard finite strip method. Inverse transforms of the transformed variables will lead to the final solutions.

This approach has been applied by Booker to consolidation analysis. The dissipation of the excess pore pressure leading to the time-dependent settlement behaviour of soils is described by the theory put forward by Biot. In terms of transformed variables, the state equations for a plane-strain problem can be written in term of displacement (U, W) and pore pressure (P) as:

$$\frac{\partial}{\partial z}\left(\frac{f}{2}\frac{\partial \overline{U}}{\partial z}\right) - a\alpha^2\overline{U} - c\alpha\frac{\partial \overline{W}}{\partial z} - \frac{\partial}{\partial z}\left(\frac{f\alpha}{2}\overline{W}\right) + \alpha\overline{P} = 0 \tag{12.36a}$$

$$\frac{\partial}{\partial z}(c\alpha\overline{U}) + \frac{f\alpha}{2}\frac{\partial \overline{U}}{\partial z} + \frac{\partial}{\partial z}\left(d\frac{\partial \overline{W}}{\partial z}\right) - \alpha^2\frac{f}{2}\overline{W} - \frac{\partial \overline{P}}{\partial z} = 0 \tag{12.36b}$$

$$\alpha\overline{U} + \frac{\partial \overline{W}}{\partial z} + \alpha^2\frac{\kappa_H}{\gamma_w s}\overline{P} - \frac{\partial}{\partial z}\left(\frac{\kappa_V}{\gamma_w s}\frac{\partial \overline{P}}{\partial z}\right) = 0 \tag{12.36c}$$

where the bar denotes a Laplace transformed variable,
κ_H and κ_V are the horizontal and vertical permeabilities,
a, c, d and f are the material constants in terms of the effective stresses. γ_w is the density of water.

It can be shown readily that the functional for a typical layer is:

$$\Phi = \int\left[\frac{1}{2}\,\overline{E}^T\mathbf{D}\,\overline{E} + \overline{P}\,\overline{E} - \frac{1}{2\gamma_w s}\,\overline{H}^T\mathbf{K}\,\overline{H}\,\right]dz \tag{12.37}$$

where $\overline{E} = (\overline{E}_{xx}\ ,\ \overline{E}_{zz}\ ,\ \overline{G}_{xz})$; $\overline{H} = (i\alpha\overline{P}\ ,\ \frac{\partial \overline{P}}{\partial z})$

$$\mathbf{D} = \begin{bmatrix} a & c & 0 \\ c & d & 0 \\ 0 & 0 & \frac{f}{2} \end{bmatrix} \quad ; \quad \mathbf{K} = \begin{bmatrix} \kappa_H & 0 \\ 0 & \kappa_V \end{bmatrix}$$

The independent variables are ($\overline{U}\ \overline{W}\ \overline{P}$), and they can be approximated by polynomials in terms of their nodal parameters.

Minimization of the functional leads to:

$$\begin{bmatrix} K_i & -L_i^T \\ -L_i^T & -\frac{\phi_i}{s} \end{bmatrix} \begin{Bmatrix} \overline{\Delta}_i \\ \overline{Q}_i \end{Bmatrix} = \begin{Bmatrix} \overline{R}_i \\ 0 \end{Bmatrix} \tag{12.38}$$

In the case of linear interpolation, the submatrices are defined as:

$$K_i = \begin{bmatrix} a_i\alpha^2\Delta z_i\boldsymbol{mm}^T + \frac{f_i}{2\Delta z_i}\boldsymbol{ll}^T & \alpha c_i\boldsymbol{ml}^T - \frac{f_i}{2}\boldsymbol{lm}^T \\ \alpha c_i\boldsymbol{lm}^T - \frac{f_i}{2}\boldsymbol{ml}^T & \frac{f_i}{2}\alpha^2\Delta z_i\boldsymbol{mm}^T + \frac{d_i}{\Delta z_i}\boldsymbol{ll}^T \end{bmatrix} \tag{12.39}$$

$$L_i = [\alpha\Delta z_i\boldsymbol{mm}^T \quad \boldsymbol{ml}^T]\ ;\ \phi_i = \left(\frac{\kappa_{Hi}}{\gamma_w}\right)\alpha^2\,\Delta z_i\,\boldsymbol{mm}^T + \left(\frac{\kappa_{Vi}}{\gamma_w}\Delta z_i\right)\boldsymbol{ll}^T$$

$$\boldsymbol{m} = \left[\frac{1}{2}\ \ \frac{1}{2}\right]^T ;\ \boldsymbol{l} = [-1,\ 1]^T\ ;\ \Delta z_i = z_{i+1} - z_i$$

a_i , c_i , d_i , f_i , κ_{Hi} and κ_{Vi} are the material constants for the i-th strip.

The characteristic matrices for the whole system can be formed by applying the continuity conditions which include continuity of displacement, pore pressure, surface traction and flow.

For given surface tractions ($\sigma_{xz} = f_x$, $\sigma_{zz} = f_z$) acting on the surface of the layered system, the boundary conditions can be defined as in Table 12.6.

Inverting the Laplace transform, we have the solution for the problem.

To demonstrate the application of the method, Booker[14] analyzed a two-layered soil system under the action of a strip loading (Figure 12.6). The consolidation curve is shown in Figure 12.7. In the consolidation curve, the degree of consolidation, U, at time τ is defined as:

$$U = \frac{u_z(t) - u_z(0)}{u_z(\infty) - u_z(0)} \tag{12.40}$$

TABLE 12.6
Boundary conditions for consolidation problem.

Base (z = H)	Rough - rigid Smooth - rigid	$U = 0$; $W = 0$ $S_{xz} = 0$; $W = 0$
	Permeable Impermeable	$P = 0$ $\frac{\partial P}{\partial z} = 0$
Surface (z = 0)	Specified surface traction	$S_{zz} = F_z = N$ $S_{xz} = F_x$
	Permeable Impermeable	$P = 0$ $\frac{\partial P}{\partial z} = 0$

where $S_{xz} = -\frac{f}{2}\left(\frac{\partial U}{\partial z} - \alpha W\right)$; $S_{zz} = P - c\alpha U - d\,\frac{\partial W}{\partial z}$

$$(F_x\ F_z) = \frac{1}{2\pi}\int_{-\infty}^{+\infty} (if_x\, ,\, f_z)\, e^{-i\alpha x}\, dx$$

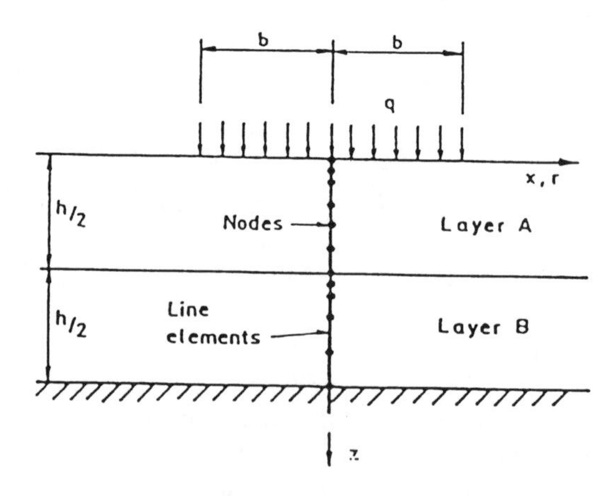

Figure 12.6 Two-layered soil system for consolidation analysis.[15]

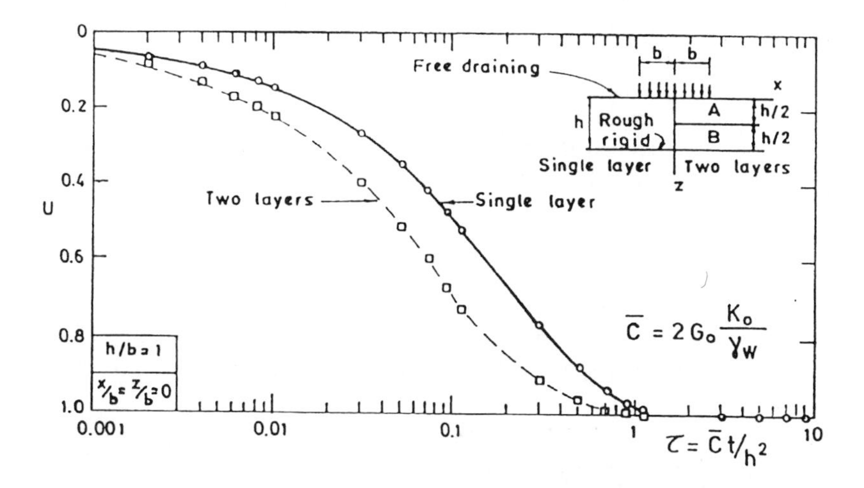

Figure 12.7 Consolidation curve for two-layered soil under strip loading.[15]

Creep and consolidation may also be coupled, and, therefore, it is also necessary to examine primary and secondary consolidation.[19] The results demonstrate that the method can be used to predict the complete range of consolidation behaviour for soils.

12.7 TRANSPORT PROBLEMS

A solution for transport problems by the transformation technique was reported by Rowe and Booker.[39] The state equation for an isotropic homogeneous layer in which the pore fluid velocity is uniform is:

$$\nabla^T(M_D\nabla c) - v^T\nabla c = (1 + \frac{\rho\kappa}{n})\frac{\partial c}{\partial t} \qquad (12.41)$$

$$M_D = \begin{bmatrix} \tilde{D}_{xx} & & \\ & \tilde{D}_{yy} & \\ & & \tilde{D}_{zz} \end{bmatrix} \quad ; \quad v = [v_x, v_y, v_z]^T$$

where c is the concentration of the dispersing substance, n is the effective porosity of the clay. $\tilde{D}_{xx}, \tilde{D}_{yy}$ and $\tilde{D}_{zz}$ are the coefficients of hydrodynamic dispersion in the respective directions. v is the seepage velocity vector; ρ is the bulk density of solid and κ is the distribution coefficient.

Laplace transform in the time domain and double Fourier transform in the x and y direction are then carried out to express the state equation in terms of the transformed variables. The relation between the flux and concentration can then be easily formed for each layer.

As usual, the matrix for the whole system can be formed by incorporating the continuity conditions at the interface such that the concentration is continuous and the net vertical flux is zero. In addition, boundary conditions have to be imposed at the surface and the base of the stratum accordingly in order to provide solution for the problem.

Examples of typical landfills were analyzed by adopting the model and satisfactory results are obtained in all cases.

REFERENCES

1. Gerrard, C. M. and Harrison, W. J., Stress and displacements in a loaded orthorhombic half space, *CSIRO Aust Div App Geom, Tech Pap No 9, 1970.*
2. Gerrard, C. M. and Harrison, W. J., The analysis of a loaded half space comprised of anisotropic layers, *CSIRO Aust Div App Geom, Tech Pap No 10, 1971.*
3. Gerrard, C. M. and Wardle, L. J., Solutions for point loads and generalised circular loads applied to a cross-anisotropic half space, *CSIRO Aust Div App Geom, Tech Pap No 13, 1973.*
4. Harrison, W. J., Wardle, L. J. and Gerrard, C. M., Computer programmes for circle and strip loads on layered anisotropic media, *CSIRO Aust Div App Geom, Geomechanics Computing Programme No 1, 1972.*
5. Wardle, L. J. and Gerrard, C. M., Computer programmes for multiple complex circle and strip loads on layered anisotropic media, *CSIRO Aust Div App Geom, Geomechanics Computing Programme No 2, 1972.*
6. Gerrard, C. M., Stresses and displacements in layered cross-anisotropic elastic systems, *Proc of 5th Aust-New Zealand Conf Soil Mech Found Eng*, 205, 1967.
7. Booker, J. R., The consolidation of a finite layer subject to surface load, *Int J of Solids and Struct*, 10, 1053-1065, 1974.
8. Small, J. C. and Booker, J. R., Finite layer analysis of layered elastic materials using a flexibility approach, Part 1 - strip loadings, *Int J for Num Meth of Eng*, 20, 1025-1037, 1984.
9. Small, J. C. and Booker, J. R., Finite layer analysis of layered elastic materials using a flexibility approach, Part 2 - circular and rectangular loadings, *Int J for Num Meth of Eng*, 23, 959-978, 1986.
10. Rowe, R. K. and Booker, J. R., The behaviour of footings on a non-homogeneous soil mass with crust - Part 1 - strip footings, *Canadian Geot Journal*, 18, 250-264, 1982.
11. Rowe, R. K. and Booker, J. R., The behaviour of footings on a non-homogeneous soil mass with crust - Part 2 - circular footings, *Canadian Geot Journal*, 18, 65-279, 1982.
12. Rowe, R. K. and Booker, J. R., Analysis of non-homogeneous soils, *J of Eng Mech, ASCE*, 108, 115-132, 1982.

13. Small, J. C. and Booker, J. R., Analysis of layered elastic viscoelastic materials, *Proc of the 8th Australasian Conf on Mech of Struct and Materials*, Newcastle, Australia, 28.1-28.6, 1982.
14. Booker, J. R. and Small, J. C., Finite layer analysis of viscoelastic layered materials, *Int J for Num and Analy Meth in Geom*, 10, 415-430, 1986.
15. Booker, J. R. and Small, J. C., Finite layer analysis of consolidation, Part 1, *Int J for Num and Analy Meth in Geom*, 6, 151-172, 1982.
16. Booker, J. R. and Small, J. C., Finite layer analysis of consolidation, Part 2, *Int J for Num and Analy Meth in Geom*, 6, 173-194, 1982.
17. Vardoulakis, I. and Harnpattanapanich, T., Numerical Laplace-Fourier transform inversion technique for layered-soil consolidation problems: I. Fundamental solutions and validation, *Int J for Num and Analy Meth in Geom*, 10, 347-365, 1986.
18. Harnpattanapanich, T. and Vardoulakis, I., Numerical Laplace-Fourier transform inversion technique for layered-soil consolidation problems: II. Gibson soil layer, *Int J for Num and Analy Meth in Geom*, 11, 103-112, 1986.
19. Small, J. C. and Booker, J. R., Finite layer analysis of primary and secondary consolidation, *Proc of the 4th Int Conf Num Meth Geom*, Edmonton, Canada, 365-372, 1982.
20. Booker, J. R. and Small, J. C., A method of computing the consolidation behaviour of layered soils using direct numerical inversion of Laplace transforms, *Int J for Num and Analy Meth in Geom*, 11, 363-380, 1987.
21. Small, J. C. and Zhang, B. Q., Consolidation of clays subjected to three-dimensional embankment loadings, *Int J for Num and Analy Meth in Geom*, 15, 857-870, 1991.
22. Booker, J. R. and Carter, J. P., Analysis of a point sink embedded in a porous elastic half space, *Int J for Num and Analy Meth in Geom*, 10, 137-150, 1986.
23. Booker, J. R. and Carter, J. P., Elastic consolidation around a point sink embedded in a half-space with anisotropic permeability, *Int J for Num and Analy Meth in Geom*, 11, 61-77, 1987.
24. Booker, J. R. and Small, J. C., The behaviour of an impermeable flexible raft on a deep layer of consolidating soil, *Int J for Num and Analy Meth in Geom*, 10, 311-327, 1986.
25. Small, J. C. and Zhang, B. Q., Finite layer analysis of the behaviour of a raft on a consolidating soil, *Int J for Num and Analy Meth in Geom*, 18, 237-251, 1994.
26. Lee, C. Y. and Small, J. C., Finite layer analysis of laterally loaded piles in cross-anisotropic soils, *Int J for Num and Analy Meth in Geom*, 15, 785-808, 1991.
27. Lee, C. Y. and Small, J. C., Finite layer analysis of axially loaded piles, *J of Geot Eng, ASCE*, 117, 1706-1722, 1991.
28. Ta, L. D. and Small, J. C., Analysis of piled raft systems in layered soils, *Int J for Num and Analy Meth in Geom*, 20, 57-72, 1996.

29. Southcott, P. H. and Small, J. C., Finite layer analysis of vertically loaded piles and pile groups, *Comp and Geot*, 18, 47-63, 1996.
30. Rowe, R. K. and Booker, J. R., The elastic displacements of single and multiple underream anchors in a Gibson soil, *Geotechnique*, 31, 125-142, 1981.
31. Small, J. C. and Booker, J. R., The time-deflection behaviour of a rigid under-reamed anchor in a deep clay layer, *Int J for Num and Analy Meth in Geom*, 11, 269-281, 1987.
32. Small, J. C. and Booker, J. R., The behaviour of layered soil or rock containing a decaying heat source, *Int J for Num and Analy Meth in Geom*, 10, 501-519, 1986.
33. Booker, J. C., Small, J. C. and Balaam, N. P., Application of microcomputers to the analysis of three dimensional problems in Geomechanics, *Proc of the 1st Int Conf, Engineering Software for Microcomputer*, Venice, Italy, Apr 2-5, 1984, 279-287, 1984.
34. Booker, J. R. and Small, J. C., Finite layer analysis of settlement creep and consolidation using microcomputers, *Proc of the 5th Int Conf on Num Meth in Geom*, Nagoya, 1-5 April, 1985, 3-18, 1985.
35. Small, J. C. and Wong, H. K. W., The use of integral transforms in solving three dimensional problems in geomechanics, *Comp and Geot*, 6, 199-216, 1988.
36. Booker, J. R., Carter, J. P., Small, J. C., Brown, P. T. and Poulos, H. G., Some recent applications of numerical methods to geotechnical analysis, *Comp and Struct*, 31, 81-92, 1989.
37. Rowe, R. K. and Booker, J. R., A novel technique for the analysis of 1D pollutant migration, *Proc of the Int Conf on Num Meth for Transient and Coupled Problems*, Venice, 699-702, 1984.
38. Rowe, R. K. and Booker, J. R., The analysis of pollutant migration in a non-homogenous soil, *Geotechnique*, 34, 601-612, 1984.
39. Rowe, R. K. and Booker, J. R., 1D pollutant migration in soils of finite layer, *J of Geot Eng, ASCE*, 111, 497-499, 1985.
40. Rowe, R. K. and Booker, J. R., Finite layer technique for calculating three-dimensional pollutant migration in soil, *Geotechnique*, 36, 205-214, 1986.
41. Rowe, R. K. and Booker, J. R., An efficient analysis of pollutant migration through soil, *Numerical methods for transient and coupled problem*, John Wiley & Sons Ltd, 13-41, 1987.
42. Booker, J. R. and Rowe, K. R., One dimensional advective-dispersive transport into a deep layer having a variable surface concentration, *Int J for Num and Analy Meth in Geom*, 11, 131-141, 1987.
43. Rowe, R. K. and Booker, J. R., A semi-analytical model for contaminant migration in a regular two- or three-dimensional fractured network: conservative contaminants, *Int J for Num and Analy Meth in Geom*, 13, 531-550, 1989.

44. Rahman, M. S. and Booker, J. R., Pollutant migration from deeply buried repositories, *Int J for Num and Analy Meth in Geom*, 13, 57-51, 1989.
45. Smith, D. W. and Booker, J. R., Boundary integral analysis of transient thermoelasticity, *Int J for Num and Analy Meth in Geom*, 13, 283-302, 1989.
46. Rowe, R. K. and Booker, J. R., Contaminant migration in a regular two- and three-dimensional fractured network: reactive contaminants, *Int J for Num and Analy Meth in Geom*, 14, 401-425, 1989.
47. Rowe, R. K. and Booker, J. R., Contaminant migration through fractured till into an underlying aquifer, *Canadian Geot Journal*, 27, 484-495, 1990.
48. Rowe, R. K. and Booker, J. R., Modelling of two-dimensional contaminant migration in layered and fractured zones beneath landfills, *Canadian Geot Journal*, 28, 338-352, 1991.

APPENDIX

Program for the analysis of folded plates and box girder bridges by classical finite strip method
(The diskette for the program is available from Professor Y. K. Cheung)

```
C
C     folded plate program - parallel programming version
C     IBM Scalable POWERparallel SP2 Supercomputer System
C
      IMPLICIT DOUBLE PRECISION (A-H,O-Z)
      CHARACTER*60 TITLE
      CHARACTER*20 DATE
      DIMENSION X(50),Z(50),T(50),FORC(50),NX(20),NY(20),NL(20),FP(20)
     1,QS(4),QN(2),S(4,4),C(4,4),E(4,4),D(4,4),F(4,4),QP(4)
      COMMON P(30),ST(30,30),DIS(200),R(4,4),Q(30,4),NF(8),NB(8,4),B
      COMMON NTERM,NELEM,NP,NMOM,NDF,NDF1,NBAND,NSIZ,II,NI,BNII,BN1,BN2
      COMMON A,BB,EX,EY,PX,PY,G,DX,DY,D1,DXY,NOD(50,2),OM(600,3),DN(9)
      COMMON DISS(600)
      integer numtask,taskid,source,nbuf(4),allgrp,type,msgid
      integer nstore(500)
      real *8 store(500)
       external D_VADD
C
C     ANALYSIS OF FOLDED PLATES AND BRIDGES
C     adapted from 'FINITE STRIP METHOD IN STRUCTURAL ANALYSIS'
C     by Y. K. Cheung P216-228
C
C     VARIABLE                        DEFINITION
C
C     DIS(200)          Displacement parameter array
C     DISS(600)         Displacement array
C     FORC(50)          Distributed vertical load acting on strip
C     QN(1)             Load coefficient for Sine series variation
C     QN(2)             Load coefficient for Cosine series variation
C     FP(20)            Magnitude of concentrated load
C     NB(8,4)           Boundary condition type for U,V,W and Theta
C                       (0 = Fixed, 1 = Free)
C     NBAND             Maximum half bandwidth
C     NBOUN             Number of restrained boundary points
C     NCON              Number of nodal lines with concentrated loads
C                       or line loads
C     NDF               Number of degrees of freedom per nodal line
C     NELEM             Number of elements
C     NF(8)             Restrained boundary node numbers
C     NI                Load type (1 = Non-symmetrical, 2 = Symmetrical)
C     NL(20)            Load definition (1 for X-Load, 2 for Y-Load,
C                                         3 for Z-Load, 4 for M-Load)
C     NMOM              Number of points along a nodal line for
C                       outputting stresses and moments
C     NOD(50,2)         Node numbers of all the strips
C     NP                Number of nodal lines
C     NTERM             Number of terms
C     NX(20)            Node numbers at which concentrated loads act
C     NY(20)            Corresponding Y-position of concentrated load
C                       (0 for Line Load,1 for Concentrated Load)
C     OM(600,3)         Stress array
C     P(30)             Forward elimination working area for load
C     QN(4)             Fourier coefficients for load
C     ST(30,30)         Forward elimination working area for stiffness
C     T(50)             Thickness of strip
C     X(50),Z(50)       X and Y coordinates of nodal points
C
       call mp_environ(numtask,taskid)
       call mp_task_query(nbuf,4,3)
       allgrp=nbuf(4)
       source = 0
C
```

```
        msglen = 4*500
        nsglen = 8*500
        len1   = 8*600
        len2   = 8*600*3
C
        if(taskid.eq.source) then
C
C  in.dat   = input datafile
C  coef.dat = load coefficient file
C
      OPEN (UNIT=5,FILE='in.dat', STATUS='old')
      OPEN (UNIT=8,FILE='out.dat', STATUS='unknown')
      OPEN (UNIT=3,FILE='coef.dat',STATUS='old')
C
C     DATA INPUT
C
      READ (5,'(A60)') TITLE
      READ (5 ,'(A20)') DATE
      WRITE (8,901)
 901  FORMAT (' ANALYSIS OF FOLDED PLATES AND BRIDGES BY',/,
     1           ' CLASSICAL FINITE STRIP METHOD')
      WRITE (8,902) TITLE,DATE
 902  FORMAT (/,' TITLE : ',A60,/,' DATE : ',A20)
      WRITE (8,4)
 4    FORMAT (/,' *** INPUT DATA ***',/)
       read (5,*) (nstore(i),i=1,9)
       write(8,11) (nstore(i),i=1,9)
       nndex=9
 11   FORMAT (' NUMBER OF TERMS ...............................',I4,/,
     1 ' NUMBER OF ELEMENTS ............................',I4,/,
     1 ' NUMBER OF NODAL LINES .........................',I4,/,
     1 ' NUMBER OF RESTRAINED BOUNDARY POINTS ...........',I4,/,
     1 ' NUMBER OF POINTS ALONG A NODAL LINE',/,
     1 ' FOR OUTPUTTING STRESSES AND MOMENTS ............',I4,/,
     1 ' NUMBER OF DEGREE OF FREEDOM ...................',I4,/,
     1 ' MAXIMUM HALF BANDWIDTH ........................',I4,/,
     1 ' LOAD TYPE (1=NON-SYMMETRICAL,2=SYMMETRICAL) ....',I4,/,
     1 ' NUMBER OF NODAL LINES WITH CONCENTRATED LOADS ..',I4)
C
      ni = nstore(8)
      nterm = nstore(1)
C
        read(5,*) store(1),(store(1+i),i=1,nstore(5))
        index = 1 + nstore(5)
       WRITE (8,335) store(1),(store(1+I),I=1,nstore(5))
 335  FORMAT (/,' LENGTH OF STRIPS : ',F15.4
     1 /,' POINTS TO GIVE STRESSES AND MOMENTS : ',5F10.4)
C
      K    = 4*nstore(5)*nstore(2)
      NDF1 = nstore(6)+1
      NSIZ = nstore(7)-nstore(6)
      NEQ  = nstore(6)*nstore(3)
      NA   = 2*nstore(6)
      NCOLN = 1
      DO 123 J=1,3
      DO 123 I=1,K
 123    OM(I,J) = 0.0
C
C     READ X AND Z COORDINATES OF ALL POINTS
C
      WRITE (8,38)
 38   FORMAT (/,' CO-ORDINATES OF POINTS : ',/
```

```
     1 ,'       X          Z ')
      DO 2 I=1,nstore(3)
        READ (5,*) store(index+1),store(index+2)
       WRITE (8,35) store(index+1),store(index+2)
        index = index + 2
2       continue
 35   FORMAT (7F10.4)
C
C     READ NUMBER, NODAL NUMBERS, THICKNESS, VERTICAL DISTRIBUTED LOAD
C
      WRITE (8,31)
 31   FORMAT (/,' STRIP PROPERTIES :',
     1        /,'  NO.  NODES        THICKNESS     U.D.L.')
      DO 109 I=1,nstore(2)
        READ (5,*) NUM,nstore(nndex+1),nstore(nndex+2)
     *    ,store(index+1),store(index+2)
       WRITE (8,45) NUM,nstore(nndex+1),nstore(nndex+2)
     *    ,store(index+1),store(index+2)
       nndex = nndex + 2
       index = index + 2
 109   continue
 45   FORMAT (3I4,2F16.8)
C
C     READ BOUNDARY CONDITIONS 0=FIXED 1=FREE
C
      IF (nstore(4).NE.0) THEN
        WRITE (8,326)
 326    FORMAT (/,' BOUNDARY CONDITIONS :',
     1      /,'  NODE  U   V   W   ANGLE  (FIXITY : 1=FREE,0=FIXED)')
         do i=1,nstore(4)
        READ (5,*) nstore(nndex+1),nstore(nndex+2),nstore(nndex+3),
     *             nstore(nndex+4),nstore(nndex+5)
        write(8,19) nstore(nndex+1),nstore(nndex+2),nstore(nndex+3),
     *              nstore(nndex+4),nstore(nndex+5)
        nndex = nndex + 5
 19     FORMAT (1X,4I4,I6)
        enddo
       endif
C
C     READ CONCENTRATED OR LINE LOAD DATA
C
      IF (nstore(9).NE.0) THEN
        WRITE (6,913)
 913    FORMAT (/,' CONCENTRATED LOADS :',/,'   NX  NY  NL          FP')
        DO I=1,nstore(9)
          READ (5,*) nstore(nndex+1),nstore(nndex+2),nstore(nndex+3),
     *    store(index+1)
       WRITE (8,45) nstore(nndex+1),nstore(nndex+2),nstore(nndex+3),
     *    store(index+1)
        nndex = nndex + 3
        index = index + 1
        enddo
      endif
C
C     READ YOUNG'S MODULUS E1 E2, POISSON'S RATIO PX PY
C     AND SHEAR MODULUS G
C
      READ (5,*) store(index+1),store(index+2),store(index+3),
     *           store(index+4),store(index+5)
      WRITE (8,311)
 311  FORMAT (/,' MATERIAL PROPERTIES :',
     1        /,5X,'EX',11X,'EY',11X,'VX',11X,'VY',11X,'G')
```

```
      WRITE (8,147) store(index+1),store(index+2),store(index+3),
     *            store(index+4),store(index+5)
      index = index + 5
 147  FORMAT (5E13.6)
      DO 113 I=1,NEQ
 113  DISS(I) = 0.0
      WRITE (8,340)
 340  FORMAT (/,' TERM NO.    FOURIER COEFFICIENTS',
     1        /,'              DISPLACEMENT PARAMETERS')
C
C    broadcast data to all task
C
      ENDIF
      call mp_bcast(nstore,nsglen,source,allgrp)
      call mp_bcast(store,msglen,source,allgrp)
C
C    unpack data for all task after broadcast
C
       if(taskid.ne.source) then
       idir=taskid+20
       open(unit=idir,status='unknown',access='direct',recl=16000)
        nterm = nstore(1)
        nelem = nstore(2)
        np = nstore(3)
        nboun = nstore(4)
        nmom = nstore(5)
        ndf = nstore(6)
        nband = nstore(7)
        ni = nstore(8)
        ncon = nstore(9)
        nndex = 9
        do i=1,nelem
        do j=1,2
         nndex = nndex + 1
         nod(i,j) = nstore(nndex)
        enddo
        enddo
       if (nboun.ne.0) then
        do i=1,nboun
         nndex = nndex + 1
         nf(i) = nstore(nndex)
        do j=1,4
         nndex = nndex + 1
         nb(i,j) = nstore(nndex)
        enddo
        enddo
       endif
       if (ncon.ne.0) then
        do i=1,ncoun
         nndex = nndex + 1
         nx(i) = nstore(nndex)
         nndex = nndex + 1
         ny(i) = nstore(nndex)
         nndex = nndex + 1
         nl(i) = nstore(nndex)
        enddo
       endif
C
        a = store(1)
        index = 1
        do i =1,nmom
        index = index + 1
```

```
        dn(i) = store(index)
        enddo
        do i=1,np
        index = index + 1
        x(i) = store(index)
        index = index + 1
        z(i) = store(index)
        enddo
        do i=1,nelem
        index = index + 1
         t(i) = store(index)
        index = index + 1
         forc(i) = store(index)
        enddo
        if (ncon.ne.0) then
        do i=1,ncon
        index = index + 1
         fp(i) = store(index)
        enddo
        endif
         index = index + 1
         e1 = store(index)
         index = index + 1
         e2 = store(index)
         index = index + 1
         px = store(index)
         index = index + 1
         py = store(index)
         index = index + 1
         g  = store(index)
C
      K = 4*NMOM*NELEM
      NDF1 = NDF+1
      NSIZ = NBAND-NDF
      NEQ = NDF*NP
      NA = 2*NDF
      NCOLN = 1
      EX = E1/(1.0-PX*PY)
      EY = E2/(1.0-PX*PY)
C
      DO I=1,NEQ
      DISS(I) = 0.0
      enddo
      DO  J=1,3
      DO  I=1,K
       OM(I,J) = 0.0
      enddo
      enddo
C
      ENDIF
C
C  end of unpacking ; start for looping NTERM
C
         ii=1-ni
         itask=0
         type=1
 600    itask=itask+1
        if(itask.gt.numtask-1) itask=1
        ii=ii+ni
C
        if(taskid.eq.source) then
C
```

```
C  read load coefficients
C
         read(3,*) (qn(i),i=1,2)
         write(8,341) ii,(qn(i),i=1,2)
 341     format(1x,i4,4x,2f10.4)
         call mp_send(qn,8,itask,type,msgid)
         call mp_wait(msgid,nbytes)
         else
C
        if(taskid.eq.itask) then
          call mp_brecv(qn,8,source,type,nbytes)
        CALL INIT(NBAND,NCOLN,NDF,itask)
        NE = 0
        itape=itask+10
        rewind itape
        DO 12 I=1,NEQ
 12     DIS(I) = 0.0
        DO 8 I=1,NBAND
          P(I) = 0.0
          DO 8 J=1,NBAND
 8        ST(I,J) = 0.0
C
C       READ FOURIER LOAD COEFFICIENTS
C
 99     BNII = 3.141592654*II
        CO = BNII*BNII
        BN1 = BNII/A
        BN2 = BN1*BN1
        DO 70 LK=1,NP
          IF (LK.GT.1) GOTO 9
 1        NE = NE+1
          IF (NE-NELEM) 9,9,92
 9        IF (NOD(NE,1)-LK) 92,3,92
 3        N1 = NOD(NE,1)
          N2 = NOD(NE,2)
          XP = X(N2)-X(N1)
          ZP = Z(N2)-Z(N1)
          H = T(NE)
          B = DSQRT(XP*XP+ZP*ZP)
          BB = B*B
          DX = E1*H**3/(12.*(1.-PX*PY))
          DY = E2*DX/E1
          DXY = G*H**3/12.
          D1 = PX*E2*DX/E1
          CALL TRAN(XP,ZP)
 C
          WRITE (itape) ((R(I,J),I=1,4),J=1,4)
 C
          CALL MOMP(D,E,F,itape)
          CALL MOMS(D,E,F,itape)
          CALL FEMS(CO,A,B,BB,DX,DY,D1,DXY,D,E,F)
          CALL LOADS(QN,QS,NE,FORC,B,BB,R)
          CALL FEMP(E,H,CO)
          CALL LOADP(QN,QP,NE,FORC,B,R)
            DO 80 LL=1,2
            J1 = 2*LL-1
            J2 = J1+1
            J = NDF*(NOD(NE,LL)-LK)
            S(1,1) = QP(J1)*A
            S(2,1) = QP(J2)*A
            S(3,1) = QS(J1)*A
            S(4,1) = QS(J2)*A
```

```
            CALL MBTTM(R,S,C,4,4,1)
            DO 15 NJ=1,NDF
              JN = J+NJ
 15           P(JN) = P(JN)+C(NJ,1)
            DO 80 KK=1,2
              DO 81 K=1,NDF
              DO 81 L=1,NDF
 81             D(K,L) = 0.0
              I1 = 2*KK-1
              I2 = I1+1
              I = NDF*(NOD(NE,KK)-LK)
              D(1,1) = E(I1,J1)
              D(2,1) = E(I2,J1)
              D(1,2) = E(I1,J2)
              D(2,2) = E(I2,J2)
              D(3,3) = F(I1,J1)
              D(4,3) = F(I2,J1)
              D(3,4) = F(I1,J2)
              D(4,4) = F(I2,J2)
              CALL MBTM(D,R,C,4,4,4)
              CALL MBTTM(R,C,D,4,4,4)
              DO 5 NJ=1,NDF
                JN = J+NJ
                DO 5 MI=1,NDF
                  IM = MI+I
                  ST(IM,JN) = ST(IM,JN)+D(MI,NJ)
 5            CONTINUE
 80       CONTINUE
          GOTO 1
 92       DO 67 I=1,NCON
            IF (LK.EQ.NX(I)) THEN
              J = NL(I)
              IF (NY(I).LE.0) THEN
                IF (J.EQ.2) THEN
                  P(J) = P(J)+FP(I)*QN(2)*A
                ELSE
                  P(J) = P(J)+FP(I)*QN(1)*A
                ENDIF
              ELSE
                P(J) = P(J)+FP(I)*QN(1)
              ENDIF
            ENDIF
 67       CONTINUE
          CALL BOUN(LK,NBOUN)
          CALL SOLVE
 70     CONTINUE
        REWIND itape
        CALL BSUB
        NP2 = NDF*NP
        WRITE (6,903) ii,(DIS(I),I=1,NP2)
 903    FORMAT (' TERM = ',I5,/,(8X,4E16.8))
        CALL MOM(D,E,F,C,S,itape)
C
C     WRITE DISPLACEMENT PARAMETERS (WITHOUT PREMULTIPLIER FOR V)
C
        ENDIF
        ENDIF
        if(ii.lt.nterm) go to 600
        call mp_reduce(diss,diss,len1,source,D_VADD,allgrp)
      call mp_reduce(om,om,len2,source,D_VADD,allgrp)
C
```

```
C   output results
C
       if(taskid.eq.source) then
      WRITE (8,609)
 609  FORMAT (/,' *** OUTPUT RESULTS ***')
      WRITE (8,610) nstore(1)
 610  FORMAT (/,' NUMBER OF TERMS USED =',I4)
      K = nstore(3)*nstore(5)*nstore(6)
      WRITE (8,152)
 152  FORMAT (/,' DEFLECTIONS AND ROTATIONS (U, V, W AND THETA)')
      WRITE (8,51) (DISS(I),I=1,K)
 51   FORMAT (4E16.8)
      K = 4*nstore(5)
      DO 133 LL=1,nstore(2)
        N1 = (LL-1)*K+1
        N2 = LL*K
        WRITE (8,27)
 27     FORMAT (/,' ELEM NO. ')
        WRITE (8,28)
 28     FORMAT (' ZIGMA -X , ZIGMA -Y , ZIGMA -XY (REPEAT FOR EACH)')
        WRITE (8,29)
 29     FORMAT (' MOMENT-X , MOMENT-Y , MOMENT-XY (REPEAT FOR EACH)')
        WRITE (8,17) LL,((OM(I,J),J=1,3),I=N1,N2)
 133  CONTINUE
 17   FORMAT (I4,/,(6E12.5))
C
      ENDIF
C
      STOP
      END
C
C
      SUBROUTINE FEMP(E,H,CO)
      IMPLICIT DOUBLE PRECISION (A-H,O-Z)
      DIMENSION E(4,4)
      COMMON P(30),ST(30,30),DIS(200),R(4,4),Q(30,4),NF(8),NB(8,4),B
      COMMON NTERM,NELEM,NP,NMOM,NDF,NDF1,NBAND,NSIZ,II,NI,BNII,BN1,BN2
      COMMON A,BB,EX,EY,PX,PY,G,DX,DY,D1,DXY,NOD(50,2),OM(600,3),DN(9)
      COMMON DISS(600)
C
C     IN-PLANE STIFFNESS MATRIX E(4,4) OF STRIP
C
      E(1,1) = EX*.5*A/B+B*BN2*A*G/6.
      E(2,1) = .25*BN1*A*PX*EY-.25*BN1*A*G
      E(3,1) = -.5*A*EX/B+B*A*BN2*G/12.
      E(4,1) = .25*A*BN1*PX*EY+.25*A*BN1*G
      E(2,2) = A*B*BN2*EY/6.+.5*A*G/B
      E(3,2) = -.25*A*BN1*PX*EY-.25*A*BN1*G
      E(4,2) = A*B*BN2*EY/12.-.5*A*G/B
      E(3,3) = .5*A*EX/B+A*B*BN2*G/6.
      E(4,3) = -.25*A*BN1*PX*EY+.25*A*BN1*G
      E(4,4) = A*B*EY*BN2/6.+.5*A*G/B
      DO 20 I=1,4
      DO 20 J=1,I
        E(I,J) = E(I,J)*H
 20     E(J,I) = E(I,J)
      RETURN
      END
C
C
      SUBROUTINE FEMS(CO,A,B,BB,DX,DY,D1,DXY,D,E,F)
      IMPLICIT DOUBLE PRECISION (A-H,O-Z)
```

```
      DIMENSION D(4,4),E(4,4),C(4,4),F(4,4)
C
C     BENDING STIFFNESS MATRIX F(4,4) OF STRIP
C
      AA = A*A
      BBBB = BB*BB
      COCO = CO*CO
      BBB = B*BB
      D(1,1) = COCO*DY*B*.5/AA
      D(2,1) = COCO*BB*DY*.25/AA
      D(3,1) = -CO*B*D1+.166667*COCO*BBB*DY/AA
      D(4,1) = -1.5*CO*BB*D1+.125*COCO*BBBB*DY/AA
      D(2,2) = .166667*COCO*BBB*DY/AA+2.*CO*B*DXY
      D(3,2) = -CO*BB*D1*.5+.125*COCO*BBBB*DY/AA+2.*CO*BB*DXY
      D(4,2) = -CO*BBB*D1+.1*COCO*BBBB*B*DY/AA+2.*CO*BBB*DXY
      D(3,3) = 2.*AA*B*DX-.6666667*CO*BBB*D1
     1         +.1*COCO*BBBB*B*DY/AA+2.66667*CO*BBB*DXY
      D(4,3) = 3.*AA*BB*DX-CO*BBBB*D1+.0833333*COCO*BBBB*BB*DY/AA
     1         +3.*CO*BBBB*DXY
      D(4,4) = 6.*BBB*DX*AA-1.2*CO*BBBB*B*D1
     1         +.0714286*COCO*BBB*BBBB*DY/AA+3.6*CO*BB*BBB*DXY
      DO 5 I=1,4
      DO 5 J=1,I
        D(I,J) = D(I,J)/A
 5      D(J,I) = D(I,J)
      CALL EM(B,BB,BBB,E)
      CALL MBTM(D,E,C,4,4,4)
      CALL MBTTM(E,C,F,4,4,4)
      RETURN
      END
C
C
      SUBROUTINE MOMP(D,E,F,itape)
      IMPLICIT DOUBLE PRECISION (A-H,O-Z)
      DIMENSION D(4,4),E(4,4),F(4,4)
      COMMON P(30),ST(30,30),DIS(200),R(4,4),Q(30,4),NF(8),NB(8,4),B
      COMMON NTERM,NELEM,NP,NMOM,NDF,NDF1,NBAND,NSIZ,II,NI,BNII,BN1,BN2
      COMMON A,BB,EX,EY,PX,PY,G,DX,DY,D1,DXY,NOD(50,2),OM(600,3),DN(9)
      COMMON DISS(600)
C
C     IN-PLANE STRESS MATRIX D(4,4) OF STRIP
C
      DO 83 INDEX=1,2
        IF (INDEX.EQ.1) THEN
          X = 0.
        ELSE
          X = B
        ENDIF
        DO 82 M=1,NMOM
          Z = DN(M)/A
          H1 = DSIN(BNII*Z)
          H2 = DCOS(BNII*Z)
          D(1,1) = -H1*EX/B
          D(2,1) = -H1*PX*EY/B
          D(3,1) = (1.-X/B)*BN1*H2*G
          D(1,2) = -(1.-X/B)*BN1*H1*PX*EY
          D(2,2) = -(1.-X/B)*BN1*H1*EY
          D(3,2) = -H2*G/B
          D(1,3) = H1*EX/B
          D(2,3) = H1*PX*EY/B
          D(3,3) = X*BN1*H2*G/B
          D(1,4) = -X*BN1*H1*PX*EY/B
```

```
         D(2,4) = -X*BN1*H1*EY/B
         D(3,4) = H2*G/B
         WRITE (itape) ((D(I,J),I=1,3),J=1,4)
 82    CONTINUE
 83   CONTINUE
      RETURN
      END
C
C
      SUBROUTINE MOMS(D,E,F,itape)
      IMPLICIT DOUBLE PRECISION (A-H,O-Z)
      DIMENSION D(4,4),E(4,4),F(4,4)
      COMMON P(30),ST(30,30),DIS(200),R(4,4),Q(30,4),NF(8),NB(8,4),B
      COMMON NTERM,NELEM,NP,NMOM,NDF,NDF1,NBAND,NSIZ,II,NI,BNII,BN1,BN2
      COMMON A,BB,EX,EY,PX,PY,G,DX,DY,D1,DXY,NOD(50,2),OM(600,3),DN(9)
      COMMON DISS(600)
C
C     MOMENT MATRIX F(4,4) OF STRIP
C
      CO = BN2
      DO 83 INDEX=1,2
        IF (INDEX.EQ.1) THEN
          X = 0.
        ELSE
          X = B
        ENDIF
        DO 82 M=1,NMOM
          Z = DN(M)/A
          H1 = DSIN(BNII*Z)
          H2 = DCOS(BNII*Z)
          H3 = -H1
          D(1,1) = -CO*H3*D1
          D(2,1) = -CO*H3*DY
          D(3,1) = 0.
          D(1,2) = -CO*H3*D1*X
          D(2,2) = -CO*X*H3*DY
          D(3,2) = 2.*BN1*H2*DXY
          D(1,3) = -2.*H1*DX-CO*X*X*H3*D1
          D(2,3) = -2.*H1*D1-CO*X*X*H3*DY
          D(3,3) = 4.*BN1*H2*DXY*X
          D(1,4) = -6.*X*H1*DX-CO*X*X*X*H3*D1
          D(2,4) = -6.*X*H1*D1-CO*X*X*X*H3*DY
          D(3,4) = 6.*BN1*X*X*H2*DXY
          BBB = BB*B
          CALL EM(B,BB,BBB,E)
          CALL MBTM(D,E,F,3,4,4)
          WRITE (itape) ((F(I,J),I=1,3),J=1,4)
 82    CONTINUE
 83   CONTINUE
      RETURN
      END
C
C
      SUBROUTINE TRAN(XP,ZP)
      IMPLICIT DOUBLE PRECISION (A-H,O-Z)
      COMMON P(30),ST(30,30),DIS(200),R(4,4),Q(30,4),NF(8),NB(8,4),B
      COMMON NTERM,NELEM,NP,NMOM,NDF,NDF1,NBAND,NSIZ,II,NI,BNII,BN1,BN2
      COMMON A,BB,EX,EY,PX,PY,G,DX,DY,D1,DXY,NOD(50,2),OM(600,3),DN(9)
      COMMON DISS(600)
C
C     TRANSFORMATION MATRIX R(4,4) OF STRIP
C
```

```
      DO 1 I=1,4
      DO 1 J=1,4
 1      R(I,J) = 0.
      S = ZP/B
      C = XP/B
      R(1,1) = C
      R(2,2) = 1.
      R(3,1) = -S
      R(1,3) = S
      R(3,3) = C
      R(4,4) = 1.
      RETURN
      END
C
C
      SUBROUTINE LOADP(QN,QP,LK,FORC,B,R)
      IMPLICIT DOUBLE PRECISION (A-H,O-Z)
      DIMENSION QN(2),QP(4),FORC(50),R(4,4)
C
C     IN-PLANE NODAL FORCES QP(4) DUE TO DISTRIBUTED LOADS
C
      FOR = FORC(LK)*R(1,3)
      QP(1) = QN(1)*FOR*B*.5
      QP(2) = 0.
      QP(3) = QP(1)
      QP(4) = 0.
      RETURN
      END
C
C
      SUBROUTINE LOADS(QN,QS,LK,FORC,B,BB,R)
      IMPLICIT DOUBLE PRECISION (A-H,O-Z)
      DIMENSION QN(2),QS(4),FORC(50),R(4,4)
C
C     OUT-OF-PLANE NODAL FORCES QS(4) DUE TO DISTRIBUTED LOADS
C
      FOR = FORC(LK)*R(1,1)
      DO 26 KK=1,2
        I = 2*(KK-1)
        QS(I+1) = QN(1)*FOR*B*.5
        IF (KK.EQ.1) THEN
          QS(I+2) = QN(1)*BB*FOR/12.
        ELSE
          QS(I+2) = -QN(1)*BB*FOR/12.
        ENDIF
 26   CONTINUE
      RETURN
      END
C
C
      SUBROUTINE BOUN(LK,NBOUN)
      IMPLICIT DOUBLE PRECISION (A-H,O-Z)
      COMMON P(30),ST(30,30),DIS(200),R(4,4),Q(30,4),NF(8),NB(8,4),B
      COMMON NTERM,NELEM,NP,NMOM,NDF,NDF1,NBAND,NSIZ,II,NI,BNII,BN1,BN2
      COMMON A,BB,EX,EY,PX,PY,G,DX,DY,D1,DXY,NOD(50,2),OM(600,3),DN(9)
      COMMON DISS(600)
C
C     INTRODUCTION OF BOUNDARY CONDITIONS
C
      DO 230 I=1,NBOUN
        IF (NF(I).EQ.LK) THEN
          DO 220 J=1,NDF
```

```
            IF (NB(I,J).EQ.0) THEN
              ST(J,J) = ST(J,J)*.1D+12
            ENDIF
 220      CONTINUE
        ENDIF
 230  CONTINUE
      RETURN
      END
C
C
      SUBROUTINE SOLVE
      IMPLICIT DOUBLE PRECISION (A-H,O-Z)
      COMMON P(30),ST(30,30),DIS(200),R(4,4),Q(30,4),NF(8),NB(8,4),B
      COMMON NTERM,NELEM,NP,NMOM,NDF,NDF1,NBAND,NSIZ,II,NI,BNII,BN1,BN2
      COMMON A,BB,EX,EY,PX,PY,G,DX,DY,D1,DXY,NOD(50,2),OM(600,3),DN(9)
      COMMON DISS(600)
C
C     FORWARD ELIMINATION
C
      CALL MATIN(ST,NDF)
      CALL STORE(ST,P,NDF)
      DO 111 J=1,NDF
      DO 111 I=1,NSIZ
        L = I+NDF
        Q(I,J) = 0.
        DO 111 K=1,NDF
          Q(I,J) = Q(I,J)+ST(K,L)*ST(K,J)
 111  CONTINUE
      DO 112 I=NDF1,NBAND
        L = I-NDF
        DO 112 K=1,NDF
          DO 113 J=NDF1,NBAND
 113        ST(I,J) = ST(I,J)-Q(L,K)*ST(K,J)
          P(I) = P(I)-Q(L,K)*P(K)
 112  CONTINUE
      DO 114 I=1,NSIZ
        K = I+NDF
        P(I) = P(K)
        P(K) = 0.
        DO 114 J=1,NSIZ
          L = J+NDF
          ST(I,J) = ST(K,L)
          ST(I,L) = 0.
          ST(K,J) = 0.
          ST(K,L) = 0.
 114  CONTINUE
      RETURN
      END
C
C
      SUBROUTINE BSUB
      IMPLICIT DOUBLE PRECISION (A-H,O-Z)
      COMMON P(30),ST(30,30),DIS(200),R(4,4),Q(30,4),NF(8),NB(8,4),B
      COMMON NTERM,NELEM,NP,NMOM,NDF,NDF1,NBAND,NSIZ,II,NI,BNII,BN1,BN2
      COMMON A,BB,EX,EY,PX,PY,G,DX,DY,D1,DXY,NOD(50,2),OM(600,3),DN(9)
      COMMON DISS(600)
C
C     BACK-SUBSTITUTION
C
      DO 30 IP=1,NP
        M = NP-IP
        CALL RDBK(ST,P,NDF)
```

```
          DO 1 I=1,NDF
          DO 1 J=NDF1,NBAND
 1          P(I) = P(I)-ST(I,J)*P(J)
          DO 2 I=1,NDF
            Q(I,1) = 0.
            DO 2 J=1,NDF
 2            Q(I,1) = Q(I,1)+ST(I,J)*P(J)
          DO 3 I=1,NDF
            P(I) = Q(I,1)
            J = NDF*M+I
 3          DIS(J) = Q(I,1)
          DO 114 I=1,NSIZ
            L = NBAND-I+1
            K = L-NDF
 114        P(L) = P(K)
 30     CONTINUE
        RETURN
        END
C
C
        SUBROUTINE MOM(D,E,F,C,S,itape)
        IMPLICIT DOUBLE PRECISION (A-H,O-Z)
        DIMENSION D(4,4),E(4,4),F(4,4),C(4,4),S(4,4)
        COMMON P(30),ST(30,30),DIS(200),R(4,4),Q(30,4),NF(8),NB(8,4),B
        COMMON NTERM,NELEM,NP,NMOM,NDF,NDF1,NBAND,NSIZ,II,NI,BNII,BN1,BN2
        COMMON A,BB,EX,EY,PX,PY,G,DX,DY,D1,DXY,NOD(50,2),OM(600,3),DN(9)
        COMMON DISS(600)
C
C       COMPUTATION OF STRESSES (D=FC), MOMENTS (D=FS) AND DISPLACEMENTS
C       DISS(600) AT SPECIFIED POINTS ALONG NODAL LINES
C       (AND DISPLACEMENTS DIS(200) AT CENTRE POINTS)
C
        NUM = 0
        DO 81 LK=1,NELEM
          READ (itape) ((R(I,J),I=1,4),J=1,4)
          DO 24 LL=1,2
            I = NDF*(NOD(LK,LL)-1)
            DO 23 J=1,NDF
              K = I+J
 23           D(J,1) = DIS(K)
            CALL MBTM(R,D,E,4,4,1)
            C(2*LL-1,1) = E(1,1)
            C(2*LL,1) = E(2,1)
            S(2*LL-1,1) = E(3,1)
            S(2*LL,1) = E(4,1)
 24       CONTINUE
          DO 25 LL=1,2
          DO 25 M=1,NMOM
            READ (itape) ((F(I,J),I=1,3),J=1,4)
            CALL MBTM(F,C,D,3,4,1)
            NUM = NUM+1
            DO 3 I=1,3
 3            OM(NUM,I) = OM(NUM,I)+D(I,1)
 25       CONTINUE
          DO 26 LL=1,2
          DO 26 M=1,NMOM
            READ (itape) ((F(I,J),I=1,3),J=1,4)
            CALL MBTM(F,S,D,3,4,1)
            NUM = NUM+1
          DO 4 I=1,3
 4            OM(NUM,I) = OM(NUM,I)+D(I,1)
 26       CONTINUE
```

```
 81    CONTINUE
       J = 1
       DO 30 K=1,NP
         DO 29 M=1,NMOM
           Z = DN(M)/A
           H1 = DSIN(BNII*Z)
           I = (K-1)*NDF+1
           DISS(J) = DISS(J)+DIS(I)*H1
           DISS(J+1) = DISS(J+1)+DIS(I+1)*DCOS(BNII*Z)
           DISS(J+2) = DISS(J+2)+DIS(I+2)*H1
           DISS(J+3) = DISS(J+3)+DIS(I+3)*H1
           J = J+NDF
 29      CONTINUE
 30    CONTINUE
       RETURN
       END
C
C
       SUBROUTINE EM(B,BB,BBB,E)
       IMPLICIT DOUBLE PRECISION (A-H,O-Z)
       DIMENSION E(4,4)
C
C      FORMULATION OF A MATRIX CONSTANT E(4,4)
C
       E(1,1) = 1.
       E(2,1) = 0.
       E(3,1) = -3./BB
       E(4,1) = 2./BBB
       E(1,2) = 0.
       E(2,2) = 1.
       E(3,2) = -2./B
       E(4,2) = 1./BB
       E(1,3) = 0.
       E(2,3) = 0.
       E(3,3) = 3./BB
       E(4,3) = -2./BBB
       E(1,4) = 0.
       E(2,4) = 0.
       E(3,4) = -1./B
       E(4,4) = 1./BB
       RETURN
       END
C
C
       SUBROUTINE MATIN(ST,N)
       IMPLICIT DOUBLE PRECISION (A-H,O-Z)
       DIMENSION ST(30,30)
C
C      INVERSION OF MATRIX
C
       DO 19 I=1,N
         Z = ST(I,I)
         ST(I,I) = 1.
         DO 60 J=1,N
 60        ST(I,J) = ST(I,J)/Z
         DO 19 K=1,N
           IF (K-I) 3,19,3
 3         Z = ST(K,I)
         ST(K,I) = 0.
         DO 4 J=1,N
 4         ST(K,J) = ST(K,J)-Z*ST(I,J)
 19    CONTINUE
```

```
      RETURN
      END
C
C
      SUBROUTINE MBTM(D,B,DB,L,M,N)
      IMPLICIT DOUBLE PRECISION (A-H,O-Z)
      DIMENSION D(4,4),B(4,4),DB(4,4)
C
C     MATRIX MULTIPLICATION DB(L,N) = D(L,M) X B(M,N)
C
      DO 110 J=1,N
      DO 110 I=1,L
        DB(I,J) = 0.
        DO 110 K=1,M
 110      DB(I,J) = DB(I,J)+D(I,K)*B(K,J)
      RETURN
      END
C
C
      SUBROUTINE MBTTM(D,B,DB,L,M,N)
      IMPLICIT DOUBLE PRECISION (A-H,O-Z)
      DIMENSION D(4,4),B(4,4),DB(4,4)
C
C     MATRIX TRANSPOSE MULTIPLICATION DB(L,N) = D(M,L) X B(M,N)
C
      DO 110 J=1,N
      DO 110 I=1,L
        DB(I,J) = 0.
        DO 110 K=1,M
 110      DB(I,J) = DB(I,J)+D(K,I)*B(K,J)
      RETURN
      END
C
C
      SUBROUTINE INIT(NBAND,NCOLN,NDF,itask)
      IMPLICIT DOUBLE PRECISION (A-H,O-Z)
      COMMON /BUFDA/ NBD,NCOL,IS,NA,LRECL,NREC,L,idir
       COMMON /BUF/X(2000)
C
C     THIS SUBROUTINE MUST BE CALLED ONCE BEFORE SUBROUTINES
C     STORE AND RDBK ARE CALLED IN ORDER TO INITIALISE THE BLOCK
C     CONTROL COUNTERS.
C
      idir=itask+20
      NA = 2000
      IS = 1
      NBD = NBAND
      NCOL = NCOLN
      LRECL = (NBD+NCOL)*NDF
      NREC = 0
      IF (LRECL-NA) 1,1,2
 1    RETURN
 2    WRITE (*,4) LRECL,NA
 4    FORMAT (' Logical Record Length of ',I6,' exceeds Buffer set at',
     1        I6)
      STOP
      END
C
C
      SUBROUTINE STORE(ST,P,NDF)
      IMPLICIT DOUBLE PRECISION (A-H,O-Z)
```

```
      DIMENSION ST(30,30),P(30)
      COMMON /BUFDA/ NBD,NCOL,IS,NA,LRECL,NREC,L,idir
       COMMON /BUF/X(2000)
C
C     STORE FORWARD ELIMINATION RESULTS IN BUFFER AREA AND WRITE ON
C     DISK IN A BLOCK OF 2000 WORDS TO SAVE TRANSFER TIME
C
C     TEST IF ROOM IN BUFFER
 5    IF ((IS+LRECL).LE.NA) THEN
C       ROOM IN BUFFER
        DO 10 I=1,NBD
        DO 10 J=1,NDF
          X(IS) = ST(J,I)
 10       IS = IS+1
        DO 15 I=1,NDF
          X(IS) = P(I)
 15       IS = IS+1
        RETURN
      ENDIF
C     NO ROOM LEFT IN BUFFER
      L = IS-1
      NREC = NREC+1
      WRITE (idir,REC=NREC) (X(J),J=1,L)
      IS = 1
      GOTO 5
      END
C
C
      SUBROUTINE RDBK(ST,P,NDF)
      IMPLICIT DOUBLE PRECISION (A-H,O-Z)
      DIMENSION ST(30,30),P(30)
      COMMON /BUFDA/ NBD,NCOL,IS,NA,LRECL,NREC,L,idir
       COMMON /BUF/X(2000)
C
C     READ FORWARD ELIMINATION RESULTS FOR BACK-SUBSTITUTION
C
C     TEST IF NEXT RECORD IN BUFFER
 10   IS = IS-LRECL
      IF (IS.GE.1) THEN
C       RECORD IS IN BUFFER
        DO 11 I=1,NBD
        DO 11 J=1,NDF
          ST(J,I) = X(IS)
 11       IS = IS+1
        DO 15 I=1,NDF
          P(I) = X(IS)
 15       IS = IS+1
        IS = IS-LRECL
        RETURN
      ENDIF
C     LAST BLOCK WRITTEN MUST BE READ
      IF (NREC.GE.1) THEN
        READ (idir,REC=NREC) (X(J),J=1,L)
        NREC = NREC-1
        IS = L+1
        GOTO10
      ENDIF
C     ILLOGICAL ERROR
      WRITE (*,100)
 100  FORMAT (' Attempt to read back too many records.')
      STOP
      END
```

Example : folded plate structure

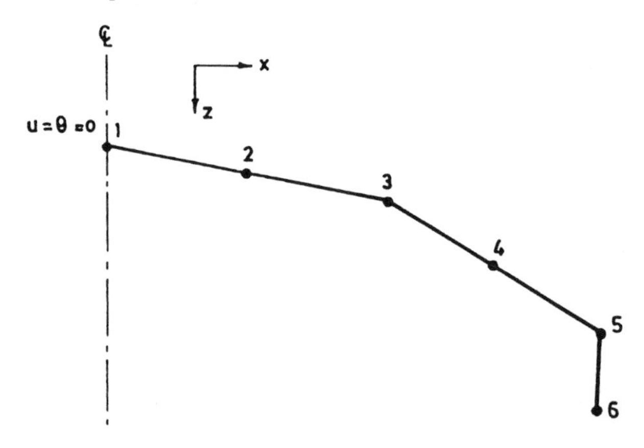

Figure A.1 Folded plate structure: division into finite strips and boundary conditions.

Nodal coordinates

Node	x(m)	z(m)
1	0.000	0.000
2	4.915	0.875
3	9.830	1.750
4	14.165	4.250
5	18.500	6.750
6	18.500	9.750

Strip properties

Strip	Left-hand node	Right-hand node	Thickness (m)	Distributed load (kN/m^2)
1	1	2	0.25	80
2	2	3	0.25	80
3	3	4	0.25	80
4	4	5	0.25	80
5	5	6	0.50	75

Material properties and length of strip

E_x	E_y	υ_x	υ_y	G	Length(m)
1	1	0	0	0.5	70

Prescribed displacements

Node	u	v	w	θ
1	0			0

INPUT FILES

IN.DAT	COEF.DAT
TEST PROBLEM - FOLDED PLATE STRUCTURE	0.6366,0.0
15TH FEBRUARY 1990	0.2122,0.0
1,5,6,1,1,4,8,1,1	0.1273,0.0
70.0,35.0	0.09095,0.0
0.0,0.0	0.07074,0.0
4.915,0.875	0.05787,0.0
9.83,1.75	0.04897,0.0
14.165,4.25	0.04244,0.0
18.5,6.75	0.03745,0.0
18.5,9.75	0.03351,0.0
1,1,2,0.25,80.0	0.03032,0.0
2,2,3,0.25,80.0	0.02768,0.0
3,3,4,0.25,80.0	0.02546,0.0
4,4,5,0.25,80.0	0.02358,0.0
5,5,6,0.5,75.0	0.02195,0.0
1,0,1,1,0	
1,1,3,0.0	
1.0,1.0,0.0,0.0,0.5	

OUTPUT FILE : OUT.DAT

```
ANALYSIS OF FOLDED PLATES AND BRIDGES BY
CLASSICAL FINITE STRIP METHOD

TITLE : TEST PROBLEM - FOLDED PLATE STRUCTURE
DATE : 15TH FEBRUARY 1990

*** INPUT DATA ***

NUMBER OF TERMS ...............................   1
NUMBER OF ELEMENTS ............................   5
NUMBER OF NODAL LINES .........................   6
NUMBER OF RESTRAINED BOUNDARY POINTS ..........   1
NUMBER OF POINTS ALONG A NODAL LINE
FOR OUTPUTTING STRESSES AND MOMENTS ...........   1
NUMBER OF DEGREE OF FREEDOM ...................   4
MAXIMUM HALF BANDWIDTH ........................   8
LOAD TYPE (1=NON-SYMMETRICAL,2=SYMMETRICAL) ....  1
NUMBER OF NODAL LINES WITH CONCENTRATED LOADS ..  1

LENGTH OF STRIPS :            70.0000
POINTS TO GIVE STRESSES AND MOMENTS :    35.0000

CO-ORDINATES OF POINTS :
     X          Z
    .0000      .0000
   4.9150      .8750
   9.8300     1.7500
  14.1650     4.2500
  18.5000     6.7500
  18.5000     9.7500

STRIP PROPERTIES :
 NO.  NODES       THICKNESS      U.D.L.
  1   1   2       .25000000     80.00000000
  2   2   3       .25000000     80.00000000
  3   3   4       .25000000     80.00000000
  4   4   5       .25000000     80.00000000
  5   5   6       .50000000     75.00000000
```

```
 BOUNDARY CONDITIONS :
  NODE  U   V   W   ANGLE  (FIXITY : 1=FREE,0=FIXED)
    1   0   1   1     0
   1   1   3        .00000000

 MATERIAL PROPERTIES :
     EX            EY            VX            VY            G
  .100000E+01  .100000E+01  .000000E+00  .000000E+00  .500000E+00

 TERM NO.     FOURIER COEFFICIENTS
              DISPLACEMENT PARAMETERS
    1          .6366     .0000

 *** OUTPUT RESULTS ***

 NUMBER OF TERMS USED =    1

DEFLECTIONS AND ROTATIONS (U, V, W AND THETA)
  -.70558770E-06  -.25448262E-01  -.65459762E+07   .16574891E-04
  -.16522476E+07  -.38507515E-01   .23541832E+07   .25261731E+07
  -.40323439E+07  -.56128076E-01   .15383234E+08   .32877995E+07
  -.12855084E+08  -.25479356E-01   .30596717E+08   .27711244E+07
  -.16545852E+08   .22534483E-02   .36960736E+08   .57525016E+06
  -.18250324E+08   .21713681E+00   .36959565E+08   .56465725E+06

 ELEM NO.
 ZIGMA -X , ZIGMA -Y , ZIGMA -XY (REPEAT FOR EACH)
 MOMENT-X , MOMENT-Y , MOMENT-XY (REPEAT FOR EACH)
   1
 -.13368E+05 -.26129E+05 -.18256E-03 -.13368E+05 -.39537E+05 -.11710E-03
 -.15197E+04 -.16902E+02 -.42339E-16  .20200E+03  .68382E+01 -.64528E-05

 ELEM NO.
 ZIGMA -X , ZIGMA -Y , ZIGMA -XY (REPEAT FOR EACH)
 MOMENT-X , MOMENT-Y , MOMENT-XY (REPEAT FOR EACH)
   2
 -.11947E+05 -.39537E+05 -.57394E-03 -.11947E+05 -.57629E+05 -.51544E-03
  .19882E+03  .68382E+01 -.64528E-05 -.59611E+03  .41574E+02 -.83983E-05

 ELEM NO.
 ZIGMA -X , ZIGMA -Y , ZIGMA -XY (REPEAT FOR EACH)
 MOMENT-X , MOMENT-Y , MOMENT-XY (REPEAT FOR EACH)
   3
 -.85004E+04 -.57629E+05 -.10496E-02 -.85004E+04 -.26160E+05 -.10079E-02
 -.62257E+03  .40233E+02 -.83983E-05  .89145E+03  .86357E+02 -.70785E-05

 ELEM NO.
 ZIGMA -X , ZIGMA -Y , ZIGMA -XY (REPEAT FOR EACH)
 MOMENT-X , MOMENT-Y , MOMENT-XY (REPEAT FOR EACH)
   4
 -.35713E+04 -.26160E+05 -.12992E-02 -.35713E+04  .23137E+04 -.12817E-02
  .88839E+03  .86357E+02 -.70785E-05  .25433E+03  .10565E+03 -.14694E-05

 ELEM NO.
 ZIGMA -X , ZIGMA -Y , ZIGMA -XY (REPEAT FOR EACH)
 MOMENT-X , MOMENT-Y , MOMENT-XY (REPEAT FOR EACH)
   5
 -.39030E+03  .23137E+04 -.44020E-03 -.39030E+03  .22294E+06 -.43905E-03
  .74207E+02  .34715E+03 -.11755E-04 -.64545E+00  .38292E+03 -.11539E-04
```

INDEX